HYDROMETALLURGICAL PROCESS FUNDAMENTALS

NATO CONFERENCE SERIES

I Ecology
II Systems Science
III Human Factors
IV Marine Sciences
V Air–Sea Interactions
VI Materials Science

VI MATERIALS SCIENCE

Volume 1 Molecular Metals
Edited by William E. Hatfield

Volume 2 Materials for Advanced Batteries
Edited by D. W. Murphy, J. Broadhead, and B. C. H. Steele

Volume 3 Adhesion in Cellulosic and Wood-Based Composites
Edited by John F. Oliver

Volume 4 Adhesion Problems in the Recycling of Concrete
Edited by Pieter C. Kreijger

Volume 5 Atomistics of Fracture
Edited by R. M. Latanision and J. R. Pickens

Volume 6 Electronic Structure and Properties of Hydrogen in Metals
Edited by P. Jena and C. B. Satterthwaite

Volume 7 Soot in Combustion Systems and Its Toxic Properties
Edited by J. Lahaye and G. Prado

Volume 8 Surface Modification and Alloying by Laser, Ion, and Electron Beams
Edited by J. M. Poate, G. Foti, and D. C. Jacobson

Volume 9 Coherence and Energy Transfer in Glasses
Edited by Paul A. Fleury and Brage Golding

Volume 10 Hydrometallurgical Process Fundamentals
Edited by Renato G. Bautista

HYDROMETALLURGICAL PROCESS FUNDAMENTALS

Edited by
Renato G. Bautista
Mackay School of Mines
University of Nevada Reno
Reno, Nevada

Published in cooperation with NATO Scientific Affairs Division
PLENUM PRESS · NEW YORK AND LONDON

Library of Congress Cataloging in Publication Data

NATO Advanced Research Institute on Hydrometallurgical Process Fundamentals (1982: Churchill College)
Hydrometallurgical process fundamentals.

(NATO conference series. VI, Materials science; v. 10)

"Proceedings of a NATO Advanced Research Institute on Hydrometallurgical Process Fundamentals, held July 25–31, 1982, in Churchill College, Cambridge University, Cambridge, United Kingdom"—T.p. verso.
Bibliography: p.
Includes index.
1. Hydrometallurgy—Congresses. I. Bautista, Renato G. II. North Atlantic Treaty Organization. Scientific Affairs Division. III. Title. IV. Series.
TN688.N38 1982 669′.028′3 84-20015
ISBN 0-306-41880-0

Proceedings of a NATO Advanced Research Institute on Hydrometallurgical Process Fundamentals, held July 25–31, 1982, in Churchill College, Cambridge University, Cambridge, United Kingdom

A Division of Plenum Publishing Corporation
233 Spring Street, New York, N.Y. 10013

Printed in the United States of America

FOREWORD

The mineral resources of the industrialized countries, especially the member nations of the North Atlantic Treaty Organization, are being depleted at such a rate that more and more of these countries are beginning to depend on ore imported from other countries. To sustain the economic and strategic well-being of these member countries, it becomes imperative that a program of developing and exploiting other non-conventional mineral resources and a conservation program where metal values from waste dumps and scrap metals and alloys are recycled must be initiated and implemented. In order to meet this challenge, new processes and technology must be available for consideration in the design and operation of the new plants. One of the possible routes of extracting the metals from their ores, especially for multimetal complex ores and very low grade ores, is by hydrometallurgical processing.

The hydrometallurgical route of metal recovery where dissolution (leaching), separation and concentration (ion exchange, solvent extraction, and membrane separation) and reduction to metal (cementation, precipitation by gaseous reduction, and electrolysis) is carried out at near ambient temperature is becoming more competitive with the conventional high temperature processes used in the smelting of metals from high grade and beneficiated ores. With increasing dependence on lower grade ores of the primary and strategically important metals such as manganese, cobalt, chromium, and the platinum group metals among others, the understanding of the above processing steps from both a chemical and physical aspect needs to be known in order that an efficient extraction technology can be designed and developed. The development of an extraction technology based on the mechanisms, kinetics, and transport phenomena of the chemical reaction system could only lead to a process that will help conserve the limited mineral resources and energy resources available in the world.

The different physical and chemical phenomena that are involved in the production of metals from their primary ores and secondary sources by hydrometallurgical methods comprised the principal topics of the NATO Advanced Research Institute on Hydrometallurgical Process Fundamentals. The measurement and acquisition of physico-chemical properties relating to their thermodynamics, chemical kinetics, mass transport and extraction equilibria, and the development and use of detailed mathematical models of the separation processes and their predictive aspects were also related topics of significance to this meeting.

This NATO Advanced Research Institute on Hydrometallurgical Process Fundamentals brought together scientists, engineers and technologists currently involved in fundamental research in the different aspects of hydrometallurgy. The attendance of these researchers whose interest and research projects are in this broad field of science and engineering provided a well-informed discussion of the problems and possible direction of future technological development.

It is hoped that this proceedings volume is a good record of the formal and informal presentations made during the institute.

The sponsorship given by the Special Programme Panel on Material Science, NATO Scientific Affairs Division, is gratefully acknowledged. The support and help given by Professor Thomas W. Chapman, Scientific Advisor, and Professor A. John Monhemius, Meeting Coordinator, in the planning of the meeting is also acknowledged. The skill and dedication of my secretary, Mrs. Janet Beer, in putting the entire volume on the word processor aided the publication of this book. To all the conferees and invited speakers goes my heartfelt thanks for making this meeting at Churchill College, Cambridge University a very successful one indeed.

Renato G. Bautista

CONTENTS

SULFIDE MINERALOGY: A REVIEW WITH SPECIAL REFERENCE TO PHASES OF INTEREST IN HYDROMETALLURGY

David J. Vaughan

Department of Geological Sciences
University of Aston in Birmingham
Birmingham B4 7ET, BRITAIN

The sulfide minerals, most particularly those containing iron, copper, nickel, cobalt, lead and zinc, are reviewed in terms of current knowledge of their crystal structures, compositional variations and physical properties. Problems concerning both structural transformations in sulfides and sulfide stoichiometry are considered. Sulfide minerals in ore deposits commonly occur in assemblages which are composed of limited groups of coexisting phases and which show characteristic mineral intergrowth textures. Such occurrences of natural sulfides are reviewed in the light of data available on the thermochemistry and phase relations in sulfide systems.

INTRODUCTION

The sulfide minerals are the major source of world supplies of a very wide range of metals and are the most important group of ore minerals. Of particular industrial importance in the context of hydrometallurgical processing are the virtually ubiquitous iron sulfides, and the extensively exploited sulfides of copper, nickel, cobalt, lead and zinc. The objectives of this paper are to briefly review the mineralogy of the industrially important sulfides, most of which have been well characterized in terms of their crystal structures, compositional limits, electrical and magnetic properties, thermochemistry and phase relations. The mineralogy and mineral chemistry of sulfides has been reviewed in recent years by Ribbe[1] and by Vaughan and Craig.[2] Also briefly discussed here are the associations and textural intergrowths occurring in the natural ores which are more fully discussed by Ramdohr[3] and by Craig and Vaughan.[4]

COMPOSITIONS AND STRUCTURES

Several of the commonest sulfide minerals were amongst the first materials to be studied by X-ray crystallography and since that time the structures of most important sulfides have been determined; a richness and diversity of structural chemistry is evident from this work. It is possible to identify groups of sulfides based on major structure-type or having key structural features in common as outlined in Table 1. These are, in many cases, the classic structures of crystalline solids such as the rocksalt structure of the galena group (Fig. 1b), the sphalerite and wurtzite forms of ZnS (Figs. 1c, d), the nickel arsenide structure (Fig. 1e) and the spinel structure of the thiospinel group (Fig. 1f). The disulfides are characterized by the presence of dianion (S-S, As-S, Sb-S, As-As, etc.) units in the structure; as well as the pyrite structure in which FeS_6 octahedra share corners along the c-axis direction, there is the marcasite form of FeS_2 in which octahedra share edges along the c-axis and variants of the marcasite structure in $FeAs_2$ and FeAsS which have shorter or alternate long and short metal-metal distances across the shared octahedral edge (see Fig. 1a). A few sulfides have layer structures, such as covellite (CuS, Fig. 1g), and a small number exhibit structures best characterized as containing rings or chains of linked atoms (e.g. realgar AsS, Fig. 1i). A diverse group referred to here as the metal-excess group comprises an unusual and di-

→

Fig. 1. Crystal structures of the major sulfides: (a) pyrite structure and the linkage of metal-sulfur octahedra along the c-axis direction in (i) pyrite FeS_2 (ii) marcasite FeS_2, (iii) loellingite $FeAs_2$ and (iv) arsenopyrite FeAsS; (b) galena (PbS) structure; (c) sphalerite (ZnS) structure; (d) wurtzite (ZnS) structure; (e) niccolite (NiAs) structure; (f) spinel (M_3S_4) structure; (g) covellite (CuS) structure; (h) cube cluster of tetrahedrally coordinated metals in the pentlandite, $(Ni, Fe)_9S_8$, structure; (i) realgar (AsS) structure. (In each case, metals are shown as the smaller or shaded spheres.)

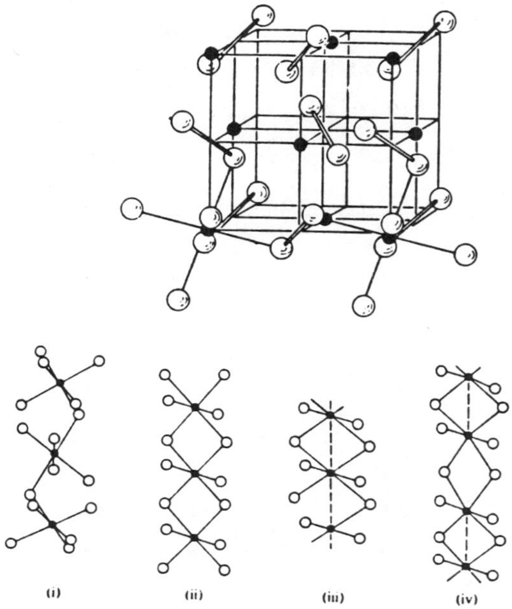

a

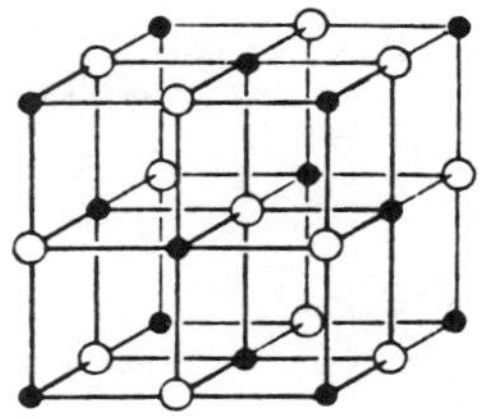

b

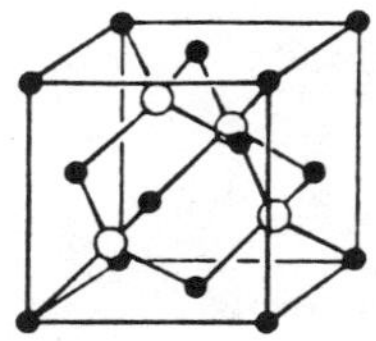

c

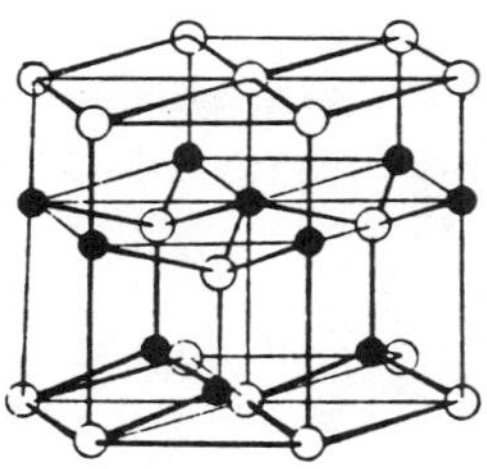

d

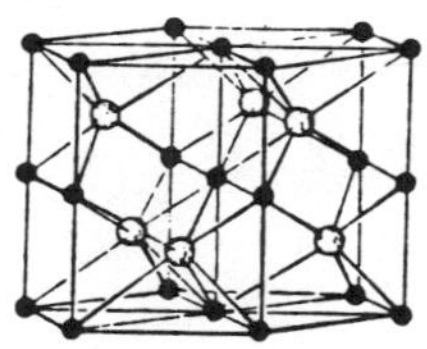

e

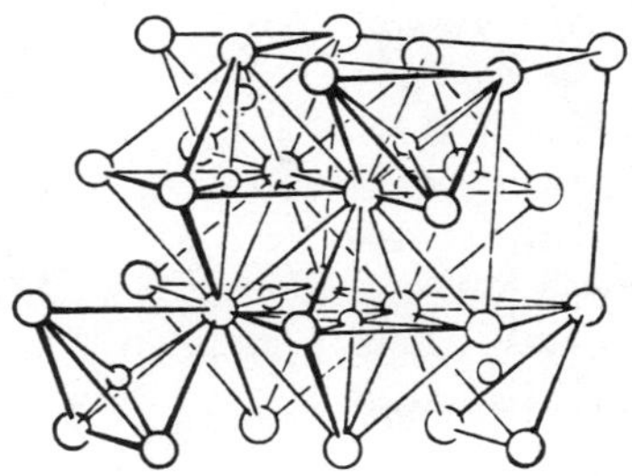

f

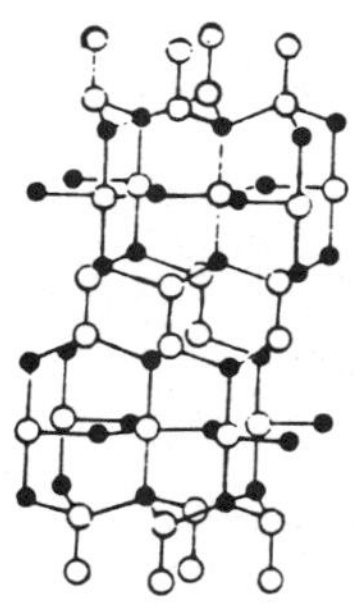

g

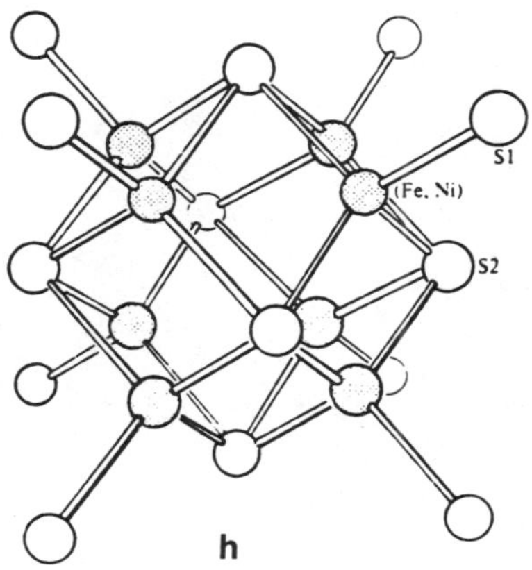

h

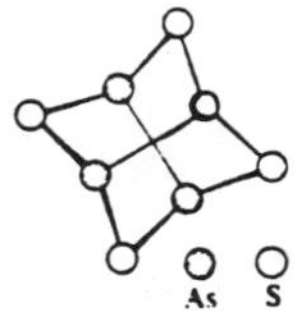

i

Table 1. Sulfide Structural Groups

1) THE DISULFIDE GROUP

Pyrite Structure	Marcasite Structure	Arsenopyrite Structure	Loellingite Structure
FeS_2 pyrite	FeS_2 marcasite	FeAsS arsenopyrite	$FeAs_2$ loellingite
CoS_2 cattierite		FeSbS gudmundite	$CoAs_2$ safflorite
			$NiAs_2$ rammelsbergite

Pyrite Structure and Marcasite Structure ↘ ↙ derived by As/S ordered substitution:

(Co,Fe)AsS cobaltite
(Ni,Co,Fe)AsS gersdorffite (I)

2) THE GALENA GROUP

PbS galena
α-MnS alabandite

3) THE SPHALERITE GROUP

Sphalerite Structure →	derived by ordered substitution →	stuffed derivatives
β-ZnS sphalerite	$CuFeS_2$ chalcopyrite	$Cu_9Fe_8S_{16}$ talnakite
CdS hawleyite	Cu_2FeSnS_4 stannite	$Cu_9Fe_9S_{16}$ mooihoekite
Hg(S,Se) metacinnabar	Cu_2ZnSnS_4 kesterite	$Cu_4Fe_5S_8$ haycockite

4) THE WURTZITE GROUP

Wurtzite Structure →	composite structure derivatives →	?further derivatives
α-ZnS wurtzite	$CuFe_2S_3$ cubanite	$Cu_3Fe_2SnS_6$ hexastannite
CdS greenockite	?$AgFe_2S_3$ argentopyrite	

↓ derived by ordered substitution

Cu_3AsS_4 enargite

5) THE NICKEL ARSENIDE GROUP

NiAs Structure →	distorted derivatives →	ordered omission derivatives
NiAs niccolite	FeS troilite	Fe_7S_8 monoclinic pyrrhotite
NiSb breithauptite	CoAs modderite	Fe_9S_{10}, $Fe_{11}S_{12}$ hexagonal pyrrhotite etc?

6) THE THIOSPINEL GROUP

Co_3S_4 linnaeite
$FeNi_2S_4$ violarite
$CuCo_2S_4$ carrollite

7) THE LAYER SULFIDES GROUP

Molybdenite Structure	Tetragonal PbO Structure	Covellite Structure
MoS_2 molybdenite	$(Fe,Co,Ni,Cr,Cu)_{1+x}S$	CuS covellite
WS_2 tungstenite		$\sim Cu_3FeS_4$ idaite

Table 1. (Cont.)

8) METAL EXCESS GROUP

Pentlandite Structure	Argentite Structure	Chalcocite Structure
$(Ni,Fe)_9S_8$ pentlandite	Ag_2S argentite	Cu_2S chalcocite ↘ ? derivative
		$Cu_{1.96}S$ djurleite
Digenite Structure →	derived by ordered substitution	Nickel Sulfide Structures
Cu_9S_5 digenite Cu_5FeS_4 bornite	Cu_7S_4 anilite	NiS millerite Ni_3S_2 heazlewoodite

9) RING OR CHAIN STRUCTURE GROUP

Stibnite Structure	Realgar Structure	Cinnabar Structure
Sb_2S_3 stibnite Bi_2S_3 bismuthinite	As_4S_4 realgar	HgS cinnabar

verse variety of structures such as that of the mineral pentlandite (Fig. 1h). Many of these compounds exhibit very short metal-metal distances suggesting they may be stabilized by metal-metal bonds.

As shown in Table 1, in many of the groups a number of minerals share the actual structure-type but there are commonly other minerals which have structures which are directly based on these "parent" structures and can be thought of as being derived from them. The relationship between derivative and parent structure may involve:

(1) Ordered substitution, e.g. the structure of chalcopyrite ($CuFeS_2$) is derived from sphalerite (ZnS) by alternative replacement of Zn atoms by Cu and Fe resulting in an enlarged (tetragonal) unit cell (see Fig. 2a).*
(2) Stuffed derivative, e.g. talnakite ($Cu_9Fe_8S_{16}$) is derived from chalcopyrite by the occupation of additional, normally empty cavities in the structure (see Fig. 2b).
(3) Ordered omission, e.g. monoclinic pyrrhotite (Fe_7S_8) is derived from the NiAs structure FeS by removal of Fe atoms leaving holes (vacancies) which are ordered (see Fig. 2c).
(4) Distortion, e.g. the troilite form of FeS is simply a distortion of the parent NiAs structure form (see Fig. 2c).

*stannite (Cu_2 $FeSnS_4$) results from further ordered substitution of half of the Fe atoms in $CuFeS_2$ by Sn (Fig. 2a).

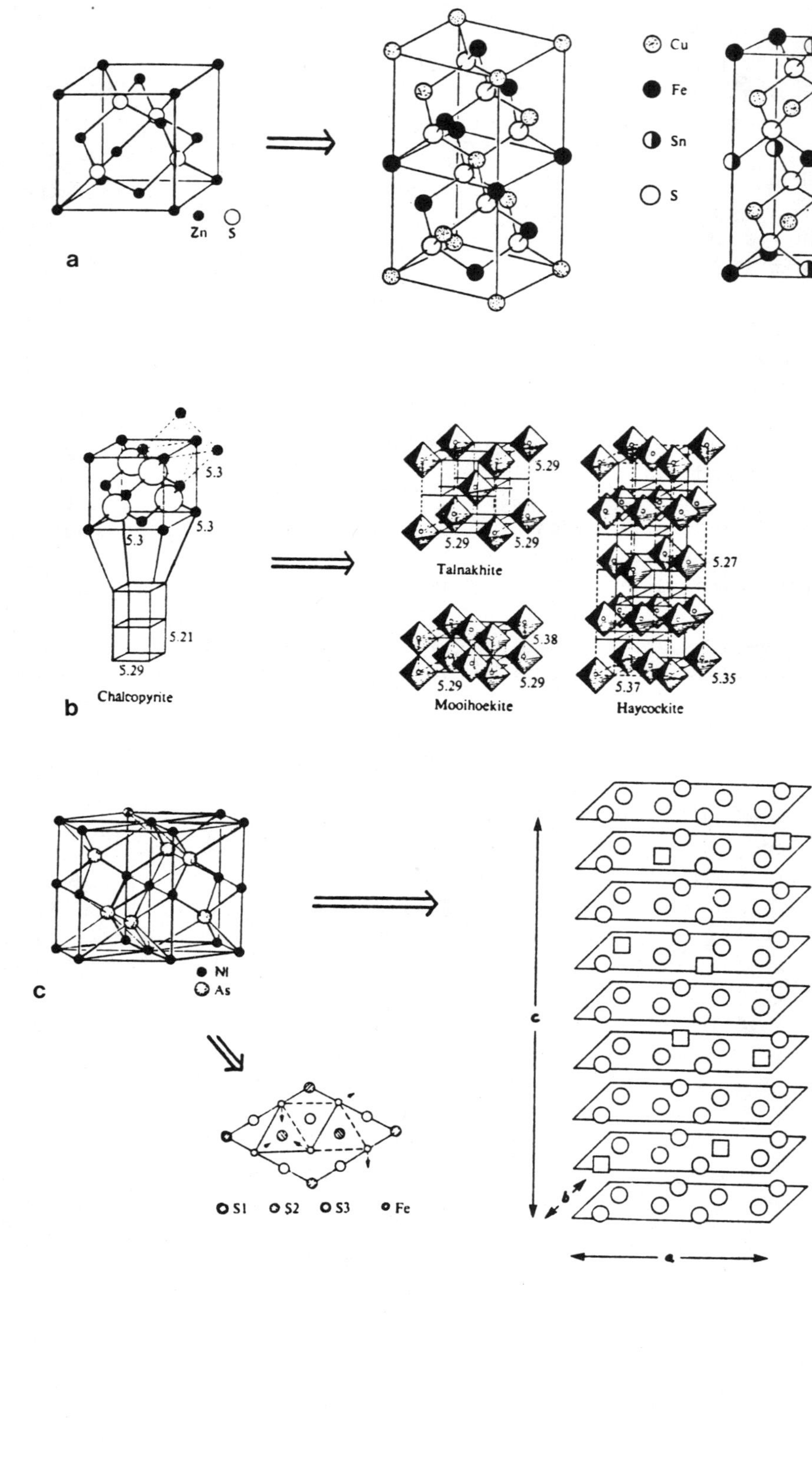
Zn
S
a
Cu
Fe
Sn
S
5.3
5.3
5.3
5.21
5.29
Chalcopyrite
b
5.29
5.29
5.29
Talnakhite
5.38
5.29
5.29
Mooihoekite
5.27
5.37
5.35
Haycockite
Ni
As
c
S1
S2
S3
Fe
c
b
a

Sometimes the relationships involved are more complex as, for example, when the resulting structure is composite and made up of slabs of the parent structure (or structures) arranged in some ordered fashion.

STRUCTURAL TRANSFORMATIONS

Many of the sulfide mineral structures described above are also related in the sense that transformations between different structures exhibited by a particular composition occur as a result of changes in conditions (commonly of temperature or pressure). Such transformations may be categorized both thermodynamically and structurally into two major types:

(1) First-order transformations usually involve breaking of bonds and changes in metal-sulfur coordination, with an abrupt increase in internal energy at the transition (Fig. 3a[i]). Examples include the transformation of HgS with a chain structure in cinnabar to a sphalerite-type structure (metacinnabar) on heating to 344°C; the transformation of CdS (hawleyite) which has a sphalerite-type structure to a rocksalt (or galena)-type structure above a pressure of 20 Kbar.

(2) Second-order transformations involve some form of disordering of the structure and, in this case, the slope of the curve

Fig. 2. Parent and derivative crystal structures in the sulfide minerals: (a) the sphalerite (ZnS) structure with the chalcopyrite ($CuFeS_2$) and stannite Cu_2FeSnS_4 structures; (b) the sphalerite and chalcopyrite unit cells with an octahedron of metals outlined within which may be an additional metal ion in the minerals talnakhite ($Cu_9Fe_8S_{16}$), mooihoekite ($Cu_9Fe_9S_{16}$) and haycockite ($Cu_4Fe_5S_8$), the arrangement of additional occupied metal sites being as shown (also shown are the dimensions of the parent sphalerite cell in Å); (c) the niccolite unit cell of high temperature FeS which has vacancies in place of Fe atoms in monoclinic pyrrhotite (Fe_7S_8) which are ordered as shown in the diagram which has vacancies represented by squares (and only Fe atom layers shown); also shown in a projection onto the basal plane are the distortions which occur in the troilite modification of FeS.

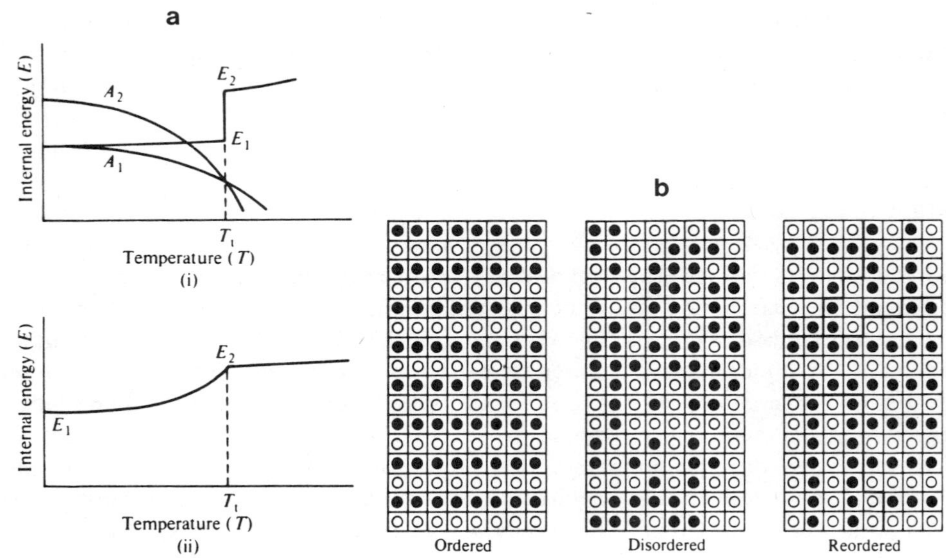

Fig. 3. Structural transformations: (a) changes in internal energy (E) and free energy (A) during (i) first-order and (ii) second-order phase transformations. (T_t = transition temperature); (b) schematic representation of order $\rightleftarrows$ disorder transformation with reordering resulting in twinning.

showing variation in internal energy is discontinuous on passing through the transition (Fig. 3a[ii]). Generally two kinds of disorder-ing may be distinguished. The first involves a distortion, and an example is the transformation from the more symmetric (hence more "disordered") true NiAs-type structure in FeS to the less symmetric troilite form. The second is a substitutional disordering in which two or more atoms (usually metals) which occupy one set of equivalent structural positions at random in the high temperature form occupy distinctly separate structural positions on cooling below the transition temperature.

The type of second-order phase transformation which involves substitutional disordering is particularly important in sulfide minerals. Although all the types of phase transformations outlined above occur at much lower temperatures (or pressures) in the sulfides than in analogous oxides, substitutional disordering transformations dominate the low-temperature crystal chemistry and phase relations of a number of important sulfide systems (e.g. Cu-S, Cu-Fe-S) and warrant further discussion.

If we consider a substance with a sulfur anion sublattice and equal numbers of two species of metal cations occupying, say, a tetrahedral site within this framework, then at low temperature the distribution of these two metals may be ordered as shown schematically in Fig. 3b. Heating a crystal of this substance above the critical transformation temperature results in breakdown of this ordered arrangement, hence producing a disordered high temperature form. On cooling again, as pointed out by Buerger,[5] although ordering may result in a single crystal as before, because ordering normally starts at many separate centers these will grow towards one another and may or may not match (see Fig. 3b). As a result, the crystal may show twinning or antiphase domains. The example chosen by Buerger to illustrate this was chalcopyrite ($CuFeS_2$) which has Cu and Fe atoms occupying alternate positions in a sphalerite-like cell (Fig. 2a) with consequent tetragonal symmetry; on heating above 557°C, disordering of Cu and Fe atoms leads to a cubic form with the sphalerite structure. It has, however, been shown subsequently that this reaction is not strictly isochemical although a slightly metal-rich "chalcopyrite" does pass through this transformation without any change in composition.[1,2]

Linked to the type of order $\rightleftarrows$ disorder transformation described above is the observation that the high temperature (disordered) phases commonly exhibit a fairly wide range of possible compositions (see Fig. 5), i.e., they form extensive fields of solid solution which shrink with falling temperature. Hence, on cooling, the high temperature solid may separate to form two or more (ordered) phases in a process of segregation or exsolution. These processes, which are very widespread in sulfide mineralogy, constitute a sort of generalized ordering transformation. An example is the segregation of pentlandite (the major ore of nickel) and pyrrhotite from the monosulfide solid solution, (MSS) of the Fe-Ni-S system (Fig. 5a):

$$(Fe, Ni)S \rightleftarrows Fe_{1-x}S + (Fe, Ni)_9S_8$$

Such processes may lead to either completely separate grains of the two phases being formed on exsolution, or the two phases may be intergrown as laths, blebs, etc., often with a clearly-defined crystallographic relationship. The crystal structures of the two phases will partly dictate the kind of texture which forms, but a crucial factor is also the time over which cooling occurs. Few kinetic studies have so far been undertaken on systems like these and most work has involved observations in the transmission electron microscope using the electron beam to heat samples through a transition temperature.[6,7]

Expressed rather simplistically, it seems that in many of the industrially important sulfide systems (e.g., Fe-S, Cu-S, Fe-Cu-S, Fe-Ni-S) metal atoms can move quite readily between positions within the sulfur sublattice framework down to relatively low temperatures. Hence, kinetic factors assume great importance in understanding the compositions and textures of the natural assemblages, a fact which only recently has been appreciated.

STOICHIOMETRY

Many metal sulfides show evidence that the elements which comprise them are not combined in a simple whole number ratio, i.e., they exhibit non-stoichiometry.

In certain cases the extent of deviation from a simple ratio is considerable. For example, the pyrrhotites are sometimes given the general formula $Fe_{1-x}S$ where $0 < x < 0.125$ and the varying compositions correspond to a varying concentration of vacancies "left" in sites which would otherwise be occupied by Fe atoms. However, in systems like these, ordering of the vacancies occurs at low temperatures and the result may be a series of stoichiometric phases of slightly different compositions. Such behavior occurs in the chromium sulfides[8] where Cr_7S_8, Cr_5S_6, Cr_3S_4 and Cr_2S_3 all have superstructures based on an NiAs cell and resulting from ordered arrangements of vacancies in alternate metal atom layers. Although Fe_7S_8 has a (monoclinic) superstructure (again based on the NiAs cell) resulting from vacancy ordering,[9] the situation in the "intermediate" or hexagonal" pyrrhotites is more complex. Some of these pyrrhotites may represent ordered phases with clearly defined compositions (Fe_9S_{10}, $Fe_{11}S_{12}$, etc.) but more complex and partial ordering in these systems may occur. One problem is certainly that free energy differences between a series of phases resulting from vacancy ordering would be very small, making a successful investigation of relationships between synthetic products very difficult or impossible.

Other examples of non-stoichiometry involve deviations from the simple ratio which are so small as to have been ignored until fairly recently. For example, galena (PbS) exhibits a range of non-stoichiometry of 0.1 atomic percent.[10] Galena is apparently stable over a wide range of values of a_{S_2} and at high a_{S_2} it has lead vacancies, whereas at low a_{S_2} there are sulfur vacancies. It has also been suggested that certain "polymorphic pairs" of minerals exhibit this type of non-stoichiometry and these reports conflict with the rigid definition of polymorphism. For example, Scott and Barnes[11] have suggested that wurtzite, formerly regarded as the high-temperature (>1020°C) polymorph of ZnS is actually sulfur-deficient relative to sphalerite. Electrical measurements indicate that zinc vacancies occur in sphalerite and sulphur vacancies in wurtzite with a total range in non-stoichiometry of about one atomic percent. Sphalerite-wurtzite equilibria would therefore be a function of a_{S_2} as well as temperature and pressure. However, another pair of minerals to which such reasoning might be applied are pyrite and marcasite and here Tossell et al[12] have offered an interpretation which is based on the reaction mechanism by which marcasite is formed (commonly in acid solution) as a metastable species relative to pyrite. An important consequence of non-stoichiometry is the effect which it has on physical (electrical, optical, hardness) and chemical properties although the latter have been less well studied.

ELECTRICAL AND MAGNETIC PROPERTIES

As well as exhibiting a richness and diversity in structural chemistry, the metal sulfides also show a tremendous range of electrical and magnetic behavior. As Table 2 indicates, whereas such non-transition metal sulfides as sphalerite and galena are diamag-

Table 2. Magnetic and Electrical Properties of Some Major Sulfide Minerals[2]

MINERAL SPECIES	MAGNETIC PROPERTIES	ELECTRICAL PROPERTIES
Sphalerite (ZnS) "Iron sphalerite" (Zn,Fe)S	diamagnetic paramagnetic	insulator (E_g~3.7eV) semiconductor (E_g~0.5eV for ~12 at % Fe)
Galena (PbS)	diamagnetic	semiconductor (n- and p-type, E_g~0.41eV)
Pyrite (FeS_2)	diamagnetic	semiconductor (n- and p-type, E_g~0.9eV)
Cattierite (CoS_2)	ferromagnetic (T_c=110K)	metallic
Chalcopyrite ($CuFeS_2$)	antiferromagnetic (T_N= 823K)	semiconductor (n-type, E_g~0.5eV)
Covellite (CuS)	diamagnetic (?)	metallic
(Monoclinic) Pyrrhotite (Fe_7S_8)	ferrimagnetic (T_c=573K)	metallic
Carrollite ($CuCo_2S_4$)	Pauli paramagnetic	metallic
Pentlandite $(Ni,Fe)_9S_8$	Pauli paramagnetic	metallic

E_g = band or "energy" gap; T_c = Curie temperature; T_N = Néel temperature.

netic, diamagnetism is also exhibited by pyrite and also marcasite (the other FeS_2 polymorph). Substitution of iron for zinc in sphalerite leads to paramagnetic behavior and many transition metal sulfides show various forms of magnetic ordering at lower temperatures including antiferromagnetism (e.g., chalcopyrite) ferromagnetism (e.g., cattierite) and ferrimagnetism (e.g., monoclinic pyrrhotite, although "hexagonal" pyrrhotites are antiferromagnetic). Other transition metal sulfides, which are metallic conductors, exhibit the weak temperature-independent paramagnetism which is known as Pauli paramagnetism. As well as the metallic conductivity occurring in many sulfides of diverse magnetic character, numerous sulfides are semiconductors and a smaller number (e.g., pure sphalerite) are insulators. Amongst semiconducting sulfides both intrinsic and impurity (or extrinsic) conduction mechanisms occur, and conduction via electrons (n-type) or via holes (p-type) is common. Such variations are often a consequence of very minor impurities being present, or of slight non-stoichiometry (e.g., galena, PbS, exhibiting a total variation in composition of 0.1 atomic percent, shows p-type conductivity in lead-deficient samples and n-type conductivity in sulfur-deficient samples).

NATURAL ASSEMBLAGES AND ORE TEXTURES

A very important feature of ore minerals in general, and of sulfides in particular, is that they frequently occur in nature grouped together in characteristic associations, with characteristic host rocks, gangue minerals and intergrowth textures.[4] For example, lead and zinc sulfides occur together in limestones (or associated sediments) in many parts of the world, copper and iron sulfides occur in certain porphyritic igneous rocks, and iron and nickel sulfides are found as massive bodies in certain types of igneous intrusions. This not only aids identification of the ores, but also enables some useful generalizations to be made with regard to assemblages and textures, although every deposit has its own unique features when studied in detail.

The natural assemblages, as well as being studied in the field and in hand specimen, have generally been studied in polished section under the microscope with further characterization of phases employing X-ray diffraction and chemical analysis using the electron microprobe.[13] The interpretation of assemblages and textures has been clarified in many cases through the study of phase equilibria in the relevant binary, ternary (or sometimes quaternary) sulfide systems. The phase chemistry and thermochemistry in turn shows a direct relationship to the crystal chemistry and crystal physics already outlined in this paper.

In a brief review such as this, it is only possible to mention a few key examples to illustrate the kinds of assemblages and textures found in natural sulfides. The following examples have been chosen to illustrate assemblages formed at high, intermediate and low temperatures; from melts, high temperature and low temperature aqueous solutions; and representing major producers of nickel, copper, cobalt, lead and zinc.

Iron-Nickel-Copper Sulfide Ores Association with Mafic and Ultramafic Igneous Rocks

Major world supplies of nickel along with significant copper, cobalt and platinum group metals are derived from these ores which occur as massive to disseminated sulfides in close association with mafic or ultramafic igneous rocks. Pyrrhotite (notably monoclinic pyrrhotite but often hexagonal types as well) is usually the major sulfide (~80%). Chalcopyrite occurs as irregular aggregates and veinlets whereas the primary nickel-bearing phase, pentlandite $(Fe,Ni)_9S_8$ can usually only be seen under the microscope and occurs either as granular material interstitial or marginal to the pyrrhotite or as "flame-like" bodies within pyrrhotite grains (see Fig. 4a). It is widely held that these ores formed by separation of an immiscible metal sulfide melt from the silicate melt of the intrusion. Segregation of such a melt led to formation of the massive ores which would have solidified at quite high temperatures (~800°C+). Changes in the ores subsequent to solidification have been clarified by experimental studies of the phase relations in the Fe-Ni-S system (see Fig. 5). As already noted in the section on structural transformations, the ores initially crystallized as a high temperature (Fe,Ni)S phase, the monosulfide solid solution (MSS). In fact this probably also accommodated substantial copper which was exsolved early on as the chalcopyrite aggregates, pentlandite was exsolved on further cooling, with the granular material probably formed at higher temperatues and the "flame-like" bodies at lower temperatures (perhaps down to as low as 100°C).

(a)

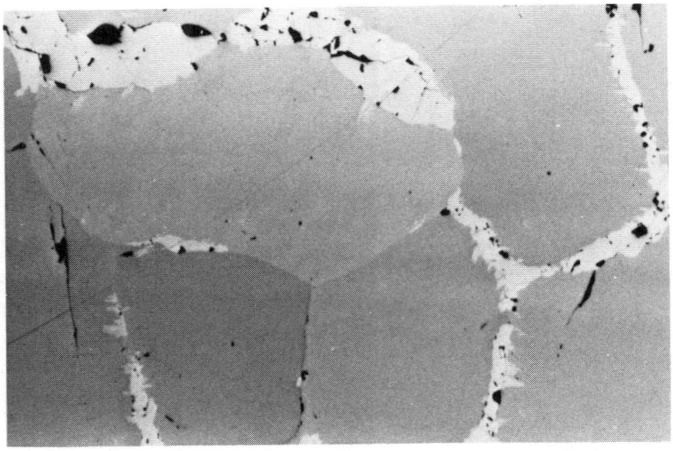

(b)

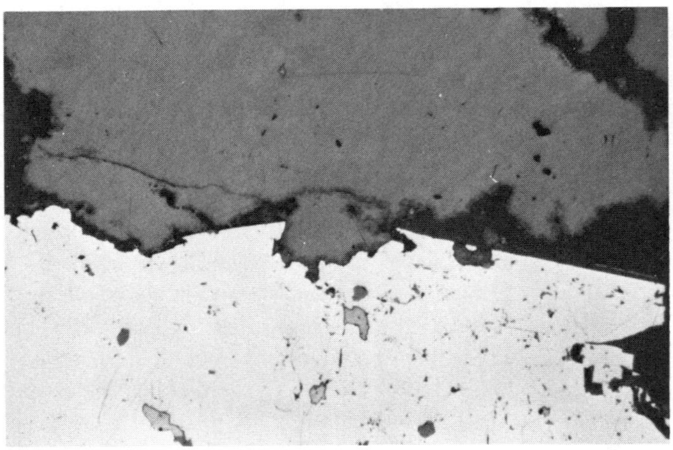

Fig. 4. Textural intergrowths occurring in some major ore assemblages involving sulfide minerals: (a) pyrrhotite ("light grey") - pentlandite ("white") exsolution textures in Sudbury type ores; (b) pyrite ("white") containing blebs of chalcopyrite with adjacent bornite and covellite rims in a porphyry copper deposit; (c) pyrite ("white") with adjacent galena showing triangular pits and spalerite with chalcopyrite blebs; (d) lenses of fine grains of chalcopyrite and pyrite; (e) sphalerite ("medium grey") with fine chalcopyrite blebs intergrown with galena and minor pyrite ("white") from a Kuroko-type deposit.

(c)

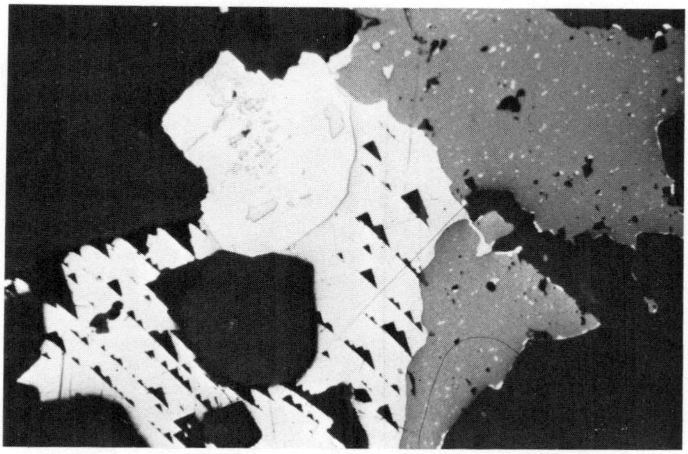

(d)

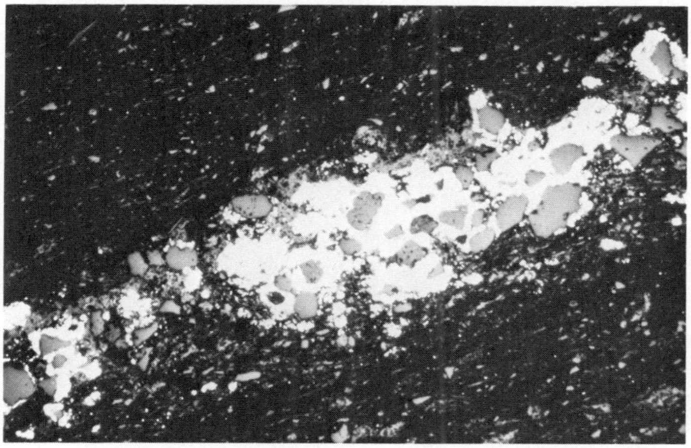

(e)

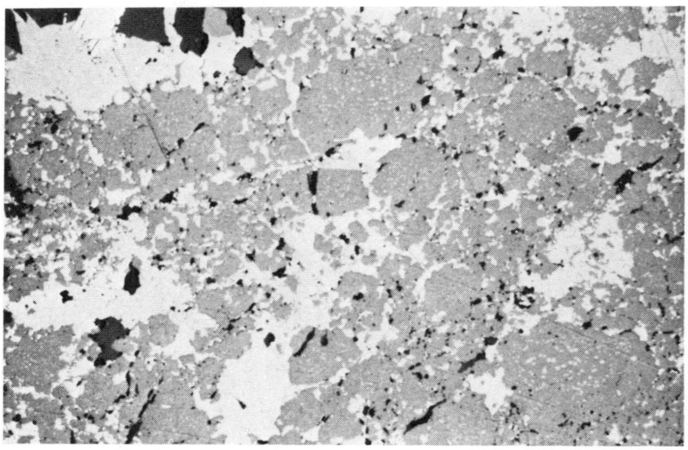

(a)

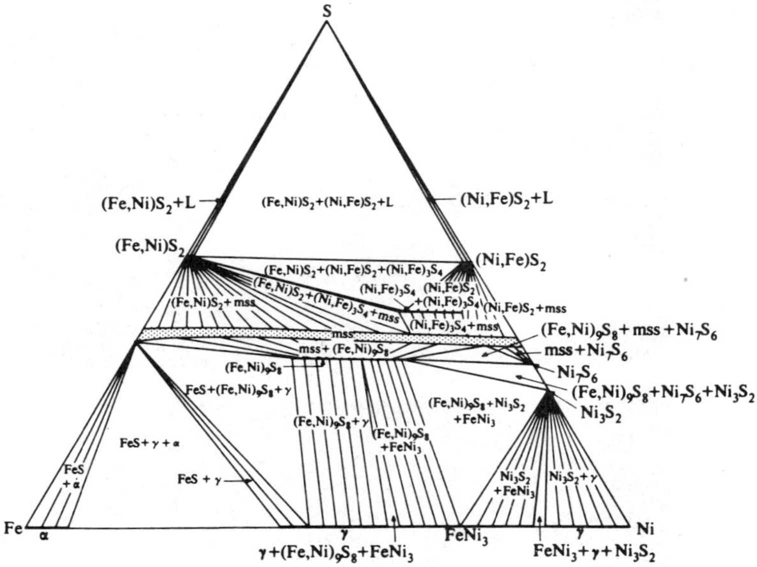
S
$(Fe,Ni)S_2+L$
$(Fe,Ni)S_2+(Ni,Fe)S_2+L$
$(Ni,Fe)S_2+L$
$(Fe,Ni)S_2$
$(Ni,Fe)S_2$
$(Fe,Ni)S_2+(Ni,Fe)S_2+(Ni,Fe)_3S_4$
$(Fe,Ni)S_2+(Ni,Fe)_3S_4+mss$
$(Ni,Fe)_3S_4$
$(Ni,Fe)S_2+(Ni,Fe)_3S_4$
$(Fe,Ni)S_2+mss$
$(Ni,Fe)S_2+mss$
$(Ni,Fe)_3S_4+mss$
mss
$(Fe,Ni)_9S_8+mss+Ni_7S_6$
$mss+(Fe,Ni)_9S_8$
$mss+Ni_7S_6$
$(Fe,Ni)_9S_8$
Ni_7S_6
$FeS+(Fe,Ni)_9S_8+\gamma$
$(Fe,Ni)_9S_8+Ni_7S_6+Ni_3S_2$
$(Fe,Ni)_9S_8+Ni_3S_2+FeNi_3$
Ni_3S_2
$(Fe,Ni)_9S_8+\gamma$
$(Fe,Ni)_9S_8+FeNi_3$
$FeS+\gamma+\alpha$
$FeS+\alpha$
$FeS+\gamma$
$Ni_3S_2+FeNi_3$
$Ni_3S_2+\gamma$
Fe
α
γ
$FeNi_3$
γ
Ni
$\gamma+(Fe,Ni)_9S_8+FeNi_3$
$FeNi_3+\gamma+Ni_3S_2$

(b)

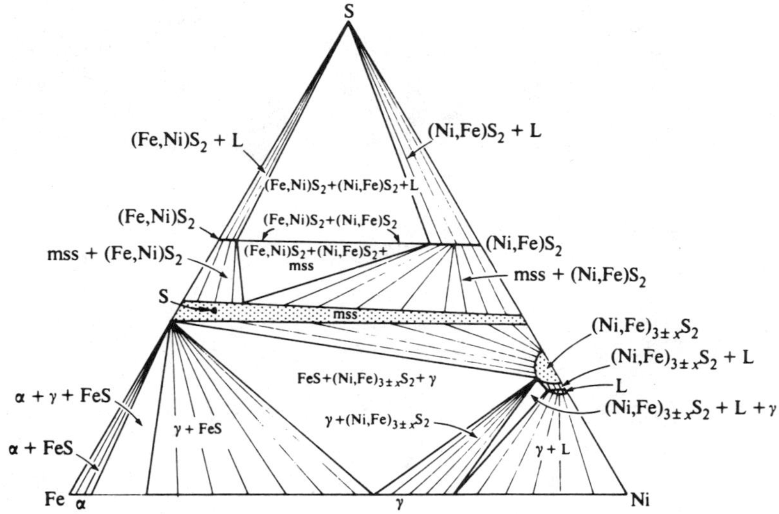
S
$(Fe,Ni)S_2 + L$
$(Ni,Fe)S_2 + L$
$(Fe,Ni)S_2+(Ni,Fe)S_2+L$
$(Fe,Ni)S_2$
$(Fe,Ni)S_2+(Ni,Fe)S_2$
$(Ni,Fe)S_2$
$mss + (Fe,Ni)S_2$
$(Fe,Ni)S_2+(Ni,Fe)S_2+mss$
$mss + (Ni,Fe)S_2$
S
mss
$(Ni,Fe)_{3\pm x}S_2$
$(Ni,Fe)_{3\pm x}S_2 + L$
$FeS+(Ni,Fe)_{3\pm x}S_2+\gamma$
L
$\alpha + \gamma + FeS$
$(Ni,Fe)_{3\pm x}S_2 + L + \gamma$
$\gamma + FeS$
$\gamma+(Ni,Fe)_{3\pm x}S_2$
$\gamma + L$
$\alpha + FeS$
Fe
α
γ
Ni

(c)

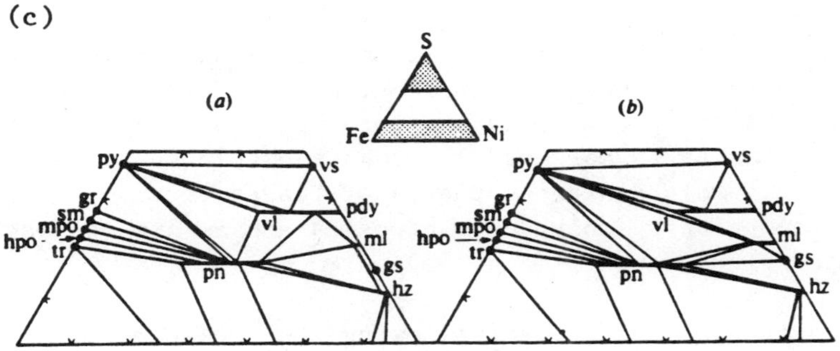

Fig. 5. Phase relations in the condensed Fe-Ni-S system:
(a) at 650°C;
(b) at 400°C;
(c) an interpretation at 25°C.

Copper (-Iron) Sulfides Associated with Porphyritic Intrusive Igneous Rocks

Substantial world supplies of copper are derived from the exploitation of the so-called "porphyry copper" deposits in which sulfides occur as veinlets and disseminated grains in, or adjacent to, porphyritic intrusive igneous rocks. Closely related deposits are also the major world source of molybdenum (as molybdenite, MoS_2) and both porphyry copper and molybdenum deposits are significant producers of silver and gold. Deposits are characteristically low grade (0.8% copper and 0.3% molybdenum respectively, on average) but large tonnage.

The sulfide ore mineral assemblages are characteristically simple, being dominated by pyrite with chalcopyrite as the major copper mineral. Bornite (Cu_5FeS_4) is present as discrete grains and as both exsolution and alteration lamellae within chalcopyrite. In near-surface ores, covellite, chalcocite ($\sim Cu_2S$) and digenite (Cu_9S_5) are commonly seen, produced by the alteration of chalcopyrite and bornite. Cubanite is more rarely seen as exsolution lamellae within chalcopyrite. Molybdenite as plates or blades occurs as a major phase in the porphyry molybdenum deposits and as a minor phase in many copper-rich examples (Fig. 4b).

These ores apparently form following the emplacement of a body of magma within 0.5 to 2 km of the earths surface. Hot aqueous solutions derived from the cooling magma and from interaction with the adjacent groundwater system precipitate the sulfide minerals within and immediately around the intrusion. These sulfides appear to have been precipitated from saline solutions at temperatures as high as 725°C. Exsolution textures involving chalcopyrite-bornite and chalcopyrite-cubanite can again be understood in terms of experimental phase relations data; in this case the central portion of the Cu-Fe-S system which at 300°C or above shows two extensive fields of solid solution (see Fig. 6).

Lead-Zinc Sulfides in Carbonate Rocks and other Sediments

Major lead and zinc producers are the ones which mostly occur in sedimentary carbonate rocks (chiefly dolomites but also limestones) and sometimes in associated sandstones, shales and conglomerates. The ores may be as veins or as more irregular bodies and have clearly been precipitated from aqueous solutions, generally below ~200°C and often below ~100°C, although details of the origins of these deposits remain controversial.

The mineral assemblages are characteristically simple with lead occurring as galena and zinc as sphalerite (although wurtzite also occurs). Pyrite and marcasite are widespread and so is chalcopyrite although it is rarely significant as a source of copper. The sphalerite generally carries only small amounts of iron or manganese subsituting for zinc but cadmium contents may be quite high. The sulfides are generally quite coarse grained and show textures resulting largely from successive precipitation of the minerals from solution (radiating crystals, banded overgrowths, etc.; see Fig. 4c).

Copper-Lead-Zinc (-Cobalt) Sulfides in Sedimentary Rocks

Another major group of copper, sometimes cobalt, and sometimes lead and zinc ores occur as sulfides disseminated within sediments; often as bodies essentially parallel with the sediment layers. Sometimes these layers are completely undisturbed as in the Kupferschiefer of Northern Europe but commonly there has been folding (e.g., Zambian Copperbelt) or even intense deformation of the sedimentary layers with consequent metamorphism of the ores and host rocks (e.g., Broken Hill and Mt. Isa, Australia). The histories of these ores are complex and consequently they are amongst the most controversial of sulfide ore deposits as regards their origin. However, there seems little doubt that the sulfides were originally formed at temperatures well below 100°C, either as part of the sediment formation process, or prior to the consolidation of the sediments.

Major minerals in the assemblages commonly include pyrite, with chalcopyrite or galena and sphalerite; pyrrhotite, bornite, chalcocite (or digenite, djurleite) occur in some ores. In those ores containing cobalt, this may be as carrollite ($CuCo_2S_4$) or as cobaltian pyrite. Except where metamorphism has led to recrystallization, the ore minerals are characteristically fine-grained and occur as aggregates or sometimes spherical bodies (See Fig. 4d). Again, in the copper-iron sulfides, textures resulting from exsolution (e.g., bornite-chalcopyrite) are observed and may be interpreted using the Cu-Fe-S system phase diagrams (Fig. 6).

Copper-Iron-Zinc Sulfide Assemblages in Volcanic Environments

The ores described above grade into examples of sulfides occurring in sequences of sediments interlayered with volcanic rocks, often the products of submarine volcanic activity. These include the so-called Kuroko-type and Besshi-type deposits of Japan, and many

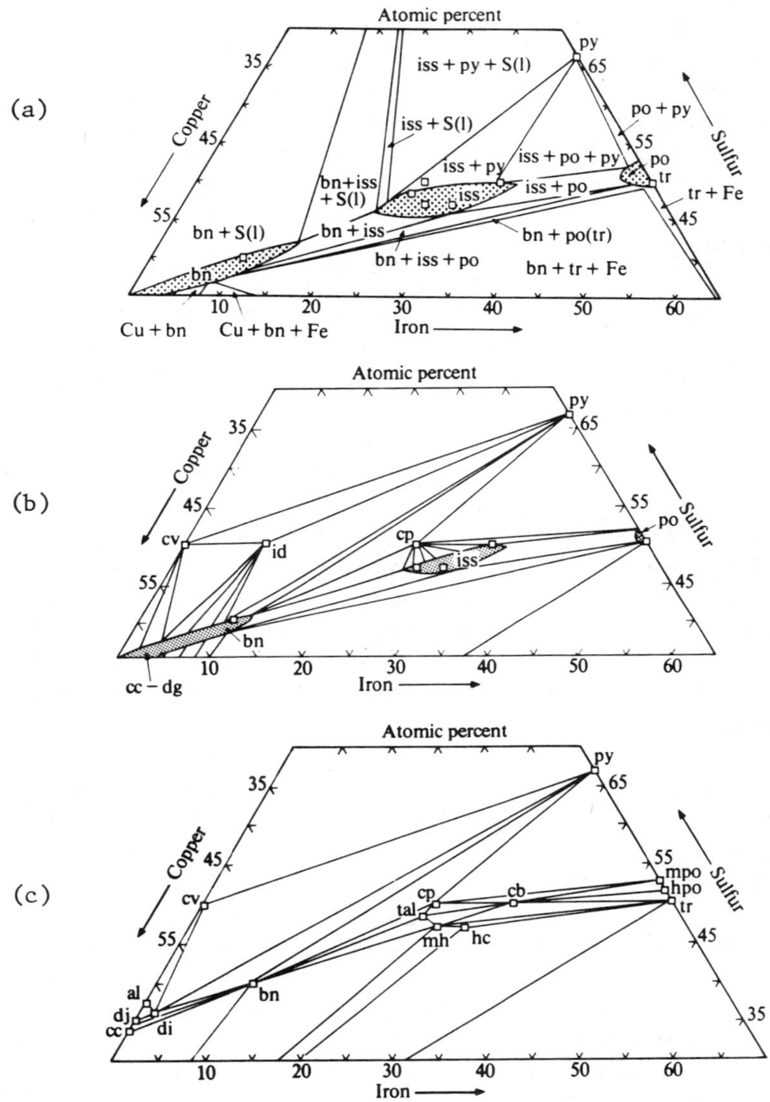

Fig. 6. Phase relations in the condensed Cu-Fe-S system:
(a) at 600°C;
(b) at 400°C;
(c) an interpretation at 25°C.

major deposits in North America and Australia. The ore minerals involved in these deposits are pyrite, chalcopyrite, shpalerite, and, in some examples, pyrrhotite or galena. Minor sulfides may include arsenic-, bismuth-, antimony- and silver-bearing phases. The textures range from massive sulfides occurring in layers to more finely disseminated material. Commonly, however, the various sulfides are intergrown on a fine scale (see Fig. 4e).

The origin of some of the deposits of this type has been clarified by observations of submarine exhalations along the crest of the East Pacific Rise.[14] Here, the ores accumulate as a "snowfall" of very fine-grained sulfides formed as hot solutions issue onto the sea-floor. Other sulfides are formed beneath the sediment surface by the introduction of hydrothermal fluids.

DISCUSSION

The crystal structures, chemical compositions and limits of solid solubility, electrical, magnetic and other physical properties of the industrially important sulfide minerals are well known. Thermochemical and phase equilibrium data are also available for binary, ternary and some quaternary systems of industrial importance (although the occurrence of persistent metastable phases means that the low temperature phase relations in some systems are still controversial). The mineralogical characteristics of natural ores have been extensively studied as regards the common assemblages and the textural relationships between phases, some of which may be interpreted in the light of the data on phase equilibria.

Aspects of sulfide mineralogy of particular importance in hydrometallurgy include the nature and mechanism of phase transformations, since solid state transformations are known to occur during leaching.[15] Also important is the phenomenon of nonstoichiometry, whether deviations from the "ideal" formula are very small or quite large. Such variations clearly affect electrical properties of the materials and also their electrochemical behavior. In determining the feasibility of hydrometallurgical processing and the efficiency of particular processes, mineral intergrowth characteristics including particle size, crystallinity, inclusions and exsolution intergrowths are all important[16] and require investigation using the techniques of ore microscopy.

A unified understanding of the mineralogy and chemistry of metal sulfides, in terms of crystal structure, solid solution limits, electrical and magnetic properties, thermochemistry and stability relations, is only possible through models of electronic structure (or "chemical bonding"). This topic forms the subject of a separate chapter.

ACKNOWLEDGEMENTS

I wish to thank Professor R. Bautista for the invitation to present this paper to the NATO Advanced Research Institute and my colleague, Dr. R. A. Ixer, for critical reading of the manuscript. Cathy Kennedy is thanked for preparation of the typescript.

REFERENCES

1. P. H. Ribbe (editor): Sulfide Mineralogy, Mineral. Soc. America Short Course Notes, Vol. 1, 1974.
2. David J. Vaughan and James R. Craig: Mineral Chemistry of Metal Sulfides, Cambridge Univ. Press, Cambridge, 1978.
3. P. Ramdohr: The Ore Minerals and their Intergrowths, Pergamon Press, Oxford, 1969 (2nd edn, 1981).
4. James R. Craig and David J. Vaughan: Ore Microscopy and Ore Petrography, Wiley-Interscience, New York, 1981.
5. M. J. Buerger: Trans. Am. Crystallogr. Assoc. 1971, Vol. 7, pp1-24.
6. A. Putnis and J. D. C. McConnell: Contrib. Mineral. Petrol. 1976, Vol. 58, pp. 127-136.
7. D. P. Kelly and D. J. Vaughan: Internat. Mineral. Assoc. Mtg., Varna, Bulgaria, 1982 Abstract.
8. F. Jellinek: Acta Crystallogr., 1957, Vol. 10, pp. 620-628.
9. M. Tokonami, K. Nishiguchi and N. Morimoto: Amer. Mineral., 1972, Vol. 57, pp. 1066-1080.
10. J. Bloem and F. A. Kroger: Z. Phys. Chem. NF, 1956, Vol. 7, pp1-14.
11. S. D. Scott and H. L. Barnes: Geochim Cosmochim Acta, 1972, Vol. 36, pp. 1275-1295.
12. J. A. Tossell, D. J. Vaughan and J. K. Burdett: Phys. Chem. Mineral., 1981, Vol. 7, pp. 177-184.
13. J. Zussman (editor): Physical Methods in Determinative Mineralogy, 2nd ed. Academic Press, New York, 1977.
14. J. Francheteau et al: Nature, 1979, Vol. 277, pp. 523-528.
15. A. R. Burkin: Mineral. Sci. Engr., 1969, Vol. 1, pp. 4-14.
16. A. P. Prosser: Proc. 9th Commonwealth Mining Metall. Congr., 1970, Vol. 3, 59-79.

ELECTRONIC STRUCTURES OF SULFIDES AND LEACHING BEHAVIOR

David J. Vaughan

Department of Geological Sciences
University of Aston in Birmingham
Birmingham B4 7ET, BRITAIN

The electronic structures of major sulfide minerals (pyrite, chalcopyrite, covellite, chalcocite, sphalerite, galena) are best understood in terms of models based on molecular orbital theory and incorporating elements of band theory. Such models may be developed from molecular orbital cluster calculations (using the SCF-X_α scattered wave method, for example) and tested and refined by comparison with the results of spectroscopic studies, particularly those employing X-ray emission and X-ray photoelectron spectroscopies. The models provide a basis for understanding the crystal chemistry and physical properties of the minerals, including their electrical behavior, and potentially can be developed to gain a fuller understanding of the surface properties of the minerals. Application of these ideas in understanding and predicting leaching behavior will be discussed with particular emphasis on the sulfides of copper and iron.

INTRODUCTION

The metal sulfide ore minerals show a great diversity in magnetic, electrical, and related physical properties which points to a richness and diversity in electronic structure (or chemical bonding) in these materials.[1,2] Sulfides must have electronic structures ranging from those with highly delocalized electron behavior of the type found in metals to the localized electron behavior characteristic of insulators. Any attempts to model the bonding in sulfides has to accommodate this diversity and an approach based largely on molecular orbital (MO) theory and incorporating elements of band theory we find to be most useful in mineral sciences. The development and testing of bonding models can be undertaken using three major approaches:

(1) Phenomenological in which available knowledge of crystal structures and properties (particularly electrical and magnetic behavior) is itself used as a basis.

(2) Experimental in which an attempt is made to determine electronic energy levels directly from data such as that obtained from optical, X-ray emission, UV or X-ray photo-electron spectroscopies.

(3) Calculational in which a first principles or more approximate method of quantum mechanical calculation is used to compute the energies and spatial distribution of electrons.

Ideally, all three approaches should be employed with emphasis on the combined use of calculation and experiment used to validate the results of calculations. In this paper, such approaches to understanding the chemical bonding in metal sulfides are applied to a number of sulfides of industrial importance, concentrating firstly on two major nontransition metal binary sulfides (ZnS, PbS), then three important transition metal binary sulfides (FeS_2, CuS, Cu_2S) and a major ternary sulfide ($CuFeS_2$). Models of electronic structure (or chemical bonding) should enable the understanding and prediction not only of sulfide crystal structures but also solid solution relations, optical and physical properties and ultimately thermodynamic stability. The potential for developing these models to gain a better understanding of leaching behavior will be considered after the models have been presented.

BINARY SULFIDES OF ZINC AND LEAD

The dominant ore mineral of zinc is the sphalerite form of ZnS which, when pure, is a diamagnetic insulator.[1] A starting point in describing the electronic structure of sphalerite might be to construct a molecular orbital energy level diagram of the type commonly seen in elementary chemistry texts, and representing the bonding in a single ZnS_4 tetrahedron cluster unit. As in Fig. 1(a), the energies of the outermost (valence) orbitals on Zn^{2+} and S^{2-} are shown qualitatively with strong overlap between zinc 4s and 4p atomic orbitals and sulfur 3s and 3p atomic orbitals resulting in a set of sigma bonding molecular orbitals (σ) stabilized with respect to the contributary atomic orbitals, and a set of correspondingly destabilized sigma antibonding molecular orbitals(σ^*). Infilling of these molecular orbitals with electrons has the σ bonding molecular orbitals

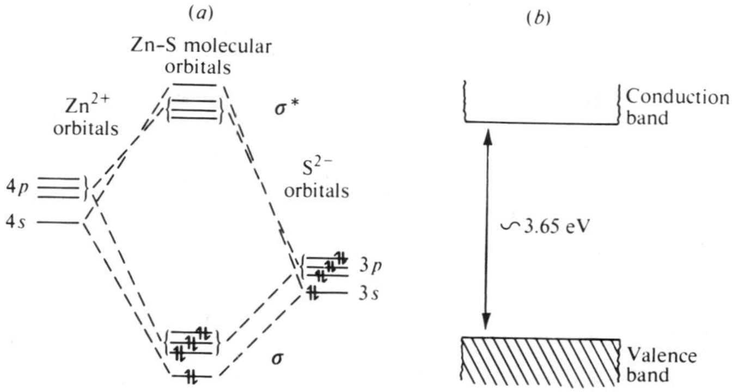

Fig. 1. Simplistic molecular orbital (MO) and band model diagrams for sphalerite: (a) schematic MO diagram for pure ZnS; (b) corresponding ZnS band model.

completely filled with electrons and the antibonding molecular orbitals (σ*) empty. Overlap between molecular orbitals on adjacent ZnS_4 tetrahedron units in a crystal of sphalerite can be considered as leading to broadening of the energy levels into bands--a σ valence band completely filled with electrons and a corresponding empty σ* conduction band. This leads to a qualitative "one electron" energy band diagram of the type shown in Fig. 1(b). The energy separation between valence and conduction bands can be measured electrically since the passage of an electric current requires that (in pure ZnS) electrons be promoted from the valence band into the conduction band so that they (or the "holes" left in the valence band) may act as carriers of the electrical charge. Both simplistic models shown in Fig. 1 correlate with a material that is an electrical insulator with a band gap of ~ 3.65eV and contains no unpaired electrons and is therefore diamagnetic.

The question now is - how can such qualitative models be tested and refined? One important approach is the experimental study of spectra, particularly X-ray emission and X-ray photoelectron (or "ESCA") spectra which involve energies capable of ejecting electrons from their orbitals and can provide a fairly complete picture of the energies of valence electrons in the solid. The valence region X-ray photoelectron spectrum of ZnS[3] shown in Fig. 2(a) arises from measuring the kinetic energies of electrons ejected on bombardment with a mono-energetic beam of X-rays. Such spectra enable direct determination of the binding energies of electrons in the solid, with all electrons giving rise to features in the spectrum. Assignment of the features in such a spectrum is clearly difficult, and one approach is

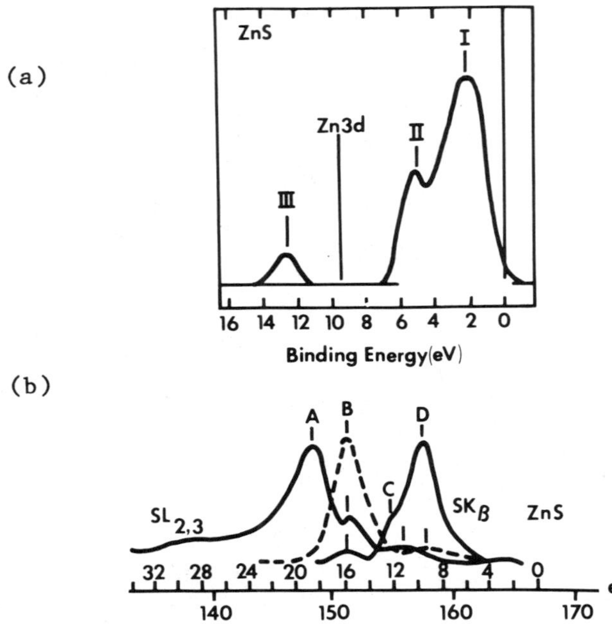

Fig. 2. (a) X-ray photoelectron spectrum of ZnS; (b) X-ray emission spectra of ZnS.

to obtain complementary data using X-ray emission spectroscopy. In this technique, bombardment of the sample with X-rays causes the ejection of more tightly bound (core) electrons and the outermost electrons drop back to fill vacancies in the core orbitals. The consequently lost energy is emitted in the form of X-rays, but this time the transitions of electrons between orbitals which can occur are governed by quantum mechanical selection rules. Hence in the example of X-ray emission spectra of ZnS[4] shown in Fig. 2(b), the $SK\beta$ spectrum arises from transitions from orbitals in the valence region with sulfur 3p character to a vacancy in the sulfur 1s orbital. The $SL_{2,3}$ spectrum arises from transitions from orbitals in the valence region with sulfur 3s character to a vacancy in the sulfur 2p orbital, and the Zn $L\alpha$ spectrum arises from transitions from dominantly zinc 3d-type orbitals to a vacancy in the zinc 2p orbital. Using the positions of peaks arising from more deeply buried (core) electrons which are likely to be little affected by bonding interactions, the energy scales on X-ray photoelectron and X-ray emission spectra can be aligned. In this way, it can be suggested that the peak at ~ 2eV binding energy in the X-ray photoelectron spectrum arises from electrons in essentially sulfur 3p-type (nonbonding) orbitals, the peak at ~ 5eV binding energy from the main zinc-sulfur bonding orbitals, and the peaks at ~ 9eV and ~ 13eV from the zinc 3d and sulfur 3s (both nonbonding) orbitals, respectively.

Clearly, interpreting spectra like these solely on the basis of qualitative models can still lead to errors, and a better approach would be to calculate the electronic structure and to compare calculation and experiment. Unfortunately, many of the molecular orbital calculational methods successfully employed to describe hydrocarbons or simple oxides give poor results for metal sulfides. One method successfully used on a variety of sulfide systems is the "SCF-Xα-Scattered Wave Cluster Method" of Slater and Johnson (or the "Xα method for short).[5,6] In this method, a "molecular" cluster fragment, taken to be a basic building block of the structure, is chosen for the calculation. In the case of ZnS, a tetrahedron with a Zn^{2+} ion coordinated to four S^{2-} ions (and hence a ZnS_4^{6-} cluster) would be an appropriate choice. The space within and around the cluster is geometrically partitioned into atomic sphere regions centered on metal and sulfur nuclei, with spheres in contact and completely enclosed within an outer sphere. Within the outer sphere but between the atomic sphere regions is the interatomic region. The electrostatic potential of the remaining atoms in the crystal lattice is approximated by a spherical shell of positive charge (the Watson Sphere) passing through the anion nuclei. The potential energy in the various regions is evaluated using electrostatics together with the Xα statistical approximation of Slater; the potential is then simplified and used to solve numerically the one-electron Schrödinger equation in each region. The resulting wave functions and their first derivatives are joined continuously through the various regions using multiple-scattered-wave theory. From this result, the spatial distribution of electron density is calculated and used to generate a new potential for the next iteration. The entire procedure is then repeated until self-consistency is attained and the result is a set of one-electron molecular orbitals characterized by their energies and electron density distributions.

An Xα method calculation has been performed on a ZnS_4^{6-} cluster by Tossell[7] and the results are shown in the energy level diagrams of Fig. 3. Shown to the left of Fig. 3 is a set of MO energy levels for the ZnS_4^{6-} cluster (labelled according to the irreducible representations of the T_d symmetry group). The calculations also provide information on the compositions of particular molecular orbitals in terms of the atomic orbitals of metal or sulfur. Thus, the $4t_2$, $1t$ and $2e$ molecular orbitals are dominantly sulfur 3p in character

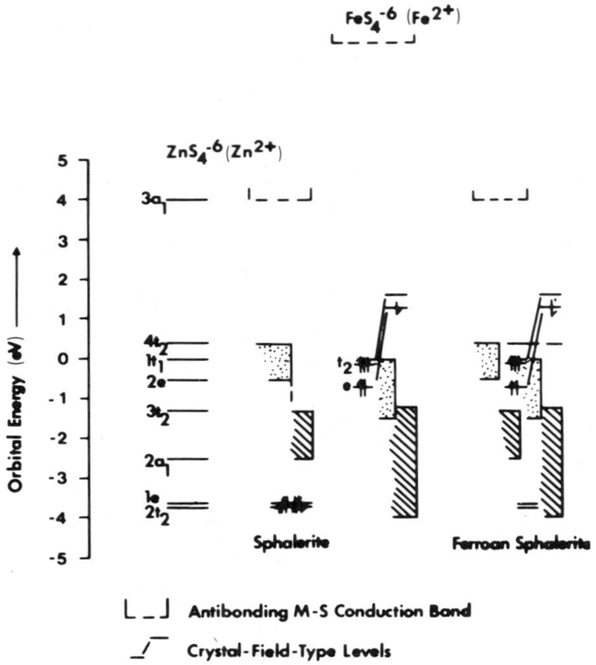

Fig. 3. Electronic structure models for sphalerite and ferroan sphalerite. Discrete energy levels are shown for the clusters ZnS_4^{6-} and FeS_4^{5-} and composite "one-electron" band model energy level diagrams for the minerals.

and nonbonding whereas the $3t_2$ and $2a_1$ orbitals are the main metal-sulfur bonding orbitals. The 1e and $2t_2$ orbitals are the dominantly metal 3d ("crystal field") type orbitals in this system. When large numbers of these tetrahedral units are brought together in a sphalerite crystal, overlap between many of these orbitals can be considered to result in formation of bands as illustrated in the simplified "one electron" band model also shown in Fig. 3.[8] Here, the highest energy band shown is the empty metal-sulfur antibonding band, beneath which is an energy gap to the filled nonbonding sulfur band. Stabilized relative to this is the main metal-sulfur bonding band, but the ("crystal-field type") Zn 3d levels are shown in this model as not forming a band but as remaining localized on the cation. Comparison of the calculated energy levels with those obtained from the study of the X-ray photoelectron and X-ray emission spectra shows that the calculations are correct in terms of the ordering and approximate separation of the MO energy levels. To facilitate comparison between calculations and spectra it is possible to use a "transition-state procedure" in which the calculation is reiterated for an electron configuration such as is found during the X-ray photoelectron or emission process.

Samples of natural sphalerite are very rarely pure ZnS and, commonly, Zn is replaced by Fe^{2+} in the tetrahedral sites. The effect of this on the electronic structure can be modelled by adding data from an Xα calculation on an FeS_4^{6-} cluster[9] as also shown in Fig. 3. The single unpaired 3d electron of Fe^{2+} results in some splitting of the valence orbital energy levels into spin-up (↑) and spin-down (↓) levels although, in this simplified diagram, it has only been shown for the crystal-field type levels since these are the levels most affected. The composite one-electron band model for ferroan sphalerite shows significant changes in the relative energies of the sulfur nonbonding band, the metal-sulfur bonding band and, in particular, the iron 3d crystal-field type levels compared to pure ZnS.

The models for sphalerite and ferroan sphalerite illustrated in Fig. 3 may be used in the interpretation (and prediction) of not only the X-ray emission and photoelectron spectra but also of properties of mineralogical interest such as spectral reflectance profiles and hardness variation.[8] For example, the overall trend to decreasing hardness and increasing cell parameter with iron substitution correlates with the destabilization inherent in the addition of the t_2 ↓ electron of Fe^{2+} and with the overall destabilization of the iron-sulfur bonding orbitals relative to zinc-sulfur. This also correlates with a decrease in thermodynamic stability with iron substitution.[8]

A similar approach can be adopted to a model of the electronic structure of galena (PbS) the properties of which have been described elsewhere.[1] The valence region X-ray photoelectron spectrum of galena is shown in Fig. 4(a).[10] The interpretation of the features in this spectrum can again be clarified with reference both to band structure calculations and cluster calculations using the "Xα method". These confirm the essential features of the simple one-electron band model shown in Fig. 4(b).[2] This shows the topmost part of the valence band as having essentially nonbonding sulfur 3p-type orbitals with the main lead (6s) - sulfur (3p) bonding orbitals about 6eV beneath the top of the valence band and the lead 5d nonbonding orbitals nearly 20eV beneath the top of the valence band.

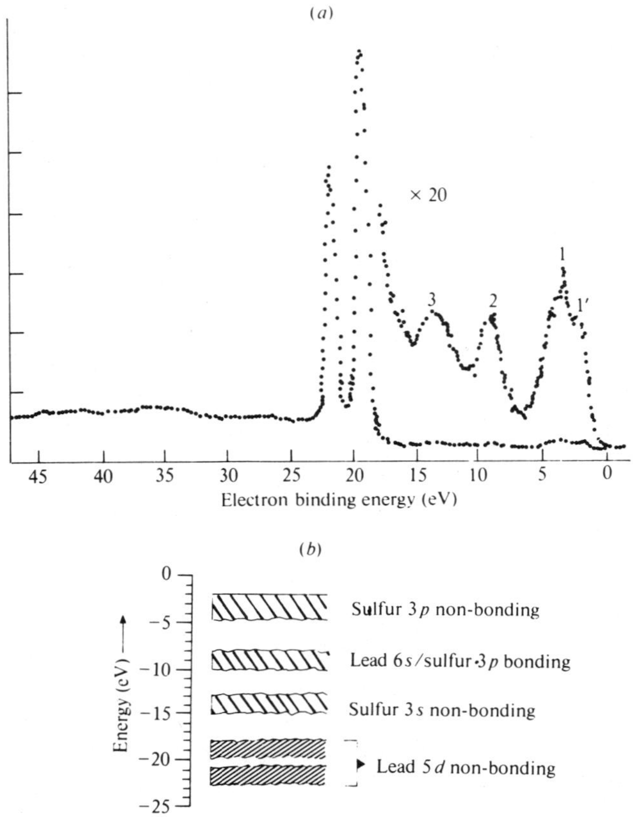

Fig. 4. (a) X-ray photoelectron spectrum for PbS; (b) simplistic band model for PbS.

BINARY SULFIDES OF IRON AND COPPER

Pyrite (FeS_2) is the most abundant of all sulfide minerals; it occurs in most sulfide ore deposits, often as a major phase.[1,2] Pyrite is a diamagnetic semiconductor with a small band gap of ~1eV and as well as containing octahedrally coordinated Fe^{2+} in the structure, there are also S-S dianion groups.[1] As in the example of sphalerite, qualitative MO and one-electron band models can be proposed on the basis of these observed properties. In the simple MO diagram of Fig. 5(a), iron-sulfur bonding is described in this transition metal system. The relative energies of outer atomic orbitals involved in bonding are qualitatively shown and overlap between s and p orbitals on metal and sulfur again regarded as forming a filled set of stabilized σ bonding orbitals and an empty (destabilized) σ* antibonding set. In pyrite, there is the additional complication of the

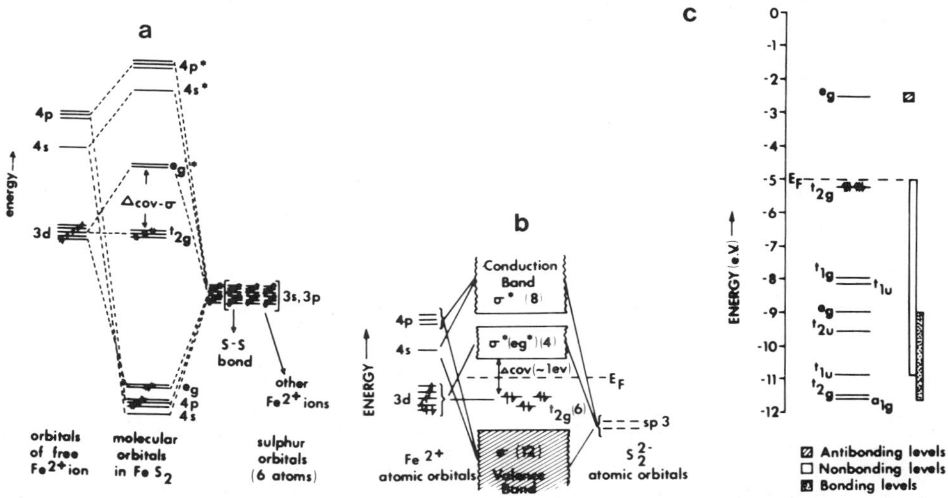

Fig. 5. Electronic structure models for pyrite: (a) molecular orbital energy level diagram; (b) schematic "one-electron" energy band diagram. Numbers in brackets refer to number of electron states per molecule available (E_f Fermi level, hatched bands are filled with electrons); (c) Energy level diagram for an FeS_6^{10-} cluster from an Xα calculation. Energy levels labelled using group theory nomenclature for the O_h symmetry group.

involvement of the 3d orbitals of iron in bonding and in this model, the 3d orbitals of e_g symmetry which are directed towards the surrounding sulfurs are shown as split by interaction with the sulfurs into a bonding e_g MO and an antibonding e_g * MO. The t_{2g}-type orbitals which are directed between the sulfur atoms surrounding the iron are shown as nonbonding orbitals and the separation in energy between the t_{2g} and e_g orbital energies ($\Delta_{cov-\sigma}$) is analogous to the "Crystal Field Stabilization Energy" of crystal field theory. As pyrite is diamagnetic, there can be no unpaired electrons in the system, so that all six of the 3d electrons of Fe^{2+} are shown spin-paired in the t_{2g} orbitals. Fig. 5(a) represents Fe-S bonding in a single FeS_6 cluster unit and, in the crystal, overlap between molecular orbitals on adjacent units causes broadening of the energy levels into bands. The simple one-electron energy band model of Fig. 5(b) illustrates this aspect. Now the σ-bonding molecular orbitals form the filled valence band with a corresponding empty σ* conduction band. The t_{2g} levels are shown here as remaining localized on the cation and not overlapping to form a band, in contrast to the empty e_g* band. The separation between filled t_{2g} levels

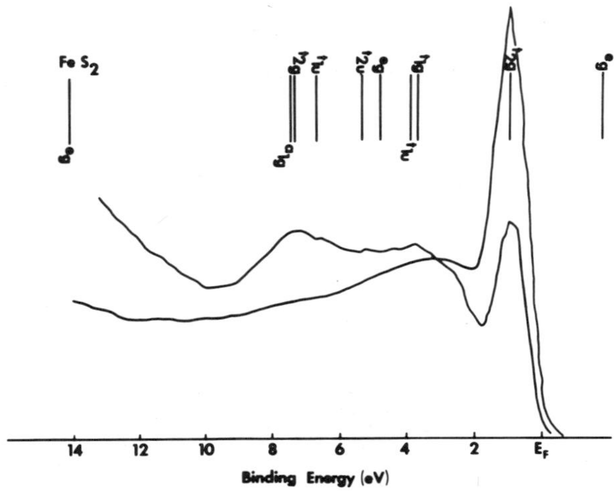

Fig. 6. UV photoelectron spectrum for pyrite with the energy levels from the Xα calculation superimposed.

and the empty e_g* band is the band gap, i.e. the energy required to promote an electron into the e_g* band which will then permit conduction of an electric current.

Again, models like these can be refined and tested through spectroscopic measurements and quantum mechanical calculations. Results of an Xα calculation on an FeS_6^{10-} cluster[13,14] are shown in Fig. 5(c) and are compared with the UV photoelectron spectrum of pyrite[15] in Fig. 6. The spectrum shows an intense feature at low binding energy which arises from the six electrons in the t_{2g} levels as predicted by the model. Roughly 3eV below this in energy are a group of orbitals (t_{1g}, t_{1u}) which are essentially S3p-type nonbonding orbitals and then immediately beneath these are the main bonding orbitals of the system (e_g, t_{2g}, a_{1g}) which have sulfur 3p and iron 3d and 4s character according to the calculations. Approximately 7eV below this is a sulfur 3s nonbonding orbital. As can be seen from Fig. 6, agreement between calculation and experiment is good. Properties of mineralogical interest such as color, reflectance, cell parameter, hardness and thermodynamic stability, not only of pyrite but also of the isostructural series FeS_2-CoS_2-NiS_2-CuS_2-ZnS_2 have been interpreted using such models.[14]

The binary copper sulfides covellite (CuS) and chalcocite (Cu_2S) have more complex crystal structures than sphalerite, galena or

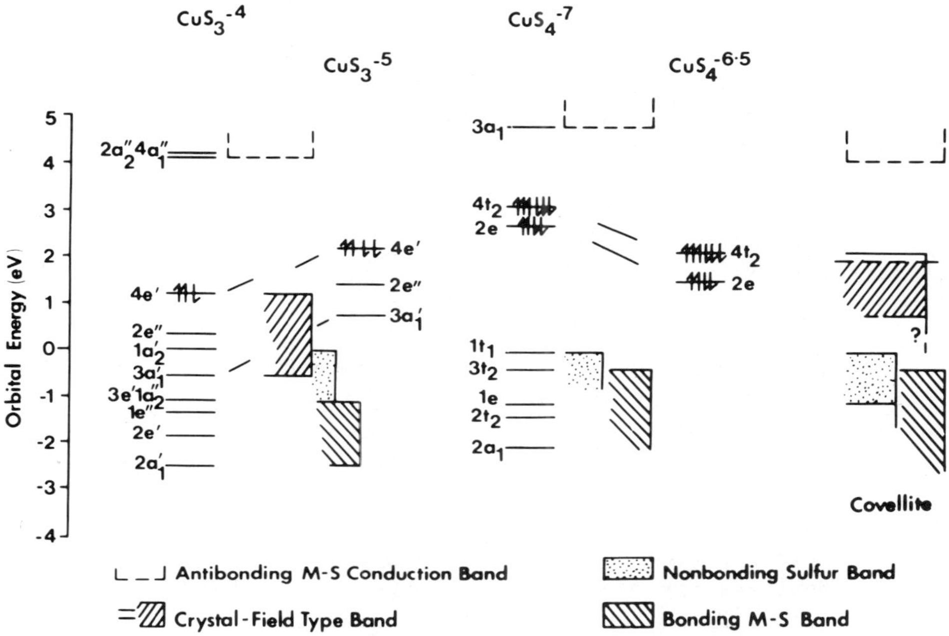

Fig. 7. Electronic structure model for covellite. Discrete energy levels are shown for clusters CuS_3^{4-}, CuS_3^{5-}, CuS_4^{7-}, $CuS_4^{6.5-}$ and a composite one-electron band model energy level diagram for the mineral.

pyrite.[1,2] In covellite, copper occurs in both triangular and roughly tetrahedral coordination and a further complexity is that two thirds of the sulfur atoms are combined in S_2^{2-} dianion units. The configuration that has been proposed for covellite is Cu_{III}^{2+} $(Cu_{IV}^{+})_2$ S^{2-} $(S_2^{2-})_2$ in which, therefore Cu^{2+} occupies triangular and Cu^{+} tetrahedral sites. Molecular orbital calculations using the Xα method have been performed on CuS_3^{4-} and CuS_4^{7-} clusters [8,15] and produced the energy level diagrams shown in Fig. 7. As in some of the earlier diagrams the compositions of particular groups of orbitals in terms of atomic orbital character have also been indicated in the simplified band models shown alongside the energy levels calculated for the isolated clusters. Again the validity of the calculations has been established by comparison with spectroscopic data.[15] However, the metallic conductivity observed in CuS[16] indicates that the highest energy orbitals containing electrons form a partly filled collective electron band. To investigate this, further calculations in which the transition state procedure was employed (transfer of half an electron from Cu^{+}_{IV} to Cu_{III}^{2+}) indicate that to form such a band, charge should flow from the $4t_2$ orbital on the tetrahedral Cu^{+} to the 4e orbital on the triangular Cu^{2+}. The resulting change in electronic structure is then represented by cluster calculations for CuS_3^{5-} and $CuS_4^{6.5-}$ also shown in Fig. 7 and these have been combined to produce the composite band model also shown. The model shows that above a nonbonding sulfur band and metal-sulfur bonding band lies an incompletely filled band dominantly of metal 3d orbital character (and minus the equivalent of one third an electron per molecule).

Chalcocite has a structure in which copper occurs in two kinds of triangular coordination.[17] The energy level diagram shown in Fig. 8 has been empirically constructed using sulfur K_β and sulfur $L_{2,3}$ X-ray emission spectra and X-ray photoelectron spectra of Cu_2S.[18] Results from the Xα method calculation on a CuS_3^{5-} cluster (Fig. 7) show good agreement with experimental data and help in interpretation of the spectra. As shown in Fig. 8, the highest occupied orbitals are predominantly copper 3d in character with a group of sulfur 3p nonbonding orbitals about 2eV below them, and then the main bonding orbitals of mixed sulfur 3p and copper 4p or 4s character.

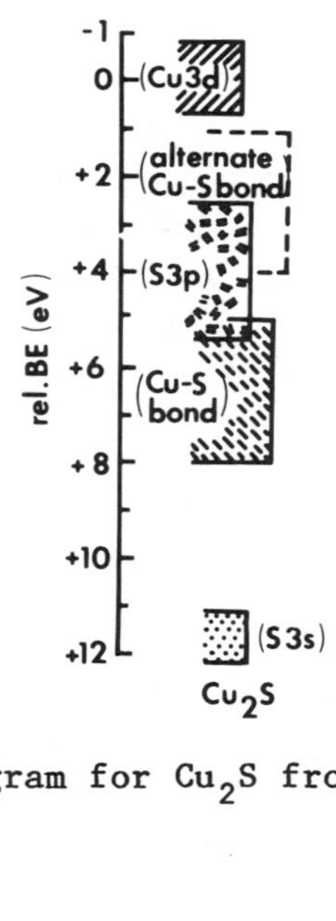

Fig. 8. Energy level diagram for Cu_2S from spectroscopic data.

COMPLEX SULFIDES - CHALCOPYRITE

Chalcopyrite ($CuFeS_2$) contains both copper and iron in tetrahedral coordination with sulfur, the crystal structure being derived from that of sphalerite by ordered substitution of Cu and Fe for Zn[1]. Although there has been some disagreement regarding the oxidation states of the cations in this structure, neutron diffraction, Mössbauer spectral and X-ray photoelectron spectral data all strongly support a configuration of Cu^{+} Fe^{3+} S_2 for chalcopyrite.[8,19]

The electronic structure of chalcopyrite has been studied using a combination of X-ray emission and X-ray photoelectron spectroscopies and Xα method calculations on the CuS_4^{7-} and FeS_4^{5-} molecular clusters.[8,18] A self-consistent band structure calculation has also been performed for $CuFeS_2$.[20] In Fig. 9, a complete set of valence region X-ray emission and X-ray photoelectron ("ESCA") spectra for chalcopyrite is shown. Assignment of the features in these spectra is possible using the results of the cluster calculations which can, in turn, be used to construct the somewhat simplified electron structure models presented in Fig. 10. In this figure are shown results for the CuS_4^{7-} and FeS_4^{5-} clusters and, between them, a composite "one-electron" band model energy level diagram for chalcopyrite.

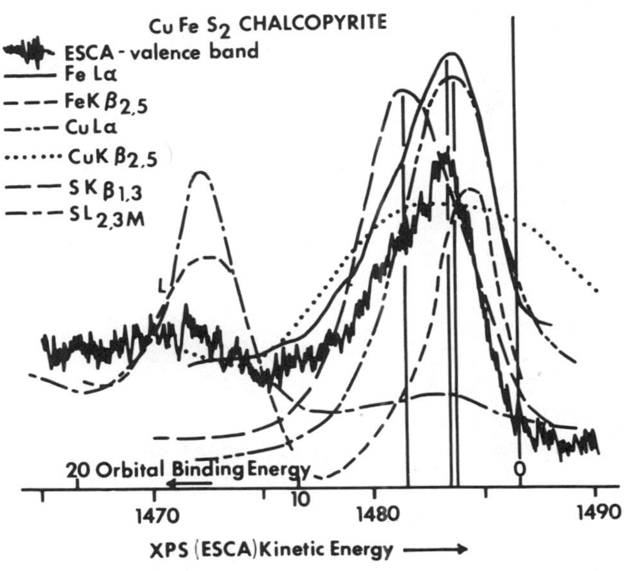

Fig. 9. X-ray emission and X-ray photoelectron ("ESCA") spectra for chalcopyrite.

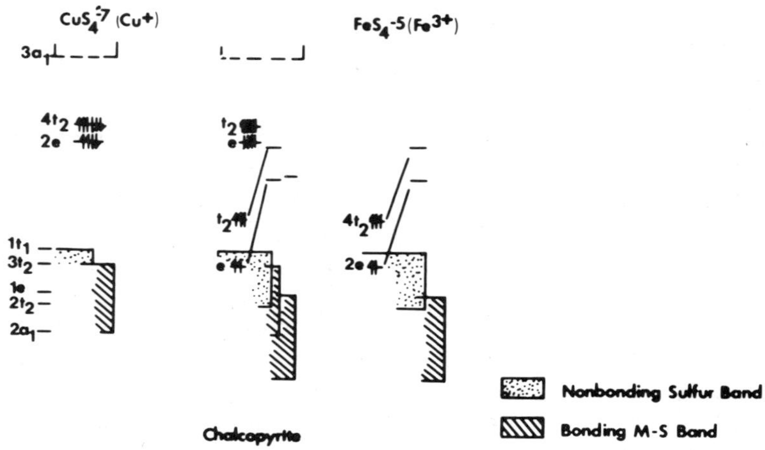

Fig. 10. Electronic structure model for chalcopyrite. Energy level diagrams are shown for the CuS_4^{7-} and FeS_4^{5-} clusters along with a composite "one-electron" band model energy level diagram for the mineral.

APPLICATIONS TO LEACHING BEHAVIOR

Electronic structure models are the key to a fundamental understanding of the physical and chemical properties of materials, including aspects of the behavior of materials during leaching. In the first part of this paper, electronic structure models have been presented for some of the industrially important metal sulfide minerals. Even though these models are based on relatively complex calculations and on spectroscopic data obtained using modern high resolution spectrometers, they are still a simplistic approach to the problem of electronic structure. They also represent the bulk properties of the pure materials and, although they form the basis for an understanding of mineral surface electronic structure and of the behavior of impurities and defects, much remains to be done in developing this aspect of the models. The same can be said regarding specific applications to leaching behavior, so that in this discussion the potential of electronic structure models in understanding leaching processes will be emphasized so as to point out some directions for future work.

The models presented in earlier sections of this article show major differences in electronic structure between the six minerals discussed. Sphalerite, which is virtually an insulator (band gap ~ 3.65eV, Fig. 1 and 3) has a topmost valence band made up of sulfur

3p nonbonding orbitals. One proposed method for the hydrometallurgical treatment of ZnS involves pressure leaching with sulfuric acid[21] which essentially proceeds according to the equation:

$$ZnS + 1/2O_2 + H_2SO_4 \rightarrow ZnSO_4 + S + H_2O \quad (1)$$

An electrochemical model proposed for this process[22] has the zinc sulfide reacting anodically according to:

$$ZnS \rightarrow Zn^{2+} + S^0 + 2e^- \quad (2)$$

An appreciable increase in the rate of dissolution was produced by introducing UV radiation. This, in line with the electronic structure model, would cause promotion of electrons from the filled valence band into the empty conduction band and result in enhanced conductivity and hence dissolution rates. In another study[23], the potential catalytic effects on this reaction of various metals in solution was investigated. Copper, in particular, had an appreciable effect and this can again be considered in the light of our models. The copper would substitute for zinc at surface sites so that the model presented for a $CuS_4{}^{7-}$ cluster (see Fig. 10) might appropriately describe the system. In fact, the oxidation state for this surface species may well be Cu^{2+} so that a partially-filled band of dominantly copper 3d orbital character (see also Fig. 7) will probably form the top of the valence region. Conductivity would be much enhanced, leading to the observed effects on kinetics of dissolution. It is interesting to note that although Ag, Hg and Pb were also tried, they had no catalytic effect. Electronic structure models of sulfides of these metals[11,18] all suggest that the topmost part of the valence region would still be made up of sulfur nonbonding orbitals forming a filled band.

Galena also has an electronic structure in which the top of the valence band is made up of sulfur 3p nonbonding orbitals (Fig. 4) and it is a semiconductor with a band gap of ~ 0.41eV[2]. An electrochemical model has also been proposed for the oxidation of galena[24] in aqueous suspensions with anodic reaction of the lead sulfide:

$$PbS \rightarrow Pb^{2+} + S^0 + 2e^-$$

and a corresponding cathodic reaction:

$$O_2 + 2H_2O + e^- \rightarrow 4OH^-$$

Galena exhibits non-stoichiometry[1] and the rate of the above reaction was shown to vary as a function of this non-stoichiometry and to be affected by the action of light. The mechanism proposed involves the chemisorption of oxygen with transfer of electrons from the PbS to leave (positive) holes (this also corresponds to breaking of Pb-S bonds and leaves S at the surface with Pb^{2+} going into solution). Because a sulfur-rich layer forms at the surface, diffusion of sulfur occurs between the surface and the interior of the crystal and this process has a controlling effect on overall reaction rate and is itself controlled by stoichiometry. The action of light on PbS (as the electronic structure model would indicate) is to promote electrons into the conduction band, simultaneously generating holes in the valence band. In a non-stoichiometric crystal, this has the effect of overriding the electronic effects of non-stoichiometry, hence the influence of this factor on reaction rates. The chemisorption of oxygen at the galena surface is another aspect of this process which could, in future, be studied by electronic structure calculations.

Pyrite is a very stable sulfide mineral as can be seen from its electronic structure (Fig. 5). The absence of electrons in the e_g* band (which would have a destabilizing effect) and the configuration whereby all six 3d electrons of iron are spin-paired in the t_{2g} orbitals is particularly stable. It is not surprising, therefore, that pyrite is one of the sulfides which it is most difficult to leach.

The copper sulfides chalcocite and covellite, and the copper-iron sulfide chalcopyrite, are perhaps the most extensively studied sulfides as regards leaching properties. Their behavior is complex, as indeed are their electronic structures. All also exhibit varying degrees of non-stoichiometry with accompanying structural complexities which may arise from ordered arrangements of additional atoms ordering of vacant sites in the structures.[1,2] Leaching reactions commonly involve successive removal of metal atoms from the structure, as in the leaching of chalcocite (Cu_2S) by ferric ion media which first produces CuS.[25] Natural leaching processes apparently lead to even more complex sequences such as djurleite ($Cu_{1.97}S$) → anilite ($Cu_{1.75}S$) → spionkopite ($Cu_{1.40}S$) → yarrowite ($Cu_{1.12}S$) → covellite (CuS).[26] The copper sulfides such as chalcocite and covellite have electronic structures (Figs. 7 and 8) in which the top of the valence band is formed by a band (sometimes partly filled) of largely copper 3d orbital character which is able to accommodate variations in metal concentration. The sulfur sublattice behaves rather as a framework within which the metal atoms are quite mobile even at near to room temperature.

The use of electronic structure models to understand aspects of the behavior of sulfide minerals during leaching should lead to a better appreciation of the mechanisms of certain leaching reactions. It should also lead to a clearer understanding of the importance in leaching of such factors as mineral stoichiometry and mineral purity, and also of the action of catalysts. Much remains to be done in applying the electronic structure models now becoming available to such industrial problems.

ACKNOWLEDGEMENTS

I wish to thank Professor R. Bautista for the invitation to present this paper to the NATO Advanced Research Institute. Dr. S. Guy of the Dept. of Minerals Engineering and Dr. M. Hayes of the Chemistry Department, both at Birmingham University, are thanked for critical comments on the manuscript. Cathy Kennedy is thanked for preparation of the typescript and Beverley Parker for drafting certain illustrations.

REFERENCES

1. D. J. Vaughan: this volume pp.
2. D. J. Vaughan and J. R. Craig: Mineral Chemistry of Metal Sulfides, Cambridge Univ Press, Cambridge, 1978.
3. L. Ley, R. A. Pollak, F. R. McFeely, S. P. Kowalczyk and D. A. Shirley: Phys. Rev. B, 1974, vol. 9, pp. 600-621.
4. E. P. Domashevskaya, V. A. Terekhov, L. N. Marshakova, Ya. A. Ugai, V. I. Nefedov and N. P. Sergushin: J. Electron. Spect., 1976, vol. 9, pp. 261-
5. J. C. Slater and K. H. Johnson: Phys, Rev. B, 1972, vol. 5, pp. 844-853.
6. K. H. Johnson: Adv. Quantum Chem., 1973, vol. 7, pp. 143-185.
7. J. A. Tossell, Inorg. Chem., 1977, vol. 16, pp. 2944-2949.
8. D. J. Vaughan and J. A. Tossell: Canad. Mineral., 1980, vol. 18, pp. 157-163.
9. D. J. Vaughan, J. A. Tossell and K. H. Johnson: Geochim. Cosmochim Acta, 1974, vol. 38, pp. 993-1005.
10. F. R. McFeely, S. Kowalczyk, L. Ley, R. A. Pollak and D. A. Shirley: Phys. Rev. B, 1973, vol. 7, pp. 5228-5236.
11. L. A. Hemstreet: Phys. Rev. B, 1975, vol. 11, pp. 2260-2270.
12. J. R. Craig and D. J. Vaughan: Ore Microscopy and Ore Petrography, Wiley-Interscience, New York, 1981.
13. J. A. Tossell: J. Chem. Phys., 1977, vol. 66, pp. 5712-5719.

14. D. J. Vaughan: Bull. Mineral, 1978, vol. 101, pp. 484-497.
15. J. A. Tossell: Phys. Chem. Minerals, 1978, vol. 2, pp. 225-236.
16. R. T. Shuey: Semiconducting Ore Minerals, Elsevier, Amsterdam, 1975.
17. H. T. Evans Jr: Nature, 1971, vol. 232, pp. 69-70.
18. J. A. Tossell and D. J. Vaughan: Inorg. Chem., 1981, vol. 20, pp. 3333-3340.
19. J. A. Tossell, D. S. Urch, D. J. Vaughan and G. Wiech: Chem. Phys. Letts., 1982, vol. 77, pp. 77-82.
20. T. Hamagima, T. Kambara, K. I. Gondaipa and T. Oguchi: Phys. Rev. B, 1981, vol. 24, pp. 3349-3353.
21. F. Habashi: Proc. 13th Internat. Mineral Processing Congr. Warsaw, 1981, Part A, pp. 902-935.
22. F. E. Pawlek: J. S. Afr. Inst. Min. Metall, 1969, pp. 632-654.
23. T. R. Scott and N. F. Dyson: Trans. Met. Soc. AIME, 1969, vol. 242, pp. 1815-1821.
24. P. Eadington and A. P. Prosser: Trans. Inst. Min. Metall., 1969, vol. 78, pp. C74-C82.
25. J. E. Dutrizac and R. J. C. MacDonald: Minerals Sci. Engng., 1974, vol. 6, pp. 59-100.
26. R. J. Goble: Canad. Mineral., 1981, vol. 19, pp. 583-591.

HETEROGENEOUS RATE PROCESSES IN THE LEACHING OF BASE METAL SULFIDES

Milton E. Wadsworth

Department of Metallurgical Engineering
University of Utah
Salt Lake City, Utah 84112

Fundamental rate processes in sulfide leaching are reviewed. Electrochemical reactions of importance in dissolution and solid-state conversion reactions are presented and emphasis is placed upon copper and lead sulfides. Parameters important in characterizing potential new hydrometallurgical processes using anodic and cathodic reactions are reviewed.

INTRODUCTION

The semi-conducting properties of many important commercial minerals gives rise to a variety of electrochemical processes important in many hydrometallurgical systems. Koch[1] has reviewed the electrochemistry of sulfide minerals and Shuey[2] and Vaughan and Craig[3] the electronic properties of both oxide and sulfide minerals.

Electrochemical reactions of importance in hydrometallurgy may be viewed in terms of the character of charge and mass transfer. In this presentation three types of coupled systems will be considered. These are:

1) Mixed potential corrosion couples
2) Mixed potential galvanic couples
3) Charge transfer through passivation layers

The concept of mixed potential is important in many anodic and cathodic processes since even a modest voltage may have an important influence on the observed kinetics where charge transfer is involved.

MIXED POTENTIALS IN ELECTROCHEMICAL PROCESSES

Electrochemical processes are unique in that the solid may assume a uniform potential throughout, providing ohmic resistance is negligible. Consequently the chemical reaction may be influenced over relatively long distances on an atomic scale. The potential may directly affect the kinetics of the reaction, if slow discharge is involved, and may also serve to stabilize intermediate solid phases. If one or more electrode reactions occur simultaneously a mixed potential may result. Vetter[4] has discussed the concept of additive partial electrode processes important in mixed electrode systems. Concepts developed in the field of corrosion are translatable to the electron conducting solids in hydrometallurgical systems, with coupled interfacial charge transfer. At the mixed potential, the sum of all anodic currents, ΣI_a, is equal to the negative sum of all cathodic currents, ΣI_c, such that

$$\Sigma I_a = -\Sigma I_c$$

where $I_a = i_a A_a$ amd $I_c A_c$. The terms A_a and A_c are the respective anodic and cathodic areas, and i_a and i_c are the corresponding current densities. As in corrosion processes, two types of mixed potential regimes may be operative (i.e. corrosion and galvanic couples). The corrosion cell type involves a solid electron conducting phase having both anodic and cathodic reactions occurring on the same surface; i.e. $A_a = A_c$. A galvanic couple is operative where two or more solid phases are in electrical contact. In this union each solid assumes either anodic or cathodic behavior and has its own separate surface area for reaction.

Individual electrode polarization curves are helpful in visualizing the concept of a mixed potential. Figure 1 depicts hypothetical polarization curves for the dissolution of a metal sulfide in the presence of a cationic oxidant N^{n+}. If MS is placed in contact with a solution containing an oxidant N^{n+} (e.g. Fe^{3+} in acid solutions or $Cu(NH_3)_4^{2+}$ in alkaline solutions) having a more positive equilbirium potential, a corrosion cell will result in the anodic oxidation of MS and the cathodic reduction of N^{n+}:

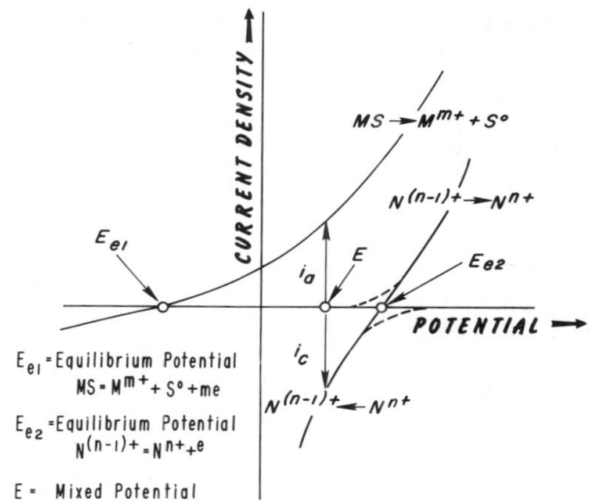

Fig. 1. Current density potential diagram showing curves for metal sulfide with rest potential E_e and redox ions with rest potential E_e forming mixed potential E.

$$MS = M^{m+} + S^{\circ} + me^{-} \quad (1)$$

and

$$N^{n+} + e = N^{(n-1)+} \quad (2)$$

The equilibrium potentials E_e and E_e for half-cell reactions (1) and (2) are illustrated in Figure 1. Equilibrium electrode potentials are independent of the properties of the electrode surface, however, electrode surface properties influence to a great extent the shape of the partial current-voltage curves. The value of the mixed potential is established by both the partial equilibrium potentials and the shape of the current-voltage curves. Hence, the mixed potential depends upon the properties of the electrode surface. Potentials more positive than the equilibrium potential drive the half-cell reaction in the net anodic direction and potentials more negative than the equilibrium potential drive the reactions in the net cathodic direction.

The net current density (i) for the system is the sum of the partial current densities i_a (anodic) and i_c (cathodic). The mixed potential (E) occurs where the net current density is zero; the values of the partial current densities therefore are equal and of the opposite sign ($i_a = -i_c$) as indicated in Fig. 1. The anodic overvoltage η_a is defined as $(E-E_{e_1})$ and is positive. The cathodic overpotential η_c is defined as $(E-E_{e_2})$ and is negative.

The electrochemical bias for charge transfer in simple reversible reactions may be expressed quantitatively by the Butler-Volmer equation. This equation relates the current density at a solid-solution of interface to the established overpotential. Applied to the specific example of a metal sulfide undergoing anodic dissolution, Eq. 1, the Butler-Volmer equation, for a single electron transfer process, is

$$i = z\mathcal{F}\vec{k}\ \exp\left[\frac{\beta\mathcal{F}E}{RT}\right] - z\mathcal{F}\overleftarrow{k}\ [M^{m+}]\ \exp\left[\frac{-(1-\beta)\mathcal{F}E}{RT}\right] \quad (3)$$

The first term in the above equation is the partial current density for the anodic (forward) direction of Eq. 1 and the second term the partial current density for the cathodic (reverse) direction, $\mathcal{F}$ is the Faraday, $\vec{k}$ and $\overleftarrow{k}$ are the rates for the forward and reverse directions, respectively, in the absence of potential, and β is the transfer coefficient. The rate $\overleftarrow{k}$ may contain a concentration term for ad-ions on the mineral surface which may change if the stoichiometry of the mineral varies. At the equilibrium potential E_{e1}, the partial current densities of Eq. 1 are equal $|\vec{i}| = |\overleftarrow{i}|$ and are represented by the exchange current density, i_o. The exchange current density is given by the following relationships

$$i_o = z\mathcal{F}\vec{k}\ \exp\left[\frac{\beta\mathcal{F}E_{e1}}{RT}\right] = z\mathcal{F}\overleftarrow{k}\ [M^{m+}]\ \exp\left[\frac{-(1-\beta)\mathcal{F}E}{RT}\right] \quad (4)$$

and it follows directly that

$$i = i_o\ \left\{\exp\left[\frac{\beta\mathcal{F}\eta}{RT}\right] - \exp\left[\frac{-(1-\beta)\mathcal{F}\eta}{RT}\right]\right\} \quad (5)$$

By definition the overvoltage η is equal to the difference between E and E_1.

If the metal sulfide is allowed to contact an aqueous solution containing an oxidant, N^{n+}, a corrosion couple involving net anodic and net cathodic reactions will result and a steady-state mixed potential will be established. The Butler-Volver expression for the net anodic, i_a, and net cathodic, i_c, current densities for reactions (1) and (2) are

$$i_a = z_a\mathcal{F}\vec{k}_a\ \exp\left[\frac{\beta_a n_a\mathcal{F}E}{RT}\right] - z_a\mathcal{F}\overleftarrow{k}_a\ [M^{m+}]\ \exp\left[\frac{-(1-\beta_a)n_a\mathcal{F}E}{RT}\right] \quad (6)$$

and

$$i_c = z_c \mathcal{F} \overleftarrow{k}_c \, [N^{(n-1)+}] \exp\left[\frac{\beta_c n_c \mathcal{F} E}{RT}\right] - z_c \mathcal{F} \overrightarrow{k}_c \, [N^{n+}] \exp\left[\frac{-(1-\beta_c) n_c \mathcal{F} E}{RT}\right] \tag{7}$$

Coupling the anodic and cathodic half cell reactions provides a means for estimating the mixed potential, although there is no explicit solution without certain assumptions. For purposes of simplification the assumption that $\beta_a \cong \beta_c = \beta$ is usually a satisfactory approximation. Also rate controlling processes most commonly involve single charge transfer steps such that n_a and n_c may be taken as unity, an assumption made here only for purposes of simplification. Accordingly Eqs. 6 and 7 may be combined giving

$$\exp\,[\mathcal{F}E/RT] = \frac{A_a z_a \overleftarrow{k}_a [M^{m+}] + A_c z_c \overrightarrow{k}_c [N^{n+}]}{A_a z_a \overrightarrow{k}_a + A_c z_c \overleftarrow{k}_c [N^{(n-1)+}]} \tag{8}$$

where E is the mixed potential. Also

$$-\frac{dn_{MS}}{dt} = \frac{A_a i_a}{z_a} \tag{9}$$

where n_{MS} refers to the moles of metal sulfide remaining at anytime t. Combining Eqs. 6, 8, and 9 results in the general charge transfer rate expression

$$-\frac{dn_{MS}}{dt} = \frac{A_a A_c z_c \overrightarrow{k}_a \overrightarrow{k}_c [N^{n+}][1-\overleftarrow{k}_a \overleftarrow{k}_c [M^{m+}][N^{(n-1)+}]/k_a k_c [N^{n+}]]}{[A_a z_a \overrightarrow{k}_a + A_c z_c \overleftarrow{k}_c [N^{(n-1)+}]]^{\beta} [A_a z_z \overleftarrow{k}_a [M^{m+}] + A_c z_c \overrightarrow{k}_c [N^{n+}]]^{1-\beta}} \tag{10}$$

It may be shown that

$$k_a k_c \, [M^{m+}][N^{(n-1)+}]/k_a k_c [N^{n+}] = \exp\,(\Delta G/RT) \tag{11}$$

where ΔG is the free energy for the reaction producing a given final product. Eq. 10 becomes

$$-\frac{dn_{MS}}{dt} = \frac{A_a A_c z_c \overrightarrow{k}_a \overrightarrow{k}_c [N^{n+}][1-\exp(\Delta G/RT)]}{[A_a z_a \overrightarrow{k}_a + A_c z_c \overleftarrow{k}_c [N^{(n-1)+}]]^{\beta} [A_a z_a \overleftarrow{k}_a [M^{m+}] + A_c z_c \overrightarrow{k}_c [N^{n+}]]^{1-\beta}} \tag{12}$$

It is apparent from Eq. 12 that for hydrometallurgical processes having a ΔG more negative than -4.0 kcal/mole ($\Delta G/RT \ll 1$) are essentially irreversible. Nevertheless products of reaction can influence the kinetics by their effect upon the associated half-cell reactions. Most sulfide and metal dissoulution processes are irreversible. Equation 12 may have several forms depending upon the relative values of the terms in the denominator. Assuming irreversibility, special cases are:

1) Mixed potential within both the anodic and cathodic Tafel regions;

$$-\frac{dn_{MS}}{dt} = (A_a\vec{k}_c)^{1-\beta}\,(A_c z_c \overleftarrow{k}_c/z_a)^{\beta}[N^{+}\,]^{\beta} \tag{13}$$

2) Mixed potential within the cathodic Tafel region only;

$$-\frac{dn_{MS}}{dt} = \frac{(A_a\vec{k}_a)^{1-\beta}\,A_c z_c \overleftarrow{k}_c [N^{n+}]}{z_a{}^{\beta}[A_a z_a \overleftarrow{k}_a [M^{m+}] + A_c z_c \overleftarrow{k}_c [N^{n+}]]^{1-\beta}} \tag{14}$$

3) Mixed potential within the anodic Tafel region only;

$$-\frac{dn_{MS}}{dt} = \frac{A_a\vec{k}_a\,(A_c z_c \overleftarrow{k}_c)^{\beta}[N^{n+}]^{\beta}}{[A_a z_a \vec{k}_a + A_c z_c \vec{k}_c [N^{(n-1)+}]]^{\beta}} \tag{15}$$

The pseudo fractional orders observed in reactions of this type result from the transfer coefficient β and fall generally in the range 0.4 to 0.6, with values commonly near 0.5.

A special case is one in which one electrode is near equilibrium, i.e. has a large exchange current density. If, for example, the cathode is near equilibrium, such that E(mixed) $\cong E_{el}$, and the anode is in its Tafel region, Eq. 12 becomes

$$-\frac{dn_{MS}}{dt} = A_a\vec{k}_a\left(\frac{\overleftarrow{k}}{\vec{k}_c}\right)^{\beta}\frac{[N^{n+}]^{\beta}}{[N^{(n-1)+}]^{\beta}} \tag{16}$$

If the cathodic reaction involves the Fe^{3+}/Fe^{2+} couple the rate of dissolution in this latter case is related to the solution E_h.

<u>Electrochemical Dissolution of Galena</u>

The electrochemical dissolution of galena in aqueous solutions has received considerable attention. The electrical conductivity of galena is unusually high, and galena is recognized as having both n- and p-type conductivity. Eadington and Prosser[10] have demonstrated

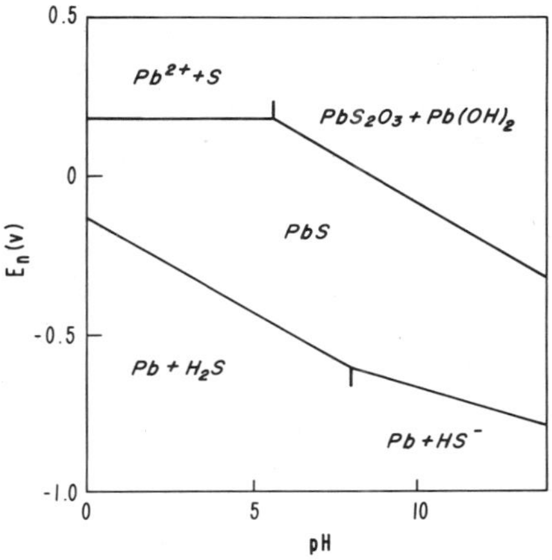

Fig. 2. Simplified E_h-pH diagram for the Pb-S system.

that the oxidation of galena is noticeably influenced by the semiconductor type. For non-stoichiometric PbS, the rate of oxidation was much higher for sulfur-rich (p-type semi-conductor) specimens than for lead-rich (n-type) specimens. In aqueous solution lead forms a large array of sparingly soluble or insoluble compounds; these materials can form surface films which mask the original electronic properties of the solid. Paul et al.[11,12] addressed the role of insoluble films as they determined the transition between active-passive behavior during the electrochemical dissolution of galena. They also developed dissolution mechanisms for various aqueous solutions.

The simplified Pourbaix diagram shown in Fig. 2 illustrates the equilibria PbS with sulfur and lead according to the following[12] reactions

$$PbS = Pb^{2+} + S^\circ + 2e \tag{17}$$

or

$$PbS + 2H^+ + 2e = Pb + H_2S \tag{18}$$

The galena electrode is acknowledged as one of the most reversible among sulfide minerals. The voltammogram[11] appearing in Fig. 3a illustrates the complex nature of the electrode kinetics. The anodic and cathodic sweeps in perchlorate solutions reveal several interesting features. During the anodic sweep there is evidence of an oxidation peak at 0.1 v which corresponds to the oxidation of H_2S to S°. The anodic dissolution of galena according to reaction (11)[12] proceeds at pH 2 without loss of activity until approximately 0.5 v.

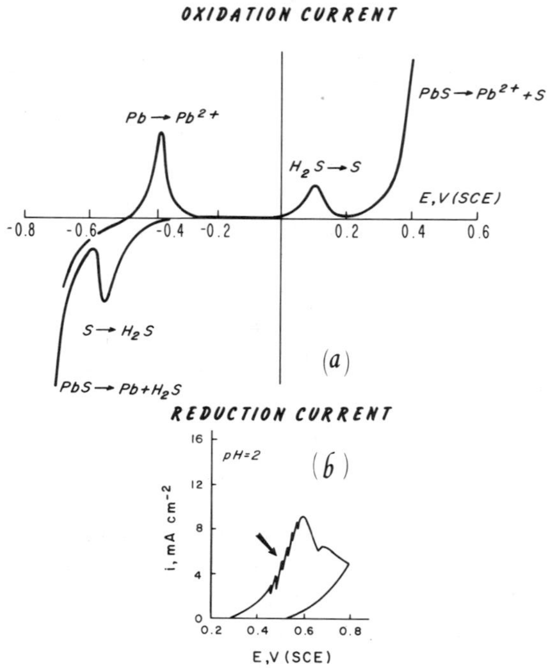

Fig. 3. a) Polarization and reduction currents for the cycl voltammetric reactions of galena in a perchlorate solution; and b) anodic dissolution of galena showing sequential passivation and oxidation.

The electrode then experiences passivation-reactivation fluctuations in the curve as denoted by the arrow in Fig. 3b. This region is a result of the sulfur layer that inhibits the diffusion of Pb^{2+} from the surface. As the concentration of Pb^{2+} increases at the galena surface there is an attendant increase in pH. The higher pH promotes the oxidation of sulfur to a higher oxidation state, and the electrode surface will be quickly passivated either by the precipitation of lead thiosulfate or the direct formation of lead thiosulfate according to

$$2\ PbS + 3H_2O = PbS_2O_3 + Pb^2 + 6H^+ + 8e \tag{19}$$

Reactivation occurs as the PbS_2O_3 dissolves as a result of the continued diffusion of lead ions from the surface through the pores of the sulfur deposit. There is a well defined passivation of the electrode surface beginning at about 0.6 v caused by the precipitation of a basic lead sulfate ($PbO \cdot PbSO_4$ or $3PbO \cdot PbSO_4$) on the galena surface. On the cathodic side of the sweep, a peak which reflects the reduction of S° to H_2S appears near -0.55 v. The peak corresponding to the oxidation of Pb° to Pb^{2+} can only occur following the reduction of galena to lead and hydrogen sulfide. It appears that the formation of lead on the galena surface does not reduce its cathodic activity. Similar cyclic-voltammetric behavior was observed for chloride and sulfide solutions. However, these systems exhibited a pronounced influence from both $PbCl_2$ and $PbSO_4$ formed from the respective solutions.

From the study described above Paul et al.[12] suggested a mechanism for the dissolution of galena in acidic perchlorate solutions. The overall anodic reaction is given in Eq. 17. Tafel slopes fail to support this simple two electron-transfer reaction. Therefore the oxidation of galena is assumed to take place according to sequential one-electron-transfer steps,

$$PbS \rightarrow Pb^{2+} + S^- + e \qquad (20)$$

$$S^- \rightarrow S^\circ + e \qquad (21)$$

The Butler-Volmer treatment for steady state conditions ($d[S^-]/dt = 0$) yields the following equation for the net anodic current

$$i_a = \frac{2\mathcal{F}\vec{k}\ k'\ \exp[(2-\beta_a-\beta_a')\mathcal{F}E/RT]}{\overleftarrow{k}\ [Pb^{2+}]_s\ \exp[-\beta_a \mathcal{F}E/RT] + k'\ \exp[(1-\beta_a)\mathcal{F}E/RT]} \qquad (22)$$

where $[Pb^{2+}]_s$ is the concentration of lead ions at the electrode surface, $\vec{k}$ and $\overleftarrow{k}$ are the respective forward and reverse rate constants for reaction (20), k' is the specific rate of reaction (21), and the other terms are analgous to those previously defined. At sufficiently positive values of E or at low $[Pb^{2+}]$, the expression for i_a reduces to

$$i_a = 2\mathcal{F}\vec{k}\ \exp[(1-\beta_a)\mathcal{F}E/RT] \qquad (23)$$

Conversely, when the $[Pb^{2+}]$ is high and e is low,

$$i_a = \frac{2\mathcal{F}\vec{k}\ k'}{\overleftarrow{k}\ [Pb^{2+}]}\ \exp\ [(2-\beta_a')\mathcal{F}E/RT] \qquad (24)$$

Both cases cited above along with an additional condition when

$[Pb^{2+}]_s$ approaches zero have been verified experimentally. Therefore, the proposed mechanism involving consecutive one-electron-transfer reactions is an accurate description of the anodic dissolution of galena.

Chloride solution chemistry is of special interest to lead hydrometallurgy because of the high solubility of $PbCl_2$ at elevated temperatures. The U.S. Bureau of Mines has developed a ferric-chloride-brine leaching process[13,14] which results in essentially complete extraction of lead at 100°C in 15 minutes. A typical recycle solution would contain in gpl 42-Fe, 90-Na, 226-Na, 17.5-Pb, and 5-$SO_4^{=}$. The yield of sulfur as elemental sulfur was essentially complete. No detectable sulfate formation occurred from the reaction of $FeCl_3$ with PbS[14]. The sulfate in solution came from pre-oxidized mineral. The process is based upon the following steps:

Dissolution

$$PbS + 2FeCl_3 \rightarrow PbCl_2 + 2FeCl_2 + S° \qquad (25)$$

Electrolysis

$$PbCl_2 \rightarrow Pb° + Cl_2 \qquad (26)$$

Ferrous Chloride Oxidation

$$2FeCl_2 + Cl_2 \rightarrow 2FeCl_3 \qquad (27)$$

It was noted in these studies[14] that the rate of leaching was enhanced by the addition of hydrochloric acid. Figure 4 illustrates the effect of HCl additions on the rate of dissolution of a PbS concentrate. Scott and Nicol[15] and Awakura, et al.[16] have examined the non-oxidative dissolution of galena and both have noted the enhanced pH and the complex chloride ion effects. Awakura, et al., measured the activity of hydrogen ions in strong NaCl solutions and found the rate was proportional to a_{H^+} over the range 0.7 to 50. The proposed mechanism was

$$PbS + H^+ \rightarrow PbS{\cdot}H^+_{ads} \qquad (28)$$

$$PbS{\cdot}H^+_{ads} \rightarrow Pb^{2+} + HS^- \qquad (29)$$

Activation energies in the range of 12.3 to 19.0 kcal/mole were reported depending upon solution make-up and ionic strength. In ferric chloride solutions any H_2S formed would react producing

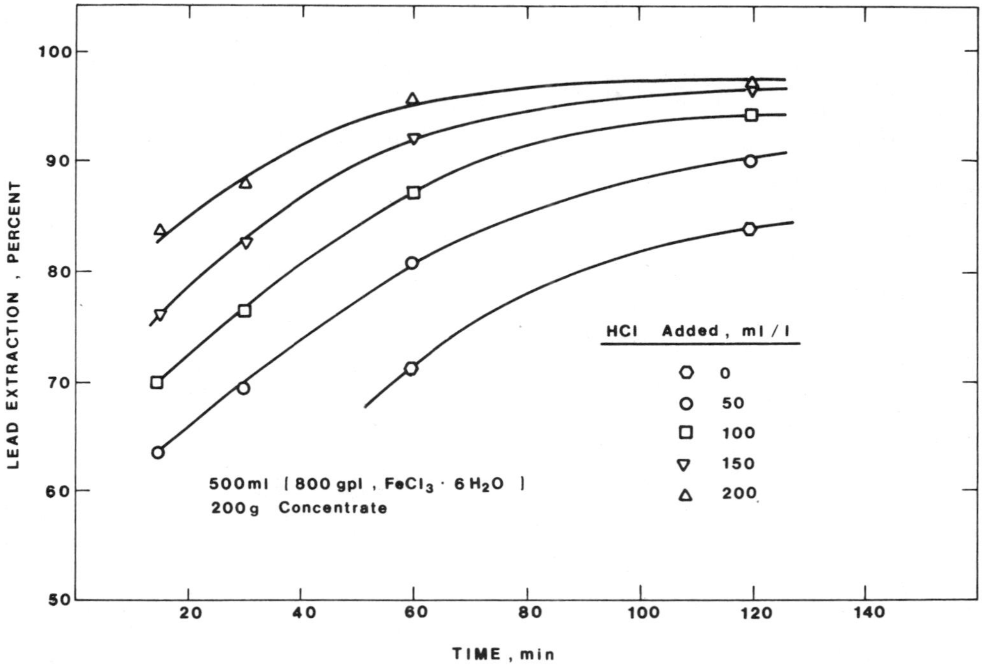

Fig. 4. Percent lead extraction for various HCl additions to $FeCl_3$ solutions. Data from U.S. Bureau of Mines.[14]

elemental sulfur according to the reaction

$$H_2S + 2FeCl_3 \rightarrow S^\circ + 2H^+ + 2Cl^- + 2FeCl_2 \tag{30}$$

The dissolution of galena by $FeCl_3$ involves both electrochemical oxidative and pH dependent non-oxidative dissolution. Recently Han[17] carried out coupled electrode and potentiodynamic measurements in strong chloride electrolytes. Figure 5 illustrates results obtained using a diaphragm and coupled PbS and graphite electrodes for the initial reaction. The mixed potential is near 340 mV (SCE) which falls on the rising portion of the anodic dissolution peak, Fig. 3(a). The initial kinetics are mixed potential controlled with the mixed potential nearer the cathode rest potential than the anodic rest potential. Palmer[18] has shown that for more extensive reaction diffusion becomes rate controlling. Figure 6 illustrates results obtained by Palmer at 50°C for 35x45 mesh galena in 4 molar NaCl and variable $FeCl_3$ concentration. The rate follows the diffusion equation

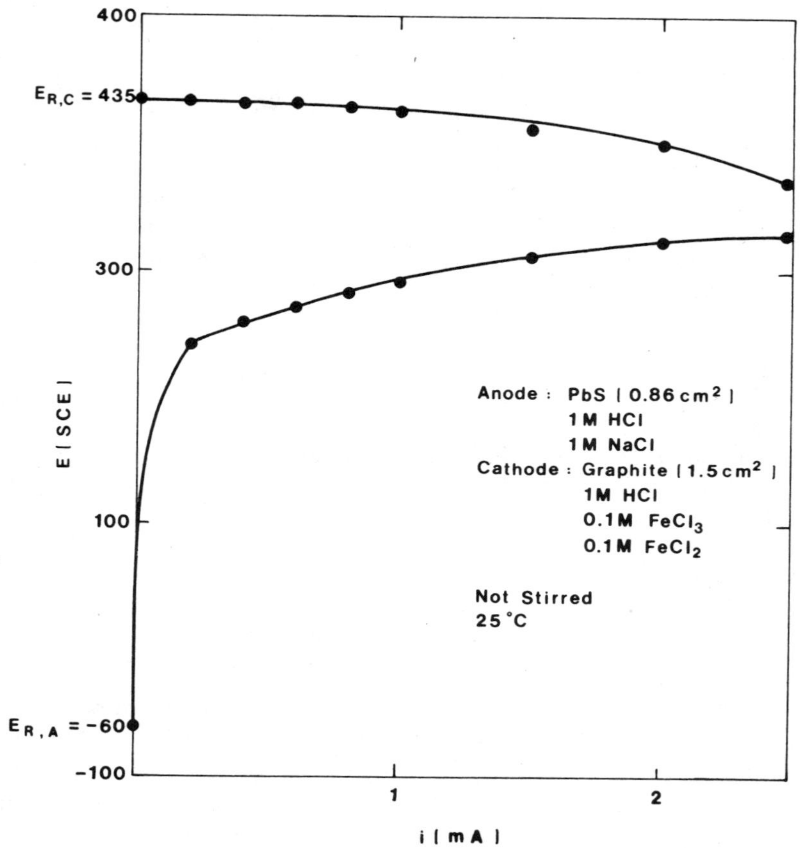

Fig. 5. Coupled cathodic (graphite) and anodic reactions (graphite/$FeCl_3$) using diaphragm cell; Han[17].

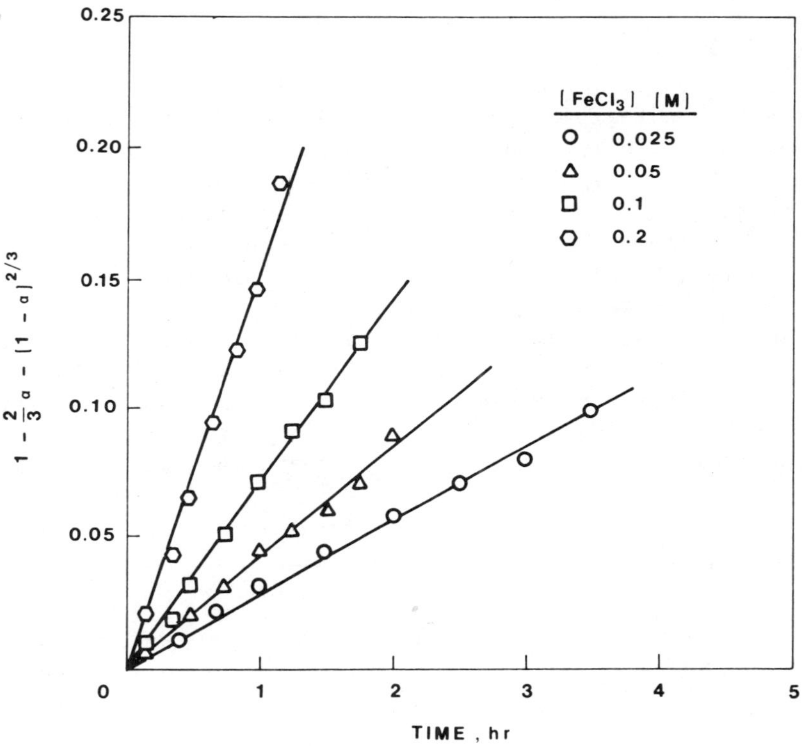

Fig. 6. Data of Palmer[18] showing geometric correlation with diffusion for dissolution of 35 x 48 mesh galena in ferric chloride and 4.0 M NaCl at 50° C.

$$1 - \frac{2}{3}\alpha - (1-\alpha)^{2/3} = \frac{k_p'}{d_o^2}\, t \tag{31}$$

where α is the fraction of lead leached, d_o is the initial average particle diameter, t is time and k_p' is the normalized rate cm^2sec^{-1} and is a function of ferric chloride concentration. Palmer also noted a very modest temperature dependence. The rate appears to involve pore diffusion through the product sulfur layer.

Han[17] has obtained potentiodynamic results for galena in strong chloride solutions. Figure 7 illustrates the effect of chloride ion concentration on the 440 mv peak for the anodic dissolution. The main peak is little influenced by chloride ion, although an after-peak is related. The main reaction is that proposed by Paul et al.[12] according to Eq. 17. The cause for the after-peak is not known at present. Han also found the anodic peak to be essentially independent of pH for pH values 2 and above. Peak height increased markedly at pH = 1. These results are similar to those reported by Paul et al.[12] To determine if H_2S in large concentration influences the anodic peak H_2S was bubbled through the solution during the initial anodic scan. As may be noted in Fig. 8 a freshly polished galena surface is strongly passivated by the H_2S. Figure 9 illustrates the effect of adding H_2S after an initial anodic scan without H_2S. As

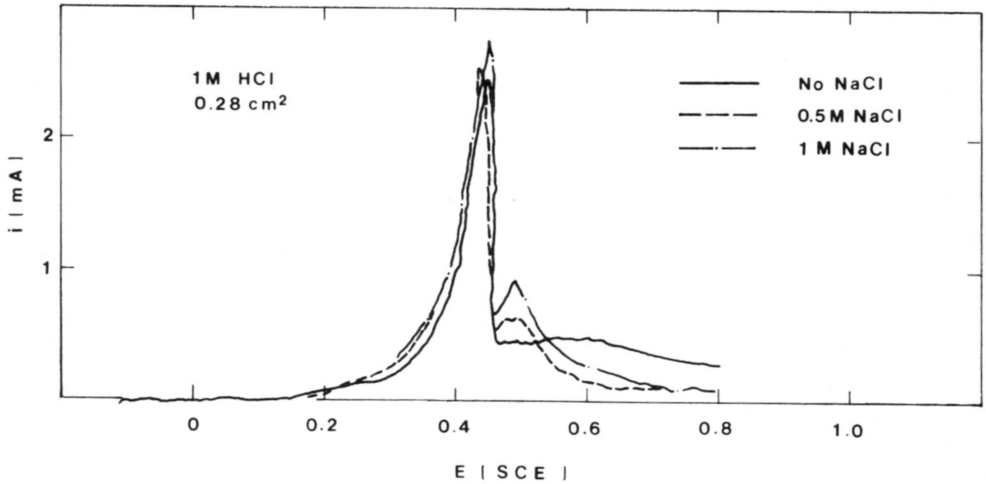

Fig. 7. Effect of chloride ion on the anode dissolution peak; Han[17]

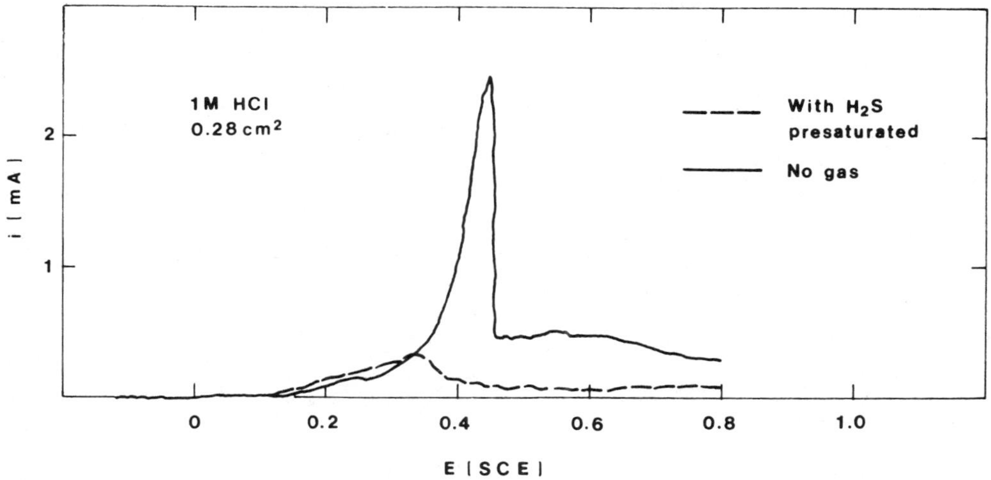

Fig. 8. Effect of H_2S on anodic dissolution peak for freshly polished galena; Han[17].

may be noted H_2S now has little influence on the anodic dissolution peak. Evidently the passivation effect of H_2S on a fresh galena surface comes from dense S° crystallization on the galena substrate. After an initial anodic scan the presence of S° creates a substrate quite different from the $H_2S \rightarrow S°$ reaction. Paul et al[11] also noted the variable character of S° related to time-dependent changes.

Second Stage Leaching of Chalcocite in Acid Ferric Sulfate Solutions

Sullivan[19] as early as 1930 identified two stages of leaching of Cu_2S. These were depicted as

$$Cu_2S + 2Fe^{3+} = Cu^{2+} + CuS + 2Fe^{2+} \qquad (32)$$

and

$$CuS + 2Fe^{3+} = Cu^{2+} + 2° + 2Fe^{2+} \qquad (33)$$

Marcantonio[20] similarly identified two stages but provided greater detail regarding the kinetics of each stage. The following half-cell reactions are consistent with the observed results

Stage Ia:

$$5Cu_2S = 5Cu_{1.8}S + Cu^{2+} + 2e^- \qquad (34)$$

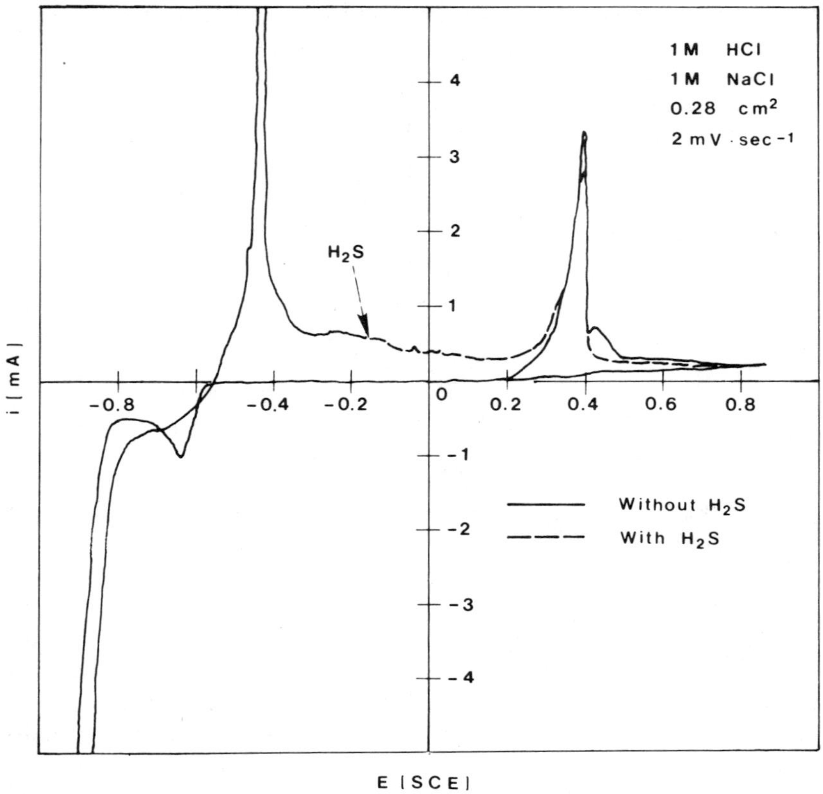

Fig. 9. Effect of H_2S addition on second cycle of the anodic dissolution of galena; Han[17].

Stage Ib:

$$\frac{5}{3}\, Cu_{1.8}\, S = \frac{5}{3}\, Cu_{1.2}S + Cu^{2+} + 2e^- \tag{35}$$

Stage II:

$$\frac{5}{6}\, Cu_{1.2}S = \frac{5}{6}\, S^\circ + Cu^{2+} = 2e^- \tag{36}$$

Stage I proceeds through a series of compounds. The sequential formation of intermediates establishes the particle potential and controls the rate of ferric ion reduction. Stage II leaching, at elevated temperatures (Eq. 36), occurs by a corrosion mechanism and

and the elemental sulfur produced does not passivate the surface. Stage I is rapidly completed at higher temperature resulting in the removal of approximately forty percent of the copper with the formation of blue remaining covellite, $Cu_{1.2}S$. The blue remaining covellite leaches at a much slower rate.

Figure 10 illustrates typical results obtained for Stage II leaching, at 75°C, for various initial Cu_2S particle sizes. The rate is independent of the initial Cu_2S particle size. The blue-remaining covellite formed in Stage I is porous, consisting of uniform columnar crystals. The rate is sufficiently slow that diffusion into the columnar array does not influence the kinetics. The rate was found to vary with [Fe III] with an apparent half-power order and the activation energy was 18.0± 1 kcal/mole. Consistent with the physical appearance of the crystals the kinetics were correlated by cylindrical crystalline geometry.

The net anodic and cathodic processes may be expressed by

$$i_a = 2\mathcal{F}\overrightarrow{k}_a \exp\left[\frac{\beta_a \mathcal{F}E}{RT}\right] - [Cu^{2+}]\,\overleftarrow{k}_a \exp\left[\frac{-(1-\beta_a)\mathcal{F}E}{}\right] \quad (37)$$

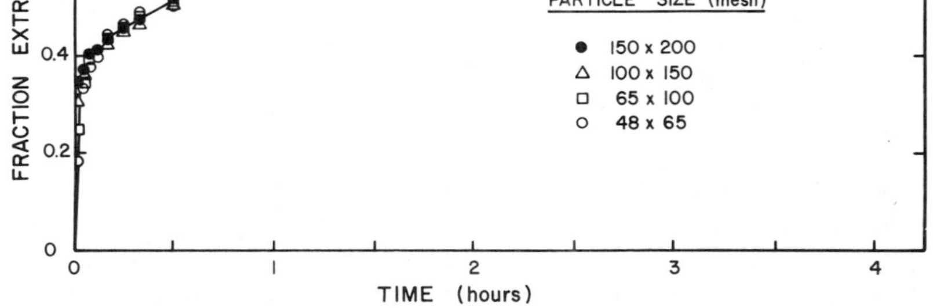

Fig. 10. Effect of particle size on the second stage dissolution of Cu_2S at 75°C and 0.12 M Fe III (Fe III/Cu = 3/1).

$$i_c = F[Fe^{2+}] \overrightarrow{k}_c \exp\left[\frac{\beta_c FE}{RT}\right] - F\overleftarrow{k}_c [Fe^{3+}] \exp\left[\frac{-(1-\beta_c)FE}{RT}\right] \qquad (38)$$

Using the relationship $d\alpha/dt = i_a A_a/2\, n_o$, where α is the fraction of total copper reacted and n_o is the total moles of copper initially added to the system, Eq. 12 for the $Cu_{1.2}S/Fe^{3+}$ system becomes

$$\frac{d\alpha}{dt} = \frac{b\, A\, \overrightarrow{k}_a\, \overleftarrow{k}_c\, [Fe^{3+}]}{n_o(1+b)\left[2b\overrightarrow{k}_\alpha + \overrightarrow{k}_o[Fe^{2+}]\right]^{1/2}\left[2b\overleftarrow{k}_a[Cu^{2+}] + \overleftarrow{k}_c[Fe^{3+}]\right]^{1/2}} \qquad (39)$$

For low concentrations of $[Cu^{2+}]$ and $Fe^{2+}]$, Eq. 39 becomes

$$\frac{d\alpha}{dt} = \frac{A\left[(b/2)\overrightarrow{k}_a\, \overleftarrow{k}_c[Fe^{3+}]\right]^{1/2}}{N_o(1+b)} \qquad (40)$$

Using shrinking core cylindrical geometry, the total crystallite surface area is given by

$$A = p\, A_o\, \left[(1 - \alpha)/(1 - \alpha_o)\right]^{1/2} \qquad (41)$$

where p contains crystallite density and appropriate geometric factors, A_o is the original surface area of the $Cu_{1.2}S$ at the beginning of Stage II, and $\alpha_o = 0.4$, the value of α at the end of Stage I. Also the concentration of ferric ion varies with α according to the relationship

$$[Fe^{3+}] = [Fe^{3+}]_o\, [1 - q\alpha]\, \gamma_3 \qquad (42)$$

where $[Fe^{3+}]_o$ is the initial ferric ion concentration, q is the stoichiometry factor, and γ_3 is the activity coefficient for ferric ion. Using these relationships Eq. 40 becomes

$$\frac{d\alpha}{dt} = \frac{p\, A_o}{n_o(1+b)} \left|\frac{b\, \overrightarrow{k}_a \overleftarrow{k}_c [Fe^{3+}]_o\, \gamma 3}{2}\right|^{1/2} \left[\frac{1 - \alpha}{1 - a_o}\right]^{1/2} \left[1 - q\alpha\right]^{1/2} \qquad (43)$$

The solid curves of Fig. 10 were calculated by numerical integration of Eq. 43.

Figures 11 and 12 illustrate respectively the effect of adding increased concentrations of Fe^{3+} and Cu^{2+}. In the latter case the rate is significantly diminished. This effect may be explained by Eq. 39. The terms containing $[Cu^{2+}]$ and $[Fe^{2+}]$ become significant

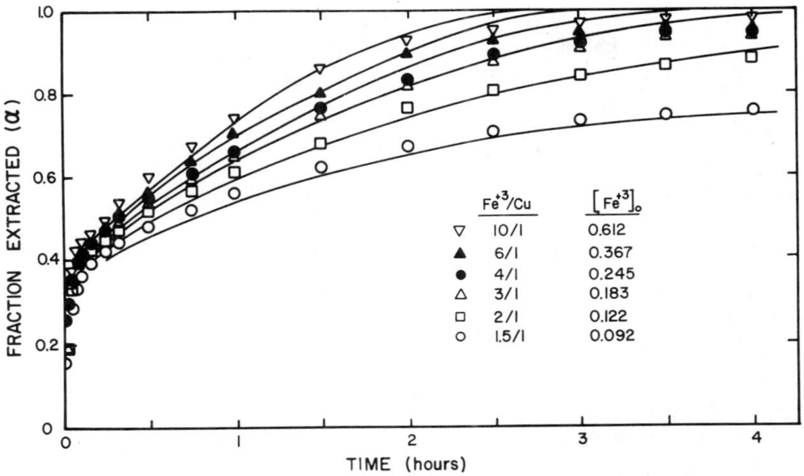

Fig. 11. Effect of ferric concentration on the second stage dissolution of 48 x 65 mesh Cu_2S at 75°C.

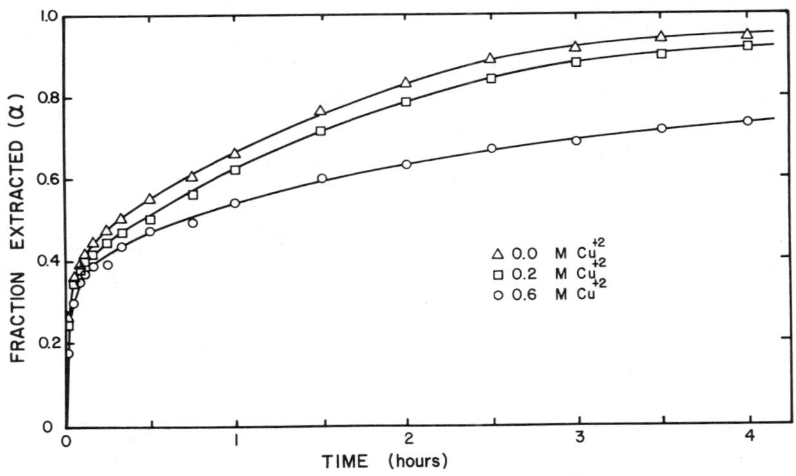

Fig. 12. Effect of cupric sulfate addition on the second stage dissolution of 48 x 65 mesh particles at 75°C and 0.24 M Fe^{+3} (Fe^{+3}/Cu = 4/1).

at high concentration levels, reducing the rate of reaction by adversely shifting the mixed potential. This result also explains the significant decrease in the rate of leaching of covellite in slurries of high pulp density and ionic concentration. Although not shown, a similar rate-depression result was obtained by adding Fe^{2+}.

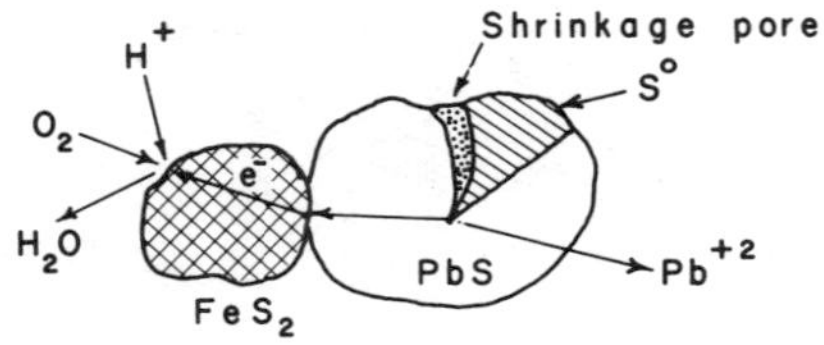

Fig. 13. Galvanic Leaching Couple for Galena in Electrical Contact with Pyrite

MIXED POTENTIAL GALVANIC COUPLES

When metallic conducting or semi-conducting minerals in electrical contact are subject to leaching, galvanic interactions may occur which can cause an enhanced corrosion of one or more members of the electro-chemically coupled mineral regime, and cathodic protection of other mineral members. The mineral with the lower "rest" potential will react anodically, whereas, cathodic behavior may be imposed on the mineral possessing the higher "rest" potential. For example pyrite possesses the highest "rest" potential of the mineral sulfides. If pyrite is brought into electrical contact with another sulfide mineral, the oxidation of the other sulfide mineral will be accelerated under oxidative leaching conditions.

Anodic Dissolution

Peters, et al.[21] have demonstrated enhanced leaching of galena in the presence of pyrite. If pyrite is brought into contact with galena in the presence of oxygen the pyrite surface acts cathodically discharging oxygen according to the reaction

$$^{1}/_{2}\ O_2 + 2H^+ + 2e^- = H_2O \tag{44}$$

and the galena dissolves anodically according to Eq. 17 releasing Pb^{2+} and forming elemental sulfur. Figure 13 illustrates the galvanic couple formed.

Another excellent example of this phenomena was recently observed in the leaching of a chalcopyrite bearing material.[22] Figure 14 illustrates schematically the polarization curve for chalcopyrite superimposed on that for pyrite. In this example log $|i|$ is plotted versus potential (SHE). Galvanic coupling between pyrite and chalcopyrite within the leaching environment will establish a mixed potential at the voltage represented by line a of

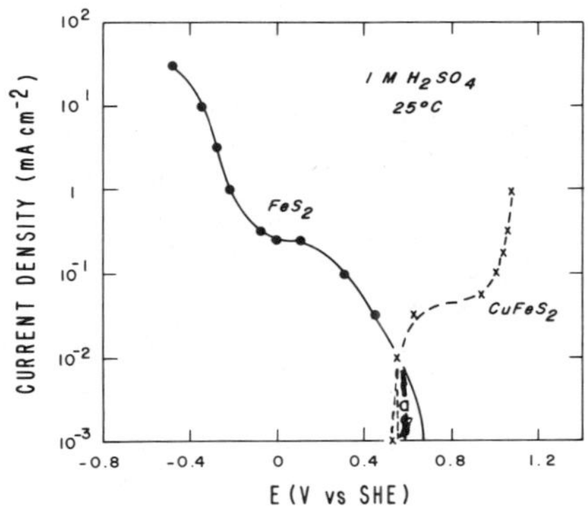

Fig. 14. Polarization curves for pyrite (FeS_2, cathodic branch) and chalcopyrite ($CuFeS_2$, anodic branch) with mixed potential a.

Fig. 14. At the mixed potential, pyrite is an inert electrode, and oxygen adsorbed on the pyrite surface is discharged cathodically according to reaction (44). Chalcopyrite reacts anodically according to the following reaction

$$CuFeS_2 = Cu^{2+} + 2S^\circ + 4e^- \quad (45)$$

The galvanic leaching mechanism is depicted in Fig. 15 showing the flow of electrons from the anodic $CuFeS_2$ particle resulting in the discharge of oxygen at the particle surface.[22]

Cathodic Conversion of Chalcopyrite

Shirts et al[23] have demonstrated that chalcopyrite will react cathodically by bringing it into physical contact with a metal having a lower potential. Three metals, Cu, Fe, and Pb, were used. The reactions observed were

$$CuFeS_2 + Cu^\circ + 2H^+ \rightarrow Cu_2S + Fe^{2+} + H_2S \quad (46)$$

$$2CuFeS_2 + Fe^\circ + 4H^+ \rightarrow Cu^\circ + 2Fe^{2+} + 2H_2S \quad (47)$$

$$2CuFeS_2 + Pb^\circ + 6H^+ \rightarrow Cu_2S + 2Fe^{2+} + Pb^{2+} + 3H_2S \quad (48)$$

$$CuFeS_2 + Pb^\circ + 4H^+ \rightarrow Cu^\circ + Fe^{2+} + Pb^{2+} + 2H_2S \quad (49)$$

Galvanic couples are readily established because of the excellent

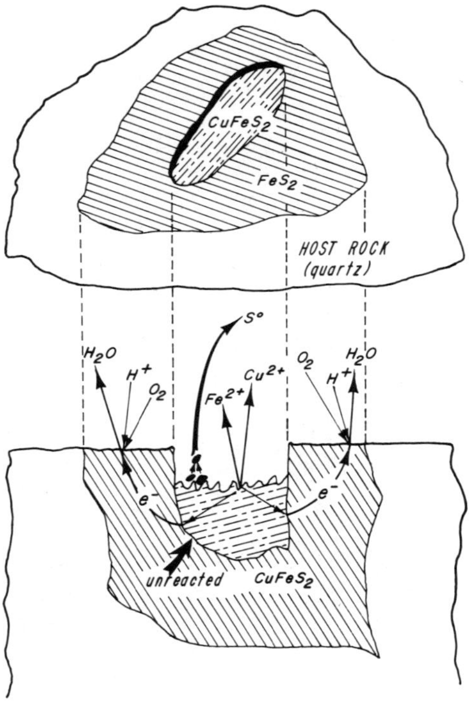

Fig. 15. Galvanic coupling of pyrite (FeS_2) and chalcopyrite showing flow of electrons from anodic $CuFeS_2$ particles to cathodic FeS_2 particle.[22]

electronic conductivity of all phases. The principal cathodic reaction is

$$2CuFeS_2 + 6H^+ + 2e^- = Cu_2S + 2Fe^{2+} + 3H_2S \qquad (50)$$

The half-cell reaction involving the copper is

$$Cu_2S + 2H^+ + 2e^- = 2Cu + H_2S \qquad (51)$$

Reaction (51), having a more negative potential than reaction (50), will react anodically. Similarly metallic iron will reduce $CuFeS_2$ to metallic copper as will metallic Pb in acid chloride solutions.

Hiskey and Wadsworth[24] measured the kinetics of the galvanic conversion of $CuFeS_2$ using copper. Figure 16 illustrates the galvanic $CuFeS_2$–Cu couple. Both the cathodic and anodic reactions produce Cu_2S which welds and establishes good electrical contact.

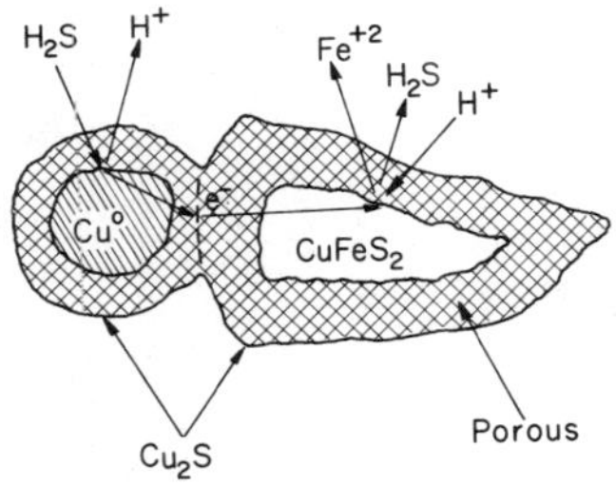

Fig. 16. Schematic representation of the $CuFeS_2$-Cu galvanic couple in sulfuric acid.

It was found that at lower temperatures, stirring disrupted the bond and retarded the reaction. At 90°C stirring had less effect on the overall kinetics. Figure 17 illustrates results obtained for an intimate mixture of copper powder and chalcopyrite concentrates. Several sizes of copper were used with 325/400 mesh chalcopyrite with excess copper, molar Cu/$CuFeS_2$ ratio of 1.5. The extent of the reaction was measured by the fraction α, indicated by the amount of iron released to the solution. The curves of Fig. 17 illustrate the importance of the copper anodic surface area. Initially the chalcopyrite surface area is larger than that of the copper and the rate is predominantly anodically controlled. As the chalcopyrite reacts, its area becomes smaller and the kinetics shift from predominantly anodic control to predominantly coathodic control. The levelling off below α = 1.0 was attributed to lack of perfect mixing of the initial solid phases. Careful blending increased the extent of reaction. In a galvanic system it is necessary to

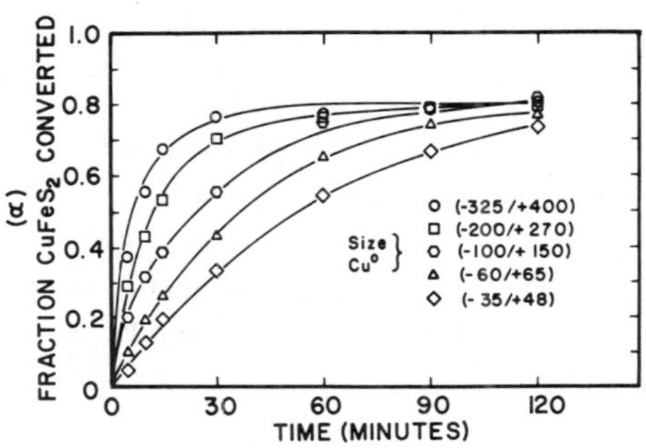

Fig. 17. Fraction reacted versus time for cathodic reduction of chalcopyrite by various sizes of metallic copper.

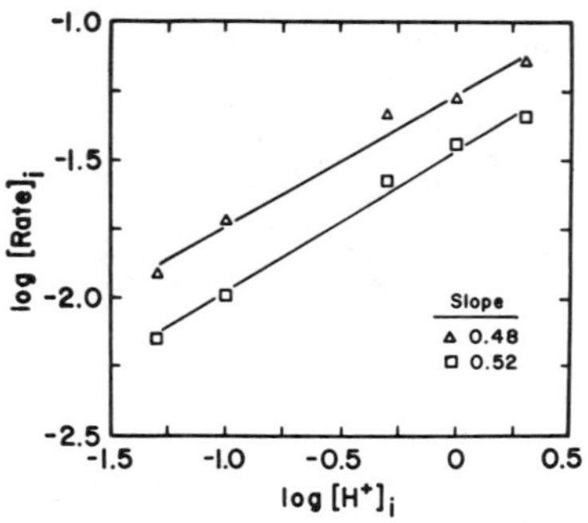

Fig. 18. Plot of Log [H^+] Versus Log [rate] to Determine the Hydrogen Ion Rate Dependence.[16] (Upper Points: 270 x 325 Mesh Cu; Lower Points: 2

normalize the data in terms of the anodic (A_a) and cathodic (A_c) surface areas by the relationships $I_a = i_a A_a$ and $I_c = i_c A_c$ where I_a and I refer to the total anodic and cathodic currents, respectively. Additional tests showed the rate of reaction to be essentially independent of [FeII] in solution. This experimental activation energy was approximately 11.5 kcal/mole. Figure 18 depicts the initial rate versus initial hydrogen ion concentration. The rate order for hydrogen ion is in the range 0.48 - 0.52 or essentially one-half. Assuming single charge transfer in the activated state, the net cathodic and anodic electrode processes may be represented by the equations

$$I_c = A_c \vec{K}_c 2F \exp[\frac{\beta_c F E_c}{RT}] - A_c \overleftarrow{k}_c a_{H^+} 2F \exp[\frac{-(1-\beta_c)FE_c}{RT}] ,$$

$$I_a = A_a \vec{k}_a \, 2F \exp[\frac{\beta_a F E_a}{RT}] - A_a \overleftarrow{k}_a a_{H^+} 2F \exp[\frac{-(1-\beta_c)FE_c}{RT}] \quad (52)$$

The rates k_c and k_a may contain the activity of H_2S. In this case a_{H_2S} is constant since H_2S continually evolves at atmospheric pressure. Also k_c may conceivably contain [Fe II] although experimental results showed little dependence on [FeII]. When steady-state currents are established E is the mixed potential such that $E_a = E_c = E$ and $I_c = -I_a$. An explicit solution for the mixed potential is not possible unless the Tafel slopes are essentially equal, or $\beta_a = \beta_c = \beta$. Using this approximation as before, Eq. 12 becomes

$$\frac{d\alpha}{dt} = \frac{a_{H^+}^{1/2} \quad A_c A_a \overleftarrow{k}_c \vec{k}_a}{(A_a \vec{k}_a + A_c \vec{k}_c)^{1/2} \; (A_a \overleftarrow{k}_a + A_c \overleftarrow{k}_c)^{1/2}} \quad (53)$$

Equation 53 is consistent with the hydrogen ion dependence and anodic control for the initial condition $A_c\vec{k}_c \gg A_a\vec{k}_a$ and $A_c\overleftarrow{k}_c \gg A_a\overleftarrow{k}_a$ observed experimentally. Similarly cathodic control results when $A_a\vec{k}_a \gg A_c\vec{k}_c$ and $A_a\overleftarrow{k}_a \gg A_c\overleftarrow{k}_c$.

Nicol[25] has demonstrated the electrochemical character of the galvanic coupling of $CuFeS_2$ with metallic copper, iron, and lead.

CHARGE TRANSFER THROUGH PASSIVATION LAYERS

Often passivation layers form as intermediate products with diffusion of charged ions or electrons rate controlling. Three examples are presented here: 1) diffusion of ions through an intermediate defect phase during the anodic dissolution of chalcopyrite; 2) high temperature passivation of chalcopyrite with diffusion of electrons through elemental sulfur as rate controlling; and 3) cathodic reduction of chalcopyrite with ionic diffusion through a bornite intermediate as rate controlling.

Anodic Dissolution of Chalcopyrite

The anodic dissolution of chalcopyrite in acidic ferric sulfate solutions occurs according to the half-cell reaction (45) forming elemental sulfur. Initially a defect chalcopyrite structure forms followed by passivation by a layer of elemental sulfur. Also, sulfate ion forms but in relatively smaller proportion than elemental sulfur. Initially ferrous ion is released much more rapidly than cupric ions[6,26,27] forming an intermediate structure according to the reaction forming an intermediate structure according to the reaction

$$CuFeS_2 \rightarrow Cu_{1-x}Fe_{1-y}S_{2-y-x} + xCu^{2+} + yFe^{2+} + (x+y)S° + 2(x+y)e^- \quad (54)$$

where y > x initially. As the reaction proceeds x approaches y. Elemental sulfur forms according to the reaction

$$Cu_{1-x}Fe_{1-y}S_{2-y-x}$$

$$(1-x)Cu^{2+} + (1-y)Fe^{2+} + (2-x-y)S° + 2(2-x-y)e \quad (55)$$

In the first step of the reaction the intermediate product serves as a passivation layer by limiting the reaction to ionic transport through the layer. Results obtained by Warren[6] are illustrated in

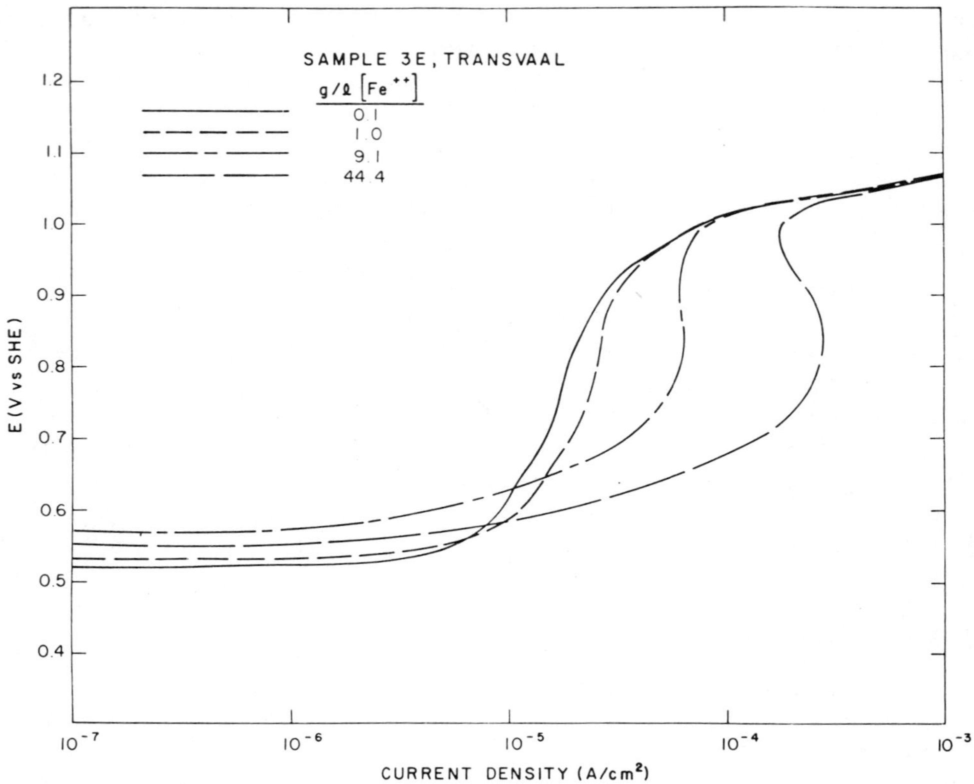

Fig. 19. Effect of ferrous ion on the anodic polarization curve of Transvaal $CuFeS_2$ in 1 M H_2SO_4, 40 mv/min, 25°C. [Cu^{++}] was held constant at 0.90 g/l.

Fig. 19 where anodic polarization curves are shown for various ferrous ion levels. The plateau region, indicating diffusional control, moves up regularly with increasing [Fe^{2+}]. The electrode is not passivated by elemental sulfur at this point as noted by Jones and Peters[27] since sulfur, being an electronic insulator, would not respond to the ferrous ion present. The increased plateau current results from the oxidation of ferrous iron to ferric iron at the electrode surface, indicating good electronic conductivity through the passivating layers. Although Warren[6] found sulfur in this region it did not appear as a dense continuous layer but occurred as patches and around etched regions of the surface. Warren observed the kinetics to be typical of ionic transport in a high potential gradient in this region. These results are in contrast to those observed by Munoz et al.[28] who found massive layers of sulfur on fine chalco-

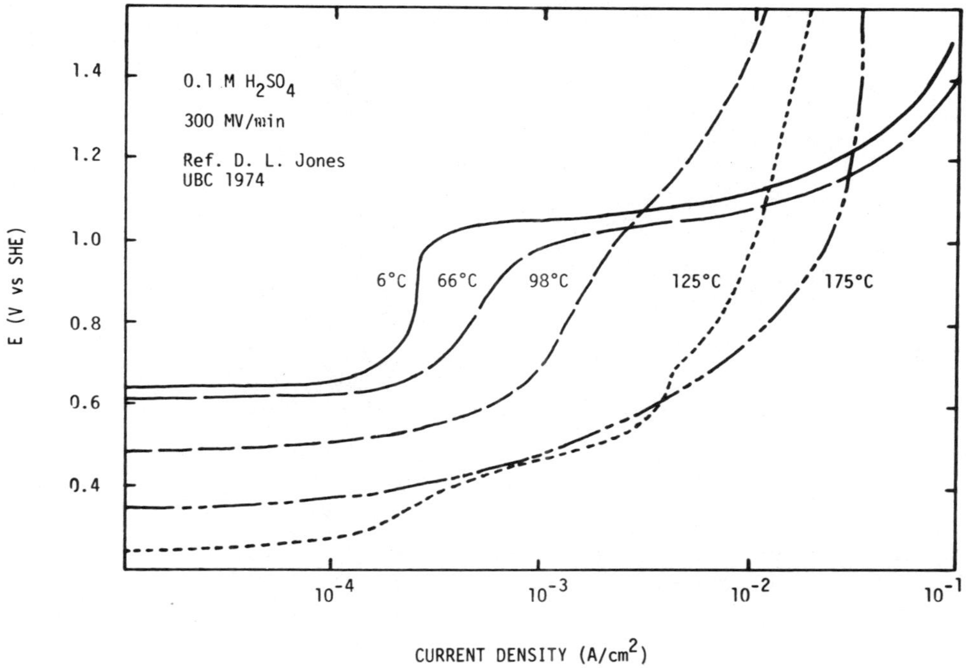

Fig. 20. Polarization curves for Craigmont chalcopyrite at various temperatures showing passivation at higher temperatures.

pyrite leached in acid ferric sulfate solutions at elevated temperatures. Under these conditions the rate of reaction was not influenced by the addition of ferrous or ferric ions to the solution. Figure 20 shows anodic polarization curves obtained by Jones[29] in sulfuric acid for various temperatures. For temperatures above 66°C a new passivation region at higher potentials is evident. In this region, at lower temperatures, the rapid increase in current for the low temperatures is caused by discharge of water resulting from transfer of electrons through electron conducting surface layers. At higher temperatures discharge of water does not occur, a result which is consistent with the formation of dense sulfur layers of low electronic conductivity.

High Temperature Passivation of Chalcopyrite

Munoz et al.[28] carried out an extensive study of the leaching of fine chalcopyrite and as noted earlier showed that the kinetics were controlled by diffusion through protective layers of sulfur. The principal overall reaction is

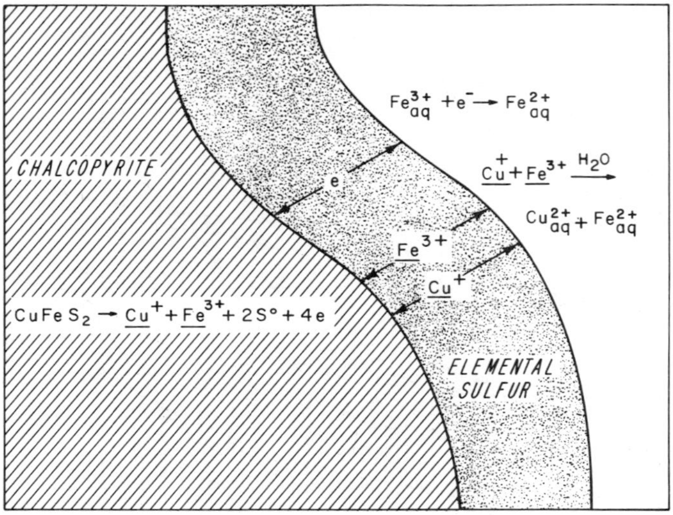

Fig. 21. Schematic illustrating transport processes in the acid ferric sulfate leaching of chalcopyrite.

$$CuFeS_2 + 4Fe^{3+} \rightarrow Cu^{2+} + 5Fe^{2+} + 2S° \qquad (56)$$

with the formation of some sulfate, particularly at high anodic overvoltages. Munoz et al.[28] explained the kinetics of leaching of monosized fractions of chalcopyrite by a model in which the main rate limiting step is electronic transport, accompanied by the outward diffusion of coupled Cu^{2+} plus Fe^{2+} ions, through the sulfur layer. Equally plausible is the outward diffusion of coupled Fe^{3+} and Cu^+. Figure 21 illustrates the transport processes proposed. The activation energy was observed to be 83.7 kJ/mol (20 kcal/mole) which compares favorably with the activation energy (23 kcal/mol) for the electronic conductivity of elemental sulfur. Figure 22 presents the rate data for three monosized fractions of chalcopyrite. The data are well correlated by the diffusion equation for spheres, for diffusion of a reactant through the product layer, Eq. 31. The rate constant k_p' (Eq. 31) is given by

$$k_p' = \frac{8\chi}{\rho} \left(\frac{D_i \varepsilon}{\tau}\right) a_i$$

where: $\frac{D_i \varepsilon}{\tau}$ = effective diffusivity, ε is the porosity and τ is the tortuosity

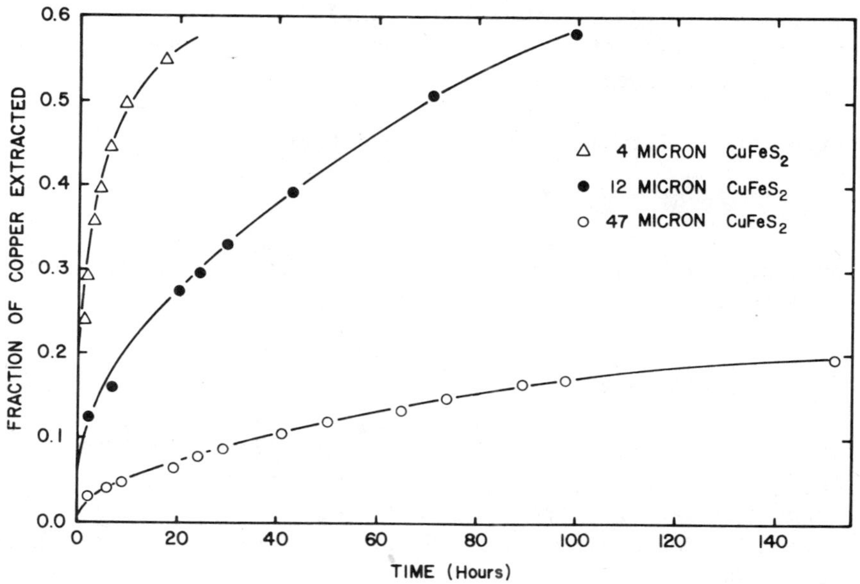

Fig. 22. A plot of copper extracted from monosize chalcopyrite particles as a function of time for 1.0 M H_2SO_4, 0.25 M $Fe_2(SO_4)_3$, 90°C, 0.5 percent solids and 1200 rpm.

a_i = activity of reacting species (electrons) at the $CuFeS_2/S°$ boundary
χ = stoichiometric factor, moles $CuFeS_2$ per mole of electrons
ρ = molar density of chalcopyrite

The experimental value of k'_p was determined to be 2.24×10^{-13} cm^2/s.

The rate of leaching was derived using Wagner's theory of oxidation which relates the intrinsic mass transport or electrical conductivity of products of reaction to rates of reaction. The flux of a charged diffusing anion, cation or electrons (mol cm^{-2} sec^{-1}) is given by

$$J_i = \frac{300}{96{,}000} \frac{t_i \sigma}{|z_i|^2 \mathcal{F}} \left[\frac{d\mu_i}{dx} - z_i \mathcal{F} \frac{d\phi}{dx}\right] \quad (57)$$

for modest voltage gradients, where

$d\mu_i/dx$ = chemical potential gradient
$d\phi/dx$ = electrical field strength

F = Faraday (23,060 cal/equiv. volt)
t_i, z_i = transport number and charge for ith species
σ = electrical conductivity

Electrical neutrality requires the condition

$$J_{\underline{Cu}^+} + 3J_{\underline{Fe}^{3+}} = J_e \tag{58}$$

The coupling equation consistent with the observed stoichiometry of the reaction is

$$J_{\underline{Cu}^+} = J_{\underline{Fe}^{3+}} = \frac{1}{A}\ \frac{dn(CuFeS_2)}{dt} \tag{59}$$

If transport of electrons through the sulfur layer is rate controlling then the transport number t_e for electrons is expected to be much smaller than the transport number $t_{Fe^{3+}}$ for $\underline{Fe}^{3+}$. Combination of Eqs. 57, 58, and 59 for the condition $t_e \ll t_{\underline{Fe}^{3+}}$, results in the general equation

$$\frac{dn}{dt}\,(CuFeS_2) = -\,A\,\frac{300}{96,500}\,\frac{t_e\sigma}{4}\,\frac{d\mu_e}{dx} \tag{60}$$

For spherical geometry Eq. 60 may be integrated, assuming quasi steady state, resulting in the equation

$$\frac{d\alpha}{dt} = \frac{3x300}{96,500}\,\frac{M}{\rho}\,\frac{t_e\sigma}{F_{d^2_o}}\,\frac{\Delta G\,(1-\alpha)^{1/3}}{1-(1-\alpha)^{1/3}} \tag{61}$$

where $\Delta G = \mu^o_e - \mu^i_e$,

μ^o_e, μ^i_e = the chemical potentials for the outer and inner boundary of the sulfur layer
ΔG = the calculated Gibbs molar free energy for the overall reaction
M,ρ = molecular weight and density of $CuFeS_2$.

Integration of Eq. 61 assuming a constant average $\overline{\Delta G}$ results in the general diffusion equation for isometric particles, Eq. 31. Equation 61 may be integrated numerically, including corrected ΔG values with time.

From k'_ρ and calculated values for ΔG at temperature, the solid lines of Fig. 22 were calculated using a best fit value of $7.6x10^{-13}$ ($ohm^{-1}cm^{-1}$) for $t_e\sigma$. This value compares to a measured value of ~10^{-13} for pure sulfur. The correlation was considered satisfactory since estimate values of the activity coefficients of Cu^{2+}(aq), Fe^{2+}(aq) and Fe^{3+}(aq) were used to calculate ΔG. Also, the sulfur produced contained impurities which have been noted to increase the electronic conductivity of sulfur. The proposed model also explains the modest response of the system to the solution concentration of feric ion.

Cathodic Reduction of Chalcopyrite

Sohn[30,31] has investigated the reduction of $CuFeS_2$ in cupric sulfate solutions using SO_2 as the reductant. The important half-cell reactions involved are:

cathodic:

$$2CuFeS_2 + 3Cu^{2+} + 4e^- = Cu_5FeS_4 + Fe^{2+} \quad (62)$$

$$Cu_5FeS_4 + 3Cu^{2+} + 4e^- = 4Cu_2S(\text{Djurleite}) + Fe^{2+} \quad (63)$$

anodic:

$$SO_2 + 2H_2O = HSO_4^- + 3H^+ + 2e^- \quad (64)$$

In the sequence of reactions, initially a defect structure was observed with a main lattice parameter near that of bornite but slightly greater. As the reaction proceeded bornite formed followed

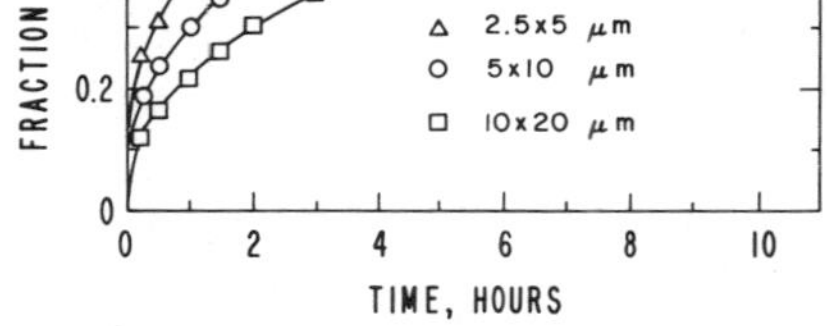

Fig. 23. A plot of iron extracted from three monosize chalcopyrite fractions as a function of time for 90°C, 0.5% solids, molor ratio of $Cu^{++}/CuFeS_2$ = 7.5 and 1200 rpm.

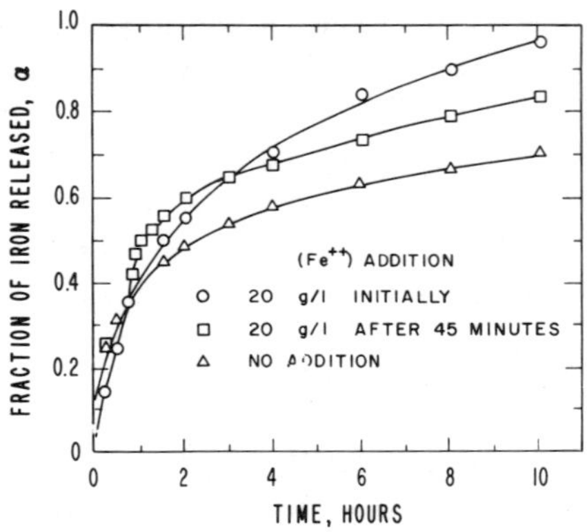

Fig. 24. Fraction of 2.5 x 5 μm chalcopyrite reduced as a function of time, showing catalytic effect of Fe^{2+}.

by djurleite, $Cu_{1.96}S$. The defect structure $Cu_{1.96}S$ (djurleite) formed in solutions of modest acidity. Solutions containing one molar or greater H_2SO_4 formed digenite, $Cu_{1.8}S$. Figure 23 illustrates the fraction of contained iron released to solution as a function of time for three monosized samples of $CuFeS_2$. The rate of conversion increased with copper concentration but was only slightly affected by the partial pressure of SO_2 gas. Ferrous iron catalyzed the reaction with significant increases in reduction rate as Fe^{2+} was added to the solution. Figure 24 illustrates results for the 2.5 μm fraction with and without added Fe^{2+} to solution. When the iron was added after 45 minutes a sharp increase in rate was noted. Warren[6] noticed a similar effect when reducing $CuFeS_2$ cathodically in the absence of cupric ion. In this case H_2S gas was evolved. Sohn[30] carried out similar cathodic reduction experiments in the presence of Cu^{2+}. The catalytic effect has been attributed to the lowered activation energy for the surface charge transfer process involving both iron and copper according to the following reactions at the electrode surface:

$$Fe^{2+}(aq) + Cu^{2+}(aq) = Cu^{+}(aq) + Fe^{3+}(aq) \tag{65}$$

$$Cu^{+}(aq) = Cu^{+}\ (\text{lattice}) \tag{66}$$

$$Fe^{3+}(aq) + e^{-} = Fe^{2+}(aq) \tag{67}$$

Figure 25 illustrates the cathodic polarization curves obtained for various Fe^{2+} solution concentrations. The sharp passivation peak formed just above 0.4 volts was attributed to the formation of an

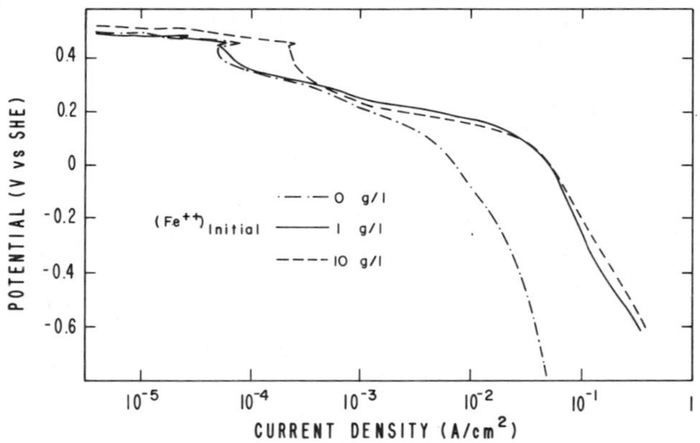

Fig. 25. Cathodic polarization curves for Transvaal chalcopyrite for various iron concentrations for 10 g/l of Cu^{++}, pH = 1.8 and 25°C.

initial defect structure. Appreciable enhancement of the current in this region resulted when the ferrous iron concentration was increased. At more negative potentials again the ferrous ion concentration increased the current although in this region only a slight increase occurred above 1 gpl Fe^{2+}. In this region $Cu_{1.96}S$ forms.

The kinetics of the process follow logarithmic kinetics, usually observed for thin tarnish films in which high voltage gradients occur. In this system thick layers of $Cu_{1.96}S$ form with a visible band of bornite which moves through the particle. Microscopic examination of a partially reacted particle clearly showed a narrow band of bornite (less than 0.5 μm in thickness) adjacent to the unreacted core of $Cu_{1.96}S$. Figure 26 illustrates the data of Fig. 23 plotted according to the logarithmic rate law.

The electrochemical nature of the process was demonstrated[30] by using a diaphragm cell with graphite electrodes. The cathodic cell contained $CuFeS_2$ and graphite powder in a solution of acidified cupric sulfate. The anode compartment was fed SO_2 gas and contained a suspension of graphite powder. Each compartment had a solid graphite electrode. The effect of ferrous iron addition was examined with separate additions to each compartment. Ferrous iron added to the catholyte increased the current markedly while addition to the anolyte decreased the current slightly.

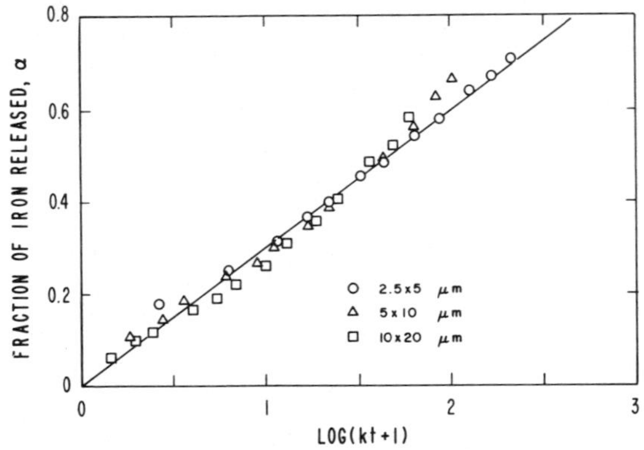

Fig. 26. A plot of the logarithmic law for the data shown in Fig. 23.

Recently Jara[32] has carried out galvanostatic experiments for the anodic dissolution of $CuFeS_2$. Figure 27 illustrates the results obtained at four separate current densities ranging from 10 to 100 microamperes. The voltage increases due to the formation of CuS(n-s). At 1.09 volts a constant voltage is maintained and is the same for the four current densities used. These results may be explained by the reactions

$$CuFeS_2 \rightarrow CuS(n\text{-}s) + Fe^{3+} + S^\circ + 3e- \quad (68)$$

$$CuS(n\text{-}s) + H_2O = CuS{\cdot}O + 2H^+ + 2e^- \quad (69)$$

$$CuS{\cdot}O = Cu^{2+} + 1/4\ SO_4^{\ 2-} + 3/4\ S^\circ + 3/2\ e^- \quad (70)$$

The plateau voltage was observed to vary with pH as suggested by Eq. 69 and the S° and $SO_4^{\ -2}$ mass balances corresponded closely to those expected in the above equations.

REFERENCES

1. G. H. Koch: Modern Aspects of Electrochemistry, No. 11, p. 211-237, Plenum Press, New York, 1975.
2. R. T. Shuey: Semiconducting Ore Minerals, Elsevier Publishing Company, New York, 1975.
3. D. J. Vaughan and J. R. Craig: Mineral Chemistry of Metal Sulfides, Cambridge University Press, New York, 1978.

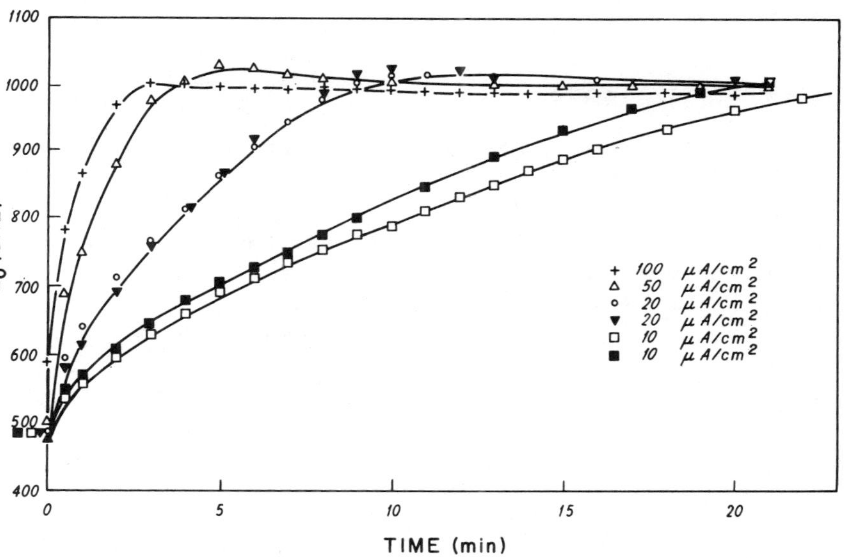

Fig. 27. Galvanostatic curves for the anodic dissolution of $CuFeS_2$ in 1.0 molar H_2SO_4.

4. T. Biegler and D. A. Swift: Electrochimica Acta, 1979, vol. 24, pp. 415-420.
5. N. Masuko and Y. Hisamatsu: Denki-Kagaku: 1963, vol. 31 pp. 907-12.
6. G. W. Warren: Ph.D. Thesis, University of Utah, Salt Lake City, Utah, 1978.
7. D. M. Chizhikov and V. N. Kovylina: Trudy Chetvertogo Soveshchaniya po Electrokhimii 1956, pp. 715-719.
8. H. Majima: Can. Met. Quart., 1969, vol. 8, pp. 269-273.
9. K. L. Vetter: Electrochemical Kinetics--Theoretical and Experimental Aspects, p. 732, Academic Press, New York, 1967.
10. P. Eadington and A. Prosser: IMM Trans., Sect. C 1969, vol. 78, pp. 74-82.
11. R. L. Paul, M. J. Nicol, J. W. Diggle and A. P. Sanders: Natn. Inst. Metall., South Africa-Report No. 1878, May 1977.
12. R. L. Paul, M. J. Nicol, J. W. Diggle, and A. P. Saunders: Electrochimica Acta, 1978, vol. 23, pp. 625-633.
13. J. E. Murphy, F. P. Haver, and M. M. Wong: U.S. Bureau of Mines RI 7913, 1974.
14. F. P. Haver and M. M. Wong: U.S. Bureau of Mines RI 8105, 1976.

15. P. D. Scott and M. J. Nicol: IMM Trans, Section C, 1976. vol. 85, pp. C40-C44.
16. Y. Awakura, S. Kamel, and H. Majima: Met Trans, 1980, vol. 11B pp. 377-381.
17. M. K. Han: unpublished research, University of Utah, Salt Lake City, Utah, 1982.
18. B. Palmer: private communication.
19. J. D. Sullivan: U.S. Bureau of Mines, Tech. Paper 473, 1930.
20. P. Marcantonio: Ph.D. Thesis, University of Utah, Salt Lake City, Utah, 1976.
21. E. Peters, I. H. Warren, and H. Veltman: First Tutorial Symposium on Hydrometallurgy, The University of Denver, 1972.
22. V. K. Berry, L. E. Murr, and J. B. Hiskey: Hydrometallurgy, 1978, vol. 3, pp. 209-326.
23. M. B. Shirts, J. K. Winter, P. A. Bloom, and G. M. Potter: U.S. Bureau of Mines RI 7953, 1974.
24. J. B. Hiskey and M. E. Wadsworth: Met. Trans. 1975, vol. 6B, pp. 183-190.
25. M. J. Nicol: IIM, 1975, vol. 84, pp. C206-C209.
26. J. P. Baur, H. L. Gibbs, and M. E. Wadsworth: U.S. Bureau of Mines, RI 7823, 1974.
27. D. L. Jones and E. Peters: "The Leaching of Chalcopyrite with Ferric Sulfate and Ferric Chloride," Extractive Metallurgy of Copper, Vol. II, J. C. Yannopoulos and J. C. Agarwal, eds., American Institute of Mining, Metallurgical and Petroleum Engineers, 1976.
28. P. B. Munoz, J. D. Miller and M. E. Wadsworth: Met. Trans. 1979, vol. 10B, pp. 149-158.
29. D. L. Jones: Ph.D. Thesis, University of British Columbia, Vancouver, British Columbia, 1974.
30. H. J. Sohn: Ph.D. Thesis, University of Utah, Salt Lake City, Utah, 1980.
31. H. J. Sohn and M. E. Wadsworth: Jour. of Metals, vol. 32, 1980, pp. 18-22.
32. J. O. Jara: M.S. Thesis, University of Utah, 1983.

THE LEACHING OF SOME ARSENIDE AND ANTIMONIDE MINERALS IN FERRIC CHLORIDE MEDIA

J. E. Dutrizac and R. M. Morrison

Mineral Sciences Laboratories, CANMET
Energy, Mines and Resources Canada
Ottawa, Ontario K1A 0G1 CANADA

The kinetics of dissolution of niccolite (NiAs), tennantite ($Cu_{12}As_4S_{13}$) and tetrahedrite ($Cu_{12}Sb_4S_{13}$) in ferric chloride-hydrochloric acid media have been investigated over the temperature range 30-95°C using both rotating disk and stirred slurry techniques. The dissolution of niccolite appears to be controlled by the aqueous diffusion of ferric ion to the mineral, and the reaction can be described by:

$$k = \frac{0.62}{n} D^{2/3} \nu^{-1/6} \omega^{1/2} C_B$$

where D is the diffusion coefficient of the ferric ion, ν is the kinematic viscosity, ω is the rotation speed and n is a number between five and seven. The apparent activation energy is 22 kJ/mole. The effects of temperature, stirring speed, ionic strength, and the concentrations of ferric ion, ferrous ion, arsenic(III) and hydrochloric acid have been evaluated. Arsenic dissolved as As^{3+} in spite of the fact that the oxidizing power of the medium was sufficient to form As^{5+}. Natural or synthetic tetrahedrite dissolves very slowly in ferric chloride media and the kinetics are essentially linear. The high activation energy, 65-80 kJ/mole together with the observed effects of ferric ion and HCl concentrations suggest that the reaction is chemically or electrochemically controlled. The reaction products are Cu^{2+} and S^o together with Sb^{5+} which precipitates in the presence of ferric ion. Tennantite dissolves in a manner similar to tetrahedrite and at approximately the same rate.

The reaction products are Cu^{2+}, S^{o} and As^{3+} which remain dissolved for the concentrations encountered.

INTRODUCTION

Arsenide and antimonide minerals are commonly present in minor amounts in many ore deposits. Niccolite (NiAs) occurs widely as a minor constituent in both base metal sulfide ores and in some uranium ores[1] where it constitutes a major source of arsenic in the processing feeds. Furthermore, niccolite is often a major mineral in Ni-Co-As vein assemblages which represent a potential source of nickel and cobalt. Tennantite ($Cu_{12}As_4S_{13}$) is commonly associated with copper and lead ores and concentrates where it is both a minor source of copper and a major arsenic contributor to the processing circuit. Tetrahedrite ($Cu_{12}Sb_4S_{13}$) is likewise an important source of copper in copper or lead feeds, but it is more significant as a source of silver and antimony in many concentrates.[2] Both tennantite and tetrahedride are remarkable for the number of minor elements (Hg, Zn, Bi, Se, etc.) they can incorporate, and these two minerals are frequently the major sources of such often unwelcome minor elements in flotation concentrates.[3] Hence, the arsenide and antimonide minerals are of dual significance in metallurgical operations: they are an important source of sought-after values of copper, nickel, cobalt and silver but they are equally a major contributor of undesirable pollutants (As, Hg, Sb, etc.) to the processing streams.

Given the dual importance of arsenide and antimonide minerals in base metal concentrates and the growing importance of hydrometallurgical processing for these feeds, it is surprising that very little is known about the behavior of niccolite, tennantite or tetrahedride in ferric chloride media. Ferric chloride solutions, with or without cupric chloride, are being widely evaluated[4-6], as a means of processing either copper or lead concentrates, and the behavior of minor elements in such processing options will directly affect their viability.

Virtually nothing has been reported for the reaction of niccolite with ferric chloride media although the reaction of this mineral with acid and oxygen has been studied.[1] The two types of study are not directly comparable, however, because of the greater difficulties involved in oxygen transport and the different valencies of arsenic encountered. A limited amount of work has been published on the dissolution of tetrahedrite or tennantite in ferric ion media, and these studies have been summarized.[7] Most previous investigations have suffered from the use of impure concentrates containing more than one copper mineral. For example, Brown and Sullivan[8] observed 10% copper dissolution from tennantite during the first 24 h of reaction, but only 1.6% additional Cu dissolution over the next 55 days. The initial rapid dissolution was attributed to impurities, with the tennantite itself dissolving very slowly. Bryner et al.[9] studied the bacterial leaching of "museum grade" tetrahedrite at room temperature, and observed only 3 mg Cu dissolution from a 1-g sample after 42 days. As far as is known, there have been no other systematic studies on the leaching of relatively pure tennantite or tetrahedrite in ferric ion media. This lack of activity is certainly more attributable to the scarcity of high-purity natural specimens and the difficulty of synthesizing tetrahedrite-tennantite than to any lack of metallurgical interest in these materials.

In the current investigation, the kinetics of dissolution of natural and/or synthetic niccolite, tennantite and tetrahedrite in acidified ferric chloride media have been determined. Efforts were made to ascertain the dissolution reaction, especially the valences of arsenic and antimony since these will directly influence the precipitation of these elements from the pregnant solutions. Both rotating disk and stirred slurry techniques were used depending on the materials under study and the rate controlling kinetic mechanism; the temperature and concentration ranges employed were chosen to encompass the conditions likely to be encountered in commercial leaching operations.

EXPERIMENTAL

Arsenides and Antimonides

The niccolite used in this study was natural high-grade ore from the south side of Sachowia Lake, Great Slave Lake area, North West Territories and was obtained from the Geological Survey of Canada. The ore consisted principally of niccolite but contained minor amounts of gersdorffite and rammelsbergite and also some gangue minerals. Some of the niccolite ore was crushed and sized into -70/+100 and -100/+150 mesh fractions which were examined with an electron microprobe. The nickel (43.8%) and arsenic (55.8%) concentrations of the niccolite itself totalled 99.6%; cobalt, iron, antimony and sulfur were not detected. The expected values for nickel monoarsenide are Ni = 43.9% and As 56.1%. Image analysis of the two mesh fractions gave weight percentages of niccolite as

Table 1. Chemical Composition of the Tennantite Grains as Determined by the Electron Microprobe.

Element	Range of Composition	Average Composition (wt%)
Cu	27.50 - 44.70	41.77
Ag	0.00 - 7.81	0.034
Fe	0.94 - 6.30	4.51
Zn	0.00 - 6.87	1.35
As	1.13 - 18.90	14.53
Sb	2.29 - 27.46	9.29
S	27.29 - 30.0	27.80
Hg	0.00 - 0.26	0.0456 (456 ppm)
Total:	98.5 - 99.5	99.32

97.2% and 97.1%, respectively. For the rotating disk work, the massive niccolite was sawed into cubes 1 cm to a side which were cast into 3 cm-resin disks and polished to 1 μm with diamond paste. The resin disks were cemented onto the ends of cast acrylic rods.

The tennantite used for this program originated from the McIntyre mine, Timmins, Ontario.[3] High grade ore samples were obtained from the 3811 slope, 3375 level, and were dry crushed and sized. The ground ore was then upgraded by flotation and magnetic means to give a concentrate containing 85% tennantite together with minor pyrite and gangue. The concentrate so produced had the following bulk composition: Cu - 36.3%, As - 8.75%, Ag - 0.10%, and Zn - 1.44%. The tennantite itself was analyzed in detail by an electron microprobe and had the composition shown in Table 1.

The As:Sb atomic ratios vary from 13.1 to 0.70, with more than 95% of the grains corresponding to tennantite (As:Sb >1). The concentrations of all elements in this mineral, but especially silver and mercury, vary greatly from grain to grain (and from sample to sample). Silver contents are generally less than 0.4 wt%, with an average of about 0.03 wt%; the highest silver content observed was 7.81 wt% for a tetrahedrite inclusion in a galena grain. Mercury contents in tennantite-tetrahedrite are generally close to the electron microprobe detection limit of 400 ppm. After the initial concentrates had been produced, the tennantite was re-cyclosized into well-defined size fractions which were used for the actual leaching experiments.

Two slightly different tetrahedrites, one iron-free and the other iron-bearing, were used in the leaching experiments, and both were synthetic. The synthesis of sintered disks of iron-free tetrahedrite, $Cu_{12.5}Sb_{4.0}S_{12.9}$, has been presented.[10] The sintered disks were 85-90% of the theoretical tetrahedrite density and were shown to be single phase. Electron microprobe analysis indicated this sulphosalt to be homogeneous and to have, within error, the analysis shown. Disks of the iron-bearing tetrahedrite were prepared in the same general manner. Appropriate amounts of the hydrogen reduced metals were placed in a silica boat and sealed under vacuum in a silica tube with sufficient sulfur to give $Cu_{11}FeSb_4S_{13}$; a twin furnace arrangement was used such that the metals were heated to 750°C while the sulfur was maintained at 400°C. The initial reaction product, which had melted, was dry-ground to -150 mesh and was then pelletized at 100,000 psi in a 1-inch die. The pellets were sealed under vacuum in a glass tube and were then sintered and homogenized for 6 weeks at 545°C. On cooling, the disks were polished through 4/0 emery paper and were then glued to the ends of plastic stirring rods for the leaching tests. The products were found to be optically and compositionally homogeneous and were shown to have the following composition in weight percent: Cu - 42.02, Fe - 3.38, Sb - 29.37, S - 25.53. These analyses correspond to $Cu_{10.9}Fe_{0.99}Sb_{3.96}S_{13.1}$.

Leaching Procedure

Reagent grade chemicals were used to prepare all the solutions which were held in thermostatted water baths. The various disk samples were rotated by a Dyna-mix stirrer whose speed was periodically trimmed with a strobe light; the same stirring assembly was used to suspend the 1-g samples of ground minerals leached in some tests. The solution volumes and concentration of reagents were such that no significant depletion of oxidants or acid occurred during the experiments. The leaching rates of the copper minerals were monitored by automatic on-line analysis for copper using a described procedure[11]; the dissolution of the nickel minerals was followed by taking pipette samples at suitable times and analyzing these for nickel by atomic absorption spectrophotometry. At various times during the experiments, the solutions were sampled for Fe^{2+}/Fe^{3+} and for As or Sb determinations using classical methods. At the end of the experiments, the elemental sulfur content was ascertained on air-dried samples by CS_2 extraction, and the residues were examined to ascertain the form and composition of any intermediate compounds which formed. The reaction stoichiometries were then deduced from the various analytical results. Solid phases were confirmed by X-ray diffraction analysis.

RESULTS AND DISCUSSION

Niccolite Dissolution

(i) Overall Reaction. Niccolite dissolves according to the reaction:

$$NiAs + nFe^{3+} \rightarrow Ni^{2+} + As^{x+} + nFe^{2+} \qquad (5)$$

The molar ratio of soluble nickel to arsenic obtained when significant amounts of niccolite (2.0 - 8.0 g) were dissolved varied from 0.95 to 1.09 indicating that niccolite dissolution is stoichiometric. No arsenide intermediate phases were detected on the surface of the NiAs as would be required for nonstoichiometric dissolution. Although a thin black film was observed on the surface of the leached samples, this was shown to be niccolite by both X-ray diffraction and chemical analyses. Presumably the film consists of very fine niccolite particles.

The value of n in Reaction 1 was calculated from the ratio of soluble nickel to ferrous ion present in the final solutions. The experimental values of n ranged from 4.6 to 5.8; when experimental errors are taken into consideration, it would appear that five electrons are withdrawn from NiAs. The form of nickel present in solution is surely a complex of nickel(II) and this yields a formal oxidation state of arsenic as +3. In acid solutions such as those used in this study, the predominant forms of arsenic[12] would be $As(OH)_3$ and $As(OH_2)_2^+$ and associated chloro-derivatives.[13] The free energy change of nickel arsenide reacting with ferric ion to form ferrous ion, nickel(II) ions and various arsenic species can be calculated using the absolute free energy for nickel arsenide[14] and the other compounds.[15] For the reaction:

$$NiAs + 5Fe^{3+} + 3H_2O \rightarrow Ni^{2+} + H_3AsO_3 + 5Fe^{2+} + 3H^+ \qquad (2)$$

the free energy change is -262 kJ/mole. For the reaction

$$NiAs + 7Fe^{3+} + 4H_2O \rightarrow Ni^{2+} + H_3AsO_4 + 7Fe^{2+} + 5H+ \qquad (3)$$

it is -300 kJ/mole. Thus, thermodynamics predict that the reaction between ferric ion and niccolite should proceed via the second reaction producing arsenic(V). However, the actual product found is arsenic(III) and Reaction 2 is appropriate. The oxidation of arsenic(III) to arsenic(V) is often under kinetic rather than thermodynamic control. Even in the presence of strong oxidizing agents such as Ce(IV), the rate of oxidation can be slow in the absence of catalysts.[16] Other workers have also reported that the oxidation of trivalent arsenic is difficult to effect with ferric ion.[17] In any event, As^{3+} seems to be the predominant arsenical reaction product and this enjoys a dynamic, if not truly thermodynamic, stability in the leach solutions.

(ii) RPM Dependence. Figure 1 shows the leaching curves (concentration of nickel in solution versus time) obtained at 40°C for various disk rotation speeds. The kinetics are linear, consistent with a system where no passivating film is created. The rate of reaction, V, derived from the leaching data is dependent upon the rotation speed although it is not directly proportional to ω^{+1}. The Levich[18] equation predicts that the rate of reaction under solution mass transport control conditions will be proportional to the square root of the disk rotation speed. Figure 2 shows that such a relationship does, in fact, exist for this reaction for $\omega \leq 1000$ rpm. Although not shown, the reaction rates measured for $\omega > 1000$ rpm were erratic; this behavior may have been due to disk "wobble" or to the onset of turbulent flow.

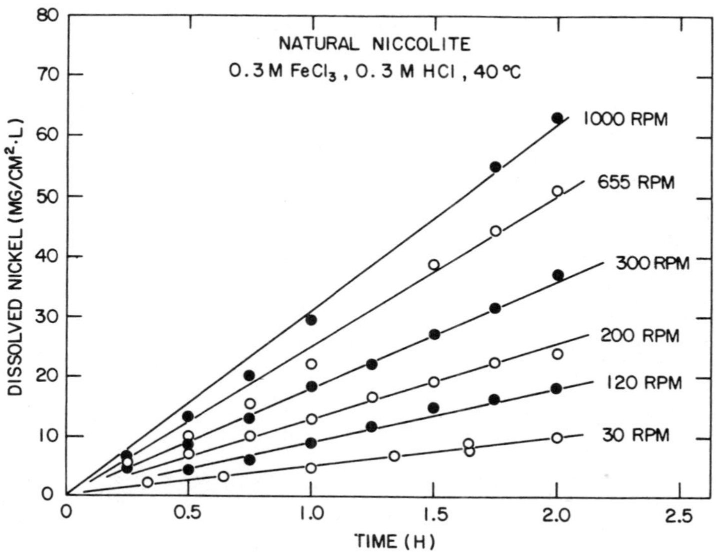

Fig. 1. Leaching curves for the dissolution of natural niccolite at various disk rotation speeds.

The Levich equation can be used to predict the limiting rate of nickel dissolution. The Levich equation:

$$j = 0.62\ D^{2/3}\nu^{-1/6}\omega^{1/2}(C_B - C_S) \qquad (4)$$

gives the mass flux of a species to or from a rotating disk. The mass flux, j, has units of $mg/cm^2.s$ when: D, the diffusion coefficient is in cm^2/s; ν, the kinematic viscosity is in cm^2/s; ω is in radians per second; and C_B and C_S, the concentration of the diffusing species in the bulk solution and at the disk surface, respectively, are in mg/cm^3. The rate of niccolite dissolution could be controlled by the mass transport of reactants (ferric ion or its complexes) to the niccolite surface, by the mass transport of reaction products (ferrous ion, nickel, or arsenic) away from the surface, or by a mixed chemical reaction-mass transport sequence.

From the magnitude of the dependence of the rate on the rotation speed, it can be concluded that mass transport of the ferric ion is the rate controlling factor. The D and ν parameters at 40°C and at a concentration of ferric ion of 0.3 moles per liter were determined. The value of $D = 7.7 \times 10^{-6}\ cm^2/s$ was calculated for ferric ion from

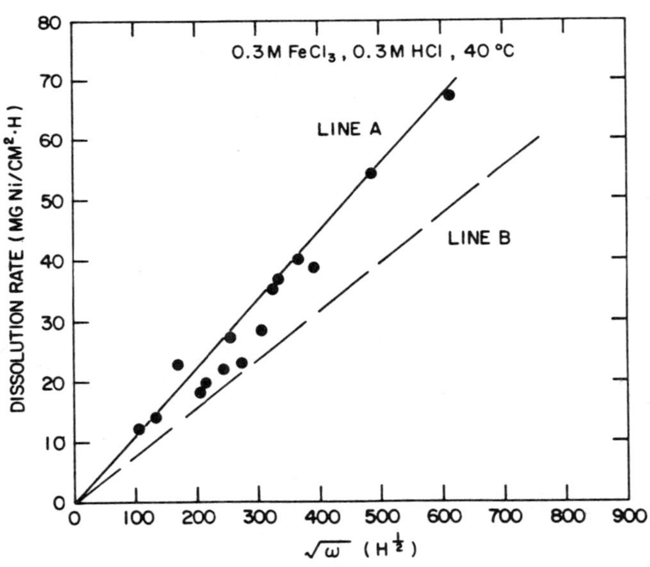

Fig. 2. Comparison of experimental and predicted leaching rates of niccolite assuming ferric ion diffusion control. Line A is for As(III) formation and Line B is for As(V) production.

the published value[19] of 4.5×10^{-6} cm^2/s at 25°C assuming an activation energy of 23 kJ/mole. The kinematic viscosity was calculated from $\nu = \eta/\rho$ where η is the viscosity of H_2O at 40°C (0.6529×10^{-2} g/cm s) and ρ, the density of the solution (approximately 1.060 g/cm^3); a value of $\nu = 6.16 \times 10^{-3}$ cm^2/s was obtained. For example, at 1000 rpm with 0.3 M $FeCl_3$ - 0.3 M HCl at 40°C, the predicted mass flux of ferric ion is 375 mg/cm^2 h. Before a direct comparison of the predicted rate of ferric ion diffusion and the observed rate of nickel dissolution can be made, however, the nature of the heterogeneous reaction must be considered.

The experimental rate is that of nickel dissolution and it is being compared to the ferric ion flux. The total ferric ion available for reaction is a fraction of the total flux if more than one oxidation step takes place at the disk surface. Also, since the rate of reaction is conveniently expressed in milligrams of nickel per square centimeter per hour, the mass flux of ferric ion must be converted via an expression reflecting both molecularity and molar/time conversion factors. Thus, the modified equation would be:

$$V_c = \frac{1}{n}\frac{58.71j}{55.85} \tag{5}$$

where j is converted to mg/cm^2 h, V_c represents the calculated rate and n is the molecularity of Reaction 1.

Line A in Fig. 2 is the curve calculated from Eq. 5 with n = 5 and assuming ferric ion diffusion control. There seems to be fairly good agreement between the experimental and calculated values. Line B in Fig. 2 is the calculated rate with n = 7 and does not give as good a correlation as Line A. This analysis of the rate dependence on n also suggests strongly that all of the oxidation steps (n = 5) take place on the disk surface. Considering the relatively imprecise nature of the rotating disk experiments and of the approximate values of D and ν used, the agreement between the calculated (n = 5) and experimental rates is excellent, and strongly argues for rate control by the diffusion of Fe^{3+} to the surface of the arsenide.

(iii) Effect of Temperature. Figure 3 shows the relationship between the leaching rate and the reaction temperature in an Arrhenius plot format. The logarithm of V is shown to be proportional to the reciprocal of the absolute temperature; linear regression gives the relationship as $\ln V = 11.5 - \frac{2600}{T}$. The apparent activation energy obtained from the slope is 22(±2) kJ/mole, a value typical of aqueous diffusion controlled processes.[7] Thus, the results of the

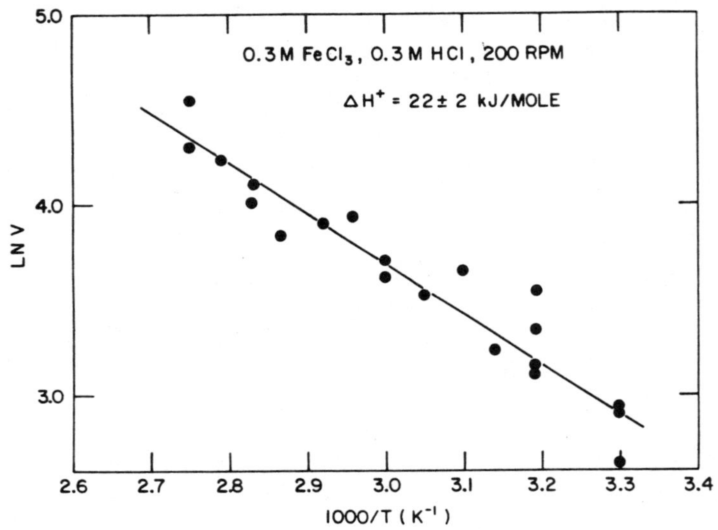

Fig. 3. Arrhenius plot for the dissolution of natural niccolite in 0.3 M $FeCl_3$ solutions.

rotation speed and temperature experiments indicate that the rate controlling step is mass transport of ferric ion to the surface of the niccolite.

(iv) Effect of Ferric Ion Concentration. The effect of various ferric ion concentrations on the rate of niccolite dissolution was also examined; Fig. 4 shows the dependence of the observed rate on the ferric ion concentration. For a mass transport controlled reaction, the rate should, as a first approximation, be proportional to the ferric ion concentration. The figure shows that for ferric concentrations less than 0.4 M, the observed rates agree fairly well with the calculated values (Line A); at concentrations greater than 0.4 M and less than 1.0 M, the rate is in fair agreement with the predicted behavior, but at higher concentrations (>1.0 M), the rate appears to decrease with increasing concentrations of Fe^{3+}. This non-ideal behavior is probably caused by mechanisms other than ferric ion diffusion becoming partly or fully rate controlling.

(v) Effect of Reaction Products. Equation 2 gives the reaction products as nickel(II), ferrous ion, acid and arsenic(III). These products could build up during a commercial leaching process, and therefore it is necessary to know if their presence in high concentrations influences the reaction kinetics.

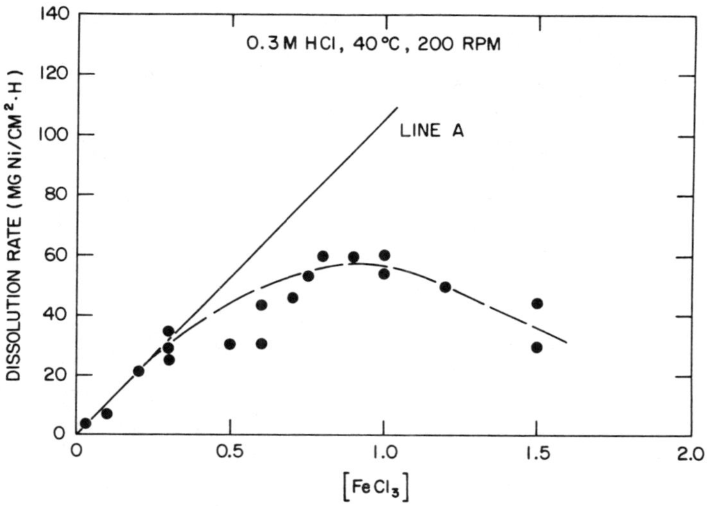

Fig. 4. Effect of the ferric chloride concentration on the rate of niccolite dissolution at 40°C.

The acid added to the leaching solution prevents the hydrolysis and precipitation of iron(III), and thus the loss of the oxidant. Previous work[20] has shown that 0.1 M acid is sufficient to prevent iron precipitation. Acid also has the potential to oxidize the mineral forming hydrogen gas. It was found, however, that hydrochloric acid alone does not dissolve niccolite to any significant extent. Also, when a large concentration of acid was added to the 0.3 M Fe^{3+} leach solution at 40°C the value of V obtained did not increase.

The addition of $FeCl_2$ to the normal 0.3 M $FeCl_3$ leaching medium did not alter the dissolution reaction at 40°C as can be seen from the data presented in Table 2. The essentially constant leaching behavior is wholly consistent with ferric ion diffusion since neither ferrous ion diffusion nor the Fe^{3+}/Fe^{2+} redox couple enter directly into the rate controlling step.

When arsenic(III) was added to the leaching media, the reaction kinetics were essentially unchanged from those observed in the normal 0.3 M Fe^{3+} - 0.3 M HCl solution as illustrated in Table 3. Higher concentrations of As^{3+} could not be studied because of precipitation problems. The lack of effect of initial concentrations of As^{3+} on the niccolite leaching rate is wholly consistent with ferric ion diffusion control.

Table 2. Effect of Ferrous Chloride Concentration on the Rate of Niccolite Dissolution in 0.3M $FeCl_3$ Media.

$[Fe^{2+}]$ (moles per liter)	V ($mg/cm^2.h$)
0.0	28.1
0.03	26.9
0.11	25.8
0.3	31.8
0.5	30.5
1.0	28.5

Table 3. Effect of As^{3+} Concentration on the Rate of Niccolite Dissolution in 0.3 M $FeCl_3$ Media.

$[As^{3+}]_{initial}$ (gpl)	V ($mg/cm^2.h$)
0	28.1
0.6	27.8
0.8	29.0

Table 4. Effect of Ionic Strength (as LiClOy) on the Rate of Niccolite Dissolution in 0.3 M $FeCl_3$ Media.

$LiClO_4$ Added (moles per liter)	V ($mg/cm^2.h$)
0	28.1
0.5	32.1
1.0	31.1
1.5	25.6

Ionic strength (varied with $LiClO_4$) was found to have no direct effect on the leaching rate as the data of Table 4 indicate. There is a suggestion that the rate declines slightly with large increases in ionic strength, but the decrease, if real, is minor and could be explained by indirect effects on the viscosity and/or Fe^{3+} diffusion coefficient.

The effect of large concentrations of dissolved nickel on the rate of niccolite dissolution could not be measured directly because of difficulties in detecting accurately small amounts of Ni against a high Ni background. Reaction 2 shows that 5 Fe^{2+} ions are produced for each Ni^{2+} formed; hence, the Fe^{2+} concentration provides a means of following the reaction in the presence of high NiCl2 concentrations. After 2 h of leaching at 40°C in 0.3 M $FeCl_3$-0.3 M HCl media, the quantity of Fe^{2+} was 129 and 141 (mg/cm^2 h) for the normal solution and for the normal solution plus 14.6 g/L Ni^{2+}, respectively. Again this result is consistent with Fe^{3+} diffusion as the rate controlling step.

Tennantite Dissolution

(i) Temperature Effect. Tennantite dissolution in 0.2 M $FeCl_3$ - 0.3 M HCl solutions was investigated as a function of temperature using 1-g samples of -50, +32 μm natural tennantite. Figure 5 shows some typical leaching curves plotted as $1-(1-\alpha)^{1/3}$ versus time where α is the fraction tennantite reacted. Several observations can be made concerning these curves. Firstly, the rate of tennantite dissolution is extremely slow even at the highest temperatures studied; at 90°C, only 15% reaction occurred after 40 h of leaching. In terms of the observed leaching rates, tennantite appears to be even more refractory than chalcopyrite.[20] Secondly, the leaching curves all extrapolate to a finite fraction reaction (~0.01) at zero time; this indicates the presence of trace amounts of more reactive copper minerals (bornite, chalcopyrite) in the natural tennantite concen-

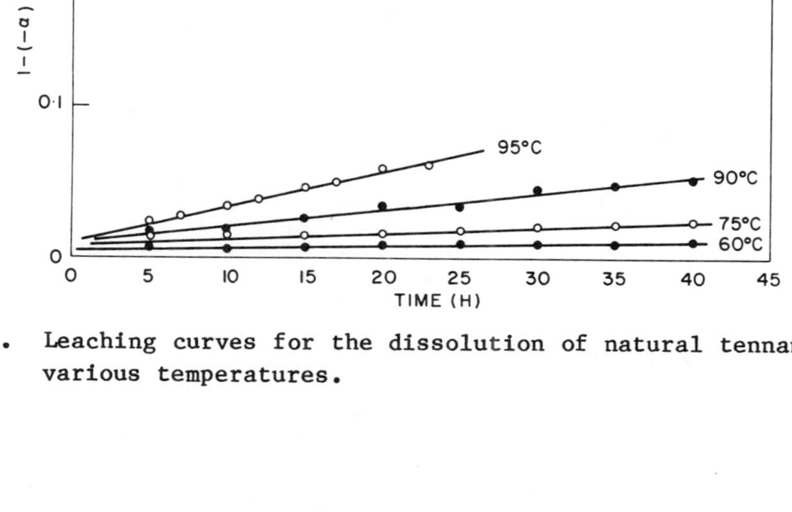

Fig. 5. Leaching curves for the dissolution of natural tennantite at various temperatures.

trate. Lastly, the leaching kinetics all appear to obey closely the $1-(1-\alpha)^{1/3}$ relationship which is characteristic of surface controlled reactions. Although elemental sulfur is formed:

$$Cu_{12}As_4S_{13} + 36Fe^{3+} + 12H_2O \rightarrow 12Cu^{2+} + 4H_3AsO_3 + 36Fe^{2+} + 13S^\circ + 12H^+ \quad (6)$$

it does not appear to be protective. The reaction stoichiometry was not thoroughly investigated, but the Fe^{2+}/Cu^{2+} ratio was found to be midway between the 3.00 value expected for As^{3+} formation and the 3.36 value demanded for As^{5+}. Given that trace amounts of SO_4 are also produced, it seems that As^{3+} is the principal arsenic product, and this is consistent with the niccolite dissolution work discussed above.

Linear rate constants were deduced from the leaching curves of Fig. 5 using the expression:

$$1-(1-\alpha)^{1/3} = a + 1_\alpha t \quad (7)$$

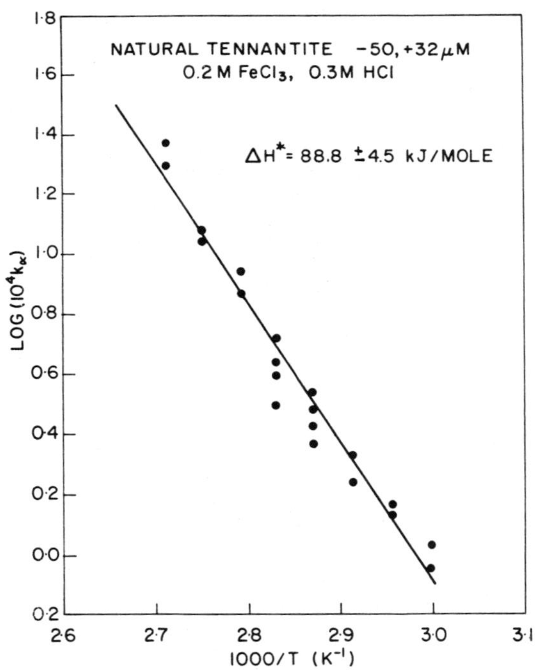

Fig. 6. Arrhenius plot for the dissolution of sized natural tennantite.

and the rate constants so deduced were displayed on an Arrhenius plot as shown in Fig. 6. The data are consistent over the temperature range 60-95°C, and suggest a single rate controlling process. The high value indicates chemical or electrochemical control. The value is considerably higher than the corresponding activation energies reported (38-63 kJ/mole) for chalcopyrite dissolution in $FeCl_3$ media,[20] and the kinetics of that reaction are also felt to be chemically or electrochemically controlled.

(ii) Effect of Tennantite Particle Size. Tennantite and tetrahedrite are very similar both structurally and compositionally and, as will be seen later, the two minerals behave nearly identically in $FeCl_3$ media. Consequently, the effect of particle size on the leaching rate was verified only for tennantite, and it was assumed that the tetrahedrite end of the $Cu_{12}As_4S_{13}$ - $Cu_{12}Sb_4S_{13}$ series would behave in the same manner. Various size fractions of the tennantite concentrate were obtained by a double cyclosizing procedure; 1-g quantities of each fraction were then leached in an excess of 0.2 M $FeCl_3$ - 0.3 M HCl solution at 95°C. Individual leaching curves, similar in form to those of Fig. 5, were obtained and rate constants were realized by fitting the data according to Eq. 7. Figure 7 shows the corresponding rate constants plotted versus 1/r, where r is the initial mean radius of the particles. If the rate of reaction is directly proportional to the surface area of the particles, a straight line passing through the origin is expected. Such behavior is, in fact, observed for leaching in ferric chloride media. Hence, it is concluded that the dissolution of both tennantite and tetrahedrite in $FeCl_3$-HCl media is directly proportional to the area of the mineral. Given the low leaching rates observed, extremely fine grinding would be required to provide commercially attractive rates for either mineral.

Tetrahedrite Dissolution

(i) Temperature Effect. Sintered disks of synthetic tetrahedrite were used to examine the kinetics of dissolution of this mineral in $FeCl_3$-HCl media; both iron-bearing and iron-free tetrahedrites were employed. Because the surface area of the disks was unaffected by small amounts of dissolution, the quantity of copper dissolved was itself indicative of the rate and there was no need to resort to shrinking core models, etc. Figure 8 illustrates the leaching curves obtained when the iron-bearing tetrahedrite was leached at various temperatures between 65 and 95°C. As was the case for tennantite, the tetrahedrite dissolves very slowly in 0.2 M $FeCl_3$ - 0.3 M HCl media. Only 30 mg Cu/cm^2 was leached after 2 days of

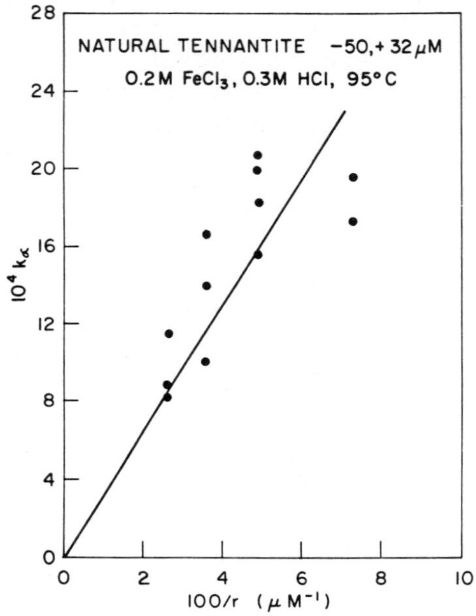

Fig. 7. Variation of the leaching rate with the mean particle size of the natural tennantite.

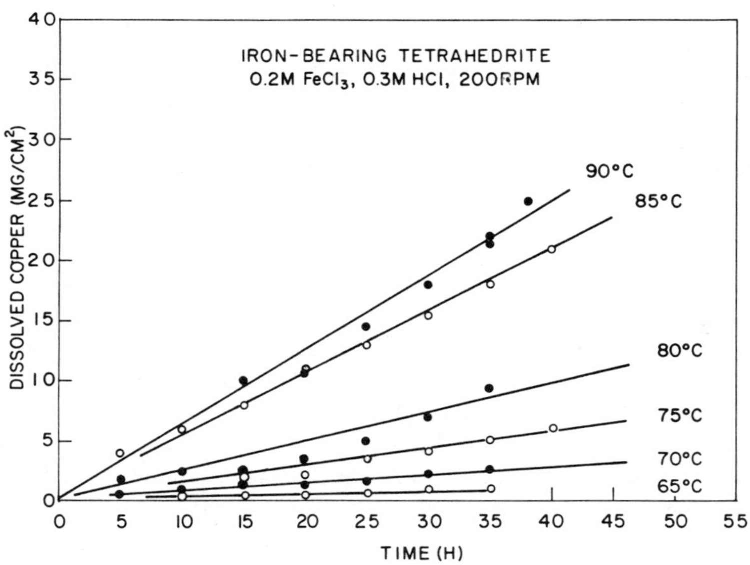

Fig. 8. Leaching curves for the dissolution of disks of synthetic iron-bearing tetrahedrite at various temperatures.

reaction at 90°C; the rates were almost immeasurably slow below 60°C. Clearly the tetrahedrite-tennantite family minerals are extremely refractory to ferric chloride, and their dissolution during a ferric chloride leach process will lag behind most copper or lead minerals, including chalcopyrite. The refractory nature of tetrahedrite-tennantite to $FeCl_3$ media probably explains why the hydrometallurgy of these minerals has tended to emphasize sulfide leach processes.[2,21] The dissolution curves presented in Fig. 8 are essentially linear and since the surface area remains largely unaffected, the kinetics are also linear as was noted previously (Fig. 5) for tennantite leaching. Rate constants were, therefore, deduced by fitting the data to an equation of the form:

$$Cu(mg/cm^2) = a + kt \tag{8}$$

where k is the linear rate constant.

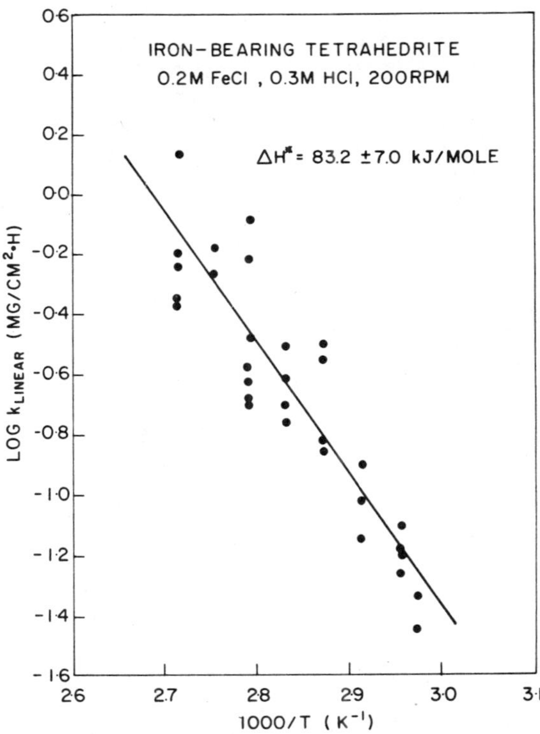

Fig. 9. Arrhenius plot for the dissolution of disks of synthetic iron-bearing tetrahedrite.

The rate data for the various tests were summarized on an Arrhenius plot as shown in Fig. 9. Although there is some scatter caused by the low amounts of copper dissolved and surface area variations from disk to disk, the data are all reasonably consistent suggesting a single rate controlling process. The data yielded an apparent activation energy of 83.2 ± 7.0 (1σ) kJ/mole.

Similar leaching tests were carried out using the iron-free tetrahedrite disks and linear kinetics were observed at all temperatures between 60 and 95°C. Because the iron-free tetrahedrite disks were slightly more porous than the iron-bearing material, the actual amount of copper dissolved was slightly greater. For example, after 20 h or leaching at 90°C in 0.2 M $FeCl_3$ media, the iron-free tetrahedrite had dissolved to the extent of 15 mg Cu/cm^2 and the iron-bearing material to 12.5 mg Cu/cm^2. Linear rate constants were also deduced for each test done with the iron-free sulfide and these data

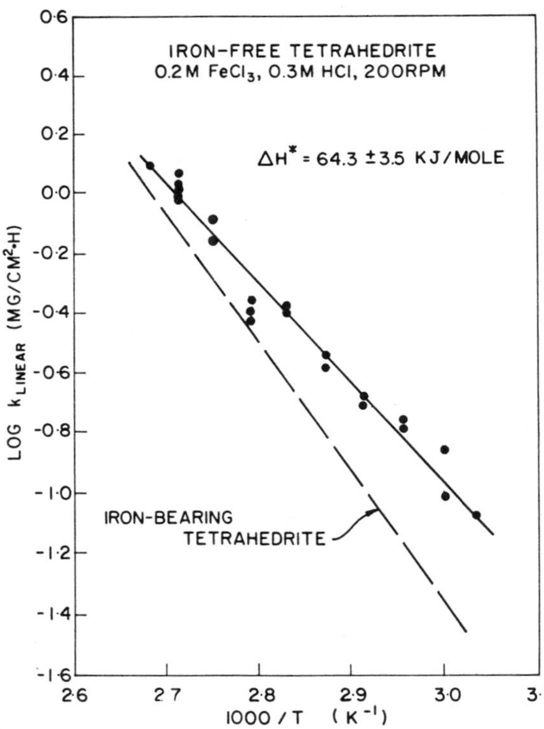

Fig. 10. Arrhenius plot for the leaching of disks of iron-free tetrahedrite and a comparison with the corresponding data for iron-bearing tetrahedrite.

are illustrated in Fig. 10 together with the smoothed curve (dashed line) for the iron-bearing tetrahedrite. The data for the iron-free sulfide are reasonably smooth over the whole temperature region, again indicating a single dissolution process. The calculated activation energy for the iron-free tetrahedrite is 64.3 ± 3.5 (1σ) kJ/mole.

The various activation energies calculated for leaching the tetrahedrite-tennantite minerals in 0.2 M $FeCl_3$ are given in Table 5. The iron-bearing tetrahedrite agrees closely with the natural tennantite and both contain about 4% Fe. Both have calculated activation energies significantly higher (~20 kJ/mole) than the iron-free tetrahedrite although the errors are such that actual differences are hard to ascertain. The data suggest that the activation energy for tetrahedrite-tennantite mineral dissolution in ferric chloride media may increase with increasing iron substitution in the mineral lattice. Additional studies employing tetrahedrites with a wider range of iron contents would be required to substantiate this hypothesis.

(ii) Ferric Chloride Concentration. The concentration of $FeCl_3$ is one of the important control parameters in any ferric chloride leaching process; increasing the $FeCl_3$ concentration is one of the easiest means (together with temperature and fine grinding) of accelerating the leaching rate. Also, the $FeCl_3$ concentration dependence is helpful for elucidating various rate controlling mechanisms. For the above reasons, the dissolution of both iron-bearing and iron-free tetrahedrites were investigated over the concentration range 0.01 - 3.0 M $FeCl_3$. All solutions contained sufficient acid (0.3 M HCl) to prevent iron precipitation, and sufficient volume was available to minimize any solution depletion effects. Figure 11 illustrates some typical leaching curves obtained for the dissolution of disks of iron-bearing tetrahedrite in ferric chloride - 0.3 M HCl media at 95°C and 200 rpm disk rotation speed. The leaching reaction is strongly affected by the $FeCl_3$ concentration, and moderately rapid rates are observed for $[FeCl_3] > 1$ M. Linear kinetics are noted at all ferric ion concentrations studied,

Table 5. Activation Energies Calculated for the Dissolution of Tetrahedrite-Tennantite Minerals in $FeCl_3$ Media.

Material	ΔH* (kJ/mole)
Natural tennantite (4.5% Fe)	88.8 ± 4.5
Iron-bearing tetrahedrite (3.4% Fe)	83.2 ± 7.0
Iron-free tetrahedrite (0% Fe)	64.3 ± 3.5

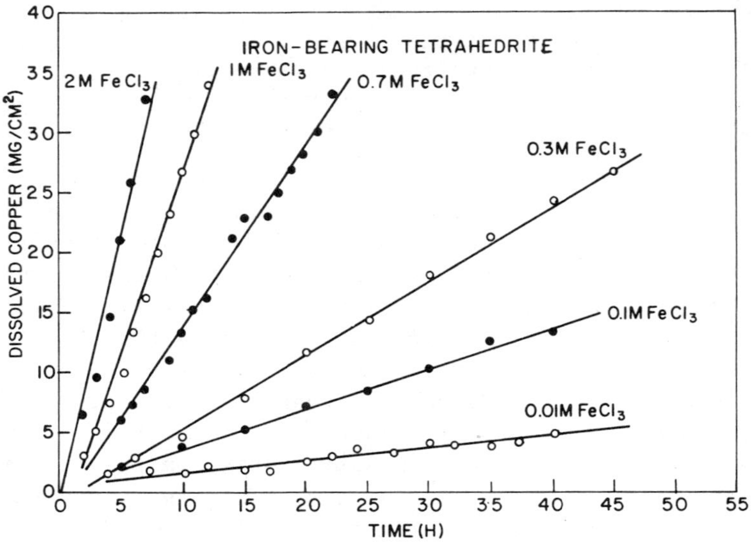

Fig. 11. Leaching curves for the reaction of disks of iron-bearing tetrahedrite in solutions containing various $FeCl_3$ concentrations.

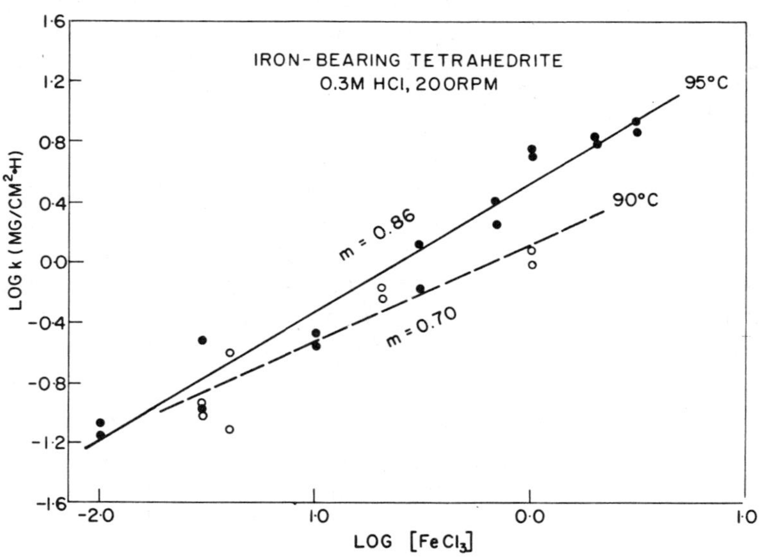

Fig. 12. The effect of the ferric ion concentration on the rate of dissolution of the iron-bearing tetrahedrite at two different temperatures.

and this supports the view that only one rate controlling process is involved under all practical conditions studied.

Linear rate constants were derived for the various leaching curves obtained at the different $FeCl_3$ concentrations and 95°C, and these data (solid line and points) are summarized in Fig. 12 together with the few values realized at 90°C (dashed line and open circles). Although there is considerable scatter, both sets of data can be reasonably well fitted by straight lines. At 95°C, the rate equation is:

$$\log k(\text{mg/cm}^2\ \text{h}) = 0.521 + 0.878 \log[\text{FeCl}_3] \qquad (9)$$

with a standard deviation of the slope of 0.05. At 90°C, the corresponding equation is:

$$\log k(\text{mg/cm}^2\ \text{h}) = 0.108 + 0.696 \log[\text{FeCl}_3] \qquad (10)$$

and the standard deviation of the slope is 0.11. It is likely that identical $FeCl_3$ dependencies exist at both 95 and 90°C, and that the calculated differences are due to experimental error, especially at the lower Fe^{3+} concentrations where the measured rates are very low.

Similar leaching experiments were also carried out using the iron-free tetrahedrite disks and, again, essentially linear kinetics

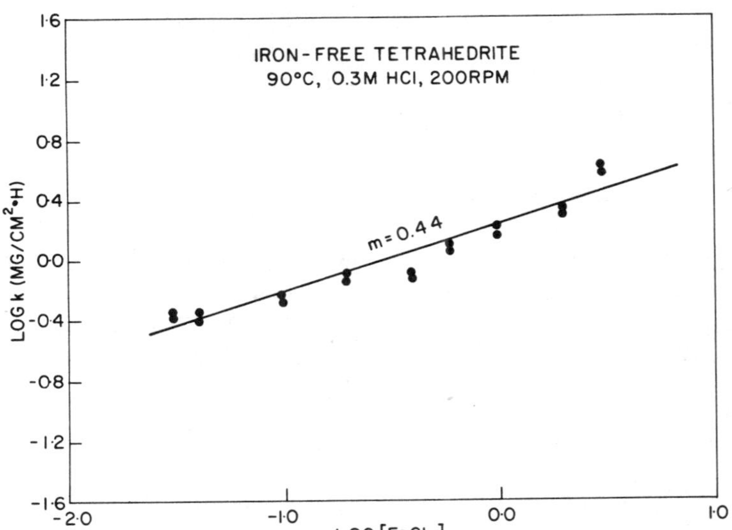

Fig. 13. The effect of the ferric ion concentration on the rate of dissolution of iron-free tetrahedrite.

were observed over the $FeCl_3$ concentration range 0.03 to 2.0 M. Linear rate constants were calculated from the individual leaching curves and these rate data are summarized in Fig. 13. The data are all fairly consistent and yield the following expression:

$$\log k(\mathrm{mg/cm^2\ h}) = 0.208 + 0.442 \log[FeCl_3] \qquad (11)$$

whose standard deviation of the slope is 0.03.

As was the case for the observed activation energies, there appears to be a significant variation of the ferric chloride concentration dependence (m in $\log k = a + m \log[FeCl_3]$) between the iron-bearing and iron-free tetrahedrites as shown in Table 6. Although all the materials exhibit a high $FeCl_3$ power dependence, there is a suggestion that this factor increases as the iron content of the tetrahedrite increases. Again, some confirmatory experiments with tetrahedrites with a broader iron concentration range would be required to confirm this trend. Ferric chloride power dependences in the range 0.4-0.8 are usually indicative of chemical or electrochemical control; such elevated power dependences are also useful on a practical scale since they permit the leaching rate to be easily increased simply by changing the solution concentration.

(iii) Effect of HCl Concentration. The overall reaction of tetrahedrite with ferric ion can be described by a reaction of the type:

$$Cu_{12}Sb_4S_{13} + 44Fe^{3+} \rightarrow 12\ Cu^{2} + 4Sb^{5+} + 13S^{\circ} + 44Fe^{2+} \qquad (12)$$

which assumes only Sb^{5+} formation. If Sb^{3+} is also formed, lesser amounts of Fe^{3+} and Fe^{2+} will be involved. The Sb^{5+} (or Sb^{3+}) almost certainly undergoes hydrolysis:

$$Sb^{5+} + 4H_2O \rightarrow H_3SbO_4 + 5H^{+} \qquad (13)$$

and Reaction 13 could occur simultaneously with Reaction 12 or in series with it. Although Reaction 12 does not involve acid directly, acid is important for preventing excessive Fe^{3+} hydrolysis; acid, of course, is directly involved in reactions of the type illustrated by

Table 6. $FeCl_3$ Power Dependence for Tetrahedrite Dissolution.

Material	Temperature	m
Iron-bearing tetrahedrite	95	0.88 $\pm$ 0.05
Iron-bearing tetrahedrite	90	0.70 $\pm$ 0.11
Iron-free tetrahedrite	90	0.44 $\pm$ 0.03

Eq. 13. Hence, the effect of acid concentration on the tetrahedrite leaching rate is of some importance and this variable was studied at both 0.3 and 0.0 M $FeCl_3$ concentrations. Conventional leaching tests were carried out and linear rate constants were calculated from each dissolution curve. Figure 14 summarizes the effect of HCl concentration on the leaching rate of iron-free tetrahedrite.

In the absence of $FeCl_3$ the leaching rate is extremely slow, almost an order of magnitude slower than in the presence of 0.3 M $FeCl_3$. The rate increases very rapidly with increasing HCl concentration at low acid levels, but is less strongly affected at the higher acid concentrations (1-3 M HCl). The direct acid attack presumably releases H_2S although the amounts are so low as to make detection difficult. The addition of ferric ion results in a marked increase in the leaching rate, and the dissolution occus via a different mechanism which forms elemental sulfur. Acid concentration is not a significant factor in the $FeCl_3$ leaching mechanism although increasing acid concentrations slightly accelerate the rate

$$k(mg/cm^2\ h) \propto [HCl]^{0.15} \tag{14}$$

The chief purpose of the acid appears to be the prevention of iron hydrolysis which, at 0.3 M $FeCl_3$ and 90°C, occurred rapidly at acid concentrations of less than 0.2 M HCl. A similar role of acid concentration has been identified for chalcopyrite[20] and enargite[22]

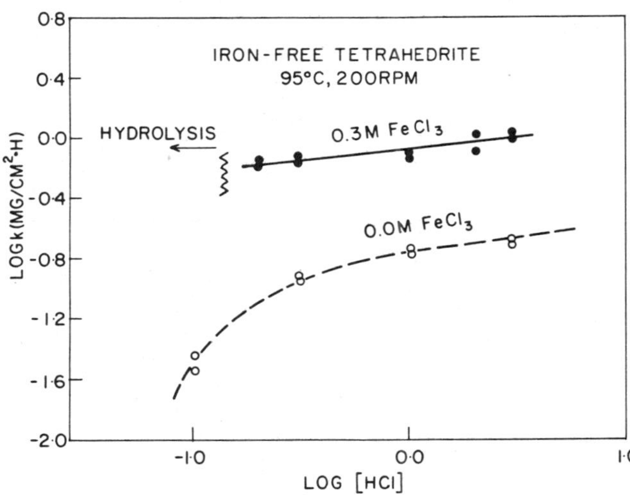

Fig. 14. Effect of HCl concentration on the rate of leaching of disks of iron-free tetrahedrite at two $FeCl_3$ concentrations.

leaching; only for the reactive sulfides like FeS or ZnS[23] is the direct acid attack mechanism significant.

(iv) <u>The Effect of LiCl, $CuCl_2$ or $FeCl_2$</u>. A commercial ferric chloride leaching process could well be carried out in the presence of substantial quantities of neutral chlorides (NaCl, $CaCl_2$, etc.) added to elevate the solution boiling point and/or to increase the solubility of AgCl, $PbCl_2$, etc. Accordingly, the effect of significant concentrations of neutral chlorides has a practical aspect as well as a fundamental interest related to ionic strength and total chloride concentration effects.

The effect of neutral chlorides/ionic strength was investigated by adding LiCl to the system; sodium and potassium salts were avoided because of their potential reaction with any sulfates formed. Also, nitrates and perchlorates which are conventionally used to vary ionic strength were avoided because of their potential oxidizing capabilities. Figure 15 shows the leaching curves obtained when the iron-bearing tetrahedrite was leached at 95°C with 0.3 M $FeCl_3$ solutions containing various concentrations of LiCl. For these particular experiments the initial room temperature pH was adjusted to 0.5 with HCl/LiOH prior to the start of the test; in some other experiments, however, the initial pH was not adjusted.

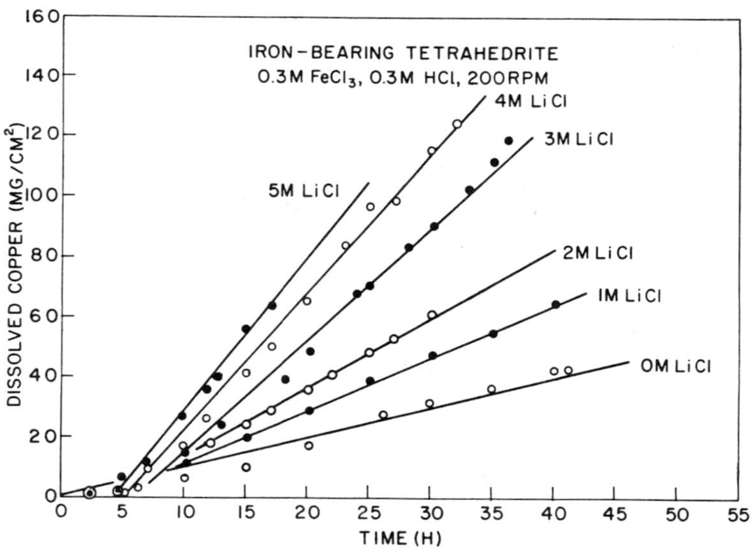

Fig. 15. Leaching curves for the dissolution of disks of iron-bearing tetrahedrite in solutions having various LiCl concentrations.

The leaching rate increases in the presence of LiCl, and the effect is fairly substantial. The addition of LiCl also causes the shape of the leaching curves to vary somewhat. In the absence of added chlorides, the dissolution commences immediately and linear kinetics are observed. When significant LiCl concentrations are present, the rate initially increases fairly slowly, at about the same rate as the 0 M LiCl test, and then accelerates rather rapidly. The "induction period" lasts 5-7 hours and seems shorter for the more concentrated media. Once the rate accelerates, linear kinetics are again observed although there is a suggestion of accelerating dissolution with increasing time. From a practical point of view, this figure indicates the desirability of leaching tetrahedrite-tennantite in high chloride media. Not only will such solutions accelerate the dissolution, but they will also help to dissolve the silver values which often accompany these minerals.

Linear rate constants were deduced from the various leaching curves obtained in the presence of LiCl and these data are summarized in Fig. 16. Because of the induction period, there is a significant error associated with the rate constants derived for solutions rich in LiCl. In spite of this variation, however, a reasonably consistent trend emerges that shows the leaching rate (after the induction period) to be directly proportional to the molar concentration of

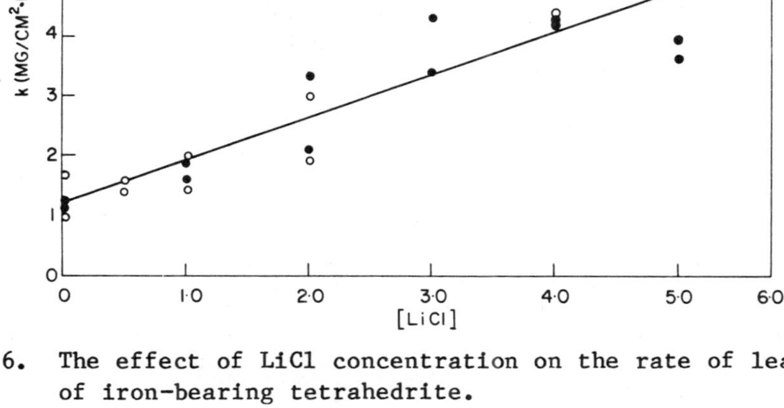

Fig. 16. The effect of LiCl concentration on the rate of leaching of iron-bearing tetrahedrite.

LiCl. There is no really significant difference between the experiments done at pH = 0.5 and those tests where the pH was not adjusted. The addition of LiCl causes minor variations in the pH, and the minimal effect of such variations on the tetrahedrite leaching rate is consistent with the data of Fig. 14.

Cupric chloride is sometimes used in conjunction with ferric chloride to accelerate leaching rates[5], and $CuCl_2$ is also a product of the dissolution of either tetrahedrite or tennantite (Eqs. 6 and 12). For these reasons, the effect of $CuCl_2$ on the tetrahedrite leaching rate is of some concern, and both high and low $CuCl_2$ concentrations were investigated at 95°C using disks of the iron-bearing tetrahedrite. All solutions contained 0.3 M $FeCl_3$ 0.3 M HCl so that cuprous chloride existed only as a transitory species. Because it is impossible to measure accurately the few ppm Cu leached from the tetrahedrite against a high copper background, the course of these experiments was monitored by periodically withdrawing pipette samples and analyzing these for total iron. The tetrahedrite used contained iron (3.38%) and this biased all the total iron concentrations positively by about 45/44 ~ 1.02:

$$Cu_{12}Sb_4S_{13} + 44Fe^{3+} \rightarrow 12Cu^{2+} + 4Sb^{5+} + 13S^0 + 44Fe^{2+} \quad (15)$$

$$Cu_{11}FeSb_4S_{13} + 44Fe^{3+} \rightarrow 11Cu^{2+} + 4Sb^{5+} + 13S^0 + 45Fe^{2+} \quad (16)$$

Figure 17 illustrates some typical tetrahedrite leaching curves obtained in the presence of high and low concentrations of $CuCl_2$ and based on the total amount of dissolved iron. All the dissolution curves are linear and all pass through the origin; i.e., there is no induction period for $CuCl_2$ additions. When the iron concentrations are corrected for disk area (~5.2 cm^2) and the reaction stoichiometry (Eq. 16), the rates are only slightly higher (~20%) than those determined by following copper. The addition of $CuCl_2$ accelerates the rate. The greatest effect is caused by additions in the range 0-100 g/L $CuCl_2$; higher concentrations seem to be without additional effect and may slightly retard the kinetics.

Linear rate constants were determined for each leaching curve and these are displayed as a function of the cupric chloride concentration in Fig. 18. Also shown on the graph is the total amount of elemental sulfur produced in each test, and this also is an indication of the rate of reaction since all the tests were run for 46-48 h. Agreement between the two calculated reaction rate indicators is good, and both show that increasing $CuCl_2$ concentrations accelerate the leaching rate.

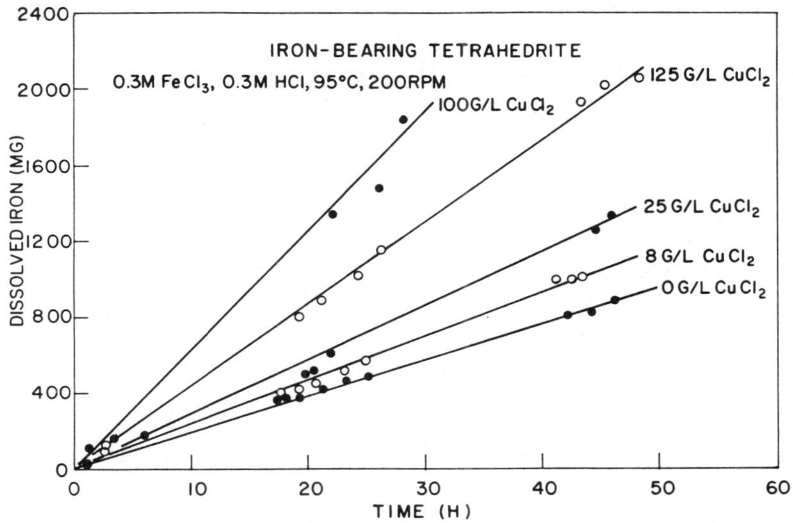

Fig. 17. Leaching curves, based on dissolved Fe^{2+}, for the dissolution of disks of iron-bearing tetrahedrite in solutions having various $CuCl_2$ concentrations.

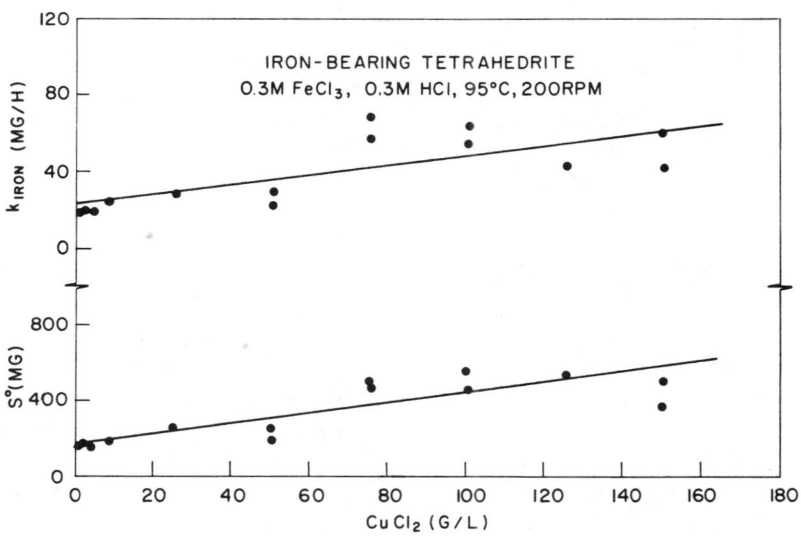

Fig. 18. The effect of the $CuCl_2$ concentration on the rate of leaching of iron-bearing tetrahedrite as determined by the ferrous ion and elemental sulfur reaction products.

$$k(mgFe^{2+}/h) = 23.053 + 0.255[CuCl_2] \quad (17)$$

and 1 M $CuCl_2$ increases the rate by 111%, and

$$S^0(mg) = 167.676 + 2.573[CuCl_2] \quad (18)$$

and 1 M $CuCl_2$ increases the sulfur yield by 153%. The effect, however, is more than would be expected by the increased chloride concentration, Fig. 16, accompanying the addition of $CuCl_2$. Although $CuCl_2$ additions are certainly beneficial, the catalytic effect of cupric ion is not particularly pronounced for tetrahedrite.

Ferrous ion is produced as a product of the tetrahedrite leaching reaction, Eq. 11, and its presence in the leaching solution could inhibit the dissolution process. To quantify the effect of $FeCl_2$ concentration, experiments were done using iron-free tetrahedrite disks at 90°C. All solutions contained 0.3 M $FeCl_3$ - 0.3 M HCl together with various concentrations of $FeCl_2$; in some experiments the pH was adjusted at room temperature to 0.5 and in other tests the natural solution pH was employed. Pipette samples and copper determinations by atomic absorption spectroscopy were employed because the $FeCl_2$-rich solutions crystallized on cooling in the feed lines of the automatic analysis apparatus. Linear copper dissolution curves were observed at all $FeCl_2$ concentrations, and linear rate constants (for Cu) were deduced from each.

The data illustrating the effect of the $FeCl_2$ concentration on the tetrahedrite leaching rate are presented in Fig. 19. Increasing the $FeCl_2$ concentration slightly increases the rate although the increase lies well within the scatter of the data:

$$k(mg\ Cu/cm^2\ h) = 0.901 + 0.143[FeCl_2] \quad (19)$$

and 1 M $FeCl_2$ increases the rate by only 16% (compare reactions 17, 18). Clearly any negative effect caused by Fe^{2+} is more than compensated for by the higher associated chloride concentration. In this regard it is interesting to compare the behavior of tetrahedrite to chalcopyrite where a similar effect is observed in ferric chloride media. It appears that the accumulation of the ferrous chloride reaction product during the course of leaching will not adversely affect the leaching rate.

(v) Inhibition by Li_2SO_4. Moderate sulfate concentrations are known to suppress significantly the leaching of chalcopyrite[20], and it is important to ascertain if similar effects exist for tetrahedrite.

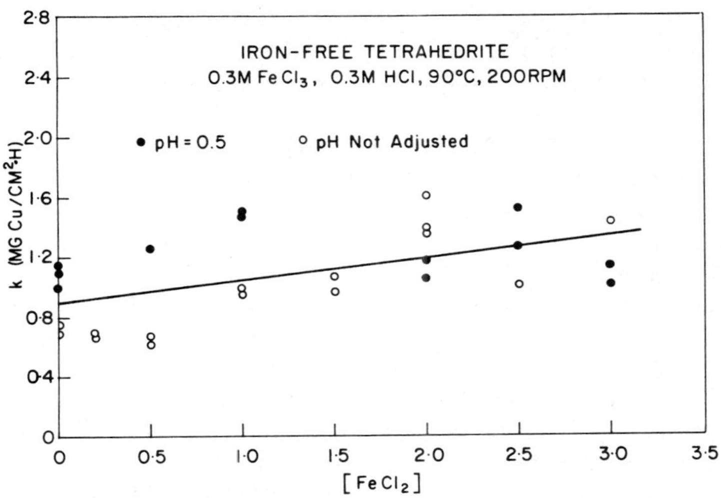

Fig. 19. Effect of $FeCl_2$ concentration on the rate of dissolution of disks of iron-free tetrahedrite.

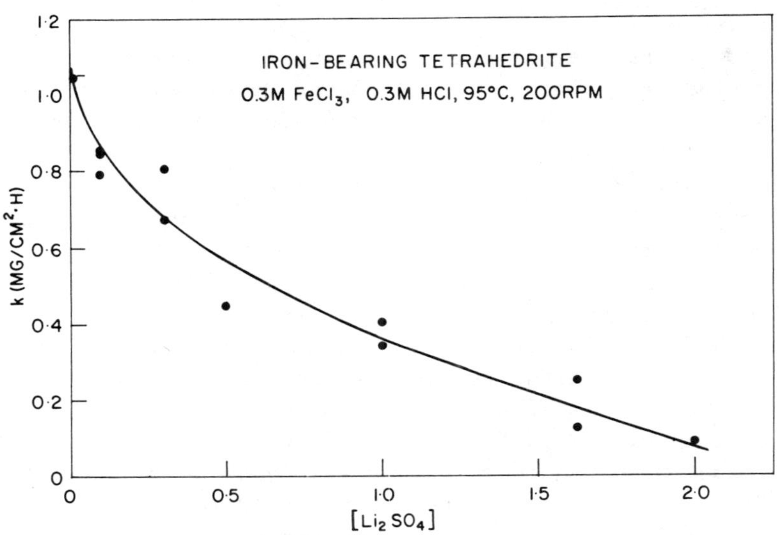

Fig. 20. The deleterious effect of sulfate ion on the rate of dissolution of tetrahedrite.

To investigate this parameter, various concentrations of Li_2SO_4 were added to the 0.3 M $FeCl_3$, 0.3 M HCl medium being used to leach iron-bearing tetrahedrite disks at 95°C. Linear copper dissolution curves were obtained for all the tests and linear rate constants were calculated from each. These data are summarized in Fig. 20.

Addition of Li_2SO_4 to the leaching medium causes the rate to fall appreciably and the effect persists with increasing concentrations. Dissolution is essentially halted in 2.0 M Li_2SO_4 solution. The inhibiting effect must be due to sulfate since lithium itself seems to be without serious negative consequences (Fig. 16). The rate of tetrahedrite dissolution is slower in sulfate media, and only minor sulfate concentrations are required to transform a chloride system effectively into a sulfate one. Similar behavior has also been observed for chalcopyrite leaching in $FeCl_3$ systems.[20] The extreme depression of the leaching reaction for $[Li_2SO_4] > 1.0$ M could in part be due to the precipitation of jarosite-type compounds on the surface of the disk. Even reagent grade Li_2SO_4 contains small amounts of K (0.03%) and Na (0.01%), and these concentrations (for large Li_2SO_4 additions) are sufficient to precipitate small amounts of a jarosite containing 4.96% K and 0.29% Na. Jarosite precipitation was observed only for 1.5 and 2.0 M Li_2SO_4 additions.

(vi) Nature of the Dissolution Reaction. The dissolution of either tetrahedrite or tennantite in ferric chloride media is very slow, and the low rates greatly complicate any interpretation of the leaching kinetics. Both minerals appear to dissolve according to linear rate laws and this was confirmed for both sized powders and massive material. The extent of dissolution never exceeded 25% and was considerably less when massive disks were employed; hence, it is not known with certainty whether the linear kinetics persist to 100% copper extraction.

Elemental sulfur is the principal sulfidic reaction product when tetrahedrite dissolves in ferric chloride media at temperatures <100°C. Elemental sulfur recoveries expressed as the molar percentage S^0 to (S^0 + sulfate sulfur) ranged from 75-98% with most determinations giving >90% S^0. The lower sulfur recoveries are possibly due to suspended sulfur in the leaching medium. Elemental sulfur is the principal species but minor sulfate is formed, likely in a parallel leaching reaction since elemental sulfur seems to be inert in $FeCl_3$ media.

$$S^{2-} + 2Fe^{3+} \rightarrow S^0 + 2Fe^{2+} \qquad (20)$$

Fig. 21. SEM photomicrograph of sulfur globules formed on iron-free tetrahedrite leached for 65 h at 95°C in 0.3 M $FeCl_3$ - 0.3 M HCl solution. White spots are an Fe-S b-O phase.

$$S^{2-} + 8Fe^{3+} + 4H_2O \rightarrow SO_4^{2-} + 8Fe^{2+} + 8H^+ \qquad (21)$$

The formation of SO_4^{2-} gives rise to increased Fe^{2+} concentrations. The bulk of the elemental sulfur seems to remain on the surface of the tetrahedrite and forms a network of interconnected globules, Fig. 21. Although the sulfur globules clearly cover the surface of the mineral, there must be sufficient porosity to permit the solution free access to the surface since linear kinetics are observed.

All the leaching solutions contained high concentrations of ferric ion at the completion of the experiments, and this ensures that copper is present as Cu^{2+}.

$$Cu^+ + Fe^{3+} \xrightarrow{\text{fast}} Cu^{2+} + Fe^{2+} \qquad (22)$$

Hence the dissolution of tetrahedrite can be expressed by:

$$Cu_{12}Sb_4S_{13} + (24 + 4x)Fe^{3+} \rightarrow 12\ Cu^{2+} + 4Sb^x + 13S^0 + (24 + 4x)Fe^{2+} \quad (23)$$

where x is the formal oxidation state of the antimony.

Various attempts were made to determine x using the molar ratios of Fe^{2+}/S^0 and Fe^{2+}/Cu^{2+} at the completion of the leaching tests, but such efforts were largely inconclusive. The Fe^{2+}/S^0 molar ratio expected for Sb^{3+} is 2.84 and for Sb^{5+} it is 3.46; measured values were within the range 2.92 - 3.74. The Fe^{2+}/Cu^{2+} molar ratio for Sb^{3+} formation is 3.36 and for Sb^{5+} it is 4.09; measured values were 3.56 - 3.80. The calculated molar ratios suggest the formation of both Sb^{3+} and Sb^{5+}, but this conclusion must be tempered by the consideration that some SO^4 is produced and that this also yields high Fe^{2+}/S^0 and Fe^{2+}/Cu^{2+} ratios.

Total antimony concentrations in the ferric chloride solutions were always less than 20 ppm (Fe^{2+} >2,000 ppm), and this indicates

Fig. 22. SEM photomicrograph showing Fe-Sb-O phase (white) on the sulfur globules formed on iron-bearing tetrahedrite (fracture surface) leached for 65 h at 95°C in 0.3 M $FeCl_3$.

that most of the Sb has precipitated. Two precipitate morphologies both shown to be Fe-Sb compounds, were identified. Figure 22 shows a thin layer of the antimony compound (white) on the surface of the elemental sulfur globules on the tetrahedrite; the view is along the fractured surface of a leached disk. The antimony precipitate is discontinuous and globular; it is clearly not protective. A small amount of the antimony precipitate is also apparent on the sulfur layer shown in Fig. 21. Although some antimony precipitates on the tetrahedrite particles, the greatest quantity forms on the vessel walls, stirrer, etc. and this material has a different morphology as is evident in Fig. 23. The samples consist of a granular mass together with a columnar type formation which may be a consequence of scraping the deposit from the vessel walls. Energy dispersion analysis showed all the precipitates to consist of Sb-Fe together with minor Cl and traces of Cu and S.

X-ray diffraction analysis of the antimony product showed it to be poorly crystallized but similar to the mineral tripuhyite ($FeSbO_4$)[24] which consists of Fe^{3+} and Sb^{5+}.[25] Two bulk chemical analyses of the antimony precipitate were obtained and these are summarized in Table 7. The first analysis seems to be contaminated

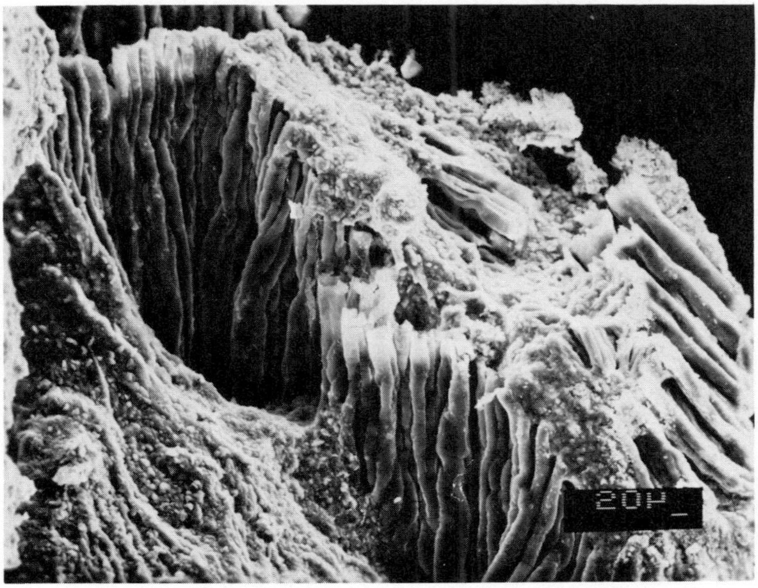

Fig. 23. SEM photomicrograph showing the Fe-Sb-O phase scraped from the walls of the glass leaching vessel.

Table 7. Chemical Analyses of the Fe-Sb Compounds Formed During Tetrahedrite Leaching.

Element	Analysis #1	Analysis #2	$FeSbO_8$
Fe	25.0	12.3	18.3
Sb	28.6	49.8	39.8
Cl	5.1	0.7	0.0

with $FeCl_3$, but the second is thought to be "pure". Unlike the idealized material, the leach product undergoes a continual 9% weight loss on heating to 800°C; the heated product is, however, well crystallized tripuhyite. The existence of a tripuhyite-like phase and the absence of soluble antimony would appear to indicate Sb^{5+} as the principal reaction product.

$$Cu_{12}Sb_4S_{13} + 44Fe^{3+} \rightarrow 12Cu^{2+} + 4Sb^{5+} + 13S^0 + 44Fe^{2+} \quad (24)$$

$$Sb^{5+} + Fe^{3+} + 4H_2O \rightarrow FeSbO_4 + 8H^+ \quad (25)$$

Presumably, an Sb(V) species forms according to Eq. 24 and this migrates into solution where it precipitates on a suitable surface. Alternatively, a Sb^{3+} species could form and migrate into solution where it would be oxidized to Sb^{5+} by ferric ion.

Unlike arsenic, antimony appears to be easily oxidized by Fe^{3+}. For example, pumping at 14 ml/h of 250 ml of acidified solution containing ~1,360 mg Sb^{3+} into 1 l of 0.3 M $FeCl_3$ solution at 95°C resulted in significant Fe^{2+} formation:

$$Sb^{3+} + 2Fe^{3+} \rightarrow Sb^{5+} + 2Fe^{2+} \quad (26)$$

Based on Fe^{2+} results, all the initial Sb^{3+} was oxidized to Sb^{5+} and was precipitated as a tripuhyite-like material. Hence, Sb^{3+} is readily oxidized to Sb^{5+} under the test conditions employed for ferric chloride leaching, and Reaction 24 probably is more important than the corresponding reaction giving Sb^{3+}.

CONCLUSIONS

The kinetics of dissolution of niccolite (NiAs), tennantite ($Cu_{12}As_4S_{13}$) and tetrahedrite ($Cu_{12}Sb_4S_{13}$) in ferric chloride-hydrochloric acid media have been investigated over the temperature range 30-95°C using both rotating disk and stirred slurry techniques.

For niccolite, the dissolution is controlled by the diffusion of ferric ion to the surface of the arsenide. The rate increases with the square root of the disk rotation speed and directly with $FeCl_3$ concentrations to about 0.4 M; the apparent activation energy is 22 kJ/mole. The dissolution is relatively insensitive to variations in ionic strength or concentrations of arsenic, ferrous ion, nickel or hydrochloric acid. Arsenic appears to dissolve as As^{3+}.

Tennantite dissolves very slowly in ferric chloride media and the kinetics are chemically or electrochemically controlled. Linear kinetics are observed together with an activation energy of 89 kJ/mole. The rate was shown to vary directly with the surface area of the tennantite particles.

Tetrahedrite also dissolves very slowly in ferric chloride media and its dissolution kinetics are also chemically or electrochemically controlled. Linear kinetics were observed and the apparent activation energy ranged from 64 to 83 kJ/mole, possibly depending on the iron content of the tetrahedrite. The dissolution rate was strongly dependent on the $FeCl_3$ concentration, but was relatively insensitive to acid concentration. The dissolution rate increased with increasing chloride concentrations and this was demonstrated for LiCl, $CuCl_2$ and $FeCl_2$. The presence of sulfate resulted in a significant reduction in the speed of leaching. The principal products of the leaching reaction were Fe^{2+}, Cu^{2+} and elemental sulfur; antimony appeared to be oxidized to Sb^{5+} which then precipitated as tripuhyite, $FeSbO_4$.

ACKNOWLEDGEMENTS

The extensive involvement of D. J. Hardy with the experimental work is gratefully recognized; P. E. Belanger characterized the products by X-ray diffraction.

REFERENCES

1. D. R. Weir and I. M. Masters: The Key Lake Uranium Process. Part I - Uranium Extraction. Paper presented at the 19th Annual Conference of Metallurgists of the CIM, Halifax, Nova Scotia, August (1980).
2. S. Dayton: Engr. Mining J., 1982, vol. 138, no. 1, pp. 78-83.
3. J. E. Dutrizac and T. T. Chen: CIM Bull., 1979, vol. 72, no. 803, pp. 201-208.
4. E. Andersen, G. H. Boe, T. Danielssen, and P. M. Finne: in Complex sulphide ores conference, Rome, October 1980, M. J.

Jones, ed., Institution of Mining and Metallurgy, London, pp. 186-192 (1980).
5. G. E. Atwood and R. W. Livingston: The CLEAR Process - A Duval Corporation Development. Paper presented at the Annual Conference of Gesellschaft Deutscher Metallhutten und Bergleute, Berlin, Sept. (1979).
6. F. P. Haver and M. M. Wong: U.S. Bureau of Mines R.I. 8105 (1976).
7. J. E. Dutrizac and R.J.C. MacDonald: Mineral. Sci. Engng., 1974, vol. 6, no. 2, pp. 59-100.
8. S. L. Brown and J. D. Sullivan: U.S. Bureau of Mines R.I. 3228 (1934).
9. L. C. Bryner, et al.: Ind. Engng. Chem., 1954, vol. 36, pp. 2587-2592.
10. J. E. Dutrizac and R.J.C. MacDonald: Mat. Res. Bull., 1973, vol. 8, pp. 961-972.
11. G. Thomas and T. R. Ingraham: Can. Metal. Quart., 1967, vol. 6, no. 2, pp. 153-165.
12. M. Pourbaix: Atlas of Electrochemical Equilibria in Aqueous Solutions, Pergamon Press, Toronto, pp. 516-523 (1966).
13. A. B. Garrett, O Holmes, and A. Laube: J. Am. Chem. Soc., 1940, vol. 62, pp. 2024-2028.
14. I. Barin and O. Knacke: Thermochemical Properties of Inorganic Substances, Springer-Verlag, New York (1973).
15. R. M. Garrels and C. L. Christ: Solutions, Minerals and Equilibria, Harper Row, New York, Appendix 2 (1965).
16. J. S. Yates and H. C. Thomas: J. Am. Chem. Soc., 1956, vol. 78, pp. 3950-3953.
17. L. M. de Smecht and Y. Berube: Chemical Stability of Arseniferous Wastes. ALUR 1973-74, IAND Publication No. 8018-000-EE-AI, Ottawa, Canada (1975).
18. V. G. Levich: Physiocochemical Hydrodynamics, Prentice-Hall, Inglewood Cliffs, New Jersey (1962).
19. A. M. Baticle, F. Perdu, and P. Vennereau: C.R. Acad. Sci. (Paris), 1962, vol. C264, pp. 12-15.
20. J. E. Dutrizac: Metal. Trans., 1981, vol. 12B, pp. 371-378.
21. W. C. Holmes: Engr. Mining J., 1944, vol. 145, no. 3, pp. 54-58.
22. J. E. Dutrizac and R.J.C. MacDonald: Can. Metal. Quart., 1972, vol. 11, no. 3, pp. 469-476.
23. J. E. Dutrizac and R.J.C. MacDonald: Metal. Trans., 1978, vol. 9B, pp. 543-551.
24. B. Mason and C. J. Vitaliano: Min. Mag., 1953, vol. 30, pp. 100-112.
25. R. J. Baker and J. G. Stevens: Rev. Chim. minerale, 1977, vol. 14, pp. 339-346.

COMPOSITION AND PHASE CHANGES DURING OXIDATIVE ACID LEACHING REACTIONS

A. R. Burkin

Department of Metallurgy and Materials Science
Imperial College
London SW7 2BP U. K.

COPPER SULFIDES

It has long been known that in nature one mineral species may be replaced by another as a result of chemical reaction. This kind of behavior is also found at times during leaching reactions in which a mineral is altered in stages before complete decomposition occurs.

One of the best known and most often studied examples is the reaction of chalcocite under acidic oxidizing conditions, when there is rapid initial dissolution of copper in the first stage of the reaction, followed by a much slower reaction which results eventually in complete dissolution of the copper present in the mineral and formation of sulfur. The solid remaining after the first stage of the reaction was, from the earliest observations of the two stage process, said to resemble covellite. Thus the reactions could be represented as

$$Cu_2S \rightarrow CuS + Cu^{2+} + 2e \qquad (1)$$

$$CuS \rightarrow S + Cu^{2+} + 2e \qquad (2)$$

and the question arises as to how Cu_2S changes into CuS and, in the case of particulate mineral, retains the original shape of the chalcocite particle.

The reaction has now been studied by many workers, usually by some kind of electrochemical technique. The use of a chalcocite electrode results in the gradual formation of partially oxidized products extending from the inner residue of unaltered chalcocite to an outside layer in contact with the solution, which may have reacted to such an extent as to have become sulfur. If, on the other hand, particulate chalcocite is leached, of particle size around 150 microns, partially leached residues have uniform composition throughout the whole of each particle by the time they are examined by x-ray powder diffraction or electron probe analysis. Thus the changes in lattice structure which occur as the composition of the leach residue changes can be easily determined.[1]

In the copper-sulfur system at low temperatures a number of different phases occur in the range of composition Cu_2S to Cu_1S, most of them having a range of non-stoichiometry although there are ranges of composition in which no single phase normally exists. Examination of the leach residues showed that there was smooth progression in the loss of copper during the reaction, with a change of phase occurring sharply when a particular composition was reached. Thus some phases continued to exist, or were formed, at compositions far outside their normal range of non-stoichiometry and this is attributed to the existence of metastable nonstoichiometry due to the reluctance of the solid to recrystallize at the low temperatures used for the leaching reaction.

COPPER-IRON SULFIDES

The low temperature phase diagram for the copper-iron-sulfur system shows that a number of different minerals occur, none apparently having a very wide range of non-stoichiometry although the natural, α-form of chalcopyrite loses sulfur on heating under inert conditions until its composition reaches $CuFeS_{1.83}$, β-chalcopyrite, at 720°C after which no more sulfur is lost. Synthetic α-chalcopyrite is leached only very slowly in acidic ferric sulfate solution, as is the naturally occurring mineral. β-chalcopyrite, however, is attacked much more rapidly, losing both copper and iron in the early stages of the reaction, although at different rates so that the copper-iron ratio in the residual solid changes.

As in the case of chalcocite, when particulate β-chalcopyrite was used, around 150 microns, partially leached materials had uniform composition throughout each particle by the time they were examined and it was found that as copper and iron were removed from the sulfur anion lattic framework certain d-spacings became smaller and the lattice collapsed when about 17.5% of the sulfur had been lost, recovering somewhat to form a structure which did not change at all on further leaching, when 31.5% of the copper, 19% of the iron and 15% of the sulfur initially present in the solid had dissolved, the latter as sulfate ions.[2]

The composition of this residual solid was $CuFe_{1.2}S_{2.3}$ but its d-spacings corresponded very closely to those of α-chalcopyrite. Its leaching rate was, however, very much faster than that of the latter. The composition of the residual solid is of interest since if sulfur is regarded as being present solely as S^{2-} ions then formal charge balance is achieved if iron is regarded as being present solely as Fe^{3+} and copper as Cu^{+}, which is also the case for α-chalcopyrite itself. It was suggested as long ago as 1954, however, that the structures of such sulfides should be considered in terms of a metallic type of bond.[3] Many of them are intrinsic semiconductors, the energy gap between the energy zone filled with electrons and the next higher zone being 0·53 eV in the case of α-chalcopyrite,[4] corresponding to a normal semiconductor. On this basis a crystal structure can be maintained despite changes in composition, for example by removal of interstitial metal ions, as long as the electron to atom ratio necessary to fill one of the Brillouin zones is maintained. It is essential that charge balance in the solid be retained, sulfide ions having a constant charge of −2 unless evidence of S_2^{2-} groups is found. Cations may be removed however, as long as the appropriate number of electrons is also removed from the Brillouin zone. This may be done by use of an oxidizing agent to accept them.

In the case of chalcocite leaching, Equations 1 and 2 could be interpreted as suggesting that oxidation of cuprous sulfide to cupric sulfide occurs followed by oxidation of sulfide to elemental sulfur. Thus the electron transfer reactions would be $2Cu(I) \rightarrow 2Cu(II)$ and S^{2-} to S° in the two stages. This would be incorrect, however, since in CuS four of the six sulfur atoms in the unit are joined to two S_2 groups while the other two sulfur atoms are isolated. All of the metal ions are Cu+ and CuS is a diamagnetic metallic conductor.

The case of acid leaching of bornite, Cu_5FeS_4 was interesting because it was possible to relate the changes which occurred in the rate of leaching as the reaction progressed to events taking place in the crystal lattice.[6] A rapid first stage of leaching ended when approximately 40% of the copper present in the bornite particles had been lost and during the first 3 minutes of the reaction the apparent

Table 1. Lattice Parameters and Compositions of Partially Leached Bornite Particles

%Cu Removed	a, Å	c, Å
0.0	10.94	21.88
7.9	10.87	21.81
14.6	10.84	21.73
27.7	10.58	21.17
30.8	10.54	21.15
37.5	10.52	21.08
43.7	10.52	20.93
57.9	10.52	20.93
80.0	10.52	20.93

activation energy was 8·8± 0·4 kJ $mole^{-1}$, clearly indicating that the rate controlling process was diffusion. Each partially leached particle (150 micron) had uniform composition and during the first stage of the reaction no iron or sulfur was lost by the solid.

Quantitative electron microprobe anlaysis of solid residues from which slightly more than 40% of the copper had been leached showed the composition to be Cu_3FeS_4 and further leaching did not alter that composition, the solid being decomposed as a whole. The solid Cu_3FeS_4 was shown by reflectivity dispersion profile measurements to be identical with the mineral idaite.

During the change from Cu_5FeS_4 to Cu_3FeS_4 x-ray powder diffraction analysis showed that no sudden phase changes occurred. There was a smooth contraction of the lattice, as was shown by the size of the unit cell. The x-ray data are shown in Table 1.

In the crystal lattice of bornite copper occurs in two different kinds of site within the framework made up from the sulfur atoms. Forty percent of the copper is in planes parallel to $(111)_{rh}$ and it was clear from a ball model that there would be no hindrance by other atoms in the structure to the free movement of this 40% of the copper atoms within their layers. This explains the very low apparent activation energy observed for the first stage of leaching of bornite. After removal of these layers of copper atoms from the structure a much simpler unit cell remains which is derived from that of chalcopyrite by replacing 4 Fe atoms of the chalcopyrite unit cell by copper atoms. Using the x-ray powder technique it proved impossible to distinguish between the d-spacing of α-chalcopyrite and the idaite leach residues.

Using specially grown single crystals of bornite x-ray diffractometer measurements were employed to identify the natural faces of the crystals and hence the $(III)_{rh}$ plane with respect to them. Crystals were then cut parallel and at several angles to this plane and mounted in araldite resin to form discs which were lightly ground and polished so as to expose a bornite surface which was parallel to the $(III)_{rh}$ plane in one case and perpendicular to it in the other.

Using the conventional rotating disc technique in 0.2 M hydrochloric acid containing 5 g/l hydrogen peroxide the rate of dissolution of copper in the first stage of the reaction was 2.51, 2.15 and 1.04×10^{-2} μg of copper/cm^2 of mineral per minute from planes cut at 90°, 70° and 45° to the $(III)_{rh}$ plane. A face cut perpendicular to that plane showed no first stage reaction.

SILVER CONTAINING COPPER SULFIDES

The chemical similarities between copper and silver can lead to substitution of silver for copper in the lattices of copper sulfides in which it sometimes occurs. If silver were present in bornite it might be expected to influence the dissolution rate of copper from it since it would affect the electronic properties of the crystal. This is one of the factors claimed to affect the chemical reactions of minerals but would be of importance only in a leaching process in which the rate-controlling step involved a chemical reaction or mass transfer through a solid phase. Such a diffusion process occurs during the first stage of the leaching of bornite. During the second stage Cu_3FeS_4 breaks down to leave elemental sulfur and under appropriate experimental conditions the rate is controlled by an electron transfer process. Thus the presence of silver could affect this stage of the process also as shown in Table 2.

Bornite containing 1.20 weight percent silver was synthesized and it was found that there was no clearly significant difference between the rate at which copper was leached from it and from pure bornite, although the trend was for slightly slower leaching of copper from the material containing silver. It was very noticeable however that very little silver appeared in solution in the latter case. The amount of silver dissolved increased with ferric iron concentration but remained very small and tended to decrease after about 1 hour.

Since copper was leached from the silver-containing bornite and the idaite formed from it reacted to give copper and iron in solution

Table 2. Leaching Data on Silver Containing Copper Sulfides

Residue	Wt % Removed		Atomic % in Residue			
	Cu	Ag	Cu	Ag	Fe	S
1	44.7	0.2	37.9	0.7	12.4	49.0
2	63.4	0.9	39.5	1.3	12.7	46.5
3	84.0	1.8	34.2	2.2	13.7	49.9
4	89.7	5.0	33.3	3.7	15.3	47.7
Cu_3FeS_4			37.5		12.5	50.0
5	97.8	1.2	19.0	9.4	16.0	55.5
6	98.8	2.2	16.3	9.6	19.6	54.6

and elemental sulfur, while silver did not appear in solution, some compound of silver must have been present in the undissolved solid. After removal of elemental sulfur using carbon disulfide some residues from which various amounts of copper had been leached were analyzed. The analysis for residues 5 and 6 show that silver can be present at 21 weight % in the residues. Agreement with the composition $(CuAg)_3FeS_4$ is quite good in residues 1 to 4, not so good in 5 and 6.

X-ray diffraction studies of the silver-containing idaite residues showed that the lines were much broader and more diffuse than for residues formed from pure bornite, but no other phase was found, after sulfur had been removed. Mixtures of pure bornite and pure silver sulfide were examined in the same instrument and 1% of silver sulfide could be easily detected.

Using electron microprobe analysis it was found that in leach residues silver concentrations were higher at the outside of the particles while copper and iron were lower than deeper within them. Pure silver sulfide leached under similar conditions to silver-containing bornite oxidized and dissolved to give much higher silver concentrations in solution than did the bornite. It was concluded therefore that in the presence of the copper-containing mineral, silver which had dissolved from the bornite was cemented out again on the mineral surface, taking up positions in a not very well crystallized idaite lattice.

Because of the significance of the absence of silver from commercial in-situ leaching operations for copper, some copper-silver minerals were studied further. Three copper-silver sulfides are known and have been found as minerals: Stromeyerite, $Cu_{1.07}Ag_{0.33}S$; McKinstryite, $Cu_{08}Ag_{1.2}S$; and Jalpaite, $Cu_{0.45}Ag_{1.55}S$.

Synthetic stromeyerite was leached over the same range of conditions that silver-containing bornite had been. The general characteristics of the reaction were that 50% of the copper present in the stromeyerite was leached out in a short time, for example 30 minutes, while the second stage was slow, for example a further 10% required another 8.5 hours. The amount of silver appearing in the solution was small, the maximum being about 0.16% of that present in the stromeyerite sample taken and, as in the case of bornite, the silver content of the solution rose initially and then fell after about 30 minutes.

X-ray diffraction studies of residues taken during the kinetic study showed that the lines of stromeyerite became progressively weaker as leaching proceded, while new lines appeared. However, after about 50% of the copper present in the original solid had dissolved, a number of the new lines had disappeared again. All of the lines observed during the earlier parts of this first stage of the leaching of stromeyerite were measured and were attributable to the published data for stromeyerite, McKinstryite, jalpaite and Ag_2S. No lines of sulfur were detected. Microscopic examination of washed, air-dried residues showed no evidence of sulfur and none was seen in mounted and polished residue specimens. In order to determine whether any of the sulfur content of the stromeyerite was lost by the solid as sulfate leaching runs were carried out in HCl - $FeCl_3$ and also in HCl - H_2O_2 solutions. The general shapes of the leaching curves were similar to those obtained using ferric sulfate and chemical analysis of solutions taken during the first stage of the reaction showed that no sulfate had been formed.

Correlation of the x-ray data on the residues from the ferric sulfate leaching experiments with the copper contents of the solutions showed that there was a sequence of reactions taking place.
I) Until 18% of the copper was in solution the residues were a mixture of stromeyerite and McKinstryite.
II) Subsequently the residues were a mixture of McKinstryite and jalpaite with the latter becoming increasingly predominant.
III) When 41.7% of the copper had been leached out of the solids the lines of silver sulfide began to appear and after 47.4% copper loss it was clear that silver sulfide was the main phase present. In this sample a number of lines measured indicated that covellite was present also.

The reaction sequence observed can be written $Ag_{0.93}Cu_{1.07}S \rightarrow Ag_{1.2}Cu_{0.8}S \rightarrow Ag_{1.55}Cu_{0.45}S \rightarrow Ag_2S + CuS$ so that in stromeyerite there is 0.93 atom of Ag to 1.0 S whereas in the final product the ratio is 2:1 and yet no sulfur or sulfate is produced. That sulfur present in stromeyerite which is not required by the sequence of reactions of solids containing silver, forms covellite and it is only the copper not required by all the solids which appears in the solution. This model for the reaction leads to the following theoretical equations, based on mass balances. For simplicity ferric and ferrous ions are not included.

$$Ag_{0.93}Cu_{1.07}S = 0\ 775\ Ag_{1.2}Cu_{0.8}S + 0.225\ CuS + 0.225\ Cu^{2+} \quad (3)$$

$$0.775Ag_{1.2}Cu_{0.8}S = 0.6Ag_{1.55}Cu_{0.45}S + 0.175\ CuS + 0.175\ Cu^{2+} \quad (4)$$

$$0.6Ag_{1.55}Cu_{0.45}S = 0.465Ag_2S + 0.135\ CuS + 0.135\ Cu^{2+} \quad (5)$$

This gives an overall reaction for the first stage of the leaching reaction of stromeyerite, again neglecting iron

$$Ag_{0.93}Cu_{1.07}S = 0.465Ag_2S + 0.535\ CuS + 0.535\ Cu^{2+} \quad (6)$$

The points on the copper extraction curve at which the various phases appear agree well with the proposed sequence of reactions except that covellite lines were not observed until about 47% of the copper was in solution. This is attributed to the facts that I) the lines of covellite found there were very diffuse and II) there was a very large number of lines present on the films from the earlier residues because of the other solids present.

The reaction mechanism proposed for the acid oxidative leaching of stromeyerite is analogous to that for chalcocite in that it proceeds by way of several intermediate phases to a terminal composition, CuS for chalcocite, Ag_2S and CuS for stromeyerite. There is however a very important difference. In the case of chalcocite, as also for bornite, a very wide range of (presumbably metastable) non-stoichiometry was observed until the terminal composition was reached. This non-stoichiometry was accompanied by a progressive shrinking of the lattice and in the case of chalcocite a phase change occurred when a particular composition was reached.

For stromeyerite on the other hand there was no evidence of line movement as leaching took place, merely the appearance of new phases. Electron microprobe studies showed the existence of silver-copper-sulfides of markedly different composition adjoining one another with no gradual change of composition between them. It is this lack of non-stoichiometry range in the silver-copper sulfides which necessitates the formation of covellite until the silver is entirely present as silver sulfide.

The literature on the crystal structures of the silver-containing sulfides relevant here suggests that it is easy for the one to transform to the next in the series containing less copper. For example the projected structure of acanthite, Ag_2S, on the (010) plane is very similar to the structure of stromeyerite projected on the (001) plane. Also the planar copper-sulfur layers parallel to the (001) plane in stromeyerite are very similar to those in covellite. Thus it seems reasonable to suggest that covellite "precipitate" because it has a different structure, having a hexagonal unit cell whereas stromeyerite is orthorhombic. Microscopical examination of residues clearly showed that covellite was being formed along preferred crystallographic directions.

NICKEL SULFIDES

The behavior of nickel-rich sulfides on acid oxidative leaching is different from that of copper-rich sulfides. Heazelwoodite, Ni_3S_2, was synthesized and leached using $HCl-FeCl_3$. When 2% of the nickel had been leached out of a sample a new phase containing 58-60% Ni could be seen at places on the edges of the particles. Using residues containing more of it, the new phase was found to be β-NiS, Millerite. Lines due to sulfur were observed when about 10% or more of the nickel present in the Ni_3S_2 had been leached out. Thus the first stage of the leaching reaction is

$$Ni_3S_2 \rightarrow NiS + Ni^{2+} \quad (7)$$

and the second reaction, the dissolution of NiS and formation of sulfur begins as soon as it is formed.

The direct formation of NiS from Ni_3S_2 with no evidence of line shifts confirmed the narrow range of stoichiometry of the solid at low temperatures and the absence of metastable non-stoichiometry of both Ni_3S_2 and β-NiS. From the nickel-sulfur phase diagram it might have been expected that Ni_7S_6 would form as an intermediate phase but although it was looked for with great care, no evidence for its presence at any stage in the reaction was found.

The apparent activation energy found for the leaching of Ni_3S_2 at the start of the reactin was about 50 kJ $mole^{-1}$, suggesting that the transformation of Ni_3S_2 to NiS took place relatively easily. The crystal structures of both solids are rhombohedral/hexagonal and the Ni -S and Ni -Ni bond distances to nearest neighbors are similar in both structures. Thus as nickel atoms are removed from Ni_3S_2 the remaining atoms need move only slighty to change into the NiS structure. However, the NiS lattice formed was considerably strained, as was shown by all its x-ray diffraction lines being broad. The structure of Ni_7S_6 is orthorhombic and since this phase did not form under the conditions used it seems that the Ni_3S_2 to Ni_7S_6 transformation is not energetically favorable.

NICKEL IRON SULFIDES

$Ni_{4.5}Fe_{4.5}S_8$ was synthesized and leached in acidic ferric chloride solutions. Both metals dissolved at the same rate within experimental error, and both sulfur and sulfate were formed, the latter due to air having free access to the reacting system. No sign of change in lattice dimensions was observed.

In nature in certain ore bodies pentlandite is oxidized during the first stages of weathering to violarite (nominally Ni_2FeS_4) with loss of metal, iron being lost more rapidly than nickel. During this oxidation the violarite is protected from more vigorous oxidation by its position below the water table. Once it is exposed the violarite is oxidized to basic nickel sulfate. In the case of natural leaching the violarite loses metal but retains the sulfur content of the pentlandite. During ferric chloride leaching the sulfide content of the mineral was oxidized to sulfur. There was no evidence of an electrochemical corrosion mechanism during the leaching when leach residues were examined, nor on examining a mounted and leached lump of the mineral. Using an electrochemical method for oxidizing synthetic pentlandite, violarite can be formed.

A monosulfide solid solution containing (wt %) Fe, 36.5; Ni, 25.0; S, 38.5%; was also synthesized. It behaved in the same way as pentlandite.

PYRRHOTITE

The pyrrhotites are the iron-sulfur phases with compositions ranging from FeS to Fe_7S_8 in the low temperature forms. Synthetic troilite, $Fe_{1.01}S$ and hexagonal pyrrhotite, $Fe_{0.86}S$, and natural monoclinic pyrrhotite were subjected to acid oxidative leaching. Each phase was identified by characteristic x-ray diffraction lines and these did not shift or appear in leach residues from a different original material. One line of the intermediate pyrrhotite is characteristic for exact composition and it did not shift at all during leaching.

REFERENCES

1. J. A. King, A. R. Burkin and R. C. H. Ferreira, Leaching and Reduction in Hydrometallurgy, A. R. Burkin, ed., Institution of Mining and Metallurgy, London (1975), pp 36-45.
2. R. C. H. Ferreira and A. R. Burkin, Leaching and Reduction in Hydrometallurgy, A. R. Burkin, ed., Institution of Mining and Metallurgy, London (1975), pp 54-66.
3. A. J. Freuh Jr., Geochim. Cosmochim. Acta, 6: 79-89 (1954).
4. I. G. Austin, C. H. L. Goodman and A. E. Pengelly, Nature, Lond., 178: 433 (1954).

THE BEHAVIOR OF IMPURITIES DURING JAROSITE PRECIPITATION

J. E. Dutrizac

Mineral Sciences Laboratories, CANMET
Energy, Mines and Resources Canada
Ottawa, Ontario K1A 0G1 CANADA

The factors affecting the incorporation of divalent base metals such as Cu^{2+} or Zn^{2+} into the alkali jarosites [$MFe_3(SO_4)_2(OH)_6$ where $M = K^+$, Na^+, NH_4^+, H_3O^+, $1/2Pb^{2+}$, Ag^+, etc.] have been determined. The extent of incorporation which is always minor, increases with increasing base metal or alkali sulfate concentration and pH, but decreases with increasing iron concentration. The divalent ions appear to replace Fe^{3+} in the jarosite lattice. A special situation exists for the major substitution of divalent ions in lead jarosite. The divalent ions replace Fe^{3+} in the jarosite structure with the charge balance being maintained by additional lead:

$$Cu^{2+} + 1/2\ Pb^{2+} = Fe^{3+}$$

Under extreme conditions, the substitutions become so large that the jarosite approaches beaverite [$PbCuFe_2(SO_4)_2(OH)_6$] in both structure and composition. Trivalent ions such as Al^{3+}, In^{3+}, Ga^{3+}, etc. also substitute for Fe^{3+}, and the conditions affecting the incorporation of several such elements have been defined. Jarosite analogs containing In^{3+} or Ga^{3+} have been produced and characterized by chemical and by X-ray diffraction analyses. Several XO_4^{2-} anions such as SeO_4^{2-} or CrO_4^{2-} substitute for sulfate, but most such species form separate non-jarosite-type precipitates. The principal factors governing the incorporation of impurities into the jarosite structure are size and ionic charge although electronic configurations are also important.

INTRODUCTION

Virtually all base metal sulfide and oxide ores contain significant amounts of iron which is present both in the sought-after minerals themselves [e.g., chalcopyrite, $CuFeS_2$ and sphalerite (Zn,Fe)S] and in discrete but often finely disseminated impurity grains of pyrite, FeS_2, hematite, Fe_2O_3, etc. Iron is an unwelcome impurity in all base metal ores and concentrates, and it often creates severe problems during the processing or refining of a particular metal. For this reason, all metallurgical circuits must include some sort of iron bleed. In pyrometallurgical operations the iron commonly reports to a slag, but in hydrometallurgical processes it is generally precipitated from solution. Futhermore, there is an operational requirement that the iron precipitate must be easily filtered and washed free of entrained pregnant solution. Significantly, simply raising the pH to hydrolyze and precipitate Fe^{3+} is not a satisfactory iron removal process since any substantial iron concentration (>1 g/L) results in a poorly settling, unfilterable precipitate which cannot be adequately washed.

During the past twenty years several innovative processes have been developed to precipitate iron in an easily filterable form. One of the more promising of these methods of eliminating iron from hydrometallurgical solutions is precipitation as a jarosite-type compound, $MFe^{3+}{}_3(SO_4)_2(OH)_6$ where M = K^+, Na^+, $NH_4{}^+$, H_3O^+, $1/2Pb^{2+}$, Ag^+, Rb^+, Tl^+ or $1/2Hg^{2+}$.[1] Figure 1 illustrates a typical manner in which jarosite precipitation can be fitted into a conventional roast-leach-electrowin circuit for zinc concentrates. The roasting step converts the contained zinc values to both ZnO and $ZnO.Fe_2O_3$, zinc ferrite. In the neutral leach circuit, the ZnO is readily dissolved and enters an essentially iron-free solution which is sent to purification and zinc electrolysis. The zinc ferrite remains largely unattacked during the neutral leach, but can be dissolved by hot, concentrated H_2SO_4 in the hot acid leach stage; this dissolves the zinc, but also much of the contained iron which must be precipitated prior to zinc purification-electrolysis. The iron is conveniently eliminated from the hot acid leach solution as a jarosite-type compound by adding a jarosite forming element (NH_4, Na, etc.) and neutralizing a portion of the excess acid. The jarosite precipitate settles well, filters readily and can be washed free of pregnant solution; the zinc remains in solution and is recycled within the circuit. Although the example given is for zinc processing, similar flowsheets can be used to reject iron from copper, cobalt, manganese or other hydrometallurgical circuits. The great advantages of jarosite precipitation are seen to be:

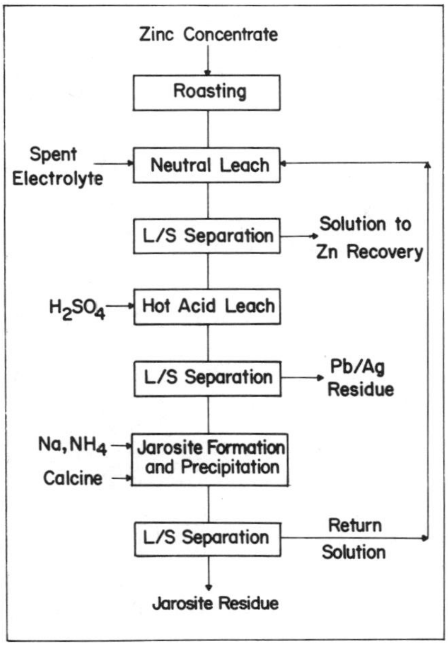

Fig. 1. Typical roast-leach-electrowin circuit for zinc concentrates that includes a jarosite precipitation operation.

- a high degree of iron rejection and the iron is precipitated in an easily filterable form;
- relatively low losses of divalent base metals into the alkali jarosite;
- the process also controls alkali and sulfate;
- operates in acidic media, thereby reducing neutralization costs;
- operates at <100°C and at P = 1 atm;
- is easy to operate and is readily integrated into conventional hydrometallurgical flowsheets.

Aside from precipitating Fe^{3+}, SO_4 and alkali ions, jarosite-type compounds also precipitate, or could precipitate, various other species including unwanted consituents like As or Ge as well as valuable elements like Pb, Ag or even Zn, Cu, Co, etc. The deportment of various "impurities" into the jarosite is of some concern to hydrometallurgists because, as Fig. 1 illustrates, the jarosite residue represents one of the major bleeds from the processing circuit. In spite of the potential importance of jarosites in eliminating impurities or trace elements from hydrometallurgical circuits, very little has been reported on the deportment of such elements during jarosite formation and few attempts have been made to optimize the impurity distributions.

Yaroslavtsev et al.[2] reported on the relative partitioning of several impurities between solution and jarosite residue when potassium jarosite was precipitated at 90°C, pH = 1.5 and Fe^{3+} = 28 g/L. These results were discussed in terms of a "capture coefficient", K, which was defined as:

$$K = \frac{\text{wt\% impurity in jarosite}}{\text{concentration of impurity in solution (g/100 cc)}}$$

The larger the value of K the greater is the tendency for the impurity element to concentrate in the jarosite precipitate. Significantly, however, Yaroslavtsev et al. did not unequivocally ascertain whether the impurity was incorporated into the jarosite lattice or whether it simply precipitated as a separate crystalline or amorphous phase. The findings of these workers are presented in Table 1 from which it is evident that the divalent base metals are not significantly precipitated into the jarosite phase. For the common divalent base metals, the reported order of co-precipitation is:

$$Co^{2+} > Cu^{2+} > Ni^{2+} > Zn^{2+} > Mn^{2+} > Cd^{2+} > Mg^{2+}$$

The higher valence elements seem to concentrate to a greater degree in the jarosite precipitate. Saarinen[3] observed that sodium jarosite precipitated from processing solutions containing Cr^{3+}, Co^{2+} or Ni^{2+} incorporated very low concentrations of these metals. Nickel concentrations in the washed jarosite residues ranged from 0.3 to 1.4 wt%; under similar precipitation conditions the corresponding Co^{2+} values were 0.5 to 1.4 wt%. Losses of Cr^{3+} into the jarosite residue were slightly higher, 0.6-1.6 wt%, at the 90°C precipitation temperature and the pH range of 1-2 employed by Saarinen.

A number of trivalent ions substitute for Fe^{3+} in the jarosite formula and these include Al^{3+}, Ga^{3+}, In^{3+} and Cr^{3+}. Aluminum has been reported[4] to substitute completely for Fe in potassium jarosite with iron being preferentially concentrated in the solid phase in acidic media. There is likely a complete solid solution series between most members of the jarosite family and their corresponding aluminum analogs. Hoeller[5] also investigated the substitution of Al^{3+} for Fe^{3+} in jarosites formed by reacting various silicate minerals with H_2SO_4; the Fe and Al released from the minerals formed jarosite-alunite-type compounds whereas the SiO_2 formed an amorphous

Table 1. Relative Partitioning of Some Impurities Between Potassium Jarosite and the Solution Phase. (Ref. 2)

Impurity Element	Initial Concentration (g/100 cc solution)	K
Zn^{2+}	3.2 - 16.3	0.20 - 0.08
Cd^{2+}	0.056 - 1.12	0.02
Cu^{2+}	0.32 - 1.60	1.0 - 0.56
Mg^{2+}	0.78 - 1.65	0.013 - 0.006
Mn^{2+}	0.275 - 1.10	0.08 - 0.07
Ni^{2+}	0.0015	0.7
Co^{2+}	0.008 - 0.02	1.2 - 0.5
Al^{3+}	0.63 - 1.4	1.4 - 1.3
As^{3+}	0.0037 - 0.187	1.6 - 0.4
As^{5+}	0.0075 - 0.15	10.6 - 7.0
Sb^{3+}	0.0007 - 0.01	11.5 - 10.0
In^{3+}	0.0011 - 0.125	12.7 - 10.5
Ga^{3+}	0.003 - 0.007	6.0 - 4.3
Ge^{4+}	0.0014 - 0.145	1.3 - 0.5
Tl^{+}	0.03 - 0.072	6.7 - 5.0
Tl^{3+}	0.03 - 0.1	6.1 - 4.7
Cl^{-}	0.337 - 0.67	0.09 - 0.08
F^{-}	0.0019 - 0.20	10.5 - 7.9

gel-like phase. The substitution of Al into either K-jarosite or Na-jarosite was favored by high temperatures and by lower acid concentrations as noted previously by Brophy et al.[4] The work of Hoeller is of only indirect significance to hydrometallurgists since the reagents used to precipitate the jarosite family compounds were indirectly derived from the acid attack of silicate rocks and minerals. The precipitation of potassium jarosite in the presence of Al^{3+} was also studied by Mar'yanchik et al.[6] Under the acidic conditions used at 170°C, the extent of aluminum precipitation was only 0-28%, again confirming the preferential precipitation of iron with respect to aluminum from acidic solutions. Aluminum-free jarosites were made provided the weight ratio Al/Fe < 0.78.

It is known that gallium analogs with Ga^{3+} replacing Fe^{3+} exist for potassium jarosite[7], sodium, rubidium and ammonium jarosites[8] as well as for hydronium jarosite[9], and that the gallium analogs form over a broad range of experimental conditions. However, the quantitative partitioning of dilute Ga^{3+} and Fe^{3+} between the solution and essentially pure iron-bearing jarosite species is not clear; this type of information would be most useful for ascertaining the behavior of minor amounts of gallium in hydrometallurgical processing solutions. Indium is also known to substitute for iron in ammonium jarosite formed at 97°C from solutions containing from 8-25 g/L Fe^{3+} and 9-29 g/L In^{3+} at pH = 1.75.[10] It appeared that both iron and indium were precipitated in about the same ratio as they were present in the aqueous phase.

Substitutions for the sulfate component of jarosite have been considerably less extensively documented. Dutrizac et al.[11] prepared selenate analogs of potassium and sodium jarosites, and showed them to be very similar to the corresponding sulfate species. The incorporation of selenium into the jarosite phase was slightly less favored than that of sulfate, but the two species tended to precipitate approximately in the same ratio as they were present in solution and they appeared to form nearly ideal solid solution series. Fersman and Shubnikova[12] have reported a natural jarosite containing 0.20 wt[5] selenite, SeO_3, but it was not established whether the selenium was actually present as SeO_3^{2-} or SeO_4^{2-}, or whether it was really incorporated into the jarosite structure at all. Breshenkov[13] also noted that low concentrations of selenium were present in natural jarosites and that the selenium concentrated in the jarosite portion of oxidized ore cappings. Chromate (CrO_4^{2-}) analogs of potassium, sodium and hydronium jarosites have been synthesized[14,15], and this implies extensive solid solution series between $MFe_3(CrO_4)_2(OH)_6$ and $MFe_3(SO_4)_2(OH)_6$ although little if anything is known about the comparative stabilities of the various chromate and sulfate analogs. Low concentrations of germanium (<50 ppm Ge) have been found in natural jarosites[16], but it is not known whether the germanium is actually present in the jarosite or if it is associated with other iron phases such as amorphous hydrogoethite.

The purpose of the present study was to expand our knowledge of jarosites by investigating in a systematic manner the substitution of various elements into jarosite-type compounds. The substitutions were studied as a function of reagent concentrations and pH; special efforts were made to ascertain the nature of the impurity phase in the precipitates. Lastly, some discussion of the significance of the data on hydrometallurgical processing is offered from time to time.

EXPERIMENTAL

Reagent-grade chemicals were used for all the jarosite syntheses. Most of the jarosites were prepared simply by heating the appropriate alkali sulfate, ferric sulfate and impurity salt (usually the sulfate) to 97°C to precipitate jarosite; the pH of the solution was adjusted initially at room temperature, and the pH was allowed to fall during the course of the experiments as acid was produced by hydrolysis. The heating was done in a titanium-baffled 2-L reaction kettle which was vigorously agitated in air.[17] At the completion of the experiment, the precipitate was filtered hot with the scrapings from the vessel walls being added to the product. The filter cake was washed with hot H_2O and then dried at 110-120°C.

Lead jarosites were, however, more difficult to synthesize, and these were prepared by reacting an excess of solid $PbSO_4$ with solutions containing ferric sulfate plus various amounts of copper sulfate or other divalent base metal sulfates in an autoclave at 130°C for 24 hours. Optimum operating conditions[18] were attained by reacting a twofold excess of $PbSO_4$ in a well-stirred baffled reactor containing 1 litre of 0.3 M Fe^{3+} as $Fe_2(SO_4)_3$ and 0.03 M H_2SO_4. Excess $PbSO_4$ was selectively leached from the jarosite precipitate by washing with four 1-litre portions of 10% ammonium acetate solution at 25°C. The jarosite residue was then filtered, water-washed, and dried at 110°C prior to chemical analysis and X-ray diffraction study. All lead-bearing jarosites were checked for residual $PbSO_4$ by sensitive Guinier-deWolff X-ray methods; separate experiments established that as little as 0.3 wt% $PbSO_4$ was faintly detectable on the Guinier patterns.

All the jarosite precipitates were analyzed by conventional analytical means; most of the metals were determined by atomic absorption spectroscopy although iron was also done by classical dichromate titration. Ammonium ion concentrations were calculated from the total nitrogen content as determined by the Kjeldahl method. Sulfate was determined by barium sulfate precipitation. Water was generally done by difference although thermogravimetry was sometimes used.

All the products were examined by sensitive Guinier-deWolff X-ray methods to try to detect impurity phases; this method can readily identify 1% of a crystalline impurity phase. Many jarosite precipitates were also examined by scanning electron microscopy (SEM) with energy analysis capabilities to try to "see" any amorphous impurity phases with distinctive morphologies which might account for the impurity elements present in the jarosites.

Some X-ray powder diffraction indexing of the jarosite phases was also carried out using a Philips 114.6 mm diameter Debye-Scherrer camera with cobalt radiation. The diffraction data were processed by the least squares computer program, PARAM[19], to refine the cell dimensions; all diffraction lines could be indexed assuming the $R\bar{3}m$ or R3m structure of jarosite. Details of the indexing procedure have been presented.[18]

RESULTS AND DISCUSSION

Divalent Base Metals in Alkali Jarosites

The type formula of jarosite family compounds $MFe_3(SO_4)_2(OH)_6$ is $KFe_3(SO_4)_2(OH)_6$ where the theoretical K composition is 7.81 wt% and the crystal-lographic space group is $R\bar{3}m$. It is well established[17] that extensive substitutions for M^+ can occur and such replacements have been reported for $M^+ = K^+$, Na^+, Rb^+, Ag^+, Tl^+, H_3O^+, NH_4^+, $1/2Pb^{2+}$ and $1/2\ Hg^{2+}$. Because jarosite-type compounds are precipitated from acidic aqueous media, virtually all natural and synthetic products appear to contain a significant degree of H_3O^+ substitution for M^+ and this substitution is commonly 15-25 atomic percent; that is, the real formula for jarosite-type compounds is:

$$K_{(1-x)}(H_3O)_xFe_3(SO_4)_2(OH)_6 \qquad x \simeq 0.15 - 0.25$$

The conditions affecting the formation of these compounds are known, and the chief factors influencing the M-site occupancy and the amount of product formed have been found to be:

- temperature,
- pH,
- concentration of M^+,
- reaction time,
- concentration of Fe^{3+},
- the presence of other cations capable of M-site occupancy.

Because jarosites are commonly precipitated in the presence of divalent base metal ions (Cu^{2+}, Zn^{2+}, Mn^{2+}, Cd^{2+}, Co^{2+}, Ni^{2+}, etc.) and since some divalent metals such as Pb^{2+} and Hg^{2+} are known to form jarosite-type compounds, it is natural to question the possible existence of other divalent jarosites and/or the nature of the incorporation of the divalent ions into the alkali jarosite structure. Initial efforts to answer these questions centered on the attempted production of hypothetical divalent base metal jarosites; e.g., $Cu_{0.5}Fe_3(SO_4)_2(OH)_6$. Attempted syntheses of such divalent base metal jarosites were done using the following conditions which are known to be near-optimum for jarosite synthesis:

- 0.1 M Fe^{3+}, 0.01 M H_2SO_4,
- 100 g/L M^{2+} as MSO_4 ($M^{2+} = Cu^{2+}$, Zn^{2+}, Mn^{2+}, Cd^{2+}, Co^{2+} or Ni^{2+})
- 97°C, 24 h,
- good agitation in air.

In every instance extensive iron hydrolysis and precipitation occurred, but the products did not resemble jarosite in color or

Table 2. Chemical and X-ray Diffraction Data for Attempted Jarosite Syntheses in the Presence of Various Divalent Ions.

M^{2+}	Product Formed	wt% M
Cu^{2+}	$Fe_4(OH)_{10}SO_4$	0.5 - 2.4
Zn^{2+}	FeO.OH	1.1
Mn^{2+}	Amorphous	0.6
Cd^{2+}	-	-
Ni^{2+}	FeO.OH	0.4
Co^{2+}	Nearly amorphous	0.8

settling properties. Table 2 illustrates the products which were identified by X-ray diffraction and chemical analyses for these syntheses. Clearly none of the divalent base metals studied is capable of forming an end-member jarosite-type compound where the divalent ion would occupy the M^+-site. In every instance a basic iron compound precipitated that contained (by substitution or adsorption) a very low concentration of the divalent base metal present in the synthesis solution.

The issue of the incorporation of divalent base metals into the jarosite lattice during the formation of alkali jarosites (Na-jarosite, NH_4-jarosite, K-jarosite) still remains, and is of paramount importance to the metallurgical industry since such alkali jarosites are commonly precipitated from iron solutions rich in Zn^{2+}, Cu^{2+}, etc. The extent of copper incorporation into sodium jarosite formed under nearly ideal conditions (97°C, 24 h, 0.3 M Fe^{3+}, 0.1 M H_2SO_4, 0.3 M Na_2SO_4) is illustrated in Fig. 2. In the absence of dissolved copper, a normal sodium jarosite containing 3.7 wt% Na (4.3% Na theoretical) with some hydronium substitution is formed. As the copper concentration in solution increases, the copper content of the jarosite also increases ultimately reaching a value of about 1.2 wt% Cu; the rate of increase is greatest at the lower copper concentrations. Significantly, the Na content, as well as the SO_4 level, of the product remains constant but the iron concentration drops steadily as the copper tenor increases; the molar increase in Cu^{2+} is not, however, exactly related to the decrease in iron although the two curves have complementary forms. The points marked by X's in Fig. 2 represent equivalent experiments carried out at constant ionic strength; for those experiments [$CuSO_4$ + $MgSO_4$] = 2.0 M whereas for the other tests, the ionic strength varied as the $CuSO_4$ concentration increased. Within the limits of experimental error, neither the sodium concentration nor the amount of copper incorporated into the jarosite are significantly affected by ionic strength.

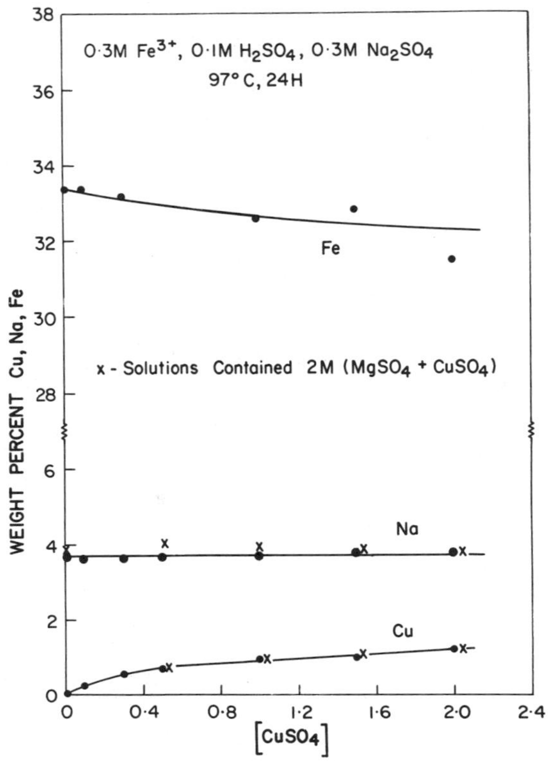

Fig. 2. The effect of dissolved $CuSO_4$ on the Cu, Na and Fe contents of sodium jarosite made at constant and varying ionic strengths.

The extent of incorporation of copper into sodium jarosite depends on the iron concentration and the initial acidity. Figure 3 compares the amount of copper incorporated into sodium jarosite as a function of the dissolved $CuSO_4$ concentration for two different iron-acid solutions. Large amounts of copper were incorporated in the products in both sets of experiments although low iron and acid concentrations clearly produce the greater degree of copper incorporation, in excess of 2 wt% Cu. As was the case for Fig. 2, the Na level remained essentially constant in these two experiments, but the iron concentration dropped progressively as copper increased.

The copper values (and a similar situation prevailed for all the divalent base metals discussed in this paper) were intimately associated with the sodium jarosite. The jarosites were thoroughly

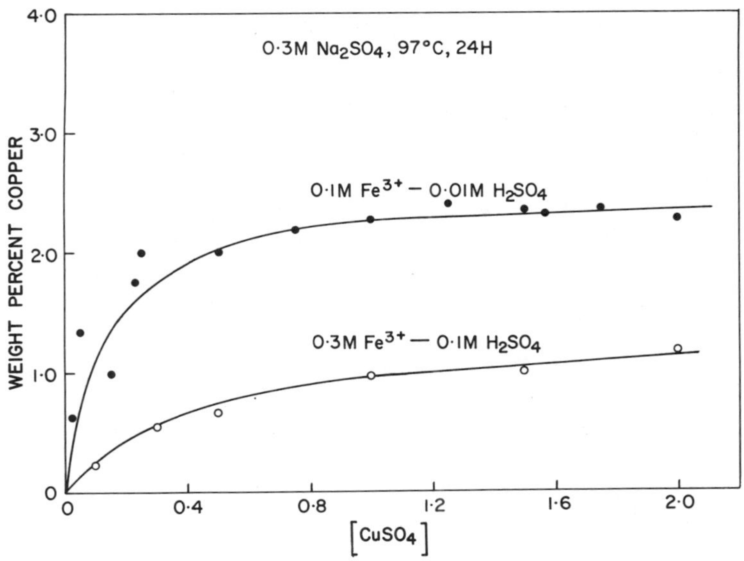

Fig. 3. The extent of copper incorporation into sodium jarosite for two different solution compositions.

washed with large volumes of hot water to eliminate entrained liquor. The success of this operation can be gauged by the fact that the jarosites made in the presence of 0–2 M $MgSO_4$ (Fig. 2) all contained less than 0.1 wt% Mg. Rewashing of the dried jarosites with additional quantities of hot water did not eliminate any significant amounts of copper. Washing with cold, dilute sulfuric acid also did not remove any copper from the jarosite. Copper-free sodium jarosite was doped with various quantities of solid copper sulfate in amounts ranging from 1% $CuSO_4$ (0.4% Cu) to 5% $CuSO_4$ (2% Cu). The mixtures were then examined by Guinier-deWolff X-ray diffraction methods which revealed that 0.4% Cu showed as a "faint trace" and that 2% Cu was present as a minor phase. Examination of many of the jarosite products which contained 0-2% Cu by Guinier-deWolff X-ray diffraction failed to detect any compound except sodium jarosite. Visual study of the products by the SEM did not reveal any anomalous phases. It would appear from these studies that the copper is actually incorporated into the jarosite lattice although the presence of amorphous copper phases cannot be ruled out entirely.

The solid points in Fig. 4 illustrate the effect of dissolved zinc sulfate on the zinc content of sodium jarosite formed at 97°C

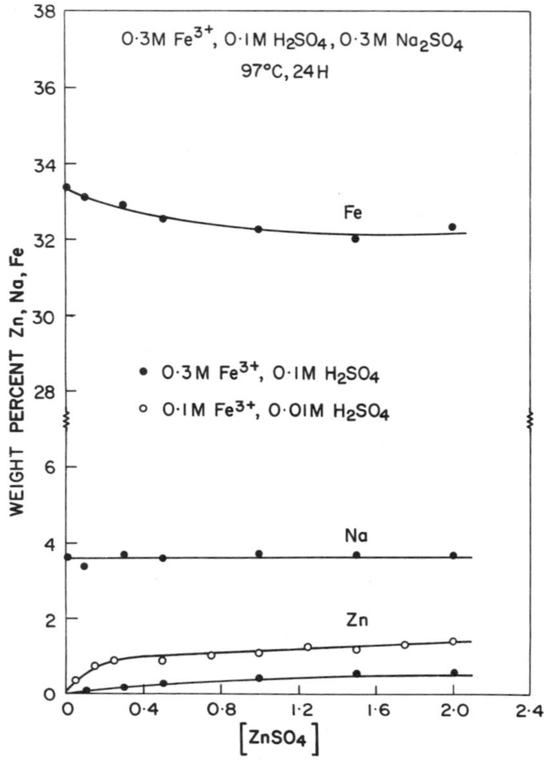

Fig. 4. The effect of dissolved $ZnSO_4$ on the Zn, Na and Fe contents of sodium jarosite made at constant and varying ionic strengths.

from solutions containing 0.3 M Fe^{3+}, 0.1 M H_2SO_4 and 0.3 M Na_2SO_4. The zinc content of the jarosite increases systematically as the amount of dissolved zinc in the synthesis solution increases. The sodium jarosite made in the presence of 2 M $ZnSO_4$ contains about 0.5 wt% Zn, significantly less than the amount of copper incorporated in the product under similar conditions (Fig. 3). This zinc was not soluble in cold dilute acid or in hot water, and no distinct zinc compounds were detectable by Guinier-deWolff X-ray diffraction analysis. As the zinc content increased, the iron concentration of the sodium jarosite decreased; consistently, more than one iron atom was lost for each zinc incorporated. The sodium, as well as the sulfate, content of the jarosite appears constant suggesting that Zn^{2+} is not replacing Na^+. The open points in Fig. 4 present the zinc content of the jarosite products formed from

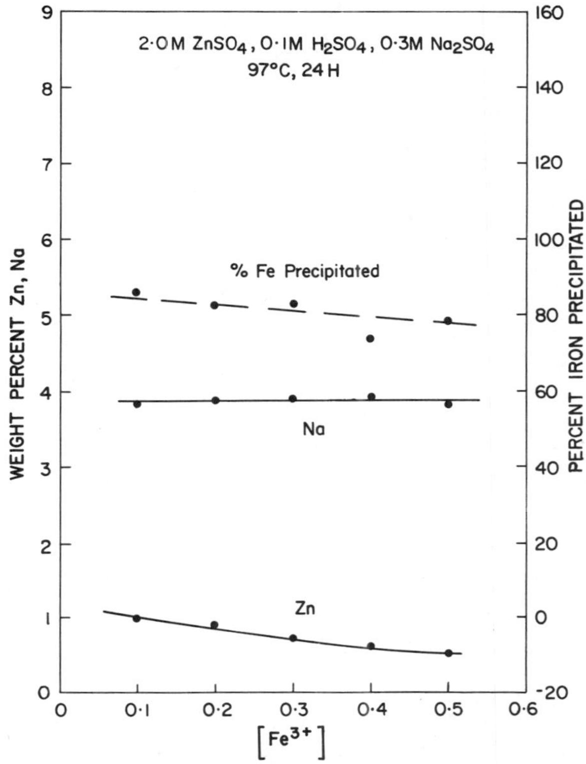

Fig. 5. The influence of initial Fe^{3+} concentration on the extent of iron precipitation and on the composition of the sodium jarosite product.

0.1 M Fe^{3+} - 0.01 M H_2SO_4 - 0.3 M Na_2SO_4 solutions under otherwise similar conditions. As was the case for copper (Fig. 3), decreasing the iron and acid concentrations serves to increase the amount of zinc actually incorporated into the precipitate.

The effect of the initial iron concentration, at a constant initial H_2SO_4 concentration of 0.1 M, on the extent of zinc loss into the jarosite was also studied, and the salient features are given in Fig. 5. Increasing the initial iron concentration from 0.1 to 0.5 M has little effect on the formation of sodium jarosite itself; the alkali and sulfate concentrations remain nearly constant and the percentage of iron precipitated drops only slightly. That is, the amount of product formed increases in a nearly linear manner with increasing initial iron concentration. The extent of zinc incorporation into the sodium jarosite, however, decreases as the

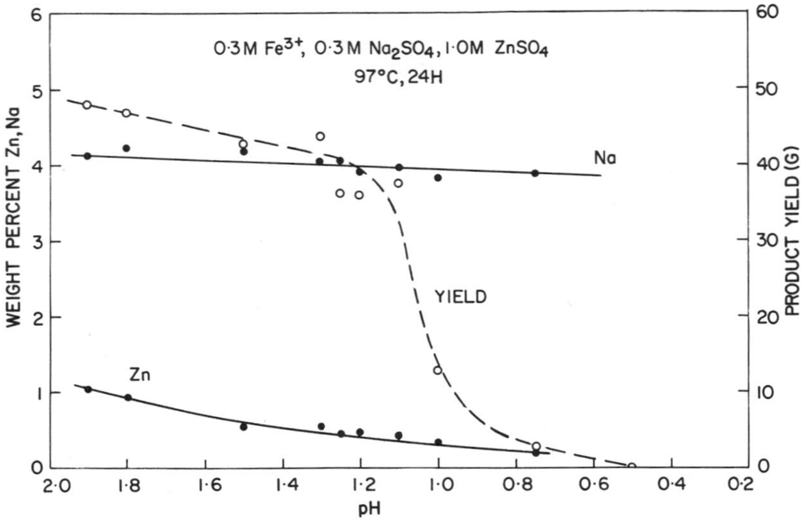

Fig. 6. The effect of initial room temperature pH of the solution on the amount of sodium jarosite formed and its compositon.

iron concentration increases falling from ~ 1% at 0.1 M Fe^{3+} to about 0.5 wt% at 0.5 M Fe^{3+}. This behavior also supports the contention that the divalent base metals are actually being incorporated into the jarosite structure. If the zinc were adsorbed or physically entrained, the zinc losses would be expected to remain constant or even increase as the iron concentration, and accordingly the volume of precipitate, increases.

Increasing the iron concentration from 0.1 to 0.3 M would be expected to decrease the zinc content of the sodium jarosite by about 28%. Hence, the variation in zinc content (>50% noted in Fig. 4 on passing from 0.1 M Fe^{3+} - 0.01 M H_2SO_4 to 0.3 M Fe^{3+} - 0.1 M H_2SO_4 cannot be due to the change in iron content only; acid concentration must also be a significant factor. Figure 6 shows the effect of the initial room temperature pH of the solution on the zinc and sodium contents of the jarosites as well as on the yield of product formed using solutions all of which contained 1.0 M $ZnSO_4$, Since the concentration of iron in the jarosite was nearly constant at 31.3 - 32.8 wt%, the product yield is essentially proportional to the fraction of iron precipitated. As the acidity increases, the amount of product formed declines. Initially the decline is gradual since the quantity of hydrolysis acid (~0.3 M) formed by the reaction

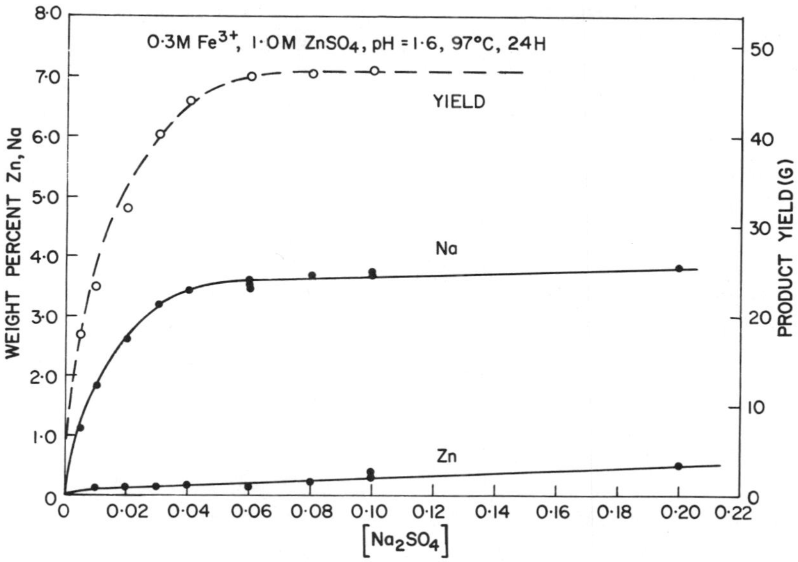

Fig. 7. The influence of Na_2SO_4 concentration on the yield of jarosite and its sodium and zinc contents.

$$Na_2SO_4 + 3Fe_2(SO_4)_3 + 6H_2O \rightarrow 2NaFe_3(SO_4)_2(OH)_6 + 6H_2SO_4 \quad (1)$$

$$K = \frac{[H_2SO_4]}{[Fe^{3+}]} \quad (2)$$

is large compared to the acid initially present at, say, pH = 1.8. Eventually, however, the acid initially present in solution becomes comparatively significant and the yield falls sharply. Finally a pH is reached where the added acid essentially suppresses Reaction 1; for sodium jarosite, this critical pH $\simeq$ 0.5. The sodium content of the jarosites formed at all pH values studied is nearly independent of acid concentration although a slight decline is observable for these particular results. Significantly, however, the extent of zinc incorporation is strongly dependent on the initial acidity. Lowering the pH from 1.9 to 0.7 reduces the amount of zinc in the sodium jarosite from 1% to ~ 0.2 wt% Zn.

Figure 7 illustrates the effect of increasing alkali concentrations in the synthesis solution on the extent of zinc incorporation into sodium jarosite as well as on the alkali content of the product and the amount of precipitate formed. Increasing the concentration of Na_2SO_4 in solution increases both the amount of product formed and its sodium content. The effect is most pronounced at the lower alkali concentrations with both the Na content of the jarosite and the amount of product formed increasing dramatically to about 0.08 M Na_2SO_4 and then essentially levelling off. The products formed at 0.01 or 0.005 M Na_2SO_4 contained less than 50 mole percent Na and are strictly sodium-bearing hydronium jarosites, $(H_3O,Na)Fe_3(SO_4)_2(OH)_6$. The zinc content of the product increases steadily and in a nearly linear manner as the Na_2SO_4 concentration increases, and this effect is known to persist to at least 0.5 M Na_2SO_4. That the zinc content of the jarosite increases when the Na content has reached a nearly constant maximum value indicates that the incorporation of divalent base metals into jarosite-type compounds is not related critically, if at all, to alkali site occupancy.

The three alkali jarosite species of concern to metallurgical operations are the sodium, ammonium and potassium members, and it is interesting to compare the relative extent of zinc incorporation into each species when all three were formed under similar conditions. The data of Table 3 illustrate the comparative degree of zinc incorporation into potassium, sodium and ammonium jarosites when the precipitates were formed at 97°C from solutions containing 2 M $ZnSO_4$, 0.1 or 0.3 M M_2SO_4 (M= K, Na or NH_4) and either 0.1 M Fe^{3+} - 0.1 M H_2SO_4 or 0.3 M Fe^{3+} - 0.1 M H_2SO_4. The extent of zinc incorporation is significantly greater for potassium jarosite, and sodium jarosite includes slightly more zinc than its ammonium analog. As has been shown above (Fig. 5 and 7), the extent of zinc incorporation decreases as the dissolved ferric ion concentration increases or as the alkali sulfate concentration decreases. Thus, in commerical zinc plants employing jarosite precipitation, minimum zinc losses will be

Table 3. Comparative Degree of Zn Incorporation into Potassium, Sodim, and Ammonium Jarosites Made at 97°C.

$[Mg_2SO_4]$	Weight Percent Zn in Jarosites	
	0.1 M Fe^{3+} - 0.1 M H_2SO_4	0.3 M Fe^{3+} - 0.1 M H_2SO_4
0.3 M K_2SO_4	2.13	-
0.3 M Na_2SO_4	0.93	-
0.3 M $(NH_4)_2SO_4$	0.80	-
0.1 M K_2SO_4	-	1.19
0.1 M Na_2SO_4	-	0.27
0.1 M $(NH_4)_2SO_4$	-	0.23

realized when the jarosite is formed from strongly acidic media with a high iron and low alkali concentration. Fortunately for the zinc industry, these are the conditions existing in the hot acid leach liquors, Fig. 1. Sodium or ammonium jarosites give the lowest zinc losses under a given set of operating conditions, and these are the species commonly employed in the industry because of their relatively lower costs.

Manganese is a common constituent of many hydrometallurgical solutions especially those in zinc processing circuits which contain several g/L of this element, mostly as Mn^{2+}. Consequently, the deportment of divalent manganese during jarosite precipitation is of some concern to hydrometallurgists. Figure 8 illustrates the effect of the initial concentration of dissolved $MnSO_4$ on the manganese content of sodium jarosite formed at 97°C from solutions containing 0.3 M Na_2SO_4 and either 0.1 M Fe^{3+} - 0.01 M H_2SO_4 or 0.3 M Fe^{3+} -

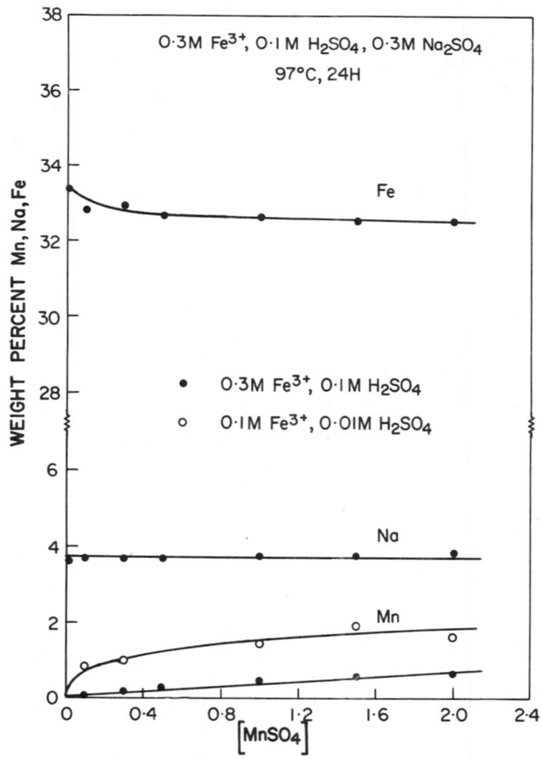

Fig. 8. The effect of $MnSO_4$ concentration on the Mn, Na, and Fe contents of sodium jarosite prepared from two different solutions.

Table. 4. The Extent of Mn Incorporation in Sodium Jarosite.

Mn in jarosite (wt %)	$[Fe^{3+}]$	$[H_2SO_4]$
0.46	0.1	0.1
0.36	0.2	0.1
0.24	0.3	0.1
0.19	0.4	0.1
0.11	0.5	0.1
0.30	0.3	0.05
0.24	0.3	0.1
0.14	0.3	0.2
0.09	0.3	0.3

0.1 M H_2SO_4. As was the case for Cu^{2+} or Zn^{2+}, the amount of Mn^{2+} incorporated into the sodium jarosite increased steadily as the manganese concentration of the solution increased, and decreased with increasing ferric ion or acid concentrations. About 2 wt% Mn was incorporated into the jarosite products made from solutions containing 2 M $MnSO_4$ and 0.1 M Fe^{3+} - 0.01 M H_2SO_4. Guinier-deWolff X-ray diffraction analysis indicated that all samples consisted only of sodium jarosite; furthermore, all the products were characteristically bright yellow indicating the absence of oxidized manganese compounds which are frequently amorphous. As the manganese content of the jarosite increased, the iron concentration fell slightly, but the sodium content of the products remained nearly constant.

As was also the case for Cu or Zn incorporation into sodium jarosite, increasing either the acid or the ferric ion concentration of the synthesis solution decreased the extent of Mn^{2+} incorporation into the precipitate. The data of Table 4 which were obtained at 97°C from solutions containing 0.3 M Na_2SO_4 and 0.3 M $MnSO_4$ are instructive. Clearly Mn^{2+} incorporation into alkali jarosite will be favored by high concentrations of dissolved $MnSO_4$ and low levels of Fe^{3+} and/or H_2SO_4.

Cadmium is commonly associated with zinc ores and concentrates and, consequently, the deportment of this element during alkali jarosite precipitation is also of considerable concern to the zinc industry. The effect of various concentrations of dissolved $CdSO_4$ on the cadmium content of the sodium jarosites formed at two different iron and acid concentrations is illustrated in Fig. 9. Clearly cadmium is not significantly incorporated into the jarosite phase; less than 0.03 wt% Cd was found in the jarosites made from solutions containing 0.3 M Fe^{3+} - 0.1 M H_2SO_4 and concentrations of

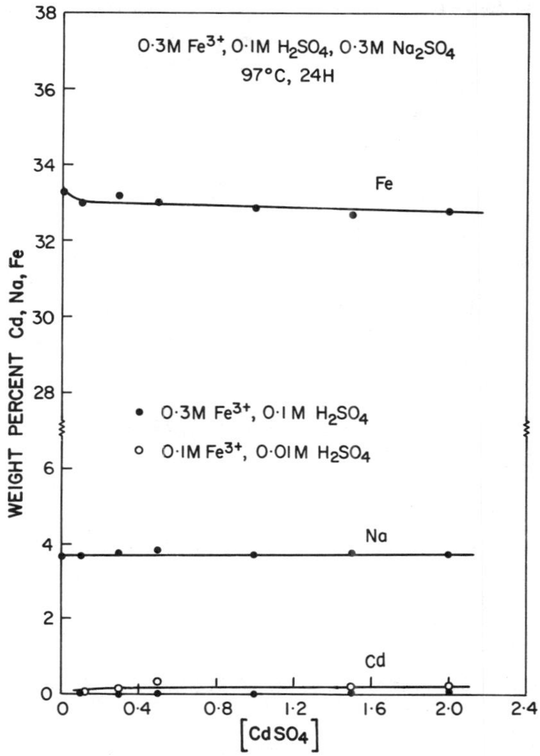

Fig. 9. The influence of $CdSO_4$ concentration on the Cd, Na, and Fe contents of the sodium jarosite made from two different solutions.

$CdSO_4$ as high as 2.0 M. Slightly greater amounts of cadmium (to 0.35 wt%) are found in the jarosites made from solutions having lower iron and acid concentrations. The sodium content of the products remains virtually constant for all $CdSO_4$ concentrations, but the iron concentration of the jarosite drops very slightly. Although part of the reduction of the iron content is probably due to the incorporation of Cd into the jarosite, a very slight iron reduction also seems to be related simply to the presence of large concentrations of MSO_4 in solution. This very minor reduction in iron content due to the presence of large amounts of MSO_4 in solution probably exists for all the divalent base metals but is normally masked by the large effect produced by the incorporation of M^{2+} into the jarosite. Given that cadmium concentrations in most solutions subjected to jarosite precipitation will likely be less

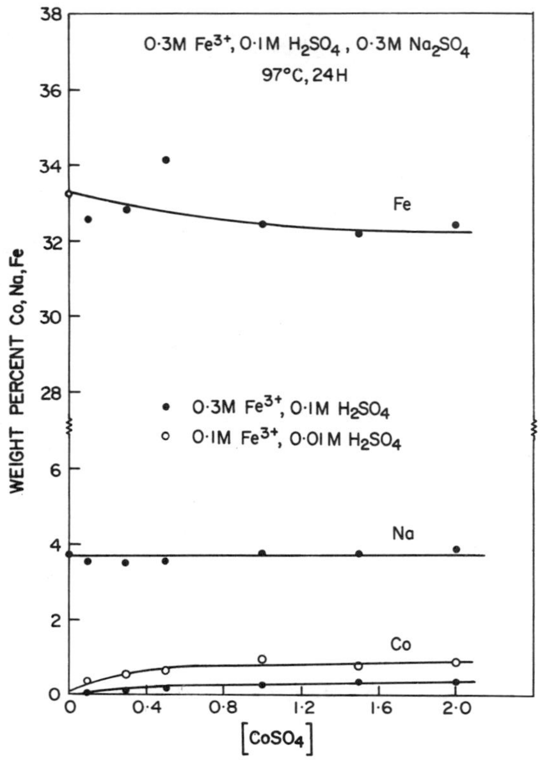

Fig. 10. The effect of various dissolved $CoSO_4$ concentrations on the composition of the sodium jarosite synthesized from two different solutions.

than 0.5 M $CdSO_4$, cadmium losses into commercial jarosite residues will be associated almost entirely with entrained liquors.

Jarosite precipitation is currently used on a limited basis in cobalt processing circuits to eliminate iron[20], and further extension of this technology in cobalt hydrometallurgy is quite likely. Hence, the deportment of cobalt during jarosite precipitation is of some commercial concern as well as being of fundamental interest. Figure 10 shows the effect of various dissolved $CoSO_4$ concentrations on the composition of the jarosite product made at 97°C under otherwise constant conditions. The alkali content of the product remains relatively constant but there is a slight decline in the iron content of the jarosite as the $CoSO_4$ concentration of the solution, and hence the Co^{2+} content of the jarosite, increases; i.e., the sum of iron

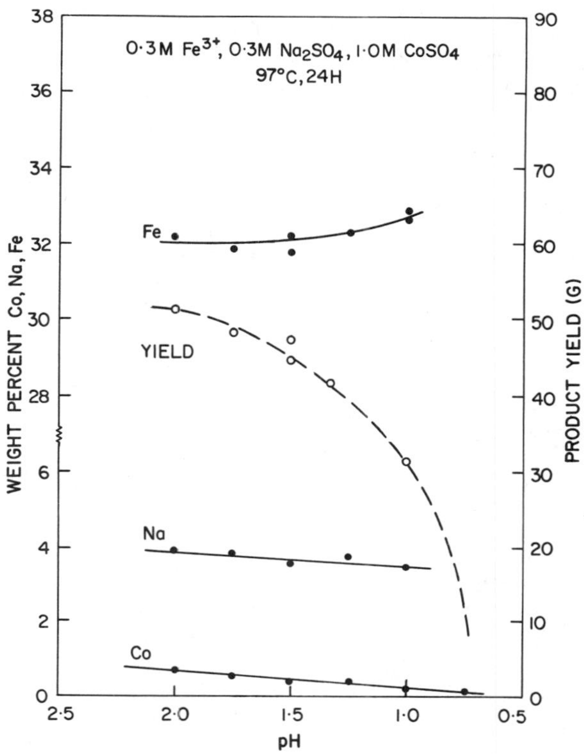

Fig. 11. The composition and yield of sodium jarosite prepared at various initial pH values but from solutions containing 1.0 M $CoSO_4$.

plus cobalt is nearly constant. The actual cobalt content of the sodium jarosite increases systematically as the $CoSO_4$ concentration of the solution increases. As was observed for the other divalent base metals, the cobalt content of the jarosite decreases as the iron and acid concentrations increase. Cobalt losses, even from very concentrated $CoSO_4$ solutions, are <0.5 wt% under the conditions of moderate acidity and iron concentration likely to be found in metallurgical operations.

Figure 11 shows the effect of the initial room temperature pH of the synthesis solution on the extent of cobalt incorporation into the sodium jarosite phase. As was observed for zinc (Fig. 6), the degree of incorporation of the divalent base metal increases as the pH increases. The sodium content of the jarosite remains constant

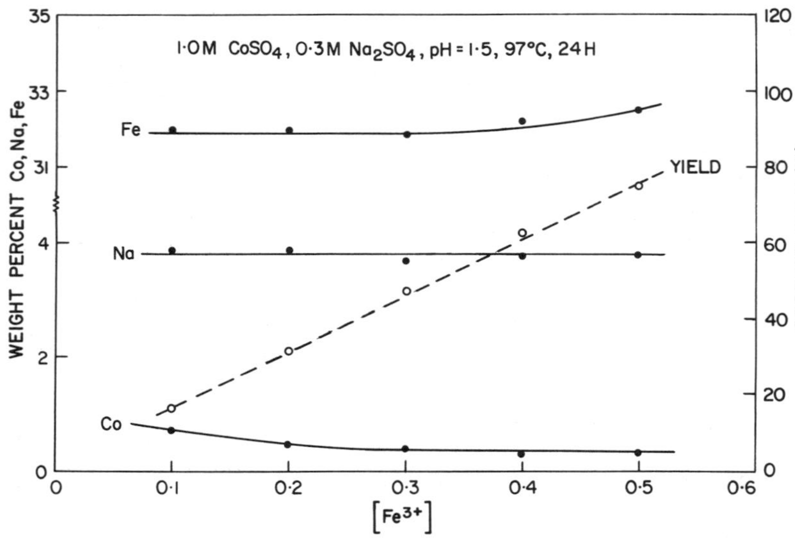

Fig. 12. The influence of initial Fe^{3+} concentration on the product yield and on the composition of the sodium jarosite made from solutions containing 1.0 M $CoSO_4$.

or, in this instance, even increases slightly as the pH and Co content increase. The iron content of these jarosites decreased slightly as the pH increased, and this is felt to be due more to the enhanced cobalt contents of the jarosites than to the variation in pH itself.

The corresponding effect of the initial ferric ion concentration on the composition and yield of the jarosite product is illustrated in Fig. 12. The cobalt content of the jarosite rises slightly as the iron concentration of the pH = 1.5 solution decreases; the iron content of the precipitate behaves in the opposite manner and rises slightly with increasing ferric ion concentration. By contrast, the alkali composition of the jarosite is nearly constant over the entire range of iron concentrations studied. All the solutions were adjusted to pH = 1.5 at room temperature prior to the actual heating of the solution. Under these conditions the amount of precipitate formed increases in an approximately linear manner with increasing iron concentration. Because the iron content of all the jarosites was ~32%, the fraction of iron precipitated was nearly the same in all the experiments.

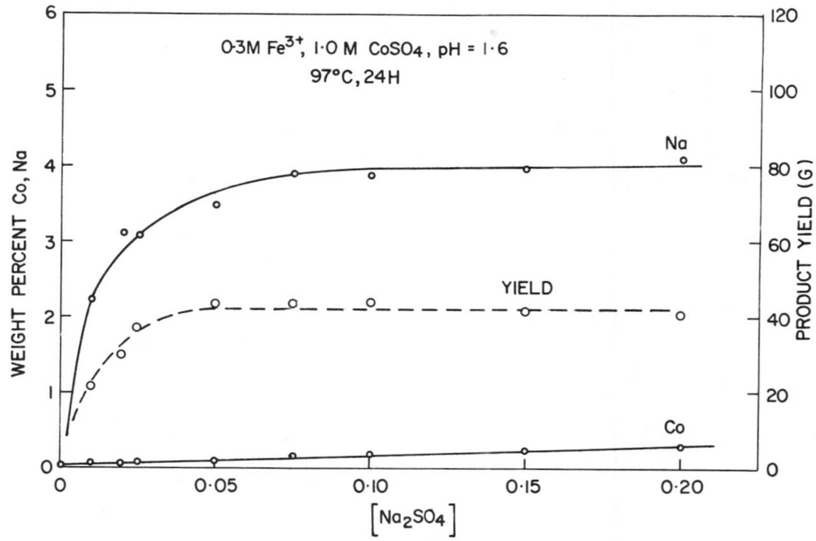

Fig. 13. The influence of Na_2SO_4 on both the yield of jarosite and its Co and Na contents for precipitation from solutions containing 1.0 M $CoSO_4$.

Dissolved alkali metal sulfate concentration is known to be an important factor affecting jarosite formation, and the influence of sodium sulfate concentration on the amount of cobalt incorporated into the jarosite is illustrated in Fig. 13. As the concentration of Na_2SO_4 in solution is raised, both the alkali content of the jarosite and the amount of cobalt incorporated both increase. The rise in sodium content is fairly dramatic with the composition changing rapidly from a sodium-rich hydronium jarosite, $(H_3O,Na)Fe_3(SO_4)_2(OH)_6$ at the low sodium sulfate concentrations to a hydronium-bearing sodium jarosite for the higher alkali concentrations. The concentration of dissolved Na_2SO_4 exerts a major influence on the sodium content of the jarosite to about 0.1 M Na_2SO_4, but thereafter, increasing sodium sulfate concentrations have negligible effect on the alkali content of the product. The alkali content of the jarosite, even for material prepared from very concentrated Na_2SO_4 solutions, never achieves the theoretical 4.3 wt% Na value of $NaFe_3(SO_4)_2(OH)_6$ and this is likely due to hydronium (H_3O^+) substitution for the Na^+ ion. The extent of cobalt incorporation into the product is fairly low for all Na_2SO_4 concentrations studied although the cobalt losses do mount progressively as the Na_2SO_4 concentration increases. The amount of product formed nearly parallels

Table 5. Comparison of the Co Loss into Potassium, Sodium and Ammonium Jarosites.

M_2SO_4	wt % Co in Jarosite
0.3 M K_2SO_4	1.35
0.3 M K_2SO_4	1.23
0.3 M Na_2SO_4	0.41
0.3 M $(NH_4)_2SO_4$	0.37
0.3 M $(NH_4)_2SO_4$	0.33

the alkali content of the jarosite itself. The weight of jarosite increases rapidly with increasing Na_2SO_4 concentrations to about 0.05 M and levels off thereafter. In general terms, the behavior of all the divalent base metals with respect to Na_2SO_4 concentration parallels that of cobalt (see Fig. 7).

It is also useful to compare the relative extent of cobalt incorporation into various alkali jarosite species of metallurgical interest. The data presented in Table 5 show the losses of cobalt into potassium, sodium and ammonium jarosites made at 97°C from solutions containing 0.3 M Fe^{3+}, 1.0 M $CoSO_4$, pH = 1.5 and 0.3 M M_2SO_4 where M = K, Na or NH_4; the data are for replicate experiments. As was observed above for zinc, the inclusion of cobalt into jarosite family minerals is most substantial for the potassium member. Sodium and ammonium jarosites give significantly lower losses of cobalt, with the amount of incorporation appearing to be slightly lower for the ammonium species. Although potassium jarosite is the most effective of all jarosite species for iron removal, it also causes the largest loss of the divalent base metals.

Nickel commonly behaves very much like cobalt and this is certainly evident during alkali jarosite formation. Comparison of the data for nickel in Fig. 14 with that for cobalt in Fig. 10 reveals many similarities. Although nickel losses into the jarosite are slightly lower than for cobalt, the extent of nickel incorporation also increases with increasing concentration of the divalent base metal sulfate, and decreases as the acid and iron concentrations increase. Sodium and iron contents of the jarosites are similar for either cobalt or nickel incorporation.

Certain trends emerge concerning the incorporation of divalent base metals into alkali jarosites, and these are enumerated below.

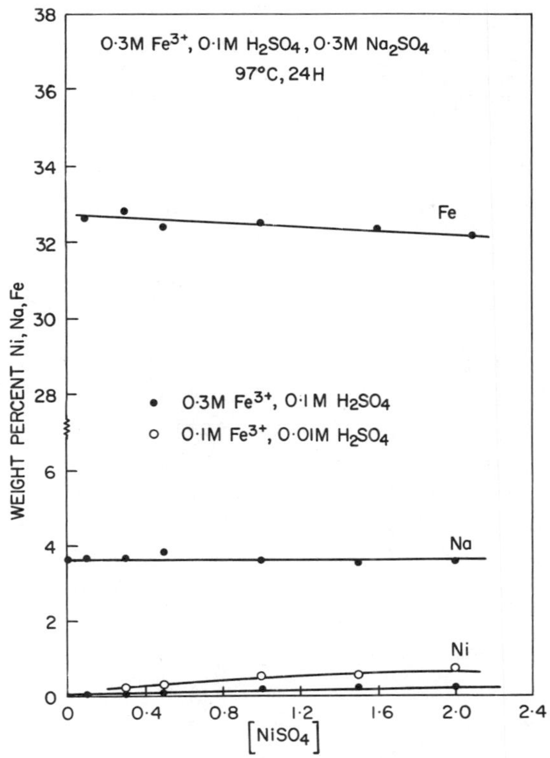

Fig. 14. The effect of $NiSO_4$ concentration on the Ni, Na and Fe contents of sodium jarosite made from two different solutions.

1. Although the extent of divalent base metal incorporation into alkali jarosites is minor (<3 wt%), it is apparently real; i.e., no second phases containing the divalent metals were detectable.

2. The degree of incorporation into the jarosite increases with increasing divalent base metal concentration and decreases with increasing ferric ion or acid concentrations.

3. Incorporation of the base metal into the alkali jarosite results in a systematic decrease in the iron content of the jarosite, but there is no systematic change in alkali content.

4. The extent of divalent base metal incorporation actually increases as the alkali concentration in solution increases.

5. Incorporation of divalent base metals into potassium jarosite is significantly greater than into sodium or ammonium jarosites; sodium jarosite incorporates slightly more of a given divalent base metal than does its ammonium analog.

The above observations suggest that the divalent base metal ions actually enter into the jarosite lattice and that they substitute for Fe^{3+}:

$$xM^{2+} + NaFe_3(SO_4)_2(OH)_6 \rightarrow Na(Fe_{3-x}M^{2+}{}_x)(SO_4)_2(OH)_6 + xFe^{3+} \qquad (3)$$

The extent of substitution according to Reaction 3 would be expected to increase with increasing concentrations of the divalent base metal in solution and to decrease with increasing Fe^{3+} concentrations. The divalent base metals are relatively difficult to hydrolyze and the extent of their incorporation into the Fe-OH sheet structure of jarosite would be expected to increase as the pH increases, again in agreement with observed data. The order of incorporation of the common divalent base metals into the jarosite structure appears to be:

$$Fe^{3+} \gg Cu^{2} > Zn^{2+} > Co^{2+} \sim Ni^{2+} \sim Mn^{2+} > Cd^{2+}$$

This order is the same as the ease of hydrolysis of metal ions; for the first hydrolysis constants:

$$M^{n+} + H_2O \rightarrow M(OH)^{(n-1)+} + H^+ \qquad (4)$$

$$K_{11} = \frac{[M(OH)^{(n-1)+}]\,[H^+]}{[M^{n+}]} \qquad (5)$$

the K_{11} values for the various metals are:

$$Fe^{3+}(10^{-3}) \gg Cu^{2+}(10^{-8}) > Zn^{2+}(10^{-9}) > Co^{2+}(6x10^{-11})$$

$$\sim Ni^{2+}(3x10^{-11}) \sim Mn^{2+}(3x10^{-11})$$

Thus the more easily the divalent base metal is hydrolyzed, the more easily it is incorporated into the jarosite lattice. Increasing solution pH values increase the extent of hydrolysis (Reaction 4) and promote the incorporation of the divalent base metals into the lattice of the alkali jarosites.

Substitution of a divalent metal ion into the Fe^{3+}-OH sheet structure of jarosite clearly introduces thermodynamic weaknesses

Table 6. Standard free energies of formation of potassium, sodium and ammonium jarosites.

Jarosite	ΔG^{o}_{298} (kcal/mole)
$KFe_3(SO_4)_2(OH)_6$	-788.6
$NaFe_3(SO_4)_2(OH)_6$	-778.4
$NH_4Fe_3(SO_4)_2(OH)_6$	-736.2

in the structure, and this is reflected by the low extent of divalent base metal incorporation both in synthetic and in natural jarosites which have presumably been equilibrated with respect to their synthesis solution. High alkali sulfate concentrations help to stabilize the jarosite (Reaction 1) and should, therefore, slightly increase the extent of base metal incorporation as apparently happens. Also, the potassium species is significantly more stable than either the sodium or ammonium jarosites[1] as indicated in Table 6. The greater the inherent stability of the jarosite, the greater it should be able to resist the disruptive factors associated with M^{2+} substitutions. Significantly, the order of thermodynamic stabilities is the same as the order of divalent base metal incorporation.

Direct crystallographic proof for the substitution of divalent metals for Fe^{3+} in the alkali jarosites is lacking because of the low degree of M^{2+} substitution and the fact that other minor chemical changes always accompany the substitution of M^{2+}. However, in the lead jarosite series, it has been forcibly argued[21] that both Cu^{2+} and Zn^{2+} substitute for Fe^{3+} eventually leading to beaverite $Pb(Cu,Fe_2)(SO_4)_2(OH)_6$. Also, the aluminum analog of beaverite, the mineral osarizawaite $Pb(Cu,Al_2)(SO_4)_2(OH)_6$, has been shown by the crystal refinement work of Giuseppetti and Tadini[22] to have Cu^{2+} substituting for Fe^{3+}. A similar situation likely holds for all members of the jarosite family.

Although Reaction 3 is balanced with respect to mass, there is a net charge imbalance; i.e., the substitution of a divalent ion for Fe^{3+} must involve compensating charge distributions. The alkali sites in sodium jarosite are not all filled with Na^{+}, and the deficiency has been assigned to hydronium substitution although substitution by hydronium ion plus minor vacancies would also be possible within the analytical errors associated with the indirect determination of H_3O^+ based on weight loss experiments. Hence, some adjustment of the charge requirements could conceivably be achieved by increasing the concentration of $(Na^+ + H_3O^+)$ in the jarosite at the expense of hypothetical vacancies. However, the consistently constant alkali content of all the products studied militates against this postulate. A more likely explanation of the incorporation of divalent base metals involves the substitution of M^{2+} for Fe^{3+} together with a corresponding reduction in the OH concentration:

$$xM^{2+} + NaFe_3(SO_4)_2(OH)_6 \rightarrow Na(Fe_{3-x}M_x^{2+})(SO_4)_2(OH)_{6-x} + xFe^{3+} + xOH^- \quad (6)$$

or

$$M^{2+} = Fe^{3+} + OH^- \quad (7)$$

The hydroxyl ion presumably combines with a proton in the acidic solution to form water, but whether this water remains in the crystal is not known. Direct and accurate determination of minor variations in the OH concentration is not easy; the addition of 1 wt% Cu to an alkali jarosite would cause a change of only 0.27 wt% in the OH content. Such differences are difficult to detect accurately in the presence of hydronium ion and/or adsorbed or bound H_2O.

Copper and Zinc Incorporation into Lead Jarosite

Figure 15 shows the effect of dissolved copper ion (as $CuSO_4$) on the lead and copper contents of lead jarosite prepared at 130°C from solutions containing 0.3 M $Fe^{3+}(SO_4)$, 0.03 M H_2SO_4, 2.0 M Li_2SO_4 and 30 g/L $PbSO_4$. The products contained <0.1 wt% Li and were subjected to a standard ammonium acetate washing procedure[18] to remove residual $PbSO_4$; the addition of neutral sulfates has been found to increase the lead content of the product significantly. In the absence of dissolved copper sulfate a conventional lead jarosite containing about 16 wt% Pb is produced; this contains slightly less than the 18.3 wt% Pb of $Pb_{0.5}Fe_3(SO_4)_2(OH)_6$, and the mass and charge difference is presumably accounted for by the substitution of H_3O^+ for Pb^{2+}. As the concentration of dissolved copper in solution increases, both the lead and copper contents of the products increase significantly. Copper contents in excess of eight weight percent are produced and such materials approach beaverite, $PbCuFe_2(SO_4)_2(OH)_6$, in both composition and structure. The incorporation of copper is thought to be due to the following type of reaction:

$$1/2\ Pb^{2+} + Cu^{2+} = Fe^{3+} \qquad (8)$$

The incorporation of an extra 1/2 Pb^{2+} is facilitated by the presence of ample vacant alkali sites in lead jarosites; i.e., one-half of the alkali sites are vacant in idealized $Pb_{0.5}Fe_3(SO_4)_2(OH)_6$. Lead-bearing jarosites, like end-member lead jarosite, also contain coupled Pb^{2+} -vacancies which allow significant divalent base metal incorporation. Charge balance is easily attained under these conditions (Reaction 8), and the copper content approaches the idealized value of beaverite together with increasing and essentially proportional lead contents.

Other divalent base metals are also readily incorporated into lead or lead-bearing jarosites because of the ease of maintaining charge balance by substituting extra lead into the vacant lattice sites. That is, the total extent of divalent substitution is governed both by solution thermodynamics and by charge balance considerations with the latter factor being of somewhat greater importance. Figure 16 shows the effect of dissolved zinc (as $ZnSO_4$) on both the zinc and lead contents of the lead jarosite prepared at 130°C from a medium containing 0.3 M Fe^{3+}, 0.03 M H_2SO_4 and 15.3 g/L $PbSO_4$. In the absence of zinc, a lead jarosite containing 12.5% Pb is pro-

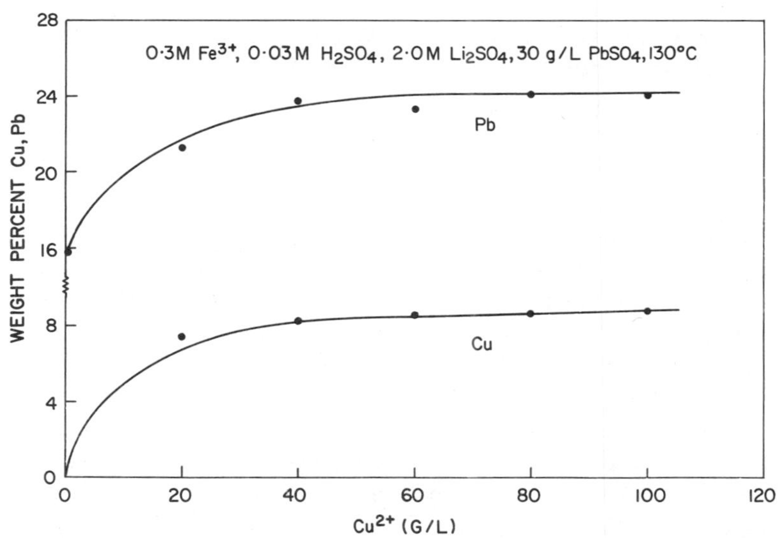

Fig. 15. The significant effect of dissolved $CuSO_4$ on the Pb and Cu contents of lead jarosite formed at 130°C.

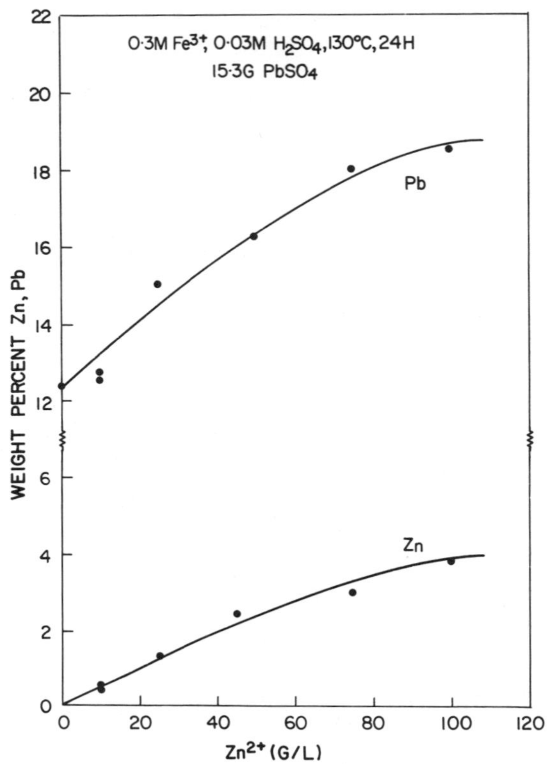

Fig. 16. The effect of dissolved $ZnSO_4$ on the Pb and Zn contents of lead jarosite formed at 130°C.

duced; the lower Pb tenor relative to that shown in Fig. 15 is due to the omission of Li_2SO_4 from the synthesis solutions. High ionic strength in sulfate media is known to increase the lead content of lead jarosite although the mechanism is not at all elucidated. As the concentration of Zn^{2+} in the synthesis solution increases, the Zn content of the lead jarosite product also increases. The extent of Zn incorporation is, however, always significantly less than for Cu^{2+} (Fig. 15); about 3-4 wt% Zn can be incorporated under optimum conditions. The increase in the lead content of the product essentially parallels that of Zn, indicating a coupled substitution to maintain charge balance:

$$1/2\ Pb^{2+} + Zn^{2+} = Fe^{3+}$$

Again the presence of ample vacant alkali sites in lead jarosite facilitates the incorporation of additional lead and, consequently, an enhanced degree of Zn^{2+} substitution.

Other divalent base metals are also incorporated into lead jarosite, and the degree of substituton is always significantly greater than for the same metal in the simple alkali jarosites. The order of inclusion into the lead jarosite is:

$$Cu^{2+} > Zn^{2+} > Co^{2+} \sim Ni^{2+} \sim Mn^{2+} > Cd^{2+}$$

This order is the same as that observed for the alkali jarosites and indicates that solution thermodynamic effects (size and electronic configuration) are also important even when the charge balance restraints are largely eliminated by the formation of lead jarosites.

Substitutions of R^{3+} Ions for Fe^{3+}

The jarosite structure is dominated by Fe^{3+} -OH sheets formed from $Fe^{3+}(OH,O)$ octahedra which are corner linked in a plane parallel to (0001). The iron can be extensively replaced by a number of R^{3+} ions such that the basic $R\overline{3}m$ structure of the jarosite family of minerals is still preserved. In many instances new R^{3+} end-member compounds are formed. Although such end-member species themselves may be of little metallurgical significance, the solid solution series with the iron members as well as the relative partitioning coefficients are of considerable concern because they directly affect impurity distributions during hydrometallurgical processing.

Aluminum is known[23] to replace iron completely in alkali jarosites, but the relative partitioning between Al^{3+} and Fe^{3+} is not quantitatively known. To help establish the relative partitioning between iron and aluminum for the potassium members of the jarosite family, experiments were carried out at 97°C from solutions containing 0.3 M K_2SO_4 and various Al^{3+} and Fe^{3+} concentrations but with the restriction that $[Al^{3+} + Fe^{3+}] = 0.3$ M. Runs were made at various initial pH values and the resulting precipitates were washed and dried prior to analysis. The relative partitioning of Al and Fe between solution and solid phases is illustrated in Fig. 17. The dashed line represents the idealized distribution curve which would be realized if iron and aluminum were both precipitated in the same molar ratio as they were present in solution. Clearly, the two elements are not equally precipitated and there is a definite tendency for Al^{3+} to concentrate in solution. The effect is particularly

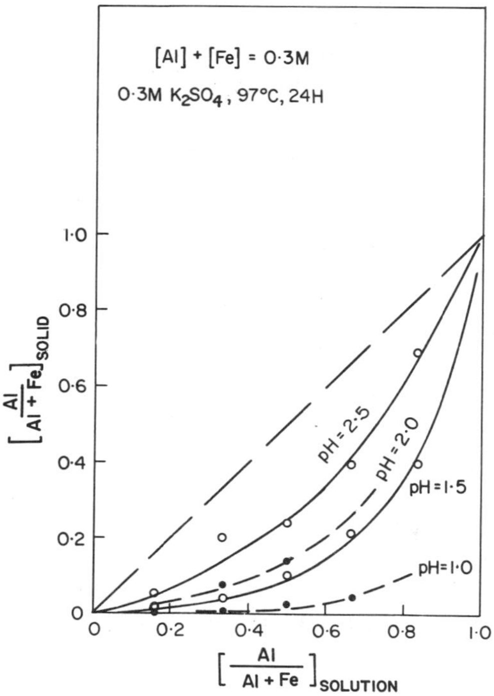

Fig. 17. Relative partitioning of Al and Fe between solutions and jarosite phases at four different initial pH values.

obvious at the lower pH values (pH 1.0–2.0) where jarosite precipitation is commonly carried out; at pH 1.5 for dilute concentrations of Al^{3+} in a ferric sulfate liquor, the relative partitioning coefficient is only about 0.1. Also, end member $KAl_3(SO_4)_2(OH)_6$ could not be prepared at pH values lower than 2.0. Clearly the aluminum analogs of jarosite family compounds will be stable only in the high pH region where Fe^{3+} will likely be unstable as a consequence of hydrolysis or at temperatures >100°C. Although jarosite precipitation from Al-bearing solutions will certainly eliminate some aluminum, this will not be an effective means of control unless high pH's and temperatures are used.

Gallium and indium are both valuable trace elements which can be present in hydrometallurgical solutions, and their behavior during jarosite precipitation is of paramount importance if attempts are to be made at recovery. End-member $KGa_3(SO_4)_2(OH)_6$ was prepared by dissolving 0.23 M Ga^{3+} (as sulfate) in 0.3 M K_2SO_4 solution, adjusting the pH to 1.6, filtering to remove the thin scum of gallium hydrolysis products, and then heating to 97°C for 24 h to precipitate the gallium analog of potassium jarosite. After drying at 110°C, it was found that 27 g of a white precipitate had been produced that had the composition shown in Table 7. The synthetic product agrees fairly closely with the hypothetical formula although it appears to contain additional K and SO_4 which were not soluble in hot water. The previously reported synthesis of this material[7] noted a significant deficiency of Ga (32.7%) and an excess of SO_4 (37.5%). Preliminary X-ray diffraction analysis confirmed that the compound had the $R\overline{3}m$ structure of jarosite, and a refinement was carried out on that basis. The interplanar d-values and visually estimated intensities of the X-ray diffraction patterns for synthetic $KGa_3(SO_4)_2(OH)_6$ are presented in Table 8 together with the corresponding data from Ref. 7. The agreement between the two sets of data is not especially good as concerns the d-spacings and the indexing (hkl) of the various lines, and the reasons for this are not immediately apparent although variations in composition between the two compounds may be partly responsible.

The cell refinement of $KGa_3(SO_4)_2(OH)_6$ yielded the lattice parameters presented in Table 9. There is fair agreement between the two reported studies of this compound. Gallium ion is about the same size as ferric ion (Ga^{3+} = 0.62 Å; Fe^{3+} = 0.64 Å), and this is reflected in the similarity in a and c parameters between the corresponding gallium and iron compounds. The a parameter of the gallium analog is slightly reduced consistent with the smaller size of Ga^{3+}; the c parameter is less affected by the size of the ion occupying the R site.[21]

Figure 18 shows the relative partitioning of gallium and iron between the solution and solid phases when a jarosite-type compound was precipitated at 97°C from a solution containing 0.4 M[Ga^{3+} + Fe^{3+}] and 0.3 M K_2SO_4 at pH =1.5. Although the total concentration of [Ga^{3+} + Fe^{3+}] was constant at 0.4 M, the relative amounts of Ga^{3+}/Fe^{3+} were varied. The figure indicates that gallium is coprecipitated with iron, but that there is a moderate tendency for the gallium to remain in solution. A solution containing 0.25 mole fraction gallium to gallium plus iron yields a solid containing 0.15 mole fraction of that element. Although the exact position of the partitioning curve is likely to vary somewhat with pH, temperature, etc., it is fairly certain that jarosite precipitation in the presence of Ga^{3+} will result in substantial losses of that element into the residue.

Table 7. Chemical Data Obtained for the Gallium Analogue of Potassium Jarosite.

Element	$KGa_3(SO_4)_2(OH)_6$	Synthetic Product
K	7.2	7.8
Ga	38.6	36.8
SO_4	35.4	39.8
OH, H_2O	18.8	15.6

Table 8. X-ray Powder Diffraction data of Synthetic $KGa_3(SO_4)_2(OH)_6$.

This work*				Tananaev et al. (Ref. 7)		
*I*est	*d* meas	*d* calc	*hkl*	*I*est	*d* meas	*hkl*
5	5.77	5.82	101	3	5.91	101
7	4.99	5.01	012	2	5.79	003
1	3.56	3.57	110	6	5.10	$10\bar{2}$
10	3.02	3.03	113	2	3.64	110
2	2.904	2.909	202	0.5	3.55	104
2	2.845	2.854	006	10	3.06	113
4	2.503	2.507	024	1	2.95	202
4	2.270	2.275	107	2	2.868	006
4	1.939	1.939	033	3	2.530	$20\bar{4}$
0.5	1.925	1.931	125	0.5	2.339	205
0.5	1.904	1.902	009	3	2.283	107
4	1.785	1.786	220	3	1.958	303
0.5	1.763	1.760	208	0.5	1.927	207
0.5	1.708	1.707	131	0.5	1.908	009
0.5	1.685	1.690	217	3	1.798	220
0.5	1.596	1.592	134	0.5	1.768	208
0.5	1.538	1.540	401	0.5	1.715	223
0.5	1.516	1.514	226	1	1.698	312
2	1.500	1.498	0.2.10	0.5	1.601	314
2	1.456	1.454	404	0.5	1.581	128

* Intensities estimated visually; indexed with *a* 7.142, *c* 17.125

Table 9. Comparison of the Lattice Parameters Obtained for the $KGa_3(SO_4)_2(OH)_6$.

Parameter (Å)	This Work	Tananaev et al. (Ref. 7)	$KFe_3(SO_4)_2(OH)_6$ (Ref. 17)
a	7.142	7.24	7.315
c	17.125	17.24	17.119

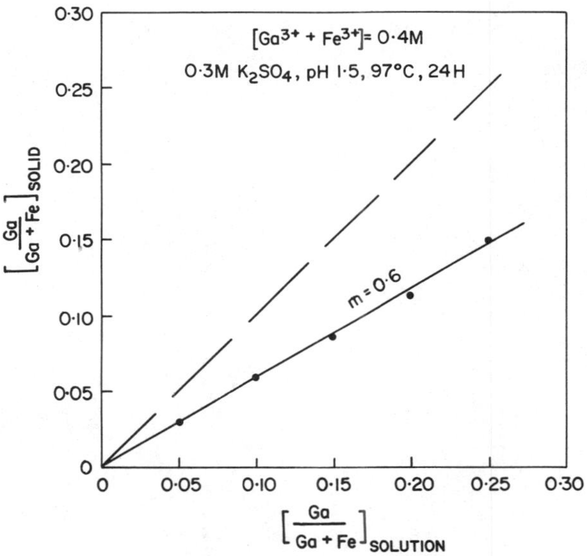

Fig. 18. Relative partitioning of Ga and Fe between solution and jarosite phases at pH = 1.5.

Indium(III) might be expected to behave in a manner similar to that of gallium and this is actually the case. End-member $KIn_3(SO_4)_2(OH)_6$ was easily precipitated at 97°C from a solution containing 0.3 M In^{3+} (as SO_4) and 0.3 M K_2SO_4 at pH = 1.5. The washed and dried product was white, and the product yield per liter of solution was 46 g. The indium analog was shown to possess a jarosite-type powder diffraction pattern and to have the composition presented in Table 10. In addition to having the jarosite-type structure and approximate composition of $KIn_3(SO_4)_2(OH)_6$, the synthetic product also possessed the characteristic blocky rhombohedral crystal habit of jarosite family compounds. Figure 19 shows a SEM photomicrograph of $KIn_3(SO_4)_2(OH)_6$ crystals with the characteristic

Table 10. Chemical Data Obtained for the Indium Analog of Potassium Jarosite.

Element	This Work	$KIn_3(SO_4)_2(OH)_6$
K	5.06	5.77
In	51.0	50.83
SO_4	29.2	28.35
OH, H_2O	14.8	15.05

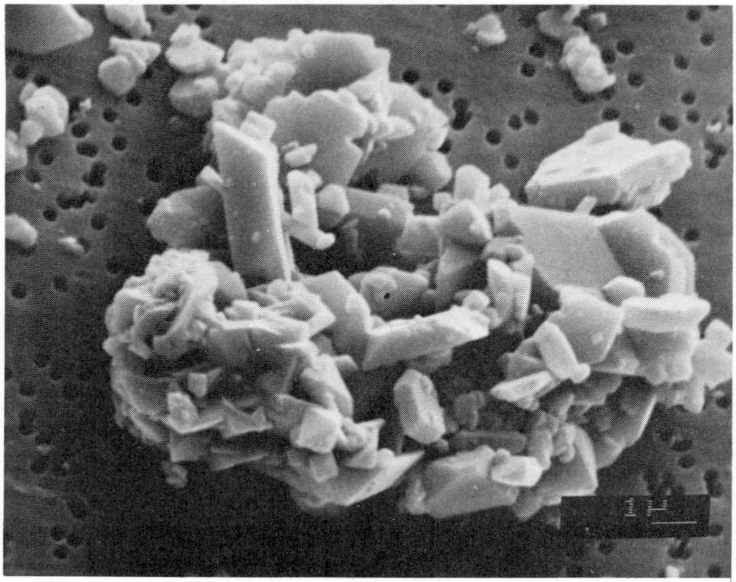

Fig. 19. SEM photomicrograph of $KIn_3(SO_4)_2(OH)_6$ crystals illustrating the characteristic jarosite habit.

shape of jarosite family minerals. It is also evident that the mean particle size is 1-5 µm, and this is also a characteristic of laboratory jarosite precipitates.

The relative partitioning of In^{3+} and Fe^{3+} between the aqueous phase and the jarosite-type precipitate is illustrated in Fig. 20. These particular tests were carried out at 97°C with solutions containing $[In^{3+} + Fe^{3+}]$ = 0.4 M and 0.3 M K_2SO_4 at pH = 1.6. Although indium precipitates with the jarosite, it, like Al^{3+} and Ga^{3+} tends to concentrate moderately in the aqueous phase. The molar distribution factor, m, is about 0.7. Bakaev et al.[10] observed that In^{3+} and Fe^{3+} were precipitated in about the same ratio as they were present in solution; i.e., m ~1.0. Indium is a potentially valuable constituent in many base metal sulfide ores, especially zinc ores, and the use of jarosite precipitation technology during the processing of such feeds will clearly result in significant indium losses to the jarosite residue.

The extent of incorporation of In^{3+} into the jarosite lattice increases as the pH increases as was also the case for Al^{3+} - Fe^{3+}

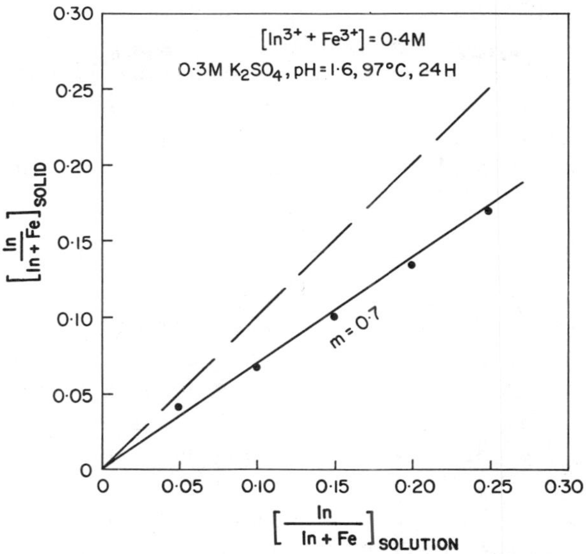

Fig. 20. Relative partitioning of In and Fe between solution and jarosite phases at pH = 1.6.

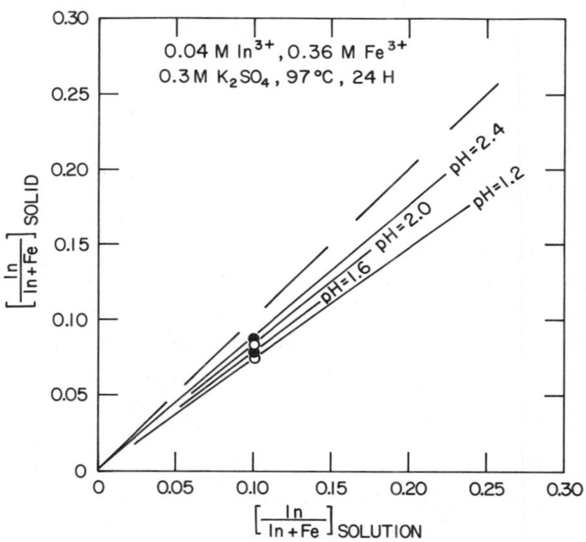

Fig. 21. The effect of initial pH on the relative partitioning of In and Fe between the jarosite phase and a solution containing a mole fraction of In/(In+Fe) = 0.1.

solutions (Fig. 17). Figure 21 shows a slight but persistent increase in the In mole fraction in the solid as the pH is raised from 1.2 to 2.4 at 97°C and for a constant In/(In+Fe) mole ratio of 0.10 in the solution phase. Comparison with the corresponding data for Al (Fig. 17) and Ga (Fig. 18) indicates that the order of preferential incorporation of the trivalent metal ions into the jarosite phase in acidic media is:

$$Fe^{3+} > In^{3+} > Ga^{3+} > Al^{3+}$$

Attempts were also made to incorporate Tl^{3+} into the jarosite phase, but no jarosite-type product was formed even at temperatures as high as 170°C. A precipitate was produced readily at room temperature, but X-ray diffraction analysis showed this to be quite different from the jarosite-type compounds. Yaroslavtsev et al.[2] indicated that Tl^{3+} was precipitated to the same extent as Tl^{+} during jarosite formation, but the nature of the precipitate was not ascertained. Based on the limited amount of work done in this study, it would have to be concluded either that the Tl^{3+} formed a non-jarosite-type precipitate or that the Tl^{3+} was reduced to Tl^{+} (with O_2 evolution) which then formed normal Tl^{+} jarosite, $TlFe_3(SO_4)_2(OH)_6$. The latter explanation could explain the similarities between Tl^{3+} and Tl^{+} that were observed by these workers.[2]

Substitutions for SO_4

Previous work[11] has shown that $SeO_4{}^{2-}$ replaces $SO_4{}^{2-}$ in either potassium or sodium jarosites, and that end-member selenate analogs exist for both jarosites. Furthermore, the solid solutions formed between the sulfate and selenate compounds were shown to be nearly ideal by both chemical and X-ray diffraction analysis; there was only a very slight tendency for the SeO_4 species to concentrate in the aqueous phase (m >0.8). Recent attempts to prepare tellurate analogs of potassium jarosite were unsuccessful. When K_2TeO_4 solution was added to a ferric nitrate medium at pH = 1.5, a precipitate formed immediately, even at room temperature. This was found to be poorly crystallized and to be similar in structure and composition to $Fe_2(TeO_4)_3.H_2O$. Possibly a tellurate-bearing jarosite could be prepared by a slow-addition technique,[17] but given the low solubility of ferric tellurate hydrate, this avenue was not pursued. Presumably any tellurate present in iron-bearing hydrometallurgical solutions will be readily precipitated, but not as a jarosite-type compound.

Efforts were also made to prepare selenite ($SeO_3{}^{2-}$) and tellurite ($TeO_3{}^{2-}$) analogs of potassium jarosite, but no jarosite-type phases were produced. The addition of either SeO_3^{2-} or TeO_3^{2-} to an acidic ferric nitrate solution resulted in the immediate precipitation of both anions as an amorphous to poorly crystalline hydrous ferric selenite or tellurite. In a hydrometallurgical circuit containing Fe^{3+}, both SeO_3^{2-} and TeO_3^{2-} will be readily precipitated, but not as a jarosite-type compound.

All zinc electrolytes contain significant concentrations of manganese(II), and part of this is oxidized to MnO_4^- at the anode during electrolysis. Work described above has elucidated the behavior of Mn^{2+} during jarosite formation, but nothing is known of the extent of substitution of MnO_4 in these compounds. Preliminary experiments indicated that large concentrations of $KMnO_4$ readily precipitated a poorly crystalline $(Mn,Fe)O_2$ in the presence of ferric ion at 97°C. Presumably the simplified reaction is

$$MNO_4 + H^+ \rightarrow MnO_2 + H_2O \tag{9}$$

Although this reaction would appear to preclude the synthesis of an end-member permanganate analog of jarosite, it does not entirely rule out the formation of a permanganate-bearing jarosite prepared from solutions containing low concentrations of permanganate in an otherwise sulfate medium. Various syntheses of potassium jarosite were carried out at 97°C from solutions containing 0.3 M Fe^{3+} (as sulfate), 0.3 M K_2SO_4, pH = 1.5 and different amounts of $KMnO_4$. The potassium member was selected since it is likely to incorporate greater amounts of any impurity. As the manganese concentration was reduced from 0.05 to 0.01 M $KMnO_4$, the manganese content of the products systematically decreased from 1.94 to 0.49 wt% Mn. Examination of all the products by sensitive Guinier-deWolff X-ray diffraction methods revealed faint traces of MnO_2 in each; also, all the jarosites were tinted light to medium brown indicating an impurity phase. Given that the MnO_2 produced by Reaction 9 is poorly crystalline, it would have to be concluded that virtually all the manganese contained in the jarosites was present as the dioxide, and that a MnO_4^- analog of jarosite or a jarosite containing significant amounts of MnO_4^- do not exist.

Most base metal ores or concentrates contain minor amounts of silica, part of which could conceivably enter the hydrometallurgical processing circuit, especially when direct leaching of ores is employed without a roasting stage. The actual ionic form of SiO_2 in solution depends on the pH, ionic strength and cations present; for this reason, all silica additions were made simply as Na_2SiO_3 to the acidic ferric sulfate media. Initial tests involved dissolving various amounts of Na_2SiO_3 in H_2O, adjusting the pH = 1.6 and then heating to 97°C to observe the solution. For SiO_2 concentrations $\geq$ 0.05 M, a silica gel formed quickly at the 97°C temperature and was amorphous to X-rays. Had a jarosite precipitation been carried out at $SiO_2 > 0.05$ M, the product would have been contaminated with silica gel which would not have been X-ray detectable; it would probably be concluded, quite erroneously, that some silicate species had been incorporated into the jarosite. No visible silica gel formation after 24 h at 97°C was, however, observed for SiO_2 concentrations of 0.03 M or less.

The possibility of incorporating some silicate into sodium jarosite was studied by forming $NaFe_3(SO_4)_2(OH)_6$ at 97°C from a medium containing 0.2 M Fe^{3+} (as SO_4) and 0.3 M Na_2SO_4 at pH = 1.6 in the presence of 0.01, 0.02 or 0.03 M Na_2SiO_3. All the water-washed products contained <0.01 wt% SiO_2, the analytical detection limit; less than 0.4% of the Si initially present in the solution was precipitated with the jarosite. Also, the pH of the initial solution was varied from 1.2 to 1.8 for a constant Na_2SiO_3 concentration of 0.03 M. In every instance the precipitate contained less than the analytical detection limit of 0.01 wt% SiO_2.

Although it is conceivable that SiO_4^{4-} might be partially incorporated into the jarosite structure, this species is prevalent only in very basic media where jarosite precipitation cannot occur. In acidic solutions, the silica is mostly present as large complex anions which do not enter the jarosite structure but can lead to amorphous silica gel formation.

Boron behaves in many ways like Si and the behavior of this element, added both as $Na_2B_4O_7 \cdot 10H_2O$ and Na_2BO_3, was studied. Attempts to precipitate a borate-bearing sodium jarosite by heating a 0.3 M Fe^{3+} (as SO_4), 0.2 M Na_2SO_4, pH = 1.6 and 0.2 M B solution to 97°C resulted in a sodium jarosite precipitate containing less than 0.008 wt% B. Variation of the initial room temperature pH from 1.8 to 1.2 did not increase the extent of borate precipitation which was consistently less than 0.2% of the amount initially present in solution. Clearly, low concentrations (<0.2 M) of borate ions in solution will not be precipitated to any significant extent during jarosite formation.

Germanium is also a valuable minor constituent of many sulfide ores and its behavior during jarosite precipitation is of some metallurgical importance. Like silicon and boron, germanium forms complex oxyanions in acidic media and there is the possibility that these could be incorporated into the jarosite. Yaroslavtsev et al.[2] observed a medium degree of incorporation of Ge^{4+} into jarosite for Ge concentrations from 0.014 to 1.45 g/L; the amount incorporated was comparable to that of aluminum (cf. Fig. 17). In the present study, sodium and potassium jarosites were synthesized at 97°C from 0.2 M Fe^{3+} (as SO_4), 0.3 M Na_2SO_4 (or K_2SO_4), pH = 1.6 media containing from 1-4 g/L GeO_2 (0.7 - 2.8 g/L Ge^{4+}). The precipitates were water washed and dried at 110°C prior to analysis. All the alkali jarosites produced in these tests contained <0.1 wt% GeO_4^{4-} (<0.07 wt% Ge). Mass balances were carried out on the synthesis systems before and after jarosite precipitation, and these confirmed that <0.05% of the germanium initially present in the solution was actually precipitated with the jarosite phase. The reason for the markedly greater degree of Ge^{4+} incorporation in the jarosites studied by Yaroslavtsev et al. is not readily explicable since their synthesis conditions were comparable to those used in the present program, and Ge^{4+} showed little evidence of hydrolysis for the concentration ranges used.

Because Pb^{2+} forms a jarosite-type compound, $Pb_{0.5}Fe_3(SO_4)_2(OH)_6$, efforts were also made to prepare a tin jarosite or a tin-bearing jarosite using either Sn^{2+} or $Na_2SnO_3 \cdot {}_3H_2O$ as the initial source of tin. Experiments based on $SnSO_4$ or $SnCl_2$ included conventional synthesis techniques as well as autoclave and slow-addition methods[17]; the tin concentration was varied from 3-50 g/L. Severe problems were encountered in the conventional and autoclave synthesis procedures with the immediate precipitation of poorly crystallized $(Sn,Fe)O_2$. At the end of the heating experiment, the product consisted of jarosite plus poorly crystallized $(Sn,Fe)O_2$. Addition of sufficient acid to retard tin hydrolysis prevented jarosite precipitation altogether. Even by the slow-addition method where no visible SnO_2 precipitation occurred, the product was shown to be contaminated with $(Sn,Fe)O_2$. Mass balance calculations indicated that about 70% of the tin initially present in the solution had precipitated. There appears to be no tin jarosite, $Sn_{0.5}Fe_3(SO_4)_2(OH)_6$.

Efforts to incorporate "stannate" in place of sulfate in the jarosite structure were also not successful. The sodium stannate dissolved only in very strongly acidic solutions where jarosite precipitation did not occur. When the pH was adjusted to a region favorable for jarosite formation, immediate precipitation of tin occurred. It seems probable that any reported tin in jarosite precipitates is the consequence of the presence of amorphous tin oxides or hydroxides.

Attempts to form a periodate analog of potassium jarosite employing ferric nitrate and KIO_4 were not successful; although a precipitate formed which contained approximately 51 wt% IO_4 and 21 wt% Fe, it was shown to consist only of poorly crystallized FeO.OH. The potassium concentration was also very low, and the product, unlike jarosite-type compounds, was very difficult to filter.

The addition of molybdate [as $(NH_4)_6Mo_7O_{24}$] to a ferric nitrate-nitric acid medium produced difficult to filter $Fe_2(MoO_4)_3$ as the only detectable product. It appears that soluble molybdenum or iodate could be extensively removed during jarosite precipitation but it is quite likely that these species are eliminated in non-jarosite form.

Table 11 summarizes the attempted replacements of SO_4 in the jarosite lattice. Germanium, boron and silicon (<0.05 M) remain almost totally in solution during jarosite precipitation under nearly optimum conditions. Tellurium (as TeO_4^{2-} or TeO_3^{2-}) and selenium (as SeO_3^{2-}) are readily precipitated as non-jarosite iron compounds, and a similar conclusion also applies to MoO_4^{2-} and IO_4^-. Manganese (MnO_4^-), tin and silicon (<0.05 M) are hydrolyzed to poorly crystalline but simple oxides. This table suggests that the extent of substitutions for SO_4 is rather limited when compared to the replacements known for iron or the alkali ions.

CONCLUSIONS

A variety of ions can substitute for the iron, and to a lesser extent for the sulfate, component of jarosite-type compounds, $MFe_3(SO_4)_2(OH)_6$ where M = K^+, Na^+, NH_4^+, 1/2 Pb_2^+, etc. In the simple alkali jarosites, the divalent base metals such as Cu, Zn, Co, Ni, Mn, etc. are incorporated to a minor extent, and these species appear to substitute for iron. The degree of base metal incorporation was found to increase with increasing base metal or alkali sulfate concentrations in solution or with rising pH, but decreased with increasing ferric ion concentrations. Potassium jarosite incorporated significantly more of a given divalent base metal than either sodium or ammonium jarosite.

Trivalent Al, In or Ga all substituted for Fe^{3+} in the jarosite structure, and complete solid solution series were observed in every instance. Based on relative partitioning data, the order of incorporation of trivalent elements into the alkali jarosites was:

Table 11. Possible Substitutions for SO_4 in Jarosite-Type Compounds

Ionic Species	Extent of Precipitation	Product Characterization
TeO_4^{2-} (0.2 M)	Precipitates with Fe^{3+}	Poorly crystallized $Fe_2(TeO_4)_3 \cdot H_2O$
SeO_3^{2-} (0.2 M)	Precipitates with Fe^{3+}	Amorphous ferric selenite
TeO_3^{2-} (0.2 M)	Precipitates with Fe^{3+}	Amorphous ferric tellurite
MnO_4^- (>0,05 M)	Precipitates with Fe^{3+}	Poorly crystalline $(Mn,Fe)O_2$
(<0.03 M)	Precipitates with jarosite	Present as poorly crystalline $(Mn,Fe)O_2$
SiO_2 (>0.05 M)	Forms silica gel at 97°C	Amorphous phase, SiO_2
(<0.03 M)	Does not precipitate	-
BO_3^{2-} or $B_4O_7^{2-}$ (<0.2 M)	Does not precipitate	-
$Ge^{4\pm}$ (0.7-2.8 g/L)	Does not precipitate	-
SnO_3^{2-} (0.025-0.1 M)	Extensive precipitation except in strongly acid media	Poorly crystallized to amorphous $(Sn,Fe)O_2$
IO_4^- (0.2 M)	Extensive Precipitation	Poorly crystallized FeO.OH containing ~50% IO_4^-
MoO_4^{2-} (0.2 M)	Extensive Precipitation	$Fe_2(MoO_4)_3$

$$Fe^{3+} > In^{3+} > Ga^{3+} > Al^{3+}$$

The extent of substitution of these elements for iron was found to rise significantly with increasing pH. It was concluded that significant losses of these trivalent elements would occur in all industrial jarosite residues.

Although SeO_4^{2-} and CrO_4^{2-} both substitute completely for the SO_4^{2-} component of jarosite-type compounds, most XO_4 species do not appear to enter significantly into the jarosite lattice. Borate and germanium did not precipitate at all, and low concentrations (<0.03 M) of silicate behaved similarly. Tellurate (TeO_4^{2-}), tellurite (TeO_3^{2-}), selenite (SeO_3^{2-}), periodate (IO_4^-) and molybdate (MoO_4^{2-}) all were readily precipitated as an iron compound, but these compounds were clearly non-jarosite-type. Permanganate (MnO_4^-), stannate (SnO_3^{2-}) and high concentrations of silicate (>0.05 M SiO_2) all precipitate as an amorphous or poorly crystallized oxide. Although these species do not extensively enter into the jarosite structure, they may nevertheless be rejected, in other forms, during jarosite precipitation.

ACKNOWLEDGEMENTS

This program could not have been carried out without the assistance of D. J. Hardy and O. Dinardo who helped to prepare the jarosites and P. E. Belanger who did the X-ray characterizations of the products. Discussions with Dr. J. L. Jambor, CANMET, were invaluable in the interpretation of parts of the work.

REFERENCES

1. J. E. Dutrizac: in Lead-Zinc-Tin '80, Cigan, J. M., Mackey, T. S. and O'Keefe, T. J., editors, AIME, New York, 1979, pp. 532-564.
2. A. S. Yaroslavtsev, L. S. Getskin, A. U. Usenov, and E. V. Margulis: Tsvetn. Metall., 1975, vol. 16, no. 4, pp. 41-42.
3. H. U. A. Saarinen: "Separating Nickel, Cobalt and Chromium from Iron in Metallurgical Products"; U.S. Patent 4,042,474, August 16, 1977.
4. G. P. Brophy, E. S. Scott, and R. A. Snellgrove: Am. Mineral., 1962, vol. 47, pp. 112-126.
5. H. Hoeller: Contrib. Mineral. Petrol. Berlin, 1967, vol. 15, no. 4, pp. 309-329.

6. L. V. Mar'yanchik, A. K. Zapol'skii, and V. S. Sazhin: Zh. Prikl. Khim., 1973, vol. 46, no. 4, pp. 918-920.
7. I. V. Tananaev, N. K. Bol'shakova, and T. I. Zazakova: Russ. J. Inorg. Chem., 1967, vol. 12, no. 2, pp. 185-188.
8. I. V. Tananaev, V. G. Kuznetsov and N. K. Bol'shakova: Russ. J. Inorg. Chem., 1967, vol. 12, no. 1, pp. 28-30.
9. G. Johansson: Arkiv Kemi, 1963, vol. 20, no. 28, pp. 343-352.
10. A. M. Bakaev, M. V. Kravets, and E. V. Margulis: Izv. Vyssh. Uchebn. Zaved., Tsvetn. Metall., 1976, vol. 76, no. 5, pp. 149-150.
11. J. E. Dutrizac, O. Dinardo, and S. Kaiman: Hydrometallurgy, 1981, vol. 6, pp. 327-337.
12. A. E. Fersman and O. M. Shubnikova: Geochemist's and Mineralogist's Companion, Lomonosov Inst. Acad. Sci., Moscow, 415 pp (1937).
13. B. K. Breshenkov: Compt. Rend. (Doklady) Acad. Sci. USSR, 1946, vol. 52, pp. 329-332.
14. A. Bounin and A. Lecerf: Comp. Rend. Ser. C., 1966, vol. 262, pp. 1782-1784.
15. D. A. Powers, G. R. Rossman, H. J. Schugar, and H. B. Gray: J. Solid State Chem., 1975, vol. 13, pp. 1-13.
16. M. F. Kulikova and N. I. Razenkova: Kora Vyvetrivaniya, 1970, vol. 11, pp. 241-245.
17. J. E. Dutrizac and S. Kaiman: Canadian Mineralogist, 1976, vol. 14, pp. 151-158.
18. J. E. Dutrizac, O. Dinardo, and S. Kaiman: Hydrometallurgy, 1980, vol. 5, pp. 305-324.
19. J. M. Steward, G. Kruger, H. L. Omman, C. Dickinson and S. R. Hall: The X-ray System of Crystallographic Programs, Univ. Maryland Computer Sci. Center, Tech. Rep. TR-192 (1972).
20. J. Aird, R. S. Celmer, and A. V. May: Proceedings, Cobalt '80, 10th Annual Hydromet. Meeting of CIM, Edmonton, Alberta (1980).
21. J. L. Jambor, and J. E. Dutrizac: Canadian Mineralogist, 1983, vol. 21, pp. 101-113.
22. G. Giuseppetti and C. Tadini: Neues Jahrb. Mineral. Monatash., 1980, pp. 401-407.
23. G. P. Brophy, E. S. Scott, and R. A. Snellgrove: Amer. Mineral., 1962, vol. 47, pp. 112-126.

COUPLED RATE CONTROL IN THE DISSOLUTION OF ELECTRON CONDUCTING MINERALS

Milton E. Wadsworth and Ting-Ke Zhong*

Department of Metallurgical Engineering
University of Utah
Salt Lake City, Utah

Coupled dissolution of metals, metal sulfides, and certain electron conducting oxides require an oxidant and in some cases both an oxidant and a complexing agent. Oxidants may be oxygen, ferric ion, peroxide or some other suitable oxidizing reagent. Typical complexing agents are cyanide ion in the case of gold, silver, and copper, ammonia/ammonium ion in the case of copper, and carbonate in the case of UO_2. Experimental methods and results obtained for the initial dissolution of chalcocite in acid ferric sulfate solutions are presented. The separate anodic and cathodic processes are quantified using an electroless configuration with graphite as an inert cathode. The method described may be used to separate electrode processes involved in the leaching of a variety of electron conducting minerals.

INTRODUCTION

The leaching of electron conducting solids in many instances is electrochemical in behavior and follows well established principles of corrosion or galvanic coupling. The response of the reaction to voltage explains the influence of locked sulfide phases, the sequential formation of intermediate phases, and unusual reaction orders observed experimentally. The dissolving particle may form a

*Present affiliation: Central South Institute of Mining and Metallurgy, Changsha, China.

mixed potential which is a unique potential where anodic and cathodic currents are equal. Leaching of several metals is influenced by rates of diffusion of both oxidant and complexing agent. In these cases the exchange current density is very large and the coupled diffusion of oxidant and complexing agent is rate determining.[1]

In this study coupled electrodes, chalcocite anode in contact with a graphite cathode in the presence of ferric sulfate, have been used to separate the various electrode processes. Investigations of the leaching and electrochemical behavior of chalcocite in acidic ferric sulfate and ammoniacal solutions have been presented by many researchers. In 1930, Sullivan[2] identified two stages of leaching of chalcocite in ferric ion solution as follows:

$$Cu_2S + 2\ Fe^{2+} = Cu^{2+} + 2\ Fe^{2+} \qquad (1)$$

and

$$CuS + 2\ Fe^{3+} = Cu^{2+} + S^0 + 2\ Fe^{2+} \qquad (2)$$

Thomas et al.[3] observed the reaction to proceed in ferric sulfate solutions according to the sequence in the first stage $Cu_2S \rightarrow Cu_{1.8}S$ (digenite) $\rightarrow Cu_{1.1}S$ (blaubleibender). Marcantonio[4,5] observed a similar sequence and developed a coupled ferrous diffusion surface discharge model to explain the results. Both studies proposed the initial kinetics involved transport control of ferric ions to the chalcocite mineral surface. Using electrochemical methods, Brennet et al.[6] observed the same sequence for anodic dissolution as Thomas et al.[3] and Marcantonio[4] observed for ferric sulfate leaching. Earlier Cavallotti and Salvago[7] observed the sequence $Cu_2S \rightarrow Cu_{1.97}S \rightarrow Cu_{1.75}S \rightarrow Cu_{1.1}S$ and more recently Koch and McIntyre[8] using a thin electrode mass balance method proposed the sequence $Cu_2S \rightarrow Cu_{1.96}S \rightarrow Cu_{1.81}S \rightarrow Cu_{1.67} \rightarrow Cu_{1.4}S \rightarrow Cu_{1.2}S$.

The purpose of this study was to define in greater detail the initial kinetic mechanism of chalcocite leaching in ferric sulfate solutions using electrochemical methods. This research departs from previous studies in that the anodic and cathodic partial reactions have been physically separated, but galvanically connected in diaphragm electrolytic cells, (D-cells), permitting the direct measurement of current under the driving force of the electrode reactions themselves. Graphite was selected as the cathode in this initial study because it could be obtained in high purity and remained inert. In addition to providing information relative to individual electrode reactions, the coupling of electrode reactions may conceivably have some future application. Such applications may include energy sources and as devices to separate interfering chemical reactions.

EXPERIMENTAL

Method

Chalcocite used in the study was synthesized using sublimed sulfur and reagent grade copper. The chalcocite particles were compacted with a cylindrical stainless steel die. Pellets were 1.31 cm in diameter and 1.348 cm^2 in surface area. Some of the experiments were performed with the full-sized disks; however, most tests were carried out using smaller chalcocite electrodes (0.0755 cm^2) prepared from the above pellets by cutting, grinding, and polishing. The small chalcocite electrodes were necessary to reduce the total current and associated IR voltage drop in the electrolytic cells. Purified graphite rods with the same surface area as the anodes were used as stationary and rotating disk cathodes. Solutions for all experiments were prepared from reagent grade chemicals and distilled, deionized water.

A schematic diagram of the experimental apparatus used is presented in Fig. 1. Two Chesapeake connectors and pulleys were used for rotating disk electrode experiments. For stationary electrodes, stirring was achieved by means of a magnetic stirrer. Experiments were carried out in a 0.5 liter electrolytic cell separated into two chambers by sintered glass diaphragms. Luggin capillaries, each with its own cooling waterjacket, were inserted into each electrode compartment as shown. Two saturated calomel standard electrodes (No. 4, Fig. 1) were inserted into the electrolyte in the body of the Luggins tube. The same electrolyte was used inside the Luggins res-

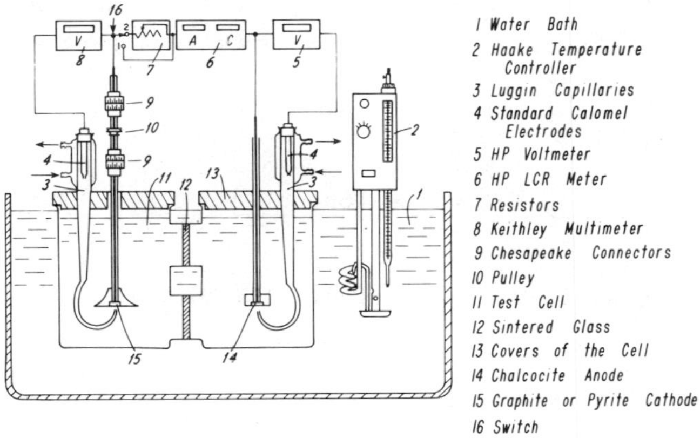

Fig. 1. Electrolytic cell used for coupled electrode measurements.

ervoir as in the cell compartment for all tests. The temperature of reference electrodes was held constant at 25°C. A Hewlett Packard 3480A voltmeter and a Keithley System 70, including 171 Digital Multimeter, 702 Scanner, and 750 Printer were used to measure the potentials. Measurements of the current and coulombs passed through the cell were carried out by means of a digitial coulumeter (Model 379, Princeton Applied Research). An external variable resistance in series was employed to fix the coupled cell current and the potentials of the anode and cathode were measured. The cell was submerged in a circulating water bath which was heated by a Haake temperature controller. For the rotating disk experiments the graphite cathode was driven with a GKH controller.

The anolyte normally contained 1 molar sulfuric acid, 0.1 molar copper sulfate and 0.5 molar magnesium sulfate as a supporting electrolyte. The catholyte contained 1 molar sulfuric acid and variable ratios of ferric and ferrous sulfate. The rest potentials of anode and cathode were measured before the cell was activated.

In order to determine initial mixed potentials, initial coupling currents, and polarization curves, two kinds of experiments were carried out. In one set of experiments the variation of initial mixed potentials and currents with electrolyte composition were obtained by direct short circuiting of the cell. Measurements were made within 30 seconds. In a second set of experiments, anodic and cathodic polarization curves were obtained by measuring the current and the anodic and cathodic potentials when the circuit of the cell was closed and the external resistance was decreased stepwise to preselected currents. All stepwise readings on a single sample were obtained within two minutes to minimize compositional changes in the anode. Mounted anode samples could be used by regrinding and polishing until the original rest potential was restored.

EXPERIMENTAL RESULTS

Effect of Anolyte and Catholyte Concentration on Observed Rest Potentials

The anodic dissolution of chalcocite may be expressed by the equations

$$\frac{1}{y-x} Cu_{2-x}S = \frac{1}{y-x} Cu_{2-y}S + Cu^{2+} + 2e^- \quad (3)$$

according to Hepel and Hepel.[9,10] Open circuit rest potentials of

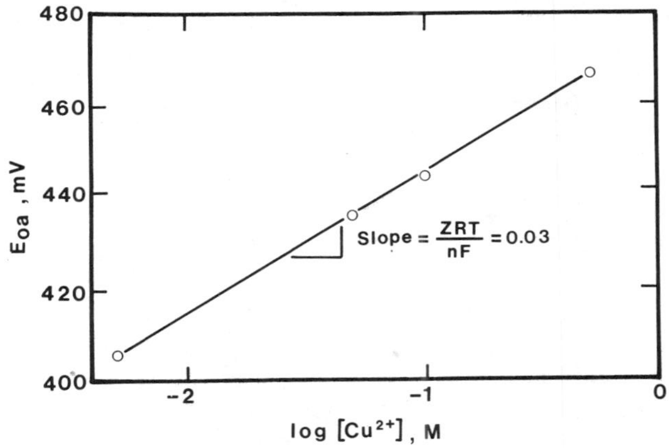

Fig. 2. Variation of anodic rest potential, E_{oa}, with copper concentration.

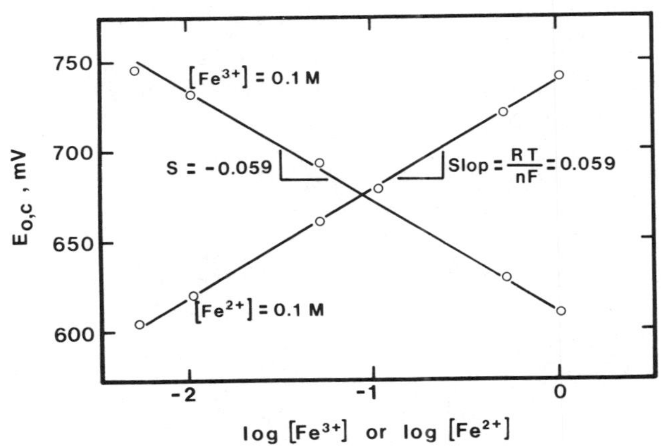

Fig. 3. Variation of cathodic rest potential, E_{oc}, with total Fe^{3+} and Fe^{2+} concentration.

the chalcocite samples used in this study are illustrated in Fig. 2 for various cupric ion concentrations. The observed values are well correlated for these solutions by the equations

$$E_{oa} = 0.476 + 0.031 \log [Cu^{2+}] \tag{4}$$

Figures 3 and 4 illustrate rest potentials for various ferrous, ferric sulfate solutions using a graphite electrode. Figure 3 shows

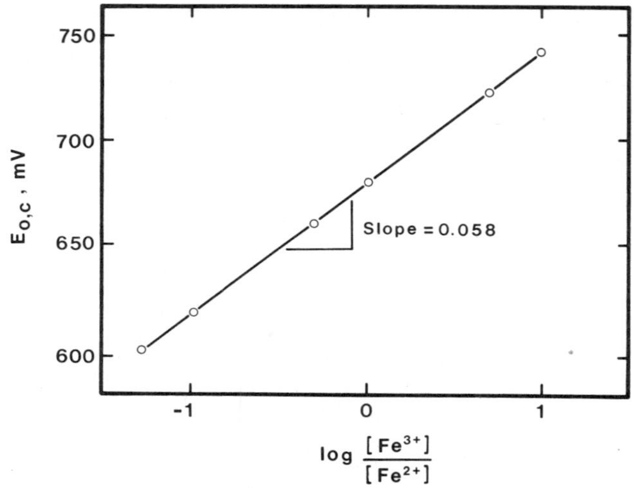

Fig. 4. Variation of cathodic rest potential, E_{oc}, with $[Fe^{3+}]/[Fe^{2+}]$ ratio.

results obtained alternately holding the ferrous and ferric sulfate concentrations constant at 0.1 M while the opposite concentration was varied. The data of Fig. 3 are plotted in Fig. 4 for the redox reaction

$$Fe^{3+} + e^- = Fe^{2+} \tag{5}$$

These results are well correlated for these solutions by the equation

$$E_{oc} = 0.680 + 0.058 \log [Fe^{3+}]/Fe^{2+}] \tag{6}$$

Mixed Potential and Mixed Current Measurements

Several experiments were carried out with the electrodes short circuited, during which time the initial mixed potential (E_m) and mixed current (i_m) were measured. Recorded values are initial values measured within 30 seconds. Initially both the catholyte and anolyte were stirred (magnetically). It was found that stirring the anolyte had no influence on the measured potentials or currents for coupled electrodes. Stirring of the catholyte had a significant and reproducible effect on both potential and current. In all subsequent figures, stirring refers to the catholyte only.

Figures 5, 6 and 7 illustrate the effect of ferric ion, ferrous ion, and cupric ion concentration on the mixed potential, E_m, and

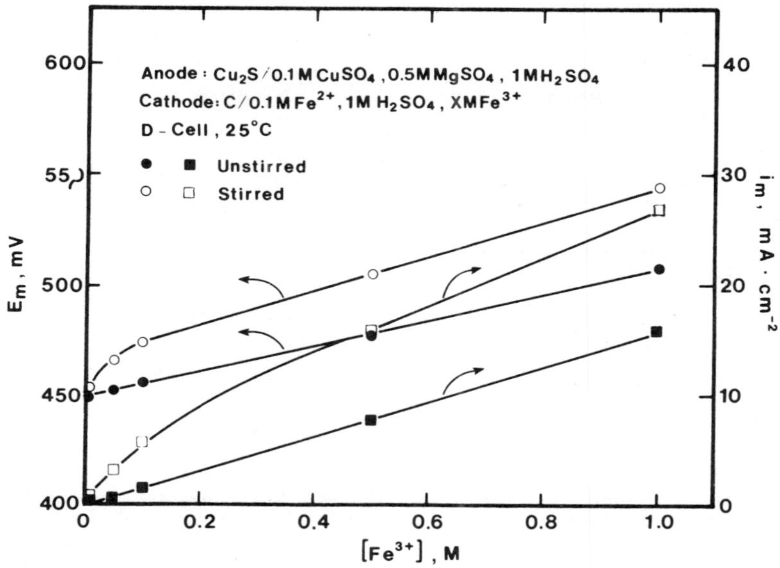

Fig. 5. Effect of $[Fe^{3+}]$ on mixed potential, E_m, and mixed current density, i_m.

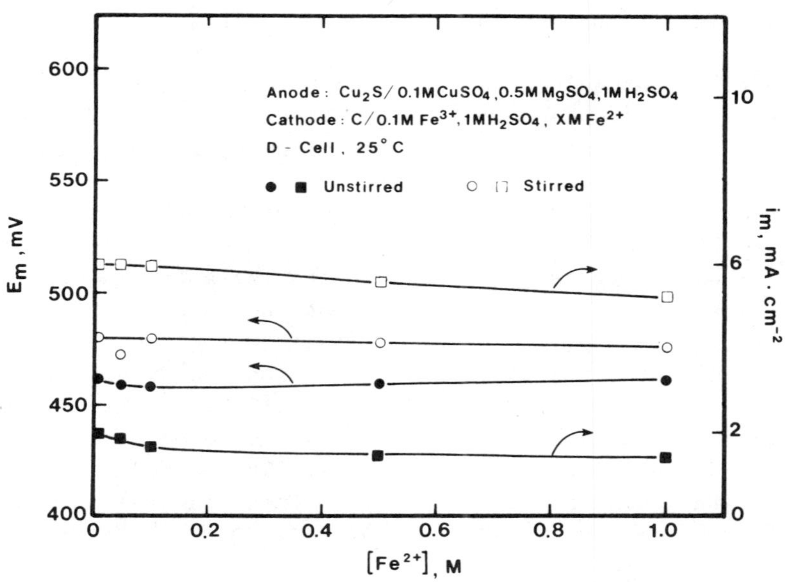

Fig. 6. Effect of $[Fe^{2+}]$ on E_m and i_m.

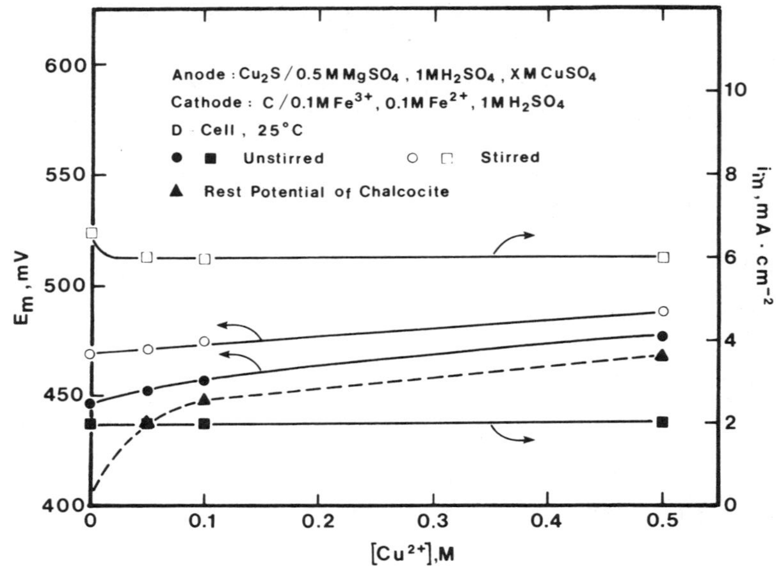

Fig. 7. Effect of $[Cu^{2+}]$ on E_m and i_m.

the mixed current density, i_m for stirred and unstirred catholyte. Increased ferric ion concentration results in an associated increase in both E_m and i_m. Variations in cupric ion concentration do not influence i_m, but a decrease in i_m is noted for increased ferrous ion concentration. Conversely, E_m is independent of ferrous ion concentration but increases slightly with cupric ion concentration.

Figure 8 illustrates the variation of E_m with time for various ferric ion concentrations. The first point on each curve corresponds to 30 seconds. Figure 9 illustrates the variation in current density with time for the same experiments. At the lower concentrations, $[Fe^{3+}]$ = 0 to 0.01 M, E_m values remain very constant and are close to the anodic rest potential E_{oa}. The anodic over voltage is less than 10 mv illustrating the high reversibility of the chalcocite electrode. The concentration of Cu^{2+} would be expected to affect the mixed potential as noted in Fig. 7. The cathodic overvoltage is very large such that back reactions involving Fe^{2+} are negligible thus affecting E_m and i_m very slightly as noted in Fig. 6. Also since the cathodic overvoltage is greater than the anodic over-voltage, the small change in E_m with $[Cu^{2+}]$ has a negligible effect upon i_m.

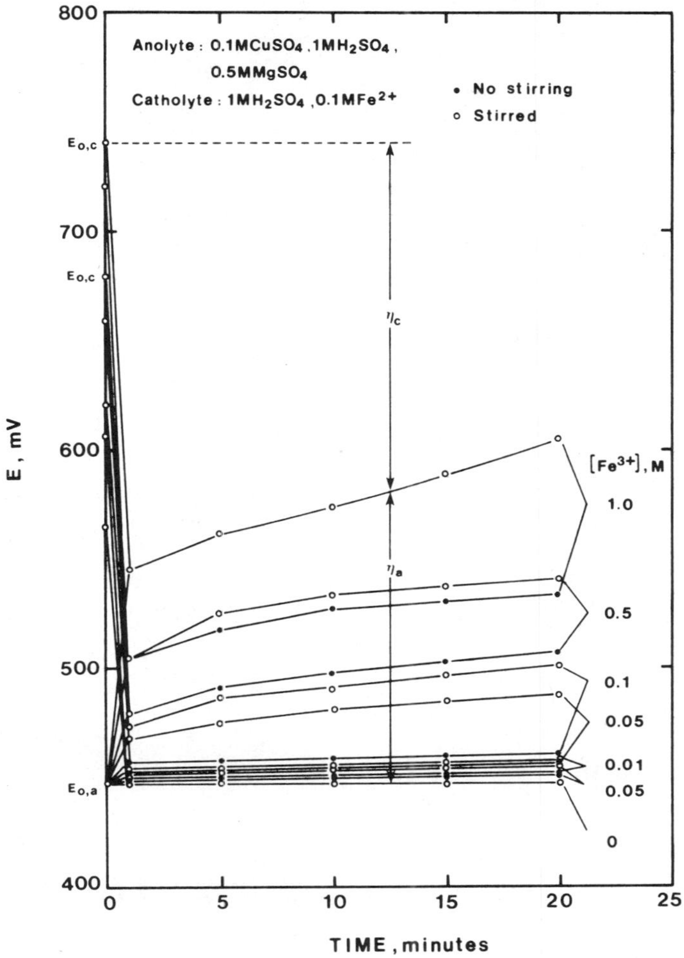

Fig. 8. Variation of mixed potential, E_m, with time for unstirred and stirred catholyte.

Of special concern in these studies is the rapid change in electrode composition that can occur as a result of voltage and total charge transferred. The total charge transferred in 30 minutes for the 0.005 and 0.01 molar Fe^{3+} solutions were 0.723 and 1.24 coulombs respectively. The total charge transferred for the 1.0 molar solution in 30 seconds was only 0.106 coulombs or approximately one-tenth that of the maximum observed for the lower concentrations. Undoubtedly as E_m increases various intermediate $Cu_{2-x}S$ phases are favored. The higher concentrations $[Fe^{3+}] \geq 0.5$ m Fe^{3+} are expected to form blue remaining covellite. This was

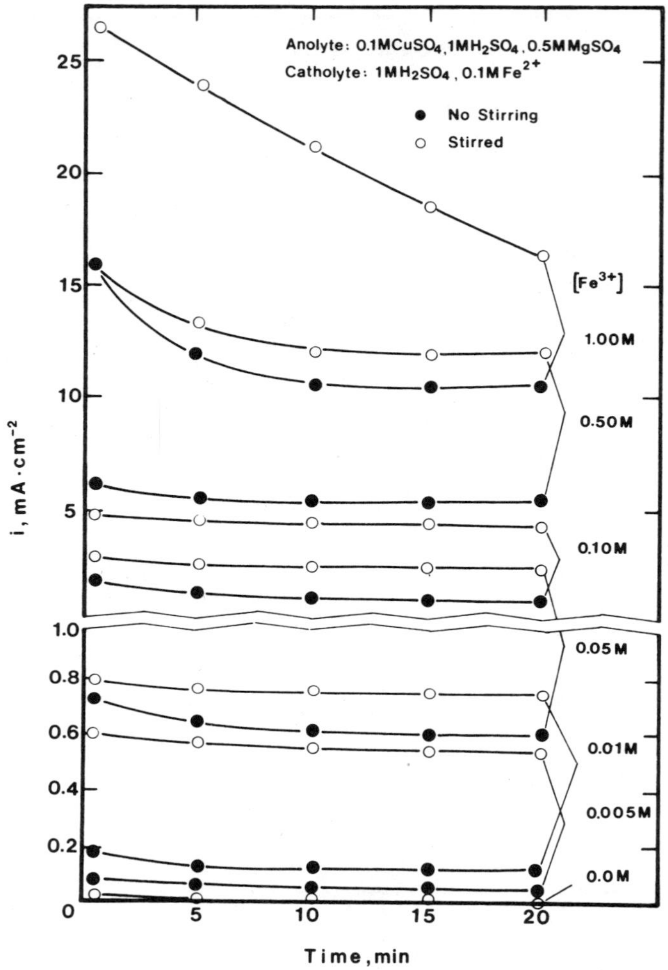

Fig. 9. Variation of mixed current density, i_m, with time for unstirred and stirred catholyte.

observed experimentally. At higher concentrations blue-remaining covellite was noted after 30 minutes. Open circuit rest potentials for $0 < [Fe^{3+}] < 0.1$ m returned to within 10 mv of the original E_{oa}, and covellite was never observed. For the two higher concentrations the open circuit potentials decreased approximately 30 to 40 mv indicating significant surface alteration. By using the 30 second points, reversibility of the anode was observed for all solution concentrations up to 0.10 molar Fe^{III}. Equation 3 remains highly reversible even for large variations in y; i.e. values approaching 0.25.

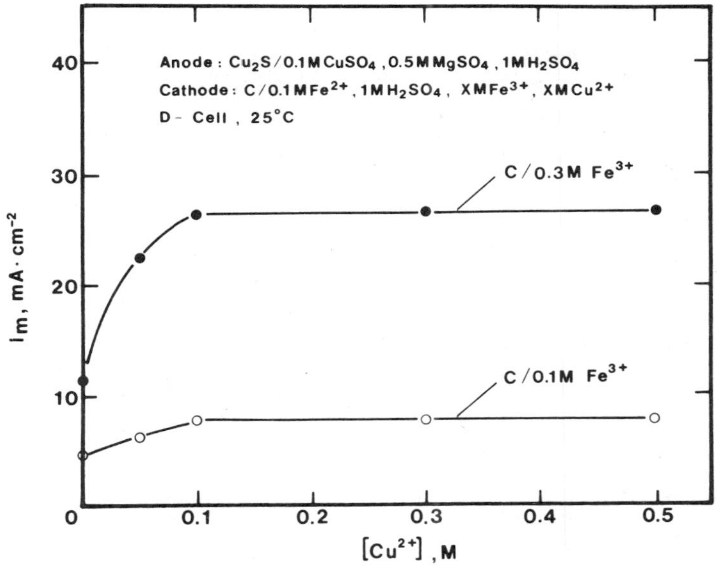

Fig. 10. Catalytic effect of cupric ion on i_m for 0.3 molar and 0.1 molar $[Fe^{3+}]$ solutions.

A separate series of experiments was carried out to see if cupric ion present in the catholyte influences the cathodic redox reactions. Figure 10 illustrates the influence of $[Cu^{2+}]$ in the catholyte on i_m. For fixed values of $[Fe^{3+}]$[0.1m and 0.3m] an enhanced initial effect was observed. Above 0.1m $[Cu^{2+}]$ increased concentrations did not increase the rate of reduction of Fe^{3+}. The catalytic effect of Cu^{2+} becomes more important for higher Fe^{3+} concentrations. Figure 11 illustrates the enhanced mixed current for constant values of Cu^{2+} in a stirred reactor for various ferric ion concentration. The mixed current density for stirred catholyte containing 0.1m Cu^{2+} is approximately double that observed in the absence of Cu^{2+}. Results observed for a solution containing 0.2m Cu^{2+} were identical to those observed for 0.1m Cu^{2+} as is indicated in Fig. 10.

The influence of electrode surface area on mixed current density is shown in Fig. 12 for various A_c:A_a ratios. The anodic area (Cu_2S) was maintained at 0.0755 cm^2 and the cathode (graphite) area was increased. The mixed current density showed a marked initial increase and then increased linearly with the A_c:A_a ratio.

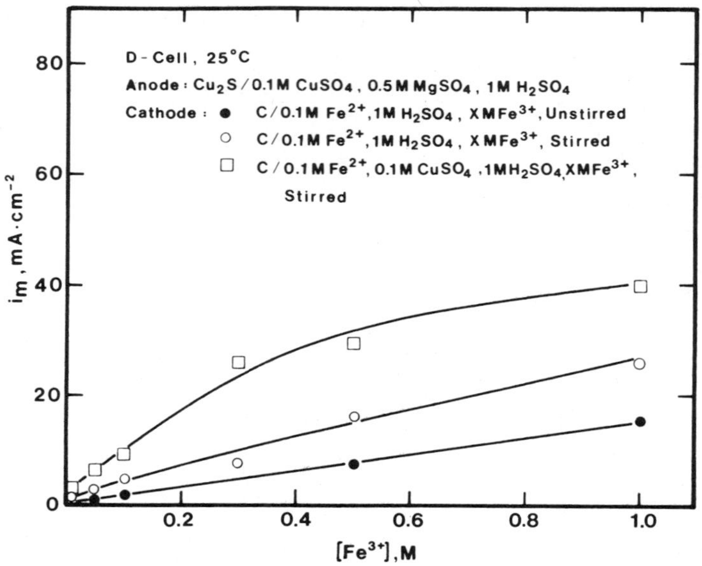

Fig. 11. Measured i_m for various Fe^{3+} concentrations with and without cupric ion for stirred catholyte.

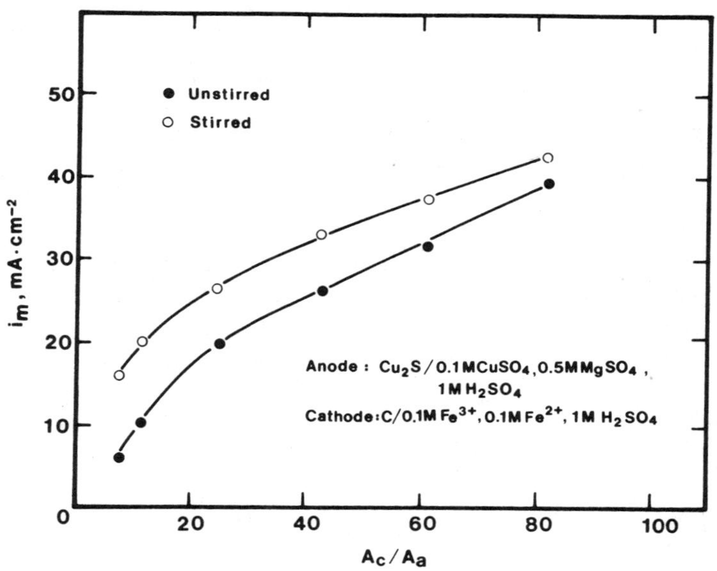

Fig. 12. Effect of electrode area showing i_m versus A_c/A_a.

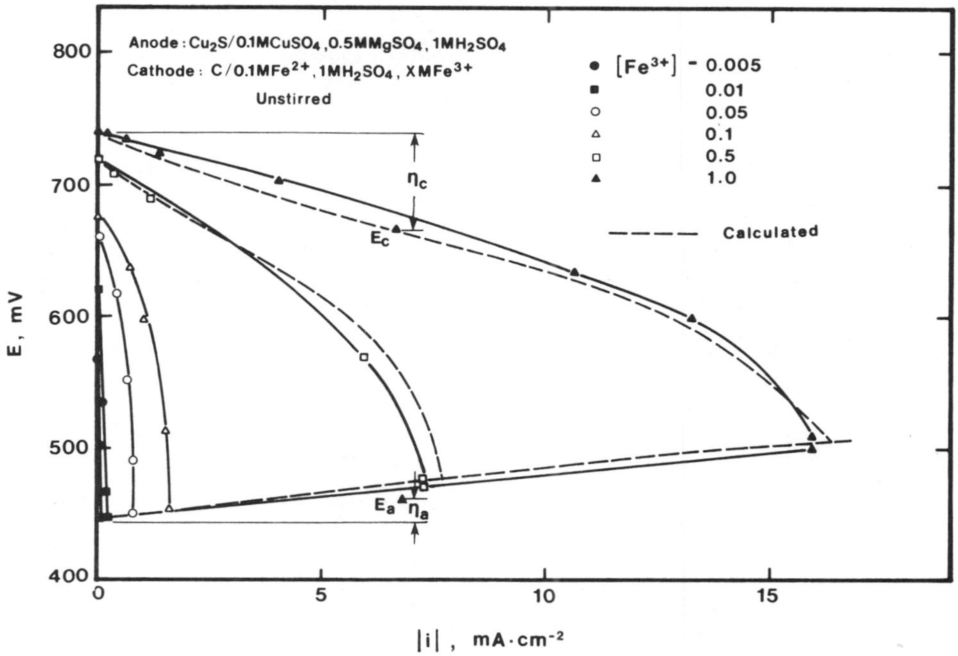

Fig. 13. Voltage-current values for coupled electrodes showing cathodic and anodic branches for unstirred catholyte and various $[\overline{Fe^{3+}}]$ concentrations.

Coupled Electrode Measurements with External Resistance

The shape of the anodic and cathodic polarization curves observed for coupled electrodes provides a means for evaluating the electrode processes separately. This was achieved by inserting progressively lower fixed external resistances.

Figure 13 illustrates results obtained for the D-cell for various ferric ion concentrations in unstirred catholyte. The $\Delta E/i$ ratio between the anodic and cathodic curves for each i value corresponds to the external resistance. The anodic branch is linear with relatively small anodic overvoltages for all values of i. Clearly the cathodic Fe^{III}/Fe^{II} processes are the predominant rate controlling kinetics processes in the coupled system. The zero current (rest potentials) are indicated for $i = 0$, and the mixed potential, E_m, and mixed current density, i_m, correspond to the point of intersection of the anodic and cathodic branches. For the short circuited

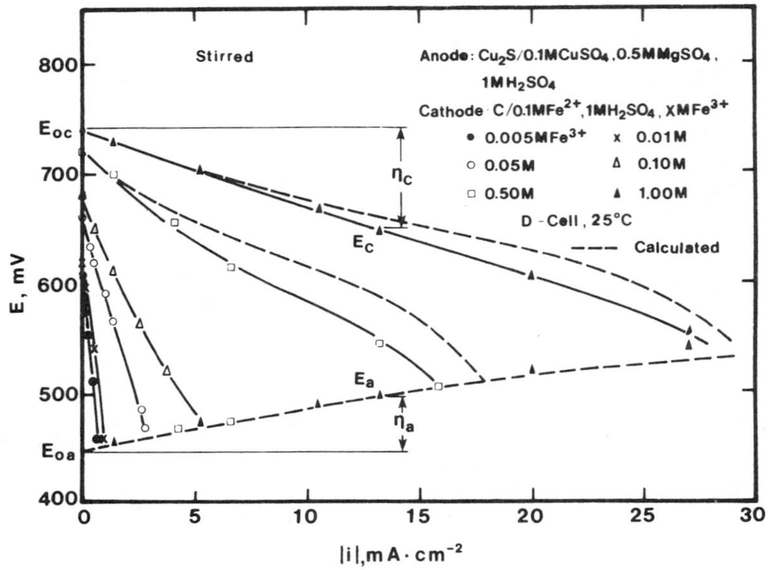

Fig. 14. Voltage-current values for coupled electrodes showing cathodic and anodic branches for magnetically stirred catholyte and various $[\overline{Fe^{3+}}]$ concentrations.

system, corresponding to the E_m and i_m intersection points, the cathodic overvoltage is always much greater than the anodic overvoltage. The shape of the anodic and cathodic branches provides a means for determining the separate electrode kinetic processes involved.

Figure 14 illustrates results obtained for the D-cell with stirred catholyte. The effect of stirring greatly influences the cathodic Fe^{III}/Fe^{II} branch. The cathodic curves are now linear and both E_m and i_m have increased greatly as noted earlier.

Because of the sensitivity of the cathode to stirring, a series of experiments was carried out using a rotating graphite cathode with the D-cell. Figure 15 illustrates the anodic and cathodic branches observed for 10 sets of experiments with rotating speeds varied from zero to 3500 rpm (367 rad sec^{-1}). In all tests the ferric ion concentration was maintained at 0.1 molar. The cathodic branches move from concave upward at low rotational speeds to convex upward at large rotational speeds. As before, the coupled electrodes display

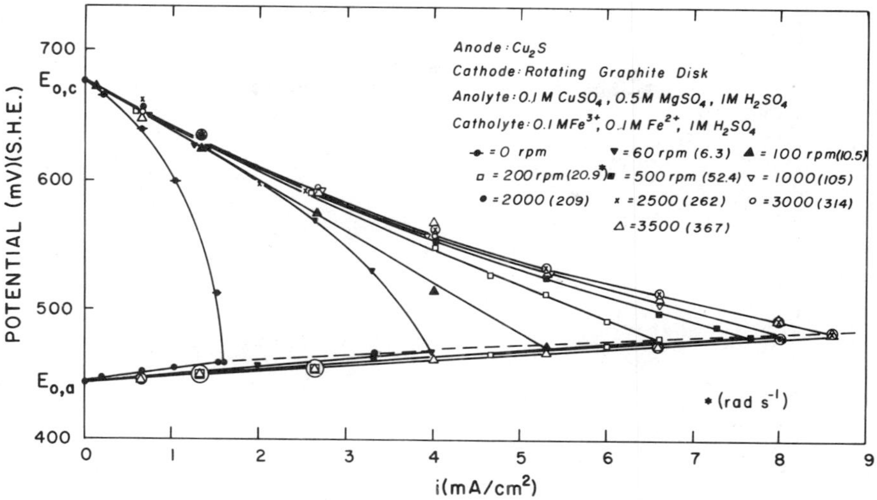

Fig. 15. Voltage-current values for coupled electrodes showing cathodic and anodic branches for a rotating disk cathode at various rotational speeds.

much larger cathodic overvoltages than anodic overvoltages. It is also apparent that the mixed current density is sensitive to rotational speed up to 500 rpm (52.4 rad sec^{-1}) and insensitive at higher rotational speeds. Figure 16 presents results of rotating disk experiments carried out at a fixed stirring speed of 2500 rpm (262 rad sec^{-1}) and with variable ferric ion concentration. Comparing these results with Fig. 14 shows that the mixed current density values measured for the rotating disk are approximately twice the values observed for the magnetically stirred catholyte. Also at the higher current densities the failure of the curves to close at zero external resistance is due to the internal ohmic resistance of the diaphragm. This was confirmed by direct measurements.

DISCUSSION

The cathodic overvoltage, η_c, for a given current density is given by

$$\eta_c = E_c - E_{oc} \tag{7}$$

where E_{oc} is the rest potential for a given catholyte and E_c is the voltage at current density i, Fig. 14. The anodic overvoltage, η_a,

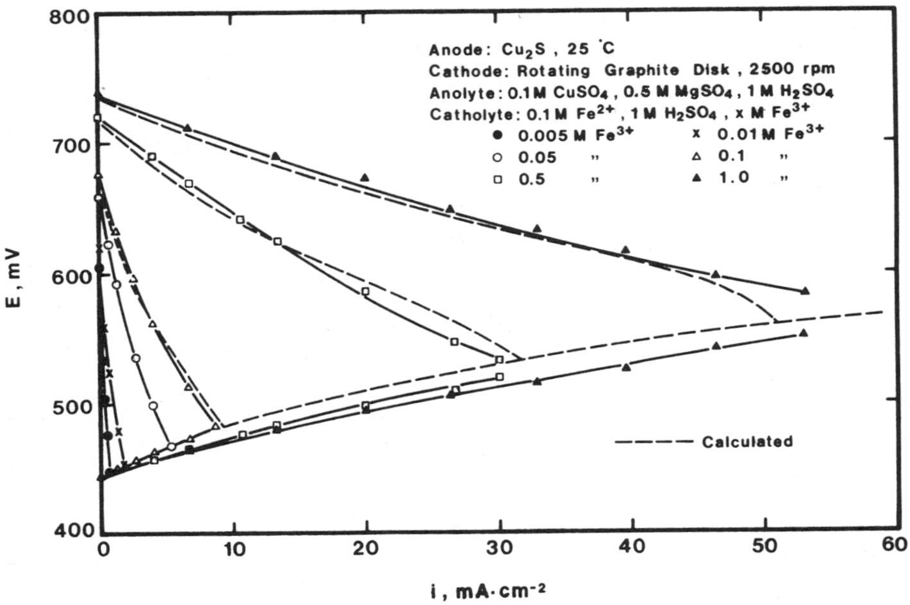

Fig. 16. Voltage-current values for coupled electrodes showing cathodic and anodic branches for a rotating disk cathode and various $[\overline{Fe^{3+}}]$ concentrations.

given by

$$\eta_a = E_a - E_{oa} \tag{8}$$

for the same current density, as shown in Figs. 13 and 14. It is apparent that the cathodic overvoltage is negative and the anodic overvoltage positive. Actually, for coupled electrodes,

$$i_a A_a = - i_c A_c \tag{9}$$

but in these studies, unless otherwise specified $A_a = A_c$.

The anodic branch for initial anodic dissolution may be expressed by the reaction

$$i_a = z_a \mathcal{F} k_a \exp[\beta_a \mathcal{F} E_a/RT] - [Cu^{2+}] z_a \mathcal{F} k'_a \exp[-(1-\beta_a)\mathcal{F} E_a/RT] \tag{10}$$

where k_a and k'_a are the specific rate constants for zero potential and k_a contains the activity coefficient for cupric ion. The anodic transfer coefficient is β_a. This equation corresponds to reaction (3) for x = 0, z_a = 2. Also Eq. 10 assumes a single charge transfer process. For a two charge process, made up of two sequential single charge transfer steps, of which only one step is rate limiting, it may be shown that[11]

$$i_a = 2i_o \exp(\beta_a F\eta_a/RT)[1-\exp(-2/F\eta_a/RT)] \qquad (11)$$

where i_o is the exchange current density for the slow step and is much less than the exchange current density for the other single electron step; also

$$i_o = Fk_a \exp(\beta_a F E_{oa}/RT) \qquad (12)$$

Assuming β_a = 0.5 calculated i_o values for the anodic branches of Figs. 13-16 were evaluated. These values for stirred, unstirred, rotating disk (variable rotation speed) and rotating disk (variable $[Fe^{3+}]$) were respectively 2.89×10^{-3}, 2.18×10^{-3}, 2.56×10^{-3}, and 3.51×10^{-3} amp cm^{-2}. Using an average value of 2.79×10^{-3} amp cm^{-2}, the dashed anodic curves shown in Figs. 13-16 were calculated.

It is apparent from the large cathodic overvoltages and the sensitivity of the cathodic reactions to solution agitation that diffusion of ferric ions is involved in the cathodic rate controlling reaction. The cathodic branches observed experimentally are well described by a steady-state kinetic model in which surface concentrations of ferrous and ferric ions are controlled by combined diffusion and surface charge transfer reactions. The consecutive steady-state processes may be represented by the equations

$$-i_c = z_c F \frac{D_3}{\delta 10^3} [\overline{Fe^{3+}}] - z_c F \frac{D_3}{\delta 10^3} [Fe^{3+}] \qquad (13)$$

$$-i_c = z_c F[Fe^{3+}]k_c \exp[-(1-\beta_c)FE_c/RT] - z_c F[Fe^{2+}]k_c' \exp[\beta_c F E_c/RT] \qquad (14)$$

$$-i_c = z_c F\frac{D_2}{\delta 10^3} [Fe^{2+}] - z_c F\frac{D_2}{\delta 10^3} [\overline{Fe^{2+}}] \qquad (15)$$

where i_c = cathodic current, and is negative
z_c = moles of charge transferred/mole of Fe^{3+}
D_3 = coefficient of diffusion for ferric ions
D_2 = coefficient of diffusion for ferrous ions
k_c and k_c' = rate constants
δ = effective thickness of the diffusion boundary

$\overline{[Fe^{3+}]}$, $\overline{[Fe^{2+}]}$ = bulk solution concentration, $m\ell^{-1}$
$[Fe^{3+}]$, $[Fe^{2+}]$ = surface solution concentration, $m\ell^{-1}$
β_c = cathodic charge transfer coefficient
and E_c = cathodic voltage when current is flowing.

In the sulfate system, ferric ion complexes such as $FeSO_4^+$ and $Fe(SO_4)_2^-$ form and generally are present in higher concentrations than uncomplexed Fe^{3+}. The concentrations $\overline{[Fe^{3+}]}$ used here actually refer to the total ferric ion in solution, $[Fe^{III}]$, since all complexes diffuse to the surface. Accordingly D_3 is an average duffusivity for Fe^{III}. The constants k_c and k_c' contain activity coefficients and equilibrium constants relating the total ferric ion concentration to uncomplexed Fe^{3+} and Fe^{2+}, and account for the deviation of E_{oc}, 298°K (25°C), 0.680 volts, Eq. 20, compared to the standard half cell potential of 0.771 volts.

For the cathodic reaction, $z_c = 1$. At steady-state the three rates of reaction (Eqs. 13, 14, and 15) are equal, resulting in the general rate expression

$$i_c = \frac{(-\mathcal{F}D_3\overline{[Fe^{3+}]}/\delta 10^3)[1 - (k_c'\overline{[Fe^{2+}]}/k_c\overline{[Fe^{3+}]})\exp(\mathcal{F}E_c/RT)]}{1+(D_3/k_c\delta 10^3)\exp[(1-\beta_c)\mathcal{F}E_c/RT + (D_3k_c'/D_2k_c)\exp(\mathcal{F}E_c/RT)} \quad (16)$$

The limiting current density for ferric ions, $i_{d,3}$, is negative,

$$i_{d,3} = -\mathcal{F}\overline{[Fe^{3+}]}D_3/\delta 10^3 = i^o_{d,3}\,[Fe^{3+}] \quad (17)$$

and the limiting current density for ferrous ions, $i_{d,2}$, is positive

$$i_{d,2} = \mathcal{F}D_2\overline{[Fe^{2+}]}/\delta 10^3 \quad (18)$$

Combining Eqs. 16-20 and assuming $\beta_c = 1/2$, results in the general expression for the cathodic current

$$i_c = \frac{|i^o_{d,3}|\overline{[Fe^{3+}]}[1 - \exp(\mathcal{F}\eta_c/RT)]}{1 + (|i^o_{d,3}|10^3/k_c\mathcal{F})\exp(\mathcal{F}E_c/2RT) + (|i_{d,3}|/i_{d,2})\exp(\mathcal{F}\eta_c/RT)} \quad (19)$$

where $|i^o_{d,3}| = \mathcal{F}D_3/\delta 10^3$ (20)

and $|i_{d,3}|/i_{d,2} = [Fe^{3+}]\,D_3/[Fe^{2+}]\,D_2$ (21)

Table 1. Calculated $i_{d,3}^{o}$ and k_c Values. (μ = ionic strength of catholyte and k_c = moles $cm^{-2}sec^{-1}$)

			Unstirred		Stirred		Rotating Disk	
$[Fe^{3+}]$	E_{oc}	μ	$i_{d,3}^{o}x10^3$	k_c	$i_{d,3}^{o}x10^3$	k_c	$i_{d,3}^{o}x10^3$	k_c
0.005	0.608	1.230	16.4	34.3	125	51.4	202	26.6
0.010	0.621	1.234	17.3	53.0	110	42.1	257	28.7
0.050	0.661	1.268	16.5	31.7	72.4	27.9	165	28.7
0.100	0.679	1.302	16.3	46.9	57.1	31.3	140	62.5
0.500	0.721	1.573	15.3	37.8	40.2	32.5	81.2	75.8
1.00	0.740	1.748	16.6	93.0	34.0	71.4	77.7	156.0

From i_c, η_c and E_c values for each of the cathodic branches in Figs. 13, 14 and 16 and assuming $D_3/D_2 \cong 1$, average $i_{d,3}^{o}$ and k_c values were determined. These values are given in Table 1. It is apparent that $i_{d,3}^{o}$ values vary for both the stirred and rotating disk experiments, but is essentially constant for the unstirred catholyte experiments. Figure 17 illustrates the variation of $i_{d,3}^{o}$ with $[\overline{Fe^{3+}}]$. The very broad variation in $[\overline{Fe^{3+}}]$ undoubtedly affects the hydrodynamics and this is apparent for the stirred electrolyte and rotating disk experiments. The variation in k_c is expected since k_c includes the equilibrium constants relating the activities of the various ferric sulfate complexes and free ferric ion.

The solid lines of Fig. 17 represent linear regression results for the data. The resulting equation are:

Rotating disk

$$|i_{d,3}^{o}| = 7.08 \times 10^{-2} [\overline{Fe^{3+}}]^{-0.26} \qquad (22)$$

Stirred

$$|i_{d,3}^{o}| = 3.31 \times 10^{-2} [\overline{Fe^{3+}}]^{-0.25} \qquad (23)$$

Unstirred

$$|i_{d,3}^{o}| = 1.62 \times 10^{-2} \qquad (24)$$

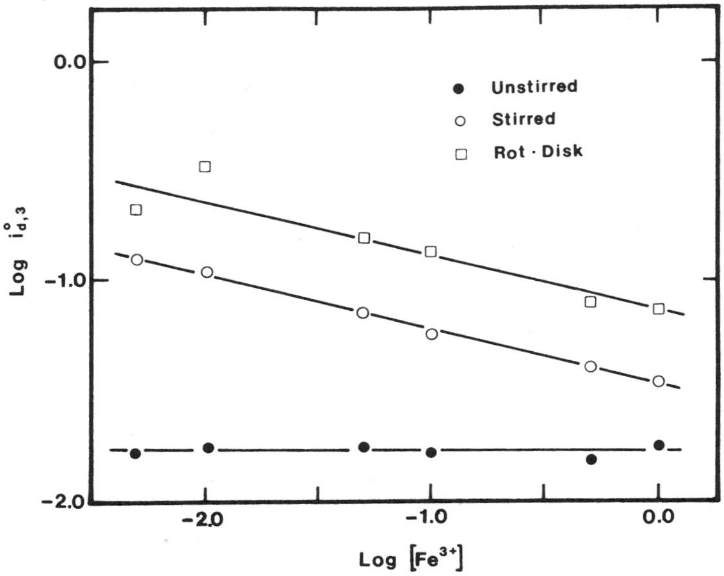

Fig. 17. Effect of [Fe^{3+}] on $i^o_{d,3}$ for unstirred and stirred catholyte and rotating disk cathode.

The ionic strengths, μ, calculated for the various catholyte concentrations are presented in Table 1. A linear regression fit of log k_c versus $\sqrt{\mu}$ values gives

$$\log k_c = -2.07 + 3.21 \sqrt{\mu} \qquad (25)$$

Using Eqs. 6, 22, 23, 24, and 25 it is possible to solve Eq. 19 for all of the cathodic curves at the various [$\overline{Fe^{3+}}$] concentrations. The dashed curves of Fig. 13 represent the calculated cathodic branches for unstirred catholyte for the two highest concentrations. Figure 18 illustrates the experimental and calculated results for the four lower concentrations. Figure 14 illustrates calculated curves for the two highest concentrations for magnetically stirred catholyte and Fig. 19 illustrates calculated curves for the four lower concentrations. Figure 16 illustrates the calculated curves for the three highest concentrations for the rotating disk experiments and Fig. 20 illustrates calculated curves for the three lower concentrations.

Equation 19 adequately explains the observed variations in E_m and i_m for various [$\overline{Fe^{3+}}$], [$\overline{Fe^{2+}}$], and [$\overline{Cu^{2+}}$] concentrations illustrated in Figs. 5, 6, and 7. Also the catalytic effect of Cu^{2+}

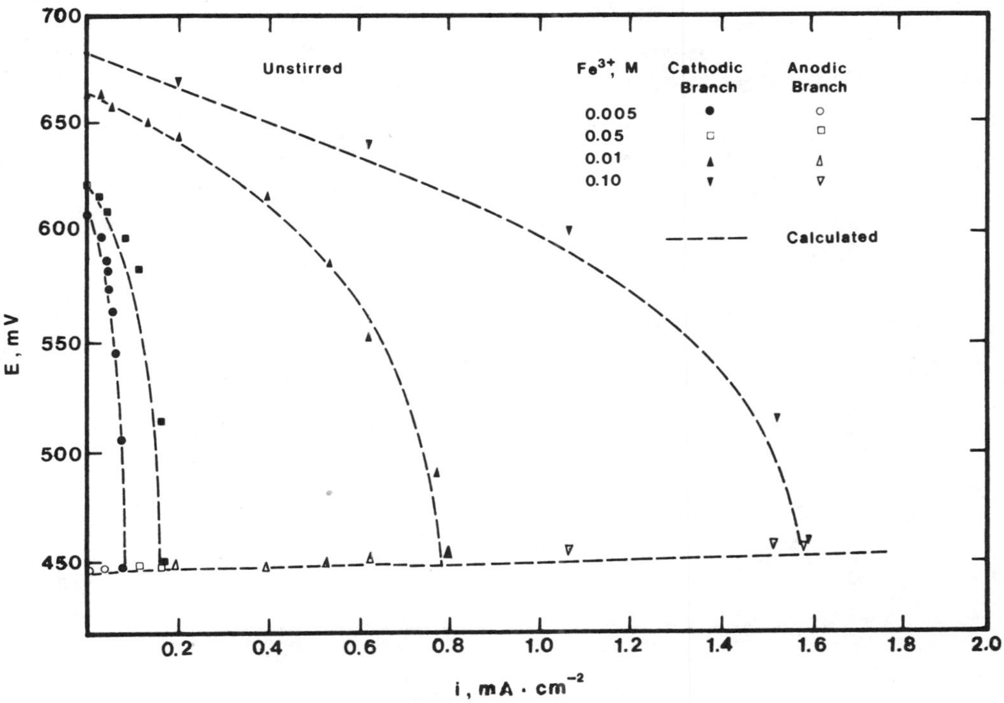

Fig. 18. Calculated cathodic and anodic branches for unstirred catholyte and various $[\overline{Fe^{3+}}]$ values.

on the rate of charge transfer for the reduction of Fe^{3+} is as expected according to Eq. 19, Figs, 10, 11. The value of k_c is expected to increase with $[Cu^{2+}]$ resulting in lower values of the second term in the denominator. Accordingly the current would increase until the second terms become small compared to unity. The current would then become independent of cupric ion concentration as observed experimentally.

CONCLUSIONS

Coupled electrode reactions may be followed separately using an electroless configuration with a diaphragm separating anolyte and catholyte. The initial kinetics for the anodic dissolution of chalcocite using a graphite-ferric sulfate couple indicates the following:

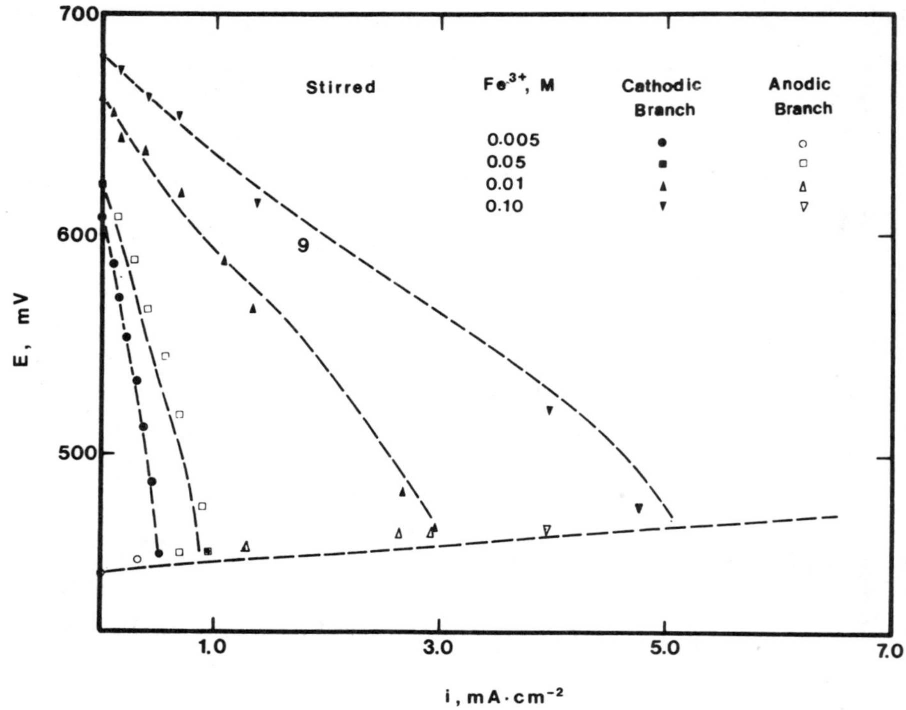

Fig. 19. Calculated cathodic and anodic branches for stirred catholyte and various [Fe^{3+}] values.

1. The anode, Cu_2S, has a high exchange current resulting in a mixed potential near the Cu_2S rest potential and is independent of degree of agitation.

2. The anode electrode reactions are well correlated by a mechanism involving two single electron transfer processes in sequence.

3. Ferric ion discharge at a coupled graphite electrode involves steady-state diffusion and surface discharge and is very sensitive to degree of agitation.

4. A single mixed kinetics, electrochemical equation may be used to correlate experimental results for unstirred, stirred, and rotating disk experiments for total Fe^{III} concentrations varying from 0.005 molar to 1.0 molar.

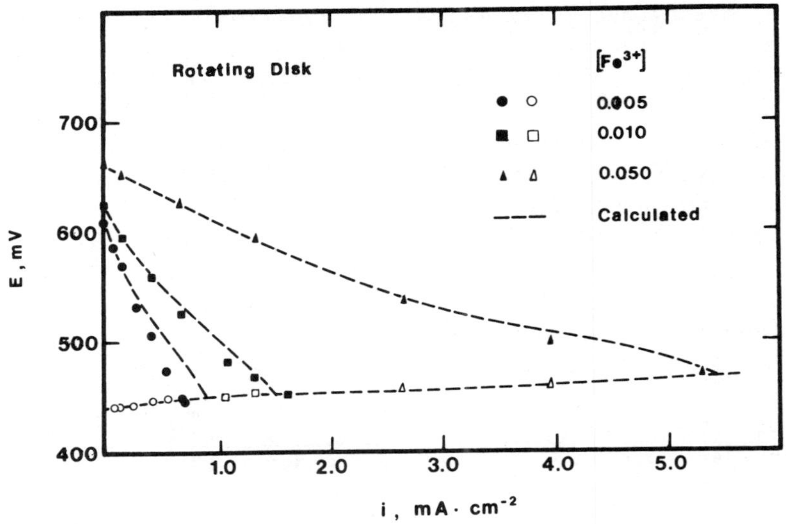

Fig. 20. Calculated cathodic and anodic branches for rotating disk cathode and various [Fe^{3+}] values.

Work at Utah is continuing using Cu_2S/FeS_2, Cu_2S/Cu_2S, $CuFeS_2/C$ and $CuFeS_2/FeS_2$ couples in both diaphragm and open cells for various temperature and concentration. These studies are providing useful information relative to the effects of galvanic coupling by locked mineral particles and on the separate electrode reactions involved.

ACKNOWLEDGEMENTS

The authors wish to thank the AMAX Foundation, Inc. which provided the financial support for this study.

REFERENCES

1. F. Habashi: Principles of Extractive Metallurgy, pp. 30-34, Gordon Breach, New York, 1970.
2. I. O. Sullivan: Chemistry of Leaching Chalcocite, U.S. Bureau of Mines, Tech. Paper 473, 1930.
3. G. Thomas, T. R. Ingraham and R. J. C. MacDonald: Can. Met. Quart., 1967, vol. 6, pp. 281-292.
4. P. J. Marcantonio: Ph.D. Thesis, Dept. of Mining, Metallurgical and Fuels Engineering, University of Utah, 1976.

5. J. B. Hiskey and M. E. Wadsworth: Electrochemical Processes in Leaching Metal Sulfides and Oxides, in: Process and Fundamental Considerations of Selected Hydrometallurgical Systems, ed. M. C. Kuhn, Soc. of Min. Engrs., AIME, New York, 303-325, 1981.
6. P. Brennet, S. Jafferali, J. Vanseveren, J. Vereecken and R. Winand: Study of the mechanism of Anodic Dissolution of Cu_2S, Met. Trans., vol. 5, pp. 127-134.
7. P. Cavallotti and G. Savalgo: Electrochim. Metal. IV, 3, 1969.
8. D. F. A. Koch and R. J. McIntyre: J. Electroanal. Chem., 1976, vol. 17, pp. 285-296.
9. M. Hepel: J. Electroanal. Chem., vol. 74, pp. 37-51.
10. M. Hepel and T. Hepel: J. Electroanal. Chem., vol. 81, pp. 161-170.
11. K. J. Vetter: Electrochemical Kinetics, Theoretical and Experimental Aspects, Academic Press, New York, 1967.

A PARTIAL EQUILIBRIUM MODEL BASED ON THE KINETICS OF DISSOLUTION AND EQUILIBRIUM IN SOLUTION OF THE UO_2-FeS_2-$Fe_2(SO_4)_3$-H_2SO_4 SYSTEM

Charles C. Allen and Renato G. Bautista*

Ames Laboratory USDOE, Mining and Mineral Resources
Institute and Department of Chemical Engineering
Iowa State University, Ames, IA 50010 USA

A mathematical model for determining the concentration changes in solution during the dissolution of a uranium dioxide-pyrite mixture has been developed using a $Fe_2(SO_4)_3$-H_2SO_4 lixiviant. When modeling such a leaching process, the two dissolution reactions are considered to be slow enough so that the reactions in solution can be assumed to be at continual equilibrium. This situation allows for the partial equilibrium approach to be utilized in modeling the overall leaching process.

The modeling process determines the equilibrium changes for all the species present in solution for every increment of solid dissolution, with the relative amount of uranium dioxide to pyrite reacting based on the experimentally-determined rates of reaction. The calculated concentration changes are obtained from simultaneously solving a series of linear equations arising from the mass action equations, a charge balance, and material balances for Fe(II), Fe(III), U(VI), and total SO_4^{2-}.

The $Fe_2(SO_4)_3$-H_2SO_4 solution was found to be an effective lixiviant provided leaching conditions were adjusted for the selective leaching of uranium dioxide. These conditions correspond to a solution with a low ferric:ferrous ratio, with the concentration of ferric ion limited by the solubility of $Fe(OH)_3$. This solubility limit is increased with increasing H_2SO_4 concentration.

*now located at Dept. of Chemical and Metallurgical Engineering
University of Nevada Reno, Reno, Nevada 89557

INTRODUCTION

There have been several major attempts to model the leaching of uranium ore under conditions applicable to in situ operations in the literature.[1-3] Langmuir[1] modeled the process utilizing a total equilibrium approach between the aqueous and solid phases. Grimes[2] analyzed the process based on a penetration model for the bacterial leaching of blasted ore deposits. Boomer[3] incorporated rate expressions for oxidant consumption and uranium production to model the leaching along streamlines within a deposit.

Since most of the uranium ore reserves in the United States are present in low grade sandstone deposits, in situ leaching has become an important method of extraction. This leaching operation involves injecting a suitable lixiviant into the ore deposit through a series of injection wells. The lixiviant then permeates through the sandstone deposit oxidizing and dissolving the various minerals within the deposit. The pregnant leaching solution is then returned to the surface for processing by a system of production wells. In the process of leaching out the uranium in the deposit, the lixiviant also dissolves many of the other gangue minerals present with the ore. These side reactions are detrimental to the overall leaching process for several reasons. Side reactions can consume oxidant and complexing agents while the porosity-permeability of the deposit can be reduced by transformation and precipitation reactions.[4] Thus optimizing the leaching conditions to selectively extract the uranium is a very important factor.

This paper outlines a chemical reaction-based model to analyze the changes which occur in the aqueous phase during the dissolution of a uranium dioxide-pyrite mixture using a $Fe_2(SO_4)_3$-H_2SO_4 lixiviant. The results obtained from this study will mainly focus on the three points: i) the effectiveness of Fe(III) ions in oxidizing uranium dioxide; ii) the changes in the concentration of the aqueous species during the leaching process; iii) conditions which will promote the selective leaching of uranium dioxide.

THE MODEL EQUATIONS

The dissolution reactions for uranium dioxide and pyrite at ambient conditions in an acidified ferric sulfate solution are given by the following expressions[5,6]:

$$UO_2 + 2Fe^{3+} \rightarrow UO_2^{2+} + 2Fe^{2+} \quad (1)$$

$$FeS_2 + 14Fe^{3+} + 8H_2O \rightarrow 15Fe^{2+} + 2SO_4^{2-} + 16H \quad (2)$$

The dissolution reactions due to oxidation by oxygen were neglected because of the limited solubility of oxygen in the lixiviant and the uncertainty in estimating the rate of oxygen replenishment in the leaching solution.

The equilibrium constants for the dissolution reactions given by Eqs. 1 and 2 at 298°K are 8.4×10^{40} and 1.34×10^{67}, respectively.[7,8] Comparison of these values with the experimental data obtained by Allen[9] indicates that the mineral dissolution reactions are far from equilibrium. The other reactions occurring during the leaching process in the aqueous phase are very rapid relative to the dissolution reactions and hence are assumed to be at continual equilibrium. This entire process can therefore be modeled utilizing a

Table 1. Homogeneous Equilibrium Reactions

$H^+ + SO_4^{2-} \rightleftarrows HSO_4^-$	(3)
$UO_2^{2+} + SO_4^{2-} \rightleftarrows UO_2SO_4^0$	(4)
$UO_2^{2+} + 2SO_4^{2-} \rightleftarrows UO_2(SO_4)_2^{2-}$	(5)
$UO_2^{2+} + 3SO_4^{2-} \rightleftarrows UO_2(SO_4)_3^{4-}$	(6)
$Fe^{3+} + H_2O \rightleftarrows Fe(OH)^{2+} + H^+$	(7)
$Fe^{3+} + SO_4^{2-} \rightleftarrows FeSO_4^+$	(8)
$FeSO_4^+ + SO_4^{2-} \rightleftarrows Fe(SO_4)_2^-$	(9)
$Fe^{3+} + HSO_4^- \rightleftarrows FeHSO_4^{2+}$	(10)
$Fe^{2+} + SO_4^{2-} \rightleftarrows FeSO_4^0$	(11)
$Fe^{2+} + HSO_4^- \rightleftarrows FeHSO_4^+$	(12)

partial equilibrium approach as outlined by Helegson[10] and successfully applied to the chalcopyrite-ferric sulfate system by Liddell and Bautista.[11-15] The overall leaching process can therefore be described by a series of partial equilibrium stages, each reversible relative to the previous step, but irreversible to the initial step.

The solution reactions occurring during the leaching process can be divided into two distinct types: i) the acid-base equilibria between SO_4^{2-}, H^+, and HSO_4^-; ii) the formation of complexes between the metal ions and either SO_4^{2-} or HSO_4^-. The solution species which will be considered in this model are given in Table 1.

The changes in concentrations of the aqueous species which which occur during the leaching process can be represented by using the definition of activity:

$$da_j = d(\gamma_j m_j) \tag{13}$$

where a_j is the activity of species j, γ_j is the activity coefficient, and m_j is the molality. The changes in concentration dm_j can also be related to the extent of reaction ξ ($0 \leq \xi \leq 1$) since the differential changes in concentration are related to the differential extent of reaction by:

$$n_j d\xi = dm_j \tag{14}$$

where n_j is the number of moles of aequeous species j reacting per liter. The sign convention which will be used have n_j positive for products and negative for reactants.

The mass action equation that can be written for each equilibrium reaction given in Table 1 can be represented by:

$$K_i = \prod_j a_j^{\nu_j} \tag{15}$$

where K_i is the equilibrium constant for reaction i and ν_j is the stoichiometric coefficient for species j. The logarithm of Eq. 15 can be differentiated with respect to ξ since the activity is a function of ξ:

$$0 = \Sigma \nu_j \frac{d \ln a_j}{d\xi} \tag{16}$$

Combining Eqs. 13, 14 and 16 gives:

$$0 = \Sigma \ \nu_j \bar{n}_j \ [\frac{1}{m_j} + \frac{d \ \ln\gamma_i}{dm_j}] \tag{17}$$

For a multicomponent electrolyte solution, the change in the ionic strength I relative to a change in the concentration of a single species is relatively small. Therefore,

$$\frac{d\gamma_j}{dm_j} = \frac{d\gamma_j}{dI}\frac{dI}{dm_j} \approx 0 \tag{18}$$

due to the high ionic strength of the solution. Thus, Eq. 17 simplifies to the linear expression

$$0 = \Sigma \ \frac{\nu_j \bar{n}_j}{m_j} \tag{19}$$

In order to obtain additional equations, material balances can be written for all species present in solution during the leaching process. These equations can be expressed by:

$$0 = \sum_k \ (\lambda_{k,j}\bar{n}_j + \lambda_{k,i}\bar{N}_i) \tag{20}$$

where $\bar{N}_i$ is the number of moles of solid i reacting per liter, $\lambda_{k,j}$ is the number of moles of k moieties present in aqueous species s_j, and $\lambda_{k,i}$ is the number of moles of k moieties present in the solid species S_i. Also, an overall charge balance can be written encompassing all species j present in solution with charge κ_j:

$$0 = \Sigma \ \kappa_j \bar{n}_j \tag{21}$$

There are a total of 17 unknowns needed to solve the model. These unknowns are:

$\bar{n}_{H^+}$, $\bar{n}_{SO_4{}^{2-}}$, $\bar{n}_{HSO_4{}^-}$, $\bar{n}_{UO_2{}^{2+}}$, $\bar{n}_{UO_2SO_4{}^0}$, $\bar{n}_{UO_2(SO_4)_2{}^{2-}}$,

$\bar{n}_{UO_2(SO_4)_3{}^{4-}}$, $\bar{n}_{Fe^{3+}}$, $\bar{n}_{Fe(OH)^{2+}}$, $\bar{n}_{FeSO_4{}^+}$, $\bar{n}_{Fe(SO_4)_2{}^-}$,

$\bar{n}_{FeSO_4{}^{2+}}$, $\bar{n}_{Fe^{2+}}$, $\bar{n}_{FeSO_4{}^0}$, $\bar{n}_{FeHSO_4{}^+}$, $\bar{N}_{UO_2}$, and $\bar{N}_{FeS_2}$.

For the equilibrium state, 10 mass action expressions of the form given by Eq. 9 can be written. Material balances for Fe(III),

Fe(II), U(VI), and total SO_4^{2-} can be written as well as an overall charge balance. Finally, the rate of dissolution for uranium dioxide and pyrite must be calculated at each increment of dissolution in order to obtain N_i values. The incorporation of actual kinetic data into the model limits the specific results obtained to the particular sample being investigated. This is because the rate of reaction for any mineral varies with the origin and formation of that mineral. However, the results from the model are general enough to encompass most ore deposits. Another disadvantage in utilizing rate data into the model is that the rate is a function of time. In order to derive useful equations, a relationship between ξ and time must therefore be derived. The necessity for an additional equation can be eliminated by reducing the number of unknowns. A covenient method of accomplishing this is to scale all the variables by scaling each unknown relative to $-\overline{N}_{UO_2}$ and then defining a new variable n_j, e.g.,

$$n_{H^+} = \frac{\overline{n}_{H^+}}{\overline{N}_{UO_2}} = \frac{\text{moles } H^+ \text{ formed}}{\text{moles } UO_2 \text{ reacting}} \; . \qquad (22)$$

The use of the minus sign preserves the sign convention introduced earlier. In a similar manner, the kinetics for the dissolution of pyrite is scaled relative to that for uranium dioxide

$$r_{FeS_2/UO_2} = \frac{r_{FeS_2}}{r_{UO_2}} = \frac{\text{moles } FeS_2 \text{ reacting}}{\text{moles } UO_2 \text{ reacting}} \approx \frac{-\overline{N}_{FeS_2}}{-\overline{N}_{UO_2}} \qquad (23)$$

The complete set of equations is given in Tables 2 and 3. There are a total of 15 linear equations and 15 unknowns, with the value of r_{FeS_2/UO_2} calculated at the start of each increment of dissolution. It should be noted that nonhomogeneous terms for the material balances in Table 3 are adaptable to account for the possibility of either uranium dioxide or pyrite becoming depleted during the course of the leaching process. For the particular case of uranium dioxide depletion, the $\overline{n}_j$ values become scaled relative to $-\overline{N}_{FeS_2}$ since $-\overline{N}_{UO_2}$ becomes zero.

Before the equations given in Tables 2 and 3 can be solved, the concentration for each aqueous species m_j must be determined at the start of the modeling process. The values for each m_j are obtained by determining the equilibrium composition of the leaching solution using the initial concentrations of $Fe_2(SO_4)_3$, $FeSO_4$, H_2SO_4, and UO_2SO_4. There are a total of 15 unknown species presented in the homogeneous equilibrium reactions given in Table 1. For each equilibrium reaction, mass action equations such as those given in Table 4 can be written. Additional equations can be obtained from the material balances for Fe(III), Fe(II), U(VI), and total SO_4^{2-}, as well as an overall charge balance (see Table 5). Thus a total of 15 equations can be written allowing the value for each m_j to be determined.

In solving for these unknowns, it is inconvenient to work directly in terms of activities since this procedure requires the use of activity coefficient calculations. To overcome this problem, apparent equilibrium constants at ionic strengths similar to those found in actual leaching operations were used in calculating the equilibrium composition.

SOLUTION OF THE MODEL EQUATIONS

The procedure used to solve this model incorporates the Newton-Raphson iterative method to solve the system of nonlinear equations given in Tables 2 and 3 and the Gauss-Jordan elimination technique to solve the system of linear equations given in Tables 4 and 5.

The initial step of the programming technique requires specifying the analytical concentrations for $Fe_2(SO)_4$, $FeSO_4$, H_2SO_4, and UO_2SO_4, as well as the apparent equilibrium constants for each mass action equation. The iterative solution for the nonlinear equations is initiated by utilizing a set of initial guesses for each m_j. The iterative process of modifying these initial guesses to obtain the exact solution is continued until either convergence is achieved or the number of specified iterations exceeds the maximum number. The criteria used for convergence was 10^{-6} for the fractional error in each unknown or equilibrium and balance equation. If convergence is obtained, the leaching solution is checked to ensure that the solubility product of $Fe(OH)_3$ (s) was not exceeded. This was accomplished by determining the solubility product $m_{Fe^{3+}} m_{(OH^-)}{}_3$ and this value was estimated with the K_{sp} value of 2.51×10^{-39} at an ionic strength of 3.0.[11]

The rate of reaction for each heterogeneous reaction was then determined using the experimentally-determined rate expressions obtained by Allen.[9] These reaction rates were then incorporated into the model to determine the nonhomogeneous coefficients for the linear equations given in Table 2.

Table 2. Mass Action Model Equations for the Homogeneous Equilibrium Reactions

$$0 = \frac{n_{H^+}}{m_{H^+}} + \frac{n_{SO_4^{2-}}}{m_{SO_4^{2-}}} - \frac{n_{HSO_4^-}}{m_{HSO_4^-}} \quad (24)$$

$$0 = \frac{n_{UO_2^{2+}}}{m_{UO_2^{2+}}} + \frac{n_{SO_4^{2-}}}{m_{SO_4^{2-}}} - \frac{n_{UO_2SO_4^0}}{m_{UO_2SO_4^0}} \quad (25)$$

$$0 = \frac{n_{UO_2^{2+}}}{m_{UO_2^{2+}}} + \frac{2\,n_{SO_4^{2-}}}{m_{SO_4^{2-}}} - \frac{n_{UO_2(SO_4)_2^{2-}}}{m_{UO_2(SO_4)_2^{2-}}} \quad (26)$$

$$0 = \frac{n_{UO_2^{2+}}}{m_{UO_2^{2+}}} + \frac{3\,n_{SO_4^{2-}}}{m_{SO_4^{2-}}} - \frac{n_{UO_2(SO_4)_3^{4-}}}{m_{UO_2(SO_4)_3^{4-}}} \quad (27)$$

$$0 = \frac{n_{Fe^{3+}}}{m_{Fe^{3+}}} - \frac{n_{Fe(OH)^{2+}}}{m_{Fe(OH)^{2+}}} - \frac{n_{H^+}}{m_{H^+}} \quad (28)$$

Table 2. (continued)

$$0 = \frac{n_{Fe^{3+}}}{m_{Fe^{3+}}} + \frac{n_{SO_4^{2-}}}{m_{SO_4^{2-}}} - \frac{n_{FeSO_4^{+}}}{m_{FeSO_4^{+}}} \qquad (29)$$

$$0 = \frac{n_{FeSO_4^{+}}}{m_{FeSO_4^{+}}} + \frac{n_{SO_4^{2-}}}{m_{SO_4^{2-}}} - \frac{n_{Fe(SO_4)_2^{-}}}{m_{Fe(SO_4)_2^{-}}} \qquad (30)$$

$$0 = \frac{n_{Fe^{3+}}}{m_{Fe^{3+}}} + \frac{n_{HSO_4^{-}}}{m_{HSO_4^{-}}} - \frac{n_{FeHSO_4^{2+}}}{m_{FeHSO_4^{2+}}} \qquad (31)$$

$$0 = \frac{n_{Fe^{2+}}}{m_{Fe^{2+}}} + \frac{n_{SO_4^{2-}}}{m_{SO_4^{2-}}} - \frac{n_{FeSO_4^{0}}}{m_{FeSO_4^{0}}} \qquad (32)$$

$$0 = \frac{n_{Fe^{2+}}}{m_{Fe^{2+}}} + \frac{n_{HSO_4^{-}}}{m_{HSO_4^{-}}} - \frac{n_{FeHSO_4^{+}}}{m_{FeHSO_4^{+}}} \qquad (33)$$

Table 3. Material and Charge Balance Model Equations

Sulfate Balance

$$C1 = n_{SO_4^{2-}} + n_{HSO_4^-} + n_{UO_2SO_4^0} + 2\,n_{UO_2(SO_4)_2^{2-}} + 3\,n_{UO_2(SO_4)_3^{4-}} + n_{FeSO_4^+} + 2\,n_{Fe(SO_4)_2^-} + n_{FeHSO_4^{2+}} + n_{FeSO_4^0} + n_{FeHSO_4^-} \quad (34)$$

U(VI) Balance

$$C2 = n_{UO_2^{2+}} + n_{UO_2SO_4^0} + n_{UO_2(SO_4)_2^{2-}} + n_{UO_2(SO_4)_3^{4-}} \quad (35)$$

Fe(III) Balance

$$C3 = n_{Fe^{3+}} + n_{Fe(OH)^{2+}} + n_{FeSO_4^+} + n_{Fe(SO_4)_2^-} + n_{FeHSO_4^{2+}} \quad (36)$$

Fe(II) Balance

$$C4 = n_{Fe^{2+}} + n_{FeSO_4^0} + n_{FeHSO_4^-} \quad (37)$$

Charge Balance

$$0 = n_{H^+} - 2\,n_{SO_4^{2-}} - n_{HSO_4^-} + 2\,n_{UO_2^{2+}} - 2\,n_{UO_2(SO_4)_2^{2-}} - 4\,n_{UO_2(SO_4)_3^{4-}} + 3\,n_{Fe^{3+}} + 2\,n_{Fe(OH)^{2+}} + m_{FeSO_4^+} - n_{Fe(SO_4)_2^-} + 2\,n_{FeHSO_4^{2+}} + 2\,n_{Fe^{2+}} + n_{FeHSO_4^+} \quad (38)$$

Table 3. (continued)

Uranium dioxide and pyrite both reacting:

$$C1 = 2\ r_{FeS_2/UO_2} \quad (39)$$

$$C2 = 1 \quad (40)$$

$$C3 = -2 - 14\ r_{FeS_2UO_2} \quad (41)$$

$$C4 = 2 + 15\ r_{FeS_2/UO_2} \quad (42)$$

Uranium dioxide reacting:

$$C1 = 0 \quad (43)$$

$$C2 = 1 \quad (44)$$

$$C3 = -2 \quad (45)$$

$$C4 = 2 \quad (46)$$

Pyrite reacting:

$$C1 = 2 \quad (47)$$

$$C2 = 0 \quad (48)$$

$$C3 = -14 \quad (49)$$

$$C4 = 15 \quad (50)$$

Table 4. Mass Action Equations and Apparent Equilibrium Constants

	K_{app}	Ionic Strength	Source	Equation
$0 = K_{1,app} - \frac{m_{H^+} m_{SO_4^{2-}}}{m_{HSO_4^-}}$	8.13×10^0	3.0	16	(51)
$0 = K_{2,app} - \frac{m_{UO_2^{2+}} m_{SO_4^{2-}}}{m_{UO_2SO_4^0}}$	7.58×10^2	2.0	17	(52)
$0 = K_{3,app} - \frac{m_{UO_2^{2+}} m^2_{SO_4^{2-}}}{m_{UO_2(SO_4)_2^{2-}}}$	7.09×10^2	2.0	17	(53)
$0 = K_{4,app} - \frac{m_{UO_2^{2+}} m^3_{SO_4^{2-}}}{m_{UO_2(SO_4)_3^{4-}}}$	2.50×10^3	2.0	17	(54)
$0 = K_{5,app} - \frac{m_{Fe(OH)^{2+}} m_{H^+}}{m_{Fe^{3+}}}$	1.07×10^{-3}	3.0	16	(55)

Table 4. (continued)

	K_{app}	Ionic Strength	Source	Equation
$0 = K_{6,app} - \frac{m_{Fe^{3+}} m_{SO_4^{+}}}{m_{FeSO_4^{+}}}$	8.51×10^{1}	3.0	16	(56)
$0 = K_{7,app} - \frac{m_{FeSO_4^{+}} m_{SO_4^{2-}}}{m_{Fe(SO_4)_2^{-}}}$	1.51×10^{0}	3.0	16	(57)
$0 = K_{8,app} - \frac{m_{Fe^{3+}} m_{HSO_4^{-}}}{m_{FeHSO_4^{2+}}}$	4.00×10^{0}	2.67	18	(58)
$0 = K_{9,app} - \frac{m_{Fe^{2+}} m_{SO_4^{2-}}}{m_{FeSO_4^{0}}}$	2.06×10^{0}	3.0	11	(59)
$0 = K_{10,app} - \frac{m_{Fe^{2+}} m_{HSO_4^{-}}}{m_{FeHSO_4^{+}}}$	1.94×10^{0}	4.0	19	(60)

Table. 5. Balance Equations for Determining Equilibrium Composition

Sulfate Balance

$$0 = C_{H_2SO_4} + C_{UO_2SO_4} + 3\,C_{Fe_2(SO_4)_3} + C_{FeSO_4} + C_{SO_4^{2-}} - m_{SO_4^{2-}} \tag{61}$$

$$- m_{UO_2SO_4^0} - 2\,m_{UO_2(SO_4)_2^{2-}} - m_{FeSO_4^0} - m_{FeHSO_4^+}$$

U(VI) Balance

$$0 = C_{UO_2SO_4^-} - m_{UO_2^{2+}} - m_{UO_2SO_4^0} - m_{UO_2(SO_4)_2^{2-}} - m_{UO_2(SO_4)_3^{4-}} \tag{62}$$

Fe(III) Balance

$$0 = 2\,C_{Fe_2(SO_4)_3} - m_{Fe^{3+}} - m_{Fe(OH)^{2+}} - m_{FeSO_4^+} - m_{Fe(SO_4)_2^-} - m_{FeHSO_4^{2+}} \tag{63}$$

Fe(II) Balance

$$0 = C_{FeSO_4} - m_{Fe^{2+}} - m_{FeSO_4^0} - m_{FeHSO_4^+} \tag{64}$$

Charge Balance

$$0 = m_{H^+} - 2\,m_{SO_4^{2-}} - m_{HSO_4^-} + 2\,m_{UO_2^{2+}} - 2\,m_{UO_2(SO_4)_2^{2-}}$$

$$- 4\,m_{UO_2(SO_4)_3^{4-}} + 3\,m_{Fe^{3+}} + 2\,m_{Fe(OH)^{2+}} + m_{FeSO_4^+} - m_{Fe(SO_4)_2^-} \tag{65}$$

$$+ 2\,m_{FeHSO_4^{2+}} + 2\,m_{FeHSO_4^{2+}} + 2\,m_{Fe^{2+}} + m_{FeHSO_4^+}$$

The linear equations given in Tables 2 and 3 are then solved to determine the corresponding n_j value for each increment of dissolution. This overall procedure requires the use of m_j, r_{FeS_2} and r_{UO_2} values at the start of the increment. The molal change is then calculated by using the relationship

$$\Delta m_j \approx n_j\,(-\bar{N}_{UO_2})\,\Delta\xi \tag{66}$$

where the number of mole of uranium dioxide, $-\bar{N}_{UO_2}$ and the value of

$\Delta\xi$ must be specified. The new concentrations are then obtained from

$$m_j \text{ (new)} = m_j \text{ (old)} + \Delta m_j \tag{67}$$

During the modeling process, the value of Δm_j can become large enough that a negative m_j (new) can arise for a particular species. This problem is automatically overcome by reducing the size of the increment $(-\overline{N}_{UO_2})\Delta\xi$ until an appropriate size is obtained which will generate only positive concentrations.

The solubility product of $Fe(OH)_3$ is then checked to ensure precipitation does not occur. If precipitation does occur, the calculations are halted. Next, the Fe(III) concentration is checked to ensure that the Fe(III) concentration is not depleted. The limiting concentration was chosen to be 10^{-4} m. This value was somewhat arbitrary but it was large enough to compensate for the error generated when computing the Fe^{3+} concentratons at very low levels.

The next step was to calculate the new rates of reaction for uranium dioxide and pyrite and recalculate the new concentrations. This overall procedure is continued until either the Fe(III) concentration is depleted or precipitation occurs.

RESULTS AND DISCUSSION

The model has been solved for the leaching process utilizing experimentally determined rate data for the dissolution of uranium dioxide and pyrite.[9] The rate of dissolution for uranium dioxide was found to be:

$$r_{UO_2} = 5.5\ [\text{Fe(III)}]^{1/2} \tag{68}$$

where the rate is given in units of μg/cm²/hr and the Fe(III) concentration in units of g/l. The dissolution of pyrite followed a Hougen-Watson model for dual site adsorption of Fe(II) and Fe(III) ions, e.g.

$$r_{FeS_2} = \frac{0.24 - 0.082\ \sqrt{\text{Fe(II)/Fe(III)}}}{1\sqrt{\text{Fe(III)}} + 0.44 + 8.1\ \sqrt{\text{Fe(II)/Fe(III)}}} \tag{69}$$

where the rate is given in the units of μg/cm²/hr and the iron concentration in g/l.

The first phase of this investigation was to determine the conditions which will promote the selective leaching of uranium dioxide

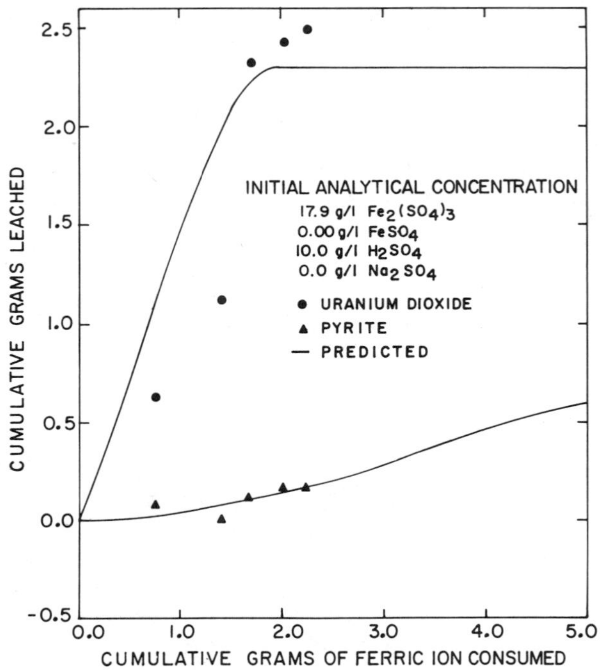

Fig. 1. Dissolution of a mixture containing 2.3 g of uranium dioxide and 97.7 g of pyrite with Fe(III) ions.

and then check these results experimentally. A mixture containing 97.7 g of pyrite and 2.3 g of uranium dioxide, similar to that used by Carlson[20], was used as a basis. Figs. 1 to 3 represent the changes within the leaching process as the initial ferric:ferrous ratio and the total iron concentration are varied, with the computer generated points given by the solid lines and the experimental points given by the circles and triangles. Figs. 1 and 2 represent the differences occurring as the initial Fe(III) concentration is decreased at a constant total iron concentration of 5.0 g/l (17.9 g/l $Fe_2(SO_4)_3$). As illustrated by these figures, when the ferric:ferrous ratio is decreased, the selectivity for uranium dioxide is increased. This is because the increased Fe(II) concentration hinders the rate of pyrite dissolution by blocking active sites on the surface. A better representation of this result is obtained by comparing Figs. 2 and 3. Both of these runs initially contained 1.0 g/l Fe(III) (3.58 g/l $Fe_2(SO_4)_3$), however the run illustrated in Fig. 2 also contains 4.0 g/l Fe(II) (10.9 g/l $FeSO_4$) present. This large excess of Fe(II) ions hinders the dissolution of pyrite to such an extent that all the uranium dioxide is selectively leached out in Fig. 2 whereas in Fig. 3 a large amount of uranium dioxide remained unoxidized.

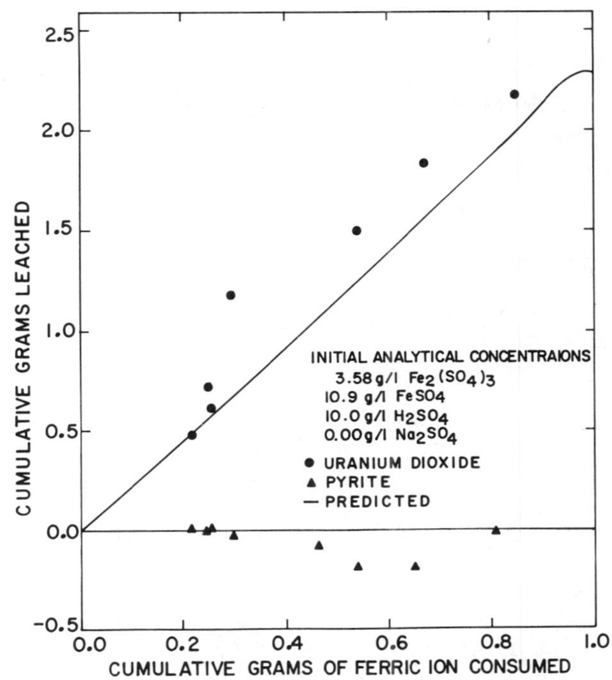

Fig. 2. Dissolution of a mixture containing 2.3 g of uranium dioxide and 97.7 g of pyrite with Fe(III) ions.

The effects on the chemical equilibrium within the aqueous phase during the leaching process due to changes in the total sulfate concentration were investigated by altering either the H_2SO_4 and Na_2SO_4 concentration. The experimental rate expressions for uranium dioxide and pyrite were not investigated sufficiently to obtain rate expressions which are dependent on the total sulfate. Thus, the range of variation in total sulfate was kept rather small.

The variation in the H^+, SO_4^{2-}, and HSO_4^- as given by Eq.(3) is given in Figs. 4 to 6. Fig. 4 represents the changes occurring while using a leaching solution having an initial composition of 5.0 g/l Fe(III) and 10.0 g/l H_2SO_4. As can be seen by this plot, the equilibria is not greatly altered during the early stages of the leaching process since uranium dioxide is the major component being leached and the corresponding dissolution reaction (see Eq. 1) does not directly affect the H^+, SO_4^{2-}, and HSO_4^- equilibria. When the uranium dioxide becomes depleted (see Fig. 1), the dissolution of pyrite causes the H^+, SO_4^{2-} and HSO_4^- concentrations to increase since these species are products of the dissolution reaction for pyrite (see Eq. 2).

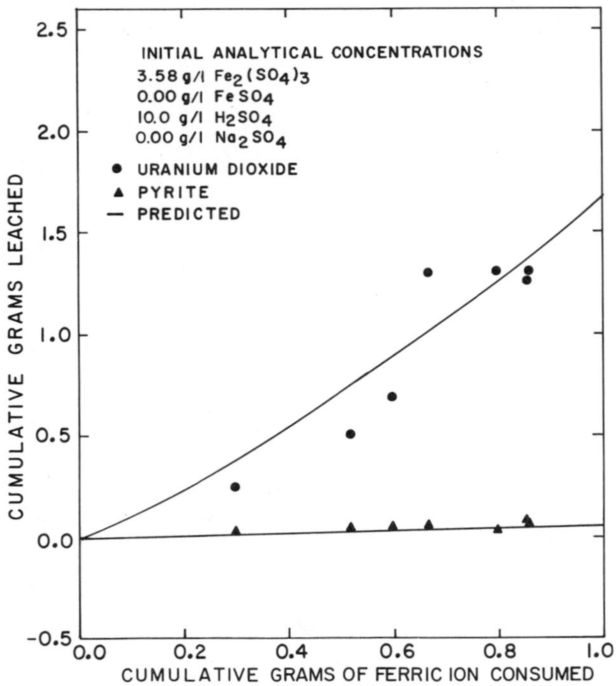

Fig. 3. Dissolution of a mixture containing 2.3 g of uranium dioxide and 97.7 g of pyrite with Fe(III) ions.

The addition of excess H_2SO_4 to give 15.0 g/l, as illustrated in Fig. 5, causes the initial H^+ and HSO_4^- concentrations to increase relative to that for the 10.0 g/l $H_2SO_4^-$ lixiviant (shown in Fig. 4). The addition of 7.24 g/l of Na_2SO_4 to a leaching solution containing 10.1 g/l H_2SO_4 produces a lixiviant with the same total sulfate concentration (23.3 g/l) as that generated using 15.0 g/l H_2SO_4 and 17.9 g/l $Fe_2(SO_4)_3$. The course of the dissolution is illustrated in Fig. 6 and indicates that excess Na_2SO_4 (7.24 gl) causes a reduction in the initial H^+ concentration and promotes the formation of SO_4^{2-} relative to HSO_4^-. However, as the dissolution of pyrite becomes more prevalent, the HSO_4^- species concentration becomes more significant and is better able to accommodate the hydrogen and sulfate ions which are generated during the dissolution of pyrite.

The variations in the U(VI) equilibrium during the leaching process are given in Figs. 7 to 9 and represent the concentration changes occurring in Eqs. 4 to 6. The range of leaching conditions are similar to those outlined for the H^+, SO_4^{2-}, and HSO_4^- equilibrium. As shown by Fig. 7, the presence of sulfate is very

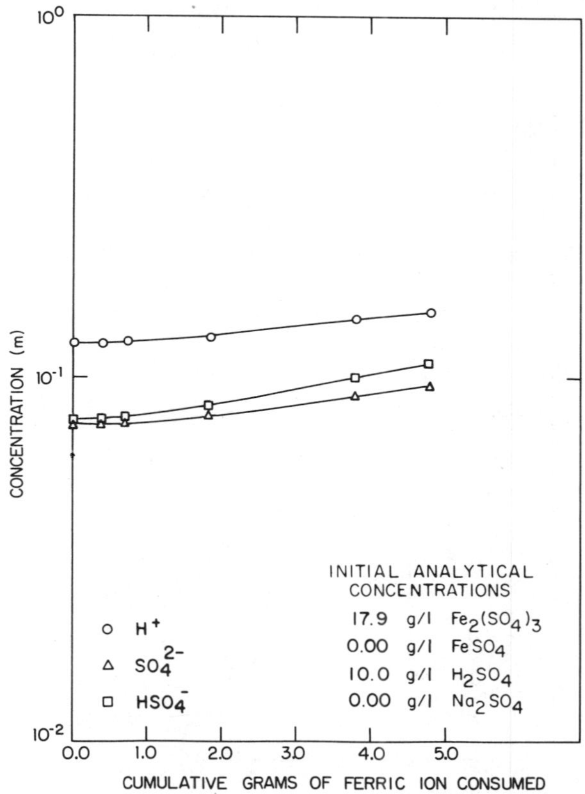

Fig. 4. Concentration change for H^+, SO_4^{2-}, and HSO_4^- during the dissolution of a mixture containing 2.3 g of uranium dioxide and 97.7 g of pyrite.

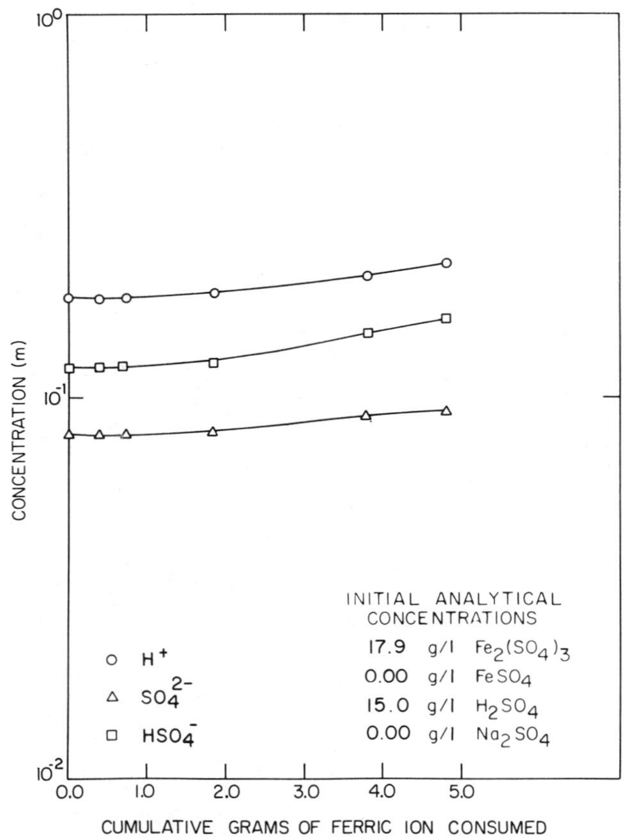

Fig. 5. Concentration change for H^+, SO_4^{2-}, and HSO_4^- during the dissolution of a mixture containing 2.3 g of uranium dioxide and 97.7 g of pyrite.

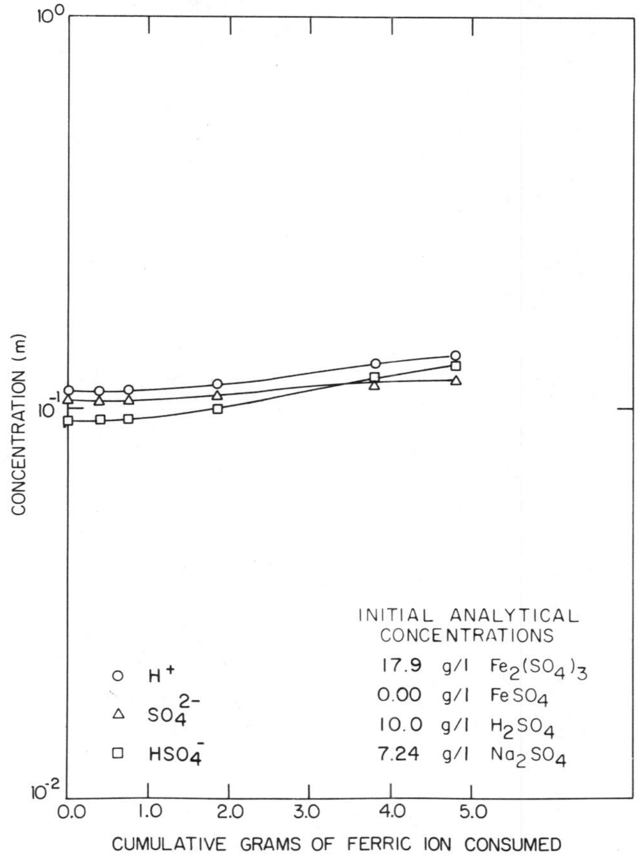

Fig. 6. Concentration changes for H^+, SO_4^{2-}, and HSO_4^- during the dissolution of a mixture containing 2.3 g of uranium dioxide and 97.7 g of pyrite.

important in the overall leaching process since over 70 percent of the soluble U(VI) exist as the $UO_2(SO_4)_2^{2-}$ species. Hence, the absence of any complexation agents in the lixiviant would greatly reduce the solubility of uranium in solution. An interesting occurrence in Fig. 7 is the presence of a maximum in the concentration for UO_2^{2+} and $UO_2SO_4^0$ at a point closely corresponding to the depletion of uranium dioxide. The maximum arises because the dissolution of pyrite, which generates sulfate ions, continually shifts the U(VI) equilibrium to the more highly sulfated $UO_2(SO_4)_2^{2-}$ and $UO_2(SO_4)_3^{4-}$ species.

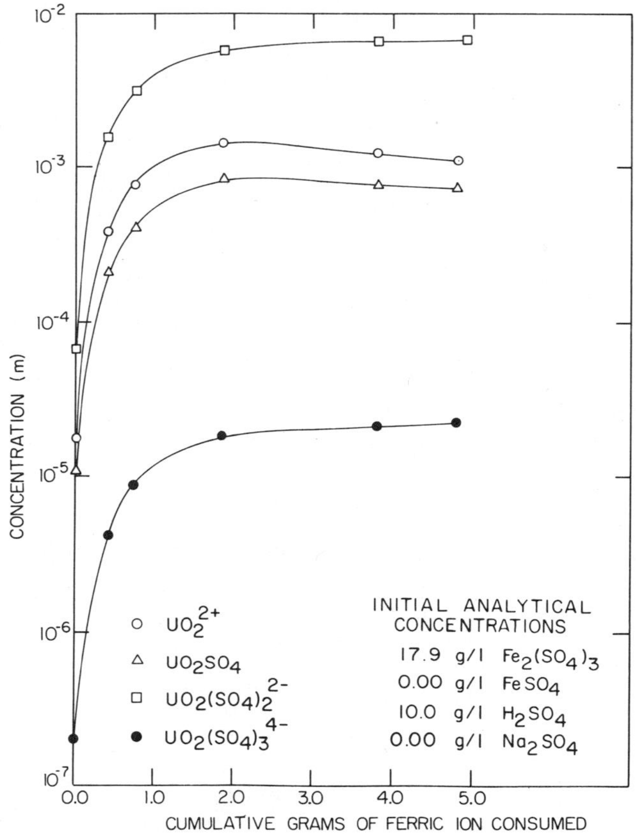

Fig. 7. Concentration changes for U(VI) species during the dissolution of a mixture containing 2.3 g of uranium dioxide and 97.7 g of pyrite.

The addition of excess H_2SO_4 to give a total of 15.0 g/l H_2SO_4 (Fig. 8) causes only a slight shift in the U(VI) equilibrium relative to the leaching solution with 10.0 g/l H_2SO_4. The concentrations of the UO_2^{2+} and $UO_2SO_4^0$ are reduced due to a shift to the more highly sulfated $UO_2(SO_4)_2^{2-}$ and $UO_2(SO_4)_3^{4-}$ species. This same effect is more pronounced in the case when excess Na_2SO_4 (7.24 g/l) is added to a solution with 10.0 g/l H_2SO_4. As illustrated in Fig. 9, when excess Na_2SO_4 instead of H_2SO_4 is added, a larger segment of the total sulfate is present as SO_4^{2-} rather than HSO_4^-. This in turn results in large driving force to form the more sulfated U(VI) species.

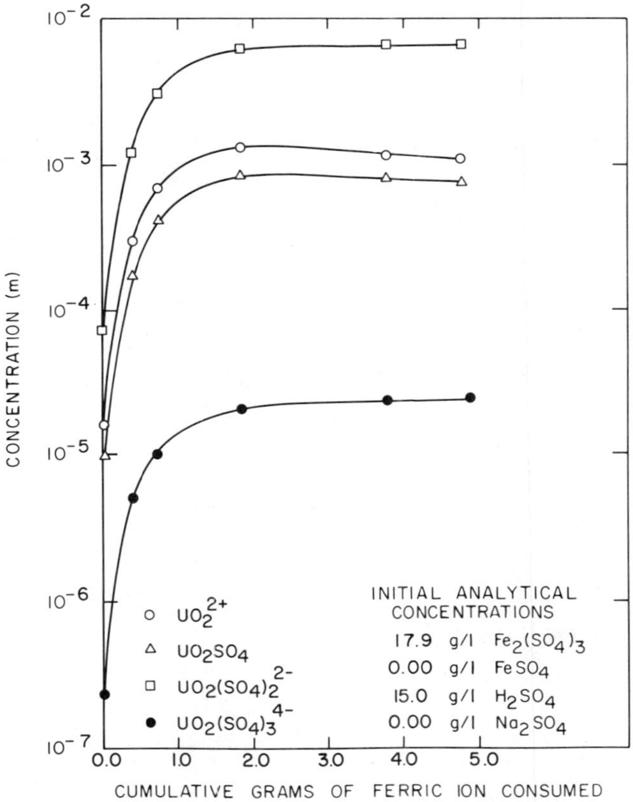

Fig. 8. Concentrations changes for U(VI) species during the dissolution of a mixture containing 2.3 g of uranium dioxide and 97.7 g of pyrite.

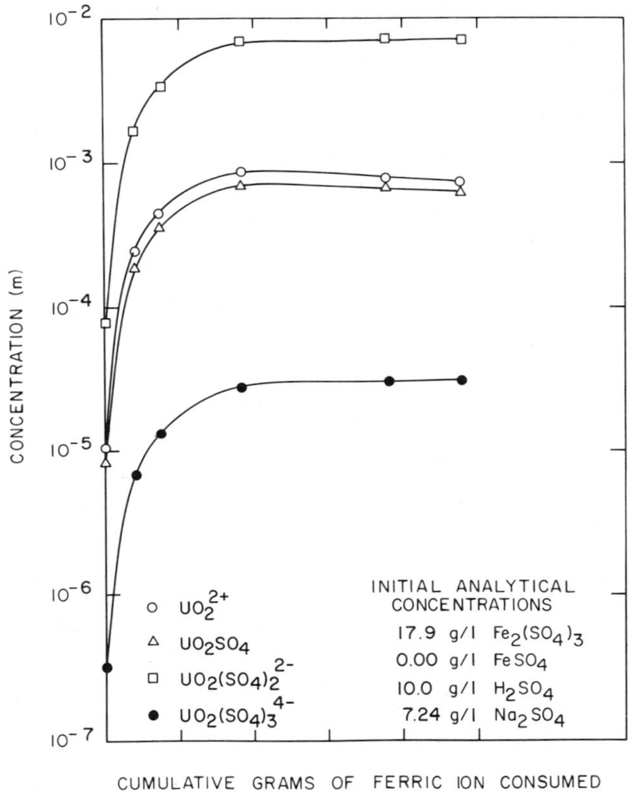

Fig. 9. Concentration changes for U(VI) species during the dissolution of a mixture containing 2.3 g of uranium dioxide and 97.7 g of pyrite.

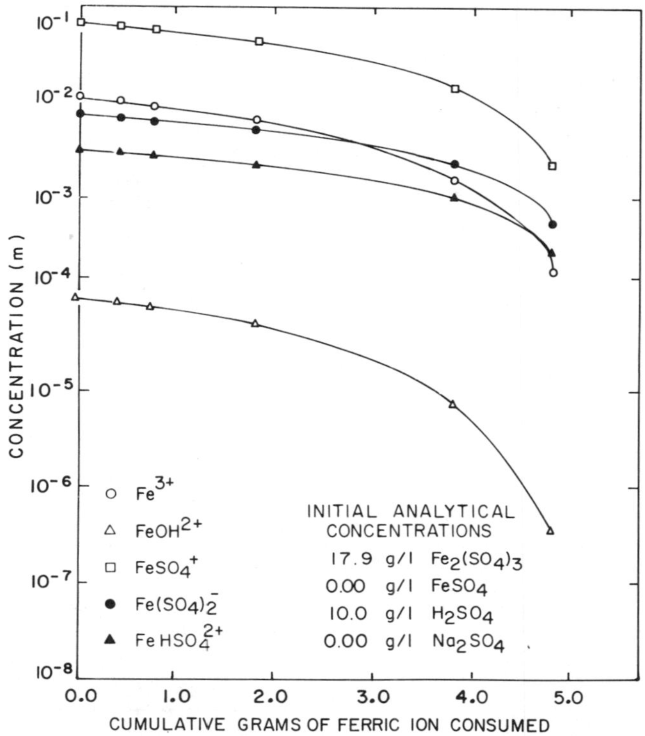

Fig. 10. Concentration changes for Fe(III) species during the dissolution of a mixture containing 2.3 g of uranium dioxide and 97.7 g of pyrite.

The variation in the Fe(III) equilibrium as given by Eqs. 7 to 10 are represented in Figs. 10 to 12. As the leaching process proceeds, the Fe(III) concentration shows a continual decline. Again, the importance of sulfate in the lixiviant can be observed since over 80 percent of the Fe(III) ions exist as $FeSO_4{}^+$. The absence of sulfate in the lixiviant will cause the Fe^{3+} species to dominate and promote the precipitation of $Fe(OH)_3$. Another factor which should be pointed out is that the Fe^{3+} species shows a much sharper decrease than the sulfated Fe(III) species. This is again due to the dissolution of pyrite shifting the Fe(III) equilibrium to the more sulfated species.

A higher concentration of H_2SO_4, 15.0 g/l, relative to 10.0 g/l as shown in Fig. 11 causes the initial $FeHSO_4{}^{2+}$ concentration to increase at a faster relative rate than the other Fe(III) species,

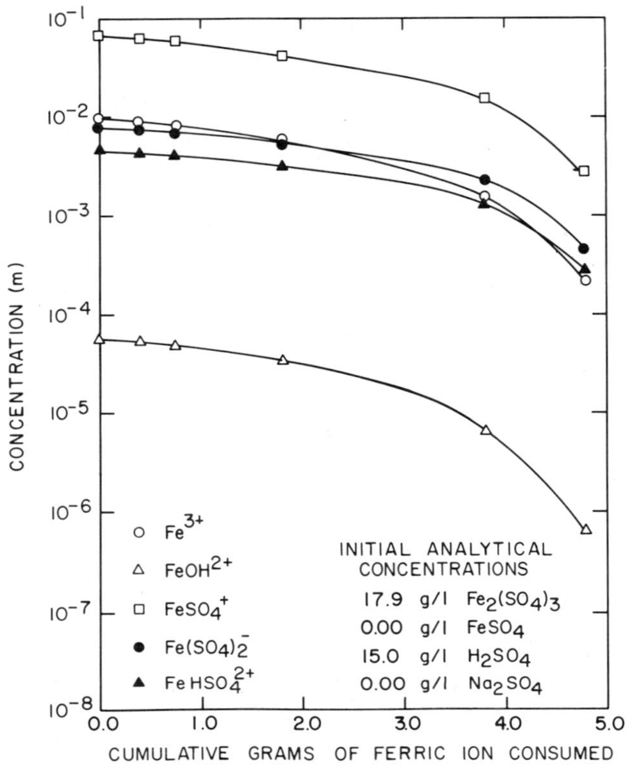

Fig. 11. Concentration changes for Fe(III) species during the dissolution of a mixture containing 2.3 g of uranium dioxide and 97.7 g of pyrite.

since the $FeHSO_4{}^{2+}$ species can accommodate both hydrogen and sulfate ions. The addition of 7.24 g/l Na_2SO_4 to a solution containing 10.0 g/l H_2SO_4 causes a shift in the Fe(III) equilibrium by promoting the formation of the more highly sulfated $Fe(SO_4)_2{}^-$ species relative to the Fe^{3+} and $FeOH^{2+}$ species.

The variations in the Fe(II) equilibrium as illustrated in Figs. 13 to 15 correspond to the equilibrium reactions given by Eqs. 11 and 12. As can be seen from these plots, the Fe(II) ions do not complex to a great extent with sulfate since the Fe^{2+} is the dominant aqueous species. The effects of pyrite oxidation in the latter part of the leaching can be seen in Fig. 13, with the $FeHSO_4{}^+$ concentration increasing faster than the $FeSO_4{}^+$ concentration. This is again

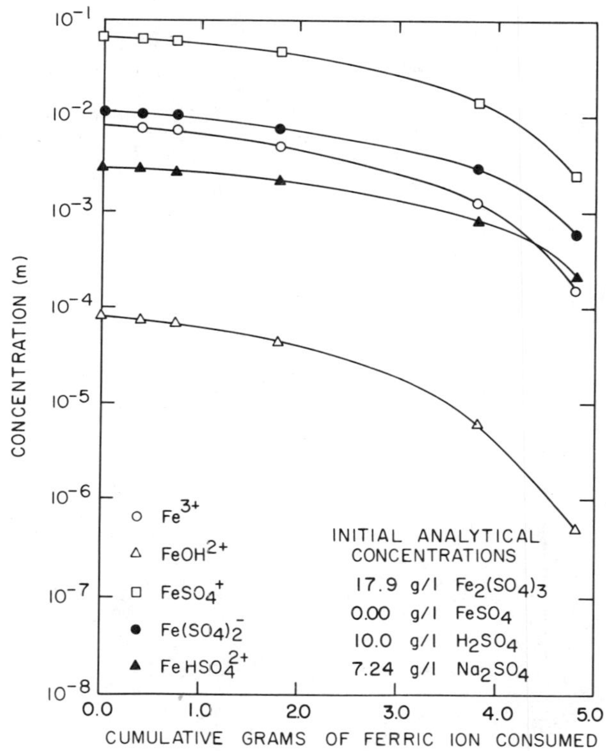

Fig. 12. Concentration changes for Fe(III) species during the dissolution of a mixture containing 2.3 g of uranium dioxide and 97.7 g of pyrite.

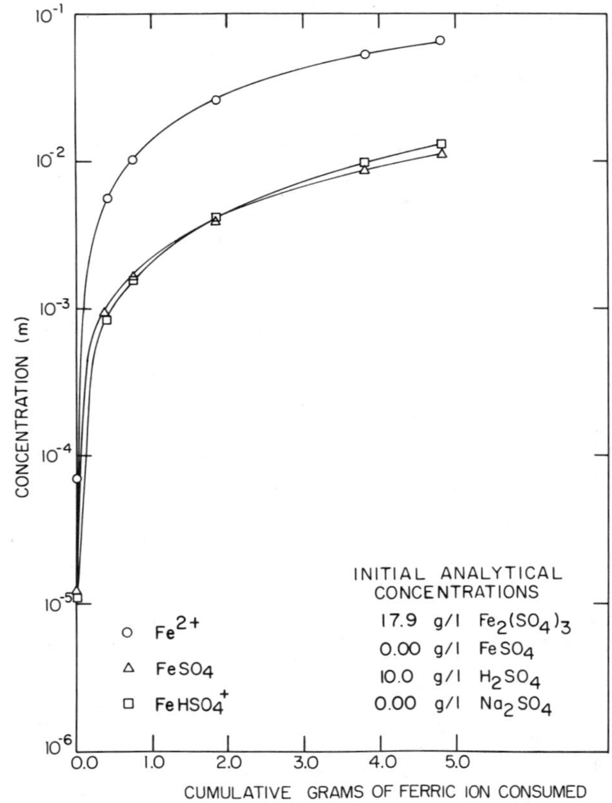

Fig. 13. Concentration changes for Fe(II) species during the dissolution of a mixture containing 2.3 g of uranium dioxide and 97.7 g of pyrite.

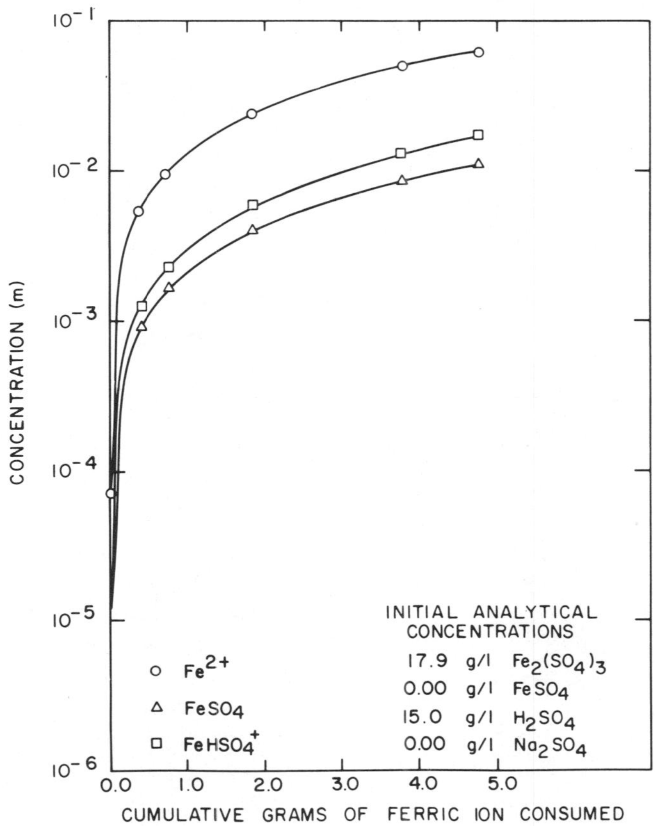

Fig. 14. Concentration changes for Fe(II) species during the dissolution of a mixture containing 2.3 g of uranium dioxide and 97.7 g of pyrite.

due to the fact that the $FeHSO_4^+$ species can accommodate the hydrolgen and sulfate generated during the dissolution of pyrite.

The addition of excess H_2SO_4 to give 15.0 g/l causes the $FeHSO_4^+$ species to dominate over the $FeSO_4^0$ species throughout the course of the leaching as shown in Fig. 14. The addition of 7.24 g/l Na_2SO_4 to a lixiviant with 10.0 g/l H_2SO_4 (Fig. 15) enhances the formation of $FeSO_4^+$ relative to the $FeHSO_4^+$ species. The changes in concentration using these conditions are similar to that for the 10.0 g/l H_2SO_4 run except that the excess sulfate delays the $FeSO_4^0/FeHSO_4^+$ inversion.

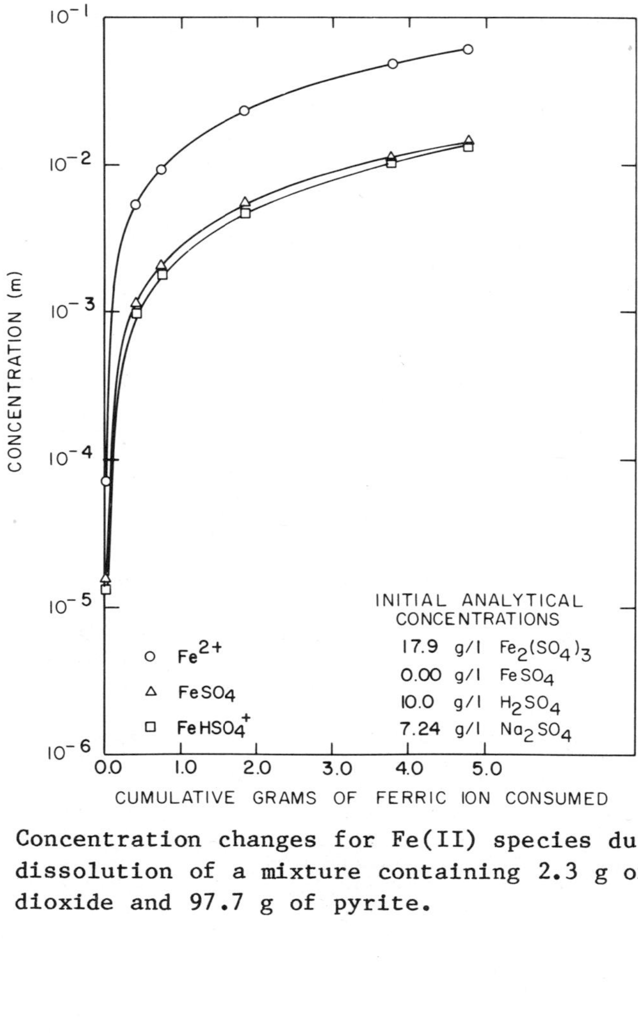

Fig. 15. Concentration changes for Fe(II) species during the dissolution of a mixture containing 2.3 g of uranium dioxide and 97.7 g of pyrite.

CONCLUSIONS

The modeling study indicated that important species present in solution for the $Fe_2(SO_4)_3$-$FeSO_4$-UO_2SO_4-H_2SO_2 systems are: H^+, SO_4^{2-}, HSO_4^-, UO_2^{2+}, $UO_2SO_4^0$, $UO_2(SO_4)_2^{2-}$, $UO_2(SO_4)_3^{4-}$, Fe^{3+}, $FeOH^{2+}$, $FeSO_4^+$, $Fe(SO_4)_2^-$, $FeHSO_4^{2+}$, Fe^{2+}, $FeSO_4^0$, and $FeSO_4^+$. The dissolution reactions for uranium dioxide and pyrite were found to be non-equilibrium processes and hence made possible the use of the partial equilibrium approach to model the overall leaching process. This type of model allowed for continual equilibrium to exist within the aqueous phase while permitting the irreversible dissolution reactions to proceed at kinetically limited rates.

The actual results obtained experimentally from the dissolution of uranium dioxide and pyrite showed fairly good agreement with those predicted by the model. These results indicate that optimal leaching conditions can be obtained by using a solution that has a low ferric:ferrous ratio. The limiting factor on the total iron concentration would be related to the maximum Fe(III) concentration which will lead to $Fe(OH)_3$ precipitation. The low ferric:ferrous ratio is desired since it inhibits the dissolution of pyrite, thus allowing the selective leaching of uranium dioxide. These results are in good agreement with those given by Laxen[21] for the leaching of South African uranium ores.

A more complete analysis on the dissolution process was not possible since the experimentally-obtained rate expressions were based only on the total Fe(III) and Fe(II) concentration. If data relating the rate of oxidation of the mineral to specific species present in solution were available, a more detailed analysis on conditions promoting the selective leaching of uranium dioxide could be determined.

The change in equilibrium occurring in the leaching solution due to changes in the composition of the initial leaching solution was analyzed relative to changes in the total sulfate composition. It was found that the presence of added sulfate tends to shift the equilibrium to the more sulfated species. Furthermore, the presence of sulfate was found to be important in increasing the solubility of both U(VI) and Fe(III) since both of these ions were mainly present as the highly sulfated species. The Fe(II) species on the other hand exhibited very little complexation with sulfate. It was also found that the leaching of pyrite had a much larger effect on the equilibrium reactions than exhibited by uranium dioxide since the dissolution of pyrite generates hydrogen and sulfate ions.

ACKNOWLEDGMENT

This research was supported in part by the U.S. Department of Energy, under Contract No. W-7405-ENG-82, by the Director for Energy Research, Office of Basice Energy Sciences, Chemical Sciences Program, WPAS-KC-03-02-02 and by the Department of Interior, Bureau of Mines under Grant No. G1106002.

REFERENCES

1. D. Langmuir: Geochim. Cosmochim. Acta, vol. 42:547, (1978).
2. M. E. Grimes: Solution Mining Symposium, Society of Mining Engineers, AIME, New York, N.Y., 338-53, 1974.
3. P. M. Boomer: Ph.D. Thesis, University of Texas, Austin, TX, 1979.
4. R. W. Potter: USGS Open File 76-835, U.S. Dept. of the Interior, Washington, D.C., 1977.
5. A. R. Amell and D. Langmuir: Open File Report 84-79, U.S. Bureau of Mines, Washington, D.C., 1978.
6. S. Sasmojo: Ph.D. Thesis, The Ohio State University, Columbus, OH, 1969.
7. W. H. Truesdall and B. F. Jones: J. Res. U.S. Geol. Surv., vol. 2:233-48, No. 2, 1974,
8. W. M. Latimer: Oxidation Potentials, 2nd ed., pp 300-4, Prentice-Hall, Inc., Englewood Cliffs, N.J., 1952.
9. C. C. Allen: M.S. Thesis, Iowa State University, Ames, IA, 1982.
10. H. C. Helgeson: Geochim. Cosmochim. Acta, vol. 32:853, 1968.
11. K. C. Liddell: Ph.D. Thesis, Iowa State Univ., Ames, IA, 1979.
12. K. C. Liddell and R. G. Bautista: Thermodynamics of Aqueous Systems with Industrial Applications, Am. Chem. Soc., Washington, D.C., pp 741-55, 1980.
13. K. C. Liddell and R. G. Bautista: Process and Fundamental Considerations of Selected Hydrometallurgical Systems, Society of Mining Engineers of AIME, New York, NY, 339-344, 1981.
14. K. C. Liddell and R. G. Bautista: Metall. Trans. B, vol. 12B:627-37, 1981.
15. K. C. Liddell and R. G. Bautista: Metall. Trans. B, vol. 14B: 5-15, (1983).
16. R. M. Smith and A. E. Martell: Critical Stability Constants, 4:1-12, 79-84, Plenum Press, New York, NY, 1976.
17. I. I. Chernyaeu, ed.: Complex Compounds of Uranium, pp. 150-171, Israel Program for Scientific Translations, Jerusalem, Israel, 1966,
18. R. S. Sapieszko, R. C. Patel, and Matijevic: J. Phys. Chem., vol. 81:1061-68, 1977.
19. C. F. Wells and M. A. Salam: J. Chem. Soc., Part A, 1968, pp 308-15.
20. R. M. Carlson, R. D. Norris, and R. Schellinger: Paper SPE 9483 55th Annual Fall Technical Conference and Exhibition of the Society of Petroleum Engineers of AIME, Dallas, TX, 1980.
21. P. A. Laxen: Unit Processes in Hydrometallurgy, Society of Mining Engineers of AIME, New York, NY, pp 143-66, 1968.

INTERFACIAL PHENOMENA IN LEACHING SYSTEMS

K. Osseo-Asare

Department of Materials Science and Engineering
The Pennsylvania State University
University Park, Pennsylvania 16802 USA

A review is presented of selected surface and colloid chemistry concepts concerned with interactions in solid/water systems. With the aid of these concepts, an attempt is made to elucidate the interfacial physico-chemical factors underlying selected hydrometallurgical leaching processes: (a) the adsorption-desorption model and the ion-transfer model of non-oxidative dissolution, (b) the effects of solid-state point defects on interfacial activity and dissolution, (c) the adsorption of hydrolyzable metal ions and how this affects metal losses in aqueous ammoniacal leaching systems, (d) interaction of aurocyanide complexes with activated carbon and carbonaceous ores, (e) the role of passive films in the dissolution of metallic phases with aqueous ammoniacal solutions, and (f) the enhancement of leaching rates via product layer modificaion with surface active agents.

INTRODUCTION

The term <u>interfacial phenomena</u> is used in recognition of the fact that the physico-chemical properties of the boundary between two phases generally deviate considerably from those of the respective homogeneous phases.[1-3] Hydrometallurgical systems are inherently heterogeneous. The three states of matter: solid, liquid and gas are all encountered and they give rise to interfaces, such as gas/liquid, liquid/liquid and solid/liquid. For example, solid/liquid interfaces are encountered during leaching, precipitation, solid/liquid separation, and electrowinning operations; liquid/liquid interfaces occur in solvent extraction systems; and solid/liquid/gas interfaces are encountered during leaching and gaseous reduction operations.

Adsorption and wetting phenomena occurring at interfaces play important roles in many aqueous metallurgical processes, e.g., leaching, flocculation of leach residues,[4-7] electrowinning,[8] hydrogen reduction of metals,[9] liquid-liquid separation,[10-12] solvent extraction,[13-15] ion flotation[16] and mass transfer from gas bubbles.[17]

In this paper, some pertinent fundamental concepts of colloid and interfacial chemistry[1-3,18-28] will be reviewed and illustrated with examples from several leaching systems. A companion paper[29] discusses the role of interfacial phenomena in hydrometallurgical solvent extraction processes.

SURFACE CHEMISTRY OF MINERAL/WATER SYSTEMS

Origin of the Surface Charge

When finely divided mineral particulates are immersed in water, they acquire a surface charge, the presence of which controls much of their interfacial behavior.[1,2,18-28] For example, when an oxide mineral is ground, the coordination bonds of the surface ions are severed. In the presence of water, there is an attempt by these exposed bonds to satisfy their coordination: the exposed cation pulls an OH^- ion, while the exposed oxygen ion pulls a proton. The result is a surface covered with a hydroxyl layer:

$$M_2O(s) + H_2O = 2MOH(s) \qquad (1)$$

It is the amphoteric dissociation of the surface hydroxyls which gives rise to the surface charge on oxides:

$$MOH(s) + H^+(aq) = MOH_2{}^+(s) \qquad (2)$$

$$MOH(s) = M^+(s) + OH^-(aq) \qquad (3)$$

$$MOH(s) = MO^-(s) + H^+(aq) \qquad (4)$$

Acid dissociation (or negative adsorption of H^+) gives negative surface sites, while basic dissociation (or positive adsorption of H+) gives positive surface sites. The magnitude of the surface

Table 1. Selected Values of the pzc's of Solid Oxides[18] and Sulfides.[20]

Material	pzc	Material	pzc
SiO_2	1.8	NiO	10.3
MnO_2	4.5	MgO	12.4
TiO_2	6.2	S°	2.0
α-Fe_2O_3	8.5	ZnS	4.0
α-Al_2O_3	9.1	PbS	3.0
CuO	9.5	FeS_2	6.2

charge (and therefore the surface potential) is controlled by H^+ and OH^- and, therefore, by analogy with reversible electrodes, these ions can be considered as the potential determining ions for mineral oxides in water.

The value of the potential determining ion activity in the aqueous solution for which the net surface charge is zero, is termed the zero point of charge, or the pzc. The pzc is a convenient means of characterizing the electrochemical properties of the solid/liquid interface. In the case of mineral oxides and sulfides, the pzc is given by the pH of the aqueous solution for which the adsorption densities of H^+ and OH^- on the mineral surface are equal. Since the surface charge on the solid depends on whether the surface hydroxyls undergo acid or basic dissociation, the relative pzc of minerals will be determined by the relative basicity or acidity of their respective surface hydroxyls. The more acidic the surface hydroxyls, the lower the pzc. Table 1 presents some representative pzc values.[18,20]

The surface potential, ψ_o, is the potential difference between the solid surface and the bulk of the solution. In general, the potential of the bulk of the solution is defined as zero, so that in the absence of specific adsorption, ψ_o is zero at the point of zero charge. The potential determining ions alter the surface potential on transfer from the solution phase to the solid surface and vice versa. At a given activity of potential determining ions a_+ and a_-, ψ_o is given by the Nernst equation:

$$\psi_o = \frac{RT}{z_+F}\ln\left(\frac{a_+}{a_+^o}\right) = \frac{RT}{z_-F}\ln\left(\frac{a_-}{a_-^o}\right) \quad (5)$$

where R is the gas constant, T is the absolute temperature, F is the Faraday constant, a_{+}^{o}, a_{-}^{o} are the activities of the potential-determining ions at the pzc, and z_{+} and z_{-} are the corresponding ionic charges (including the sign).

The Electrical Double Layer

The charging of the solid surface gives rise to a separation of electric charge such that the solid and solution acquire opposite charges at their phase boundary. This is the electrical double layer.[1,2,19] The double layer arises because in order to preserve electroneutrality, the surface charge on the solid must be balanced by ions of opposite charge, i.e., the counter ions. Of these ions, those that are adsorbed only by electrostatic attraction are called indifferent electrolyte ions; those ions which are adsorbed by other means in addition to electrostatic forces are termed specifically adsorbed ions. Figure 1 presents a schematic illustration of the interfacial charge and potential distributions associated with the electrical double layer. The interfacial region may be divided into two parts. The first consists of the layer of counter-ions directly adjacent to the solid surface; these ions are generally treated as having finite size, i.e., they are considered to reside at a distance δ (called the Stern plane) from the solid surface; the potential in the Stern plane is denoted by ψ_{δ}. Specifically adsorbed ions are located in the Stern plane. The second region of counter-ions lies beyond the Stern plane and is considered

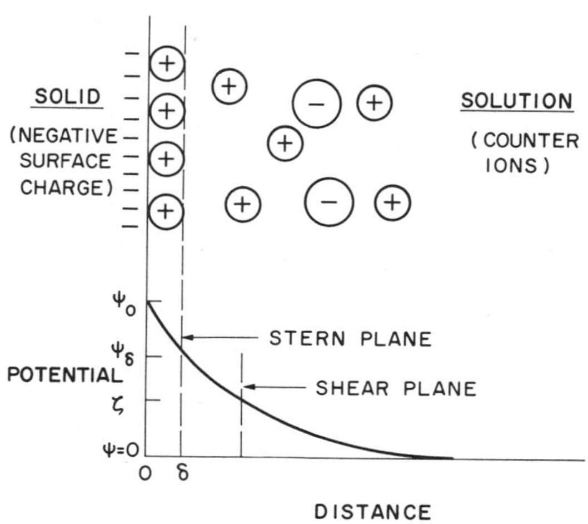

Fig. 1. The electrical double layer.

as a diffuse cloud of ions which behave as point charges. Theoretical analysis of the electrostatics of the double layer shows that the charge density in the aqueous solution decreases rapidly with the distance from the solid surface and that the potential, ψ, declines monotonically, as a functon of distance. For small surface potentials ($zF\psi_o << RT$) in the absence of specific adsorption, the following relations can be derived[1,2] for the potential (ψ) and the surface charge density (σ)

$$\psi(s) = \psi_o \exp-\kappa x \tag{6}$$

$$\sigma = \varepsilon\kappa\psi_o/4\pi \tag{7}$$

where x is the distance measured from the solid surface into the aqueous side of the interface, and κ is the reciprocal of the effective thickness of the diffuse layer. If n_i and z_i represent respectively the bulk concentration and the charge on the ith ion, then κ is given by:

$$\kappa^2 = 4\pi e^2 \sum_i n_i z_i^2/\varepsilon\kappa T \tag{8}$$

As a result of the charge on their surfaces, colloidal particles immersed in water can move under the influence of an electrical potential gradient, (E). This motion, called electrophoresis, is controlled by the zeta potential, ζ, the electrical potential at the slipping plane between the solid and liquid phases. Electrophoresis measurements therefore offer a convenient means of studying the double layer properties at the solid/water interface.[1,2,18-28] The relationship between E, ζ, and the velocity of motion (v) of the particles is derived by performing a force balance according to which the force exerted by the field E on the charged particle is balanced by the viscous drag opposing the particle movement. The final expression is of the form

$$v/E = (\varepsilon\zeta/6\pi\eta)[C_1 + C_2\zeta + C_3\zeta^2] \tag{9}$$

where η is the viscosity of the aqueous medium, and the coefficients C_1, C_2, C_3 are complicated functions of the double layer thickness ($1/\kappa$), the mobilities of the electrolyte ions, and the particle size (a). The electrophoretic mobility becomes directly proportional to the zeta potential, and also independent of particle size for the following limiting conditions:

$$\kappa a << 1, \quad v/E = \varepsilon\zeta/6\pi\eta \tag{10}$$

$$\kappa a >> 1, \quad v/E = \varepsilon\zeta/4\pi\eta \tag{11}$$

Adsorption at the Solid/Water Interface

Both ionic and nonionic species are capable of adsorbing at the solid/aqueous interface. The free energy of adsorption, ΔG_{ads}, is of the general form,

$$\Delta G_{ads} = \Delta G_{coul} + \Delta G_{sp} \tag{12}$$

The term $\Delta G_{coul}[=z_i F\psi(x)]$ represents the electrostatic energy due to the interaction of a charged adsorbing species with the double layer potential. The specific energy term, ΔG_{sp}, accounts for interactions other than coulombic and can be expressed as

$$\Delta G_{sp} = \Delta G_{solv} + \Delta G_{chem} \tag{13}$$

or

$$\Delta G_{sp} = \Delta G_{cc} + \Delta G_{cs} + \Delta G_{chem} \tag{14}$$

depending on whether the adsorbing species are inorganic ions (Eq. 13) or organic surface active agents (Eq. 14). The term ΔG_{solv} is the change in the solvation energy of an ion on moving from the bulk solution to the interfacial region (Fig. 2a). The terms ΔG_{cc} and ΔG_{cs} represent respectively, the free energy change due to hydrophobic interactions of the organic groups of surfactant molecules with each other and with the solid surface (Fig. 2b). Finally the term ΔG_{chem} accounts for the formation of chemical bonds between the solid surface and the adsorbing species. The process of adsorption is termed <u>physical</u> if the electrostatic and hydrophobic interactions constitute the driving force for adsorption. On the other hand, when the adsorbing species form chemical bonds with ions or atoms in the solid surface, the process is termed <u>chemisorption</u>.

The solvation energy term arises because the dielectric constant of the interfacial region (ε_i) is much lower than that of the bulk solution (ε_b) (Fig. 2). The solvation energy change can be expressed as:[25,28]

$$\Delta G_{solv,i} = (z_i^2 N/16\pi\varepsilon_o)[\alpha(1/\varepsilon_i - 1/\varepsilon_b + \beta(1/\varepsilon_s - 1/\varepsilon_i)] \tag{15}$$

where N is the Avogadro number, ε_o is the permittivity of free space, ε_s is the dielectric constant of the solid, and the terms α and β are functions of the radius of the inorganic ion (r_i) and the radius of the water molecule (r_w):

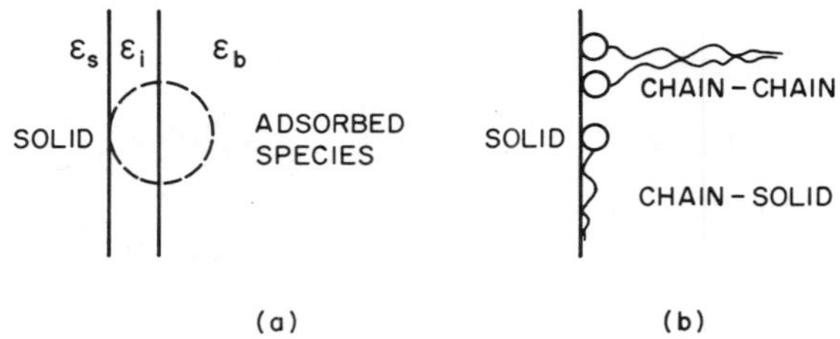

Fig. 2. Adsorption of ions at the solid/water interface.
(a) adsorption of an inorganic species in the Stern plane.
(b) hydrophobic chain-chain and chain-solid interactions.

$$\alpha = 1/(r_i + 2r_w) - r_i/2(r_i + 2r_w)^2 \tag{16}$$

$$\beta = 1/2(r_i + 2r_w) \tag{17}$$

Details of the computation of the interfacial dielectric constant are given by James and Healey.[25]

The hydrophobic bonding contributions to the specific free energy of adsorption depend on the organic chain length. The greater the number of CH_2 groups, the larger the negative value of the chain interaction terms. Thus for straight chain surfactants:

$$\Delta G_{cc} = N_c\phi_{cc} \text{ and } \Delta G_{cs} = N_c\phi_{cs} \tag{18}$$

where N_c represents the number of CH_2 groups per molecule, and ϕ_{cc} and ϕ_{cs} respectively, the molar cohesive free energy associated with each CH_2 group for chain-chain and chain-solid interactions. The free energy decrease accompanying the removal of a CH_2 group from water (i.e. chain-chain interaction) is about 1RT per mole.[19] On the other hand, if the adsorbed surfactant consists of horizontally oriented chains, the accompanying decrease in free energy is found to be about 1/2 RT per mole of CH_2 groups.[21,22] This is because even though half the chain surface is next to the solid surface, the other half is still exposed to water.

The adsorption density of a specifically adsorbed species (i.e. species located in the Stern plane) is given by

$$\Gamma_\delta = 2rC\exp(-\Delta G_{ads}/RT) \tag{19}$$

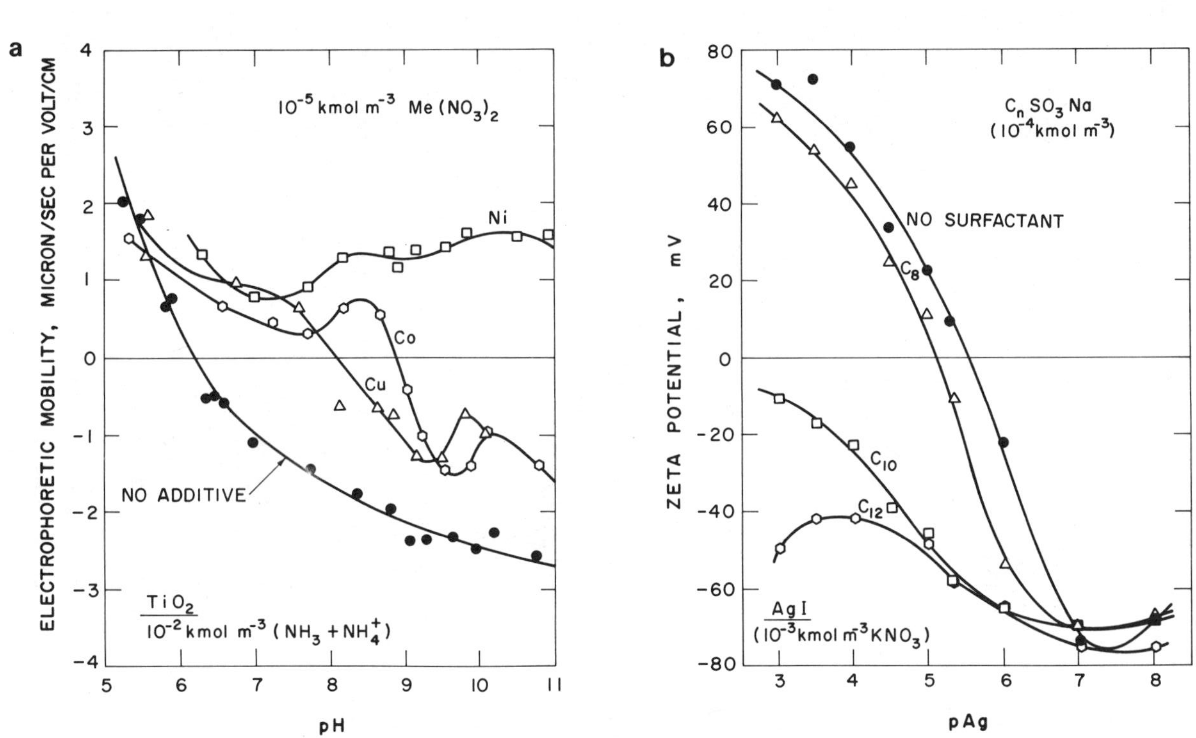

Fig. 3. (a) The effect of pH on the electrophoretic mobility of titania in the absence and presence of 10^{-5} kmol m^{-3} $Me(NO_3)_2$ (Me = Cu, Ni, Co); 10^{-2} kmol m^{-3} (NH_3 + $NH_4{}^+$).[27]
(b) The effect of pAg on the zeta potential of AgI in the absence and presence of 10^{-4} kmol m^{-3} sodium alkyl sulfonates: octadecyl (C_8), decyl (C_{10}) and dodecyl (C_{12}); 10^{-3} kmol m^{-3} KNO_3.[22]

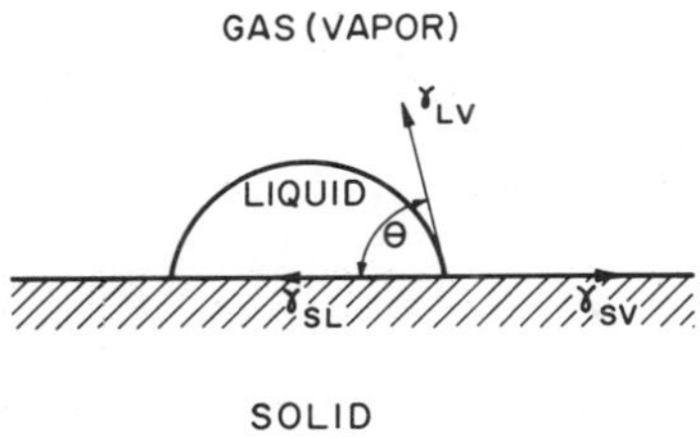

Fig. 4. Interfacial tension and contact angle.

where r represents the effective radius of an ionic species or the radius of the polar group of an organic surfactant molecule, and C is the bulk concentration of the adsorbing spcies. The magnitude as well as the sign of the double layer potential may be altered by adsorption of inorganic or organic species, as illustrated by the electrophoretic mobility and zeta potential results presented in Fig. 3.[22,27]

Interfacial Tension and Contact Angle

Interfacial tension refers to the work required to create a unit area of interface. Usually, the interfacial tension at the solid/air or liquid/air interface is referred to simply as surface tension. Consider a drop of liquid lying on a flat plate as shown in Fig. 4. The angle which the drop makes with the solid plate is termed the contact angle (θ). The equilibrium condition satisfies the relation:

$$\gamma_{VS} = \gamma_{SL} + \gamma_{LV}\mathrm{Cos}\,\theta \qquad (20)$$

Thus decreasing γ_{LV} or γ_{SL} decreases θ (for $\theta<90°$). Surface active agents are used to achieve this. When an organic compound adsorbs on a solid, it imparts some of its interfacial characteristics to the solid surface. As can be seen from Table 2, the surface tensions of organic liquids are much lower than that of water. Thus adsorption of an organic surface active agent can lead to a lowering of γ_{SL} which can in turn lead to a change in the contact angle as illustrated in Fig. 5.[19] Furthermore, as discussed in more detail elsewhere in this book[29], organic surface active agents can also decrease interfacial tensions at the liquid/gas interface as a result of adsorption.

Table 2. Surface and Interfacial Tensions of Some Liquids at 20°C.

	Liquid/Air γ(mN m^{-1})	Water/Liquid γ(mN m^{-1})
Water	72.75	
Benzene	28.85	35.0
n-actanol	27.53	8.5
n-hexane	18.43	51.1

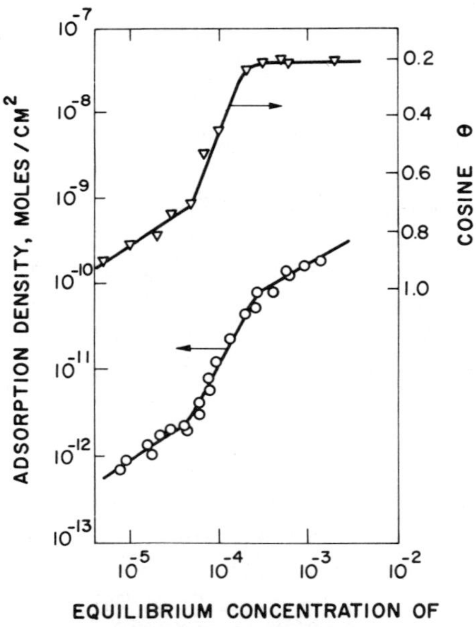

Fig. 5. Effect of the adsorption of sodium dodecyl sulfonate on the contact angle on alumina at pH 7.2.[19]

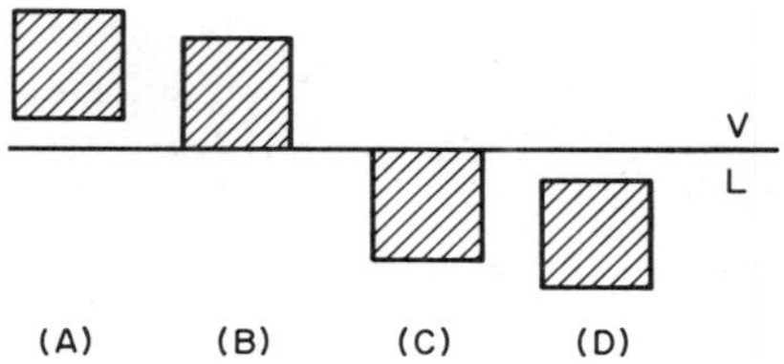

Fig. 6. Stages in the wetting process:
(A) → (B) Adhesional wetting
(B) → (C) Immersional wetting
(C) → (D) Spreading wetting

Wettability

Wetting is usually discussed quantitatively in terms of interfacial tensions and contact angles.[1,2,30] Consider a hypothetical particle in the form of a cube of unit dimensions. Figure 6 illustrates three steps leading to the complete wetting of this particle by liquid. The first stage is termed adhesional wetting (Fig. 6a) and has associated with it a work of adhesion W_a given by

$$W_a = (5\gamma_{SV} + \gamma_{SL}) - (6\gamma_{SV} + \gamma_{LV}) = \gamma_{SL}-(\gamma_{SV}+\gamma_{LV})$$

$$= -\gamma_{LV}(\cos\theta + 1) \qquad (21)$$

Equation 21 expresses the fact that there is a change in the net surface energy when an initial condition consisting of a L/V interface and a S/V interface is collapsed into a S/L interface. In the second stage, immersion wetting, the particle is wetted on all sides but one, and the work of immersion is given by:

$$W_i = (5\gamma_{SL} + \gamma_{SV}) - (5\gamma_{SV}+ \gamma_{SL}) = 4\gamma_{SL} - 4\gamma_{SV}$$

$$= -4\gamma_{LV} \cos\theta \qquad (22)$$

The final stage wherein the particle is completely surrounded by the liquid is termed spreading wetting. The corresponding work of spreading is given by

$$W_s = (6\gamma_{SL} + \gamma_{LV}) - (5\gamma_{SL} + \gamma_{SV}) = (\gamma_{SL} + \gamma_{LV})-\gamma_{SV}$$

$$= -\gamma_{LV}(\cos\theta-1) \qquad (23)$$

The net work of wetting is then the sum of W_a, W_i, and W_s.

When the value of W is negative, the process is spontaneous, while a positive W value implies that work must be done in order for the process to take place. Thus it follows that

$$W_a < 0 \text{ if } \theta < 180° \text{ or } \gamma_{SV} + \gamma_{LV} > \gamma_{SL} \quad (24)$$

$$W_i < 0 \text{ if } \theta < 90° \text{ or } \gamma_{SV} > \gamma_{SL} \quad (25)$$

$$W_s < 0 \text{ if } \theta < 0 \text{ or } \gamma_{SV} > \gamma_{SL} + \gamma_{LV} \quad (26)$$

It can be seen from Eqs. 24 to 26 that decreasing γ_{SL} should enhance wetting. As discussed above, surface active agents provide a means of altering interfacial tensions.

Inter-Particle Forces

The inter-particle forces of interest in particulate dispersions can be divided into three types: (a) attractive: London-van der Waals forces; (b) repulsive: electrical double layer forces; and (c) repulsive: adsorbed layer forces. The London-van der Waals forces are always present. Thus a dispersion will be stable only if a stronger repulsive force is available to oppose the attraction.

London-van der Waals Forces. The general attractive force present in dispersions is an electromagnetic effect. It is due to: (a) interaction between permanent dipoles of the molecules present in the particle; (b) interaction between transient induced dipoles (polarizability of an atom); and (c) interaction between permanent and induced dipoles (polarizability of a molecule).

The van der Waals interaction energy between two equal spheres is of the form

$$V_A = -\frac{A}{6} f(a,R) \quad (27)$$

where a = radius of a sphere; R = center to center distance ($=2a+H_o$); and H_o = smallest distance between the surfaces of the spherical particles.

The term A represents the Hamaker constant and for liquids has the range:

$$5 \times 10^{-13}(H_2O) \leq A(\text{erg}) \leq 4 \times 10^{-12} \ (CCl_4, \text{ highly polarizable})$$

For small distances,

$$V_A \approx -\frac{Aa}{12H_o} \tag{28}$$

Electrostatic Forces. As two charged particles of the same sign approach each other, they experience mutual repulsion because of the interaction of their double layers. In the absence of other effects, the particles therefore tend to drift away from each other. For small surface potentials, the repulsive energy V_R for two similar particles is given by:

$$V_R = \frac{\varepsilon a \psi^2}{2} \ln\{1 + \exp(-\kappa H_o)\} \tag{29}$$

Adsorbed Layer Repulsion Forces. The presence of adsorbed layers affects both the repulsive and attractive forces between particles. Let us consider two spherical particles of radius a with adsorbed layers of thickness δ. Further, let us for the sake of convenience assume that the adsorbed molecules are uncharged. In the first place, the mere presence of an adsorbed layer of thickness δ means that the effective distance of approach of the two particles has been decreased by an amount 2δ. Thus, since the electrical double layer potential decreases monotonicaly with distance, the approaching particles will meet each other at lower electrical potentials, thus leading to a decrease in the repulsive energy. The presence of an adsorbed layer also affects the magnitude of the van der Waals attractive energy. The value of V_A now assumes the form:[31]

$$V_A = -\frac{1}{12}\left\{A^*_{32}\left(\frac{a+\delta}{\Delta}\right) + A^*_{13}\left(\frac{a}{\Delta+2\delta}\right) + \frac{4a(A^*_{32}A^*_{13})^{1/2}(a+\delta)}{(\Delta+\delta)(2a+\delta)}\right\} \tag{30}$$

where the subscripts 1, 2 and 3 refer to the solid, the liquid medium and the adsorbed layer, Δ is the effective distance (i.e. from the outer surface of the adsorbed layers) between the two spherical particles, A^*_{32} is the effective Hamaker constant for particles of the adsorbed material dispersed in the liquid medium, and A^*_{13} is the effective Hamaker constant for particles of the solid dispersed in a medium of the adsorbed material. It has been demonstrated theoretically by Ottewill[31] that for particles of radius 230Å, adsorbed layers of thickness 20Å are capable of lowering the interparticle attraction at Δ=5Å to about 5% of the value obtained in the adsorbed layer-free system.

The presence of adsorbed layers also introduces additional forces including "surface", "osmotic", and "chain elasticity" effects. The surface effect has to do with the fact that the approach of two adsorbed layers imposes a restriction on the movement of the adsorbed molecules. In the thermodynamic sense, this restriction in movement is equivalent to a reduction in entropy, which constitutes an increase in the free energy of the system. Thus the particles will tend to move apart in an effort to decrease the free energy. The osmotic effect arises because when the two adsorbed layers overlap, the liquid in the overlapping volume will have an abnormally high concentration of solute molecules i.e., the adsorbed molecules. Thermodynamically, this corresponds to an increase in the osmotic pressure in the overlapping volume. This high interparticle pressure will act to separate the particles. When two particles with adsorbed layers bump into each other, the adsorbed layers may be compressed in the process. Thus if the adsorbed molecules possess some degree of elasticity, there will be a tendency for the particles to spring apart as the stress in the adsorbed layer is released. This constitutes the chain elasticity effect.

DLVO Theory of Colloid Stability. In order to treat the stability of colloidal dispersions quantitatively, all the forces acting in the system must be considered. The DLVO theory (Derjaguin, Landau, Verwey, Overbeek) considers colloid stability in terms of electrical double layer and London-van der Waals forces, i.e., the total interaction energy, V_T, is given as,

$$V_T = V_R + V_A \tag{31}$$

Curves of the total energy of interaction against the distance of interaction between two particles can be calculated by using the appropriate expressions for V_R and V_A (e.g. Eqs. 28 and 29). The characteristic features of these curves are (a) the primary minimum, which is a deep well at a small separation, (b) the secondary minimum, a shallower minimum at larger separation, and (c) the energy barrier (V_M) between these two minima; the magnitude of this barrier controls stability. The stability of a colloidal dispersion breaks down when V_M is reduced to an order of magnitude comparable to the energy of the particles associated with their Brownian motion. The parameters which determine the form of these curves are the surface potential, dielectric constant, ionic charge, concentration of ions, temperature, Hamaker constant, and particle dimension.

Dispersion of Particles in Fluids

The term dispersion describes the combination of processes which enable a powder to distribute uniformly in a liquid as discrete particles.[30] Hydrometallurgical operations, e.g. leaching, are often inefficient unless the proper degree of dispersion is maintained in the slurry supplied to the processing device. As illustrated in Fig. 7, dispersion may be treated as a three-stage process consisting of (a) wetting, (b) disaggregation, and (c) stabilization. Powders usually consist of primary particles which adhere to one another to form lumps, clusters, aggregates or agglomerates. The liquid medium must perform two functions: (a) it must coat the outside surface of the clusters; and (b) it must coat the inside surface of the clusters (this involves displacement of air from the pores of the clusters). Basically, what is taking place both in (a) and (b) is the conversion of the solid/air interface to a solid/liquid interface.

The purpose of disaggregation is to break inter-particle bonds so that all external surfaces of the primary particles may be exposed to the liquid medium. It must be noted that the fact that particles have been unstuck from each other does not necessarily mean that they are dispersed. Suspended particles are not static, they move as a result of Brownian motion; this random motion can lead to particle-particle collisions which in turn may lead to flocculation. This tendency towards flocculation results from the presence of the attractive London-van der Waals forces. The necessary repulsion may be achieved by causing the solid surface to attain electrical charges so that the particles repel each other, or by introducing adsorbed layers.

Fig. 7. Dispersion of particles in fluids.

INTERFACIAL EFFECTS IN LEACHING SYSTEMS

The Effect of the pzc on the Dissolution of Simple Oxides

As discussed above, an oxide surface acquires a hydroxyl layer (MOH(s), Eq. 1) which upon further interaction with water produces positive and negative surface sites ($[MOH_2^+(s)]$, $[MO^-(s)]$, Eqs. 2-4). With this as a basis, Warren and Devuyst[32,33] have proposed a general model to account for the dissolution of simple oxides in acidic solutions.

Let the equilibrium constant for the formation of the positive site be K_1:

$$MOH(s) + H^+(aq) = MOH_2^+(s) \qquad K_1 \tag{32}$$

An anion A^- (from the acid HA) may adsorb onto the protonated surface site:

$$MOH_2^+(s) + A^-(aq) = MOH_2A(s) \qquad K_2 \tag{33}$$

Dissolution occurs when the surface comlexes undergo desorpton reactions with rate constants k_1 and k_2 respectively:

$$MOH_2^+(s) \rightarrow MOH_2^+(aq) \qquad k_1 \tag{34}$$

$$MOH_2A(s) \rightarrow MOH_2A\ (aq) \qquad k_2 \tag{35}$$

With equations 34 and 35 as the rate determining steps, the rate of dissolution, r_M is given by

$$r_M = \frac{k_1K_1a_H}{1+K_1a_H} + \frac{k_2K_2a_Ha_A}{1 + K_1a_H} \tag{36}$$

where a_H and a_A are respectively the aqueous phase activities of the proton and the anion. When the anion adsorbs very weakly on the oxide surface, the rate of dissolution is given by the first term on the right hand side of Eq. 36. On the other hand, when the anion has a strong affinity for the oxide surface, the second term predominates. The resulting rate versus hydrogen activity profiles are illustrated schematically in Fig. 8.

In perchloric acid solutions, Warren and Devuyst[32,33] found the desorption of the protonated surface complex (Eq. 34) to be rate-

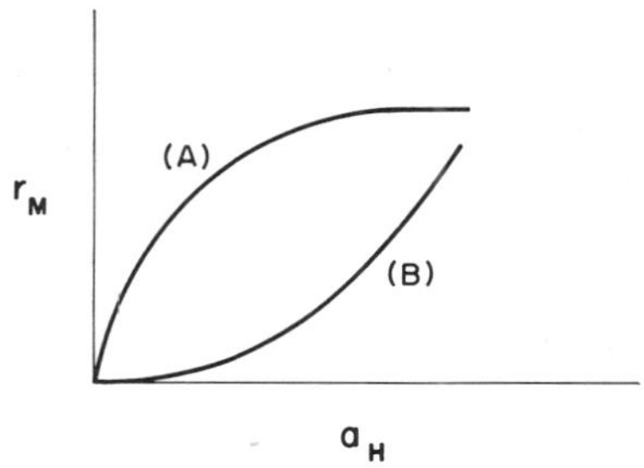

Fig. 8. Dissolution of simple oxides: rate limited by desorption of (a) protonated complex, (b) anion complex.

limiting for Cu_2O, CuO, and α-Fe_2O_3. A similar behavior was found for CuO, and α-$Al_2O_3 \cdot 3H_2O$ in hydrochloric acid solutions. On the other hand, in HCl solutions, Cu_2O and α-Fe_2O_3 dissolution rates were controlled by the desorpton of the surface anion complex (Eq. 35). The difference in $HClO_4$ and HCl behavior with respect to Cu_2O and α-Fe_2O_3 reflects the greater affinity of the smaller Cl^- anion for the oxide surface compared with the much larger ClO_4^- anion. The order of reactivity was:

$$Cu_2O > CuO > Al_2O_3 \cdot 3H_2O > Fe_2O_3 \quad (37)$$

Table 3 lists the surface adsorption constants which were obtained[32,33] when the leaching model was applied to the experimental results. In addition, this table presents values of the pzc and the stability constants for the chlorocomplexes. It can be seen that the oxides with the highest dissolution rates (Eq. 37) also have the highest pzc and K_1 values. This trend can be interpreted in terms of the surface chemistry of the oxide/water system. Let the equilibrium constant for the formation of the surface negative sites be K_3:

$$MOH(s) = MO^-(s) + H^+(aq) \qquad K_3 \quad (38)$$

The equilibrium condition between the positive and negative sites is obtained by combining Eqs. 32 and 38:

$$MO^-(s) + 2H^+ = MOH_2^+(s) \qquad K_4 = K_1/K_3 \quad (39)$$

At the pzc, the surface has no net charge and therefore the number of positive sites must equal the number of negative sites. It follows then that

$$\log K_4 = \log(K_1/K_3) = 2pzc \quad (40)$$

Table 3. Correlation Between Oxide Dissolution and Surface and Bulk Chemical Parameters.[32,33]

Oxide	Rate⁺	pzc	K_1 (Eq. 32)	K_2 (Eq. 33)	log K(Cl)*
Cu_2O	1	>11.5	19	$>2x10^4$	2.70
CuO	2	9.5	5	$6x10^{-1}$	0.40
α-$Al_2O_3 \cdot 3H_2O$	3	9.1	3	$3.3x10^{-1}$	-
α-Fe_2O_3	4	8.5	1.2	$1.4x10^{-1}$	1.48

⁺ The oxide with the highest reaction rate is indicated as number 1, the next highest as number 2, etc.

* log K values are taken from Smith and Martell (34) and are for the formation of the first chlorocomplex at zero ionic strength.

Therefore a large pzc is necessarily associated with a large value of the surface protonation constant K_1. It can be seen from Eq. 36 that if an oxide is dissolving with the desorption of the protonated complex rate-determining, then an increase in K_1 should increase the dissolution rate, in agreement with the experimental results. Table 3 also shows that Cu(I) and Fe(III) chloro complexes have the largest log K(Cl) values, and must therefore be the strongest complexes. The oxides of Cu(I) and Fe(III) are also the two oxides whose dissolution rate behavior in HCl was found to be limited by anion desorption.

Further insight into the role of surface chemical effects in dissolution processes is provided by the ion-transfer model of nonoxidative dissolution.[35-39] According to this model, the nonoxidative dissolution of ionic solids, i.e. dissolution under conditions where there is no apparent change in the oxidation state of the consituent ions, is controlled by the double layer potential.

Consider the acid dissolution of a metal hydroxide, MOH:

$$MOH(s) + H^+(aq) = M^+(aq) + H_2O \quad (41)$$

The dissolution reaction may be treated as an ion-removal process in which the cation and anion are released separately:

$$M^+(s) \rightarrow M^+(aq) \qquad k_+ \quad (42)$$

$$OH^-(s) + H^+(aq) \rightarrow H_2O \qquad k_- \quad (43)$$

where k_+ and k_- are respectively the rate constants for the cation and anion reactions. The rate of release of cations (r_+) and anions (r_-) is given by Eqs. 44 and 45 respectively:

$$r_+ = k_+\Gamma_+ \exp(\alpha_+ z_+ F\psi/RT \tag{44}$$

$$r_- = k_-\Gamma_- a_H \exp(\alpha_- z_- F\psi/RT) \tag{45}$$

where α_+ and α_- are respectively transfer coefficients for cations and anions, Γ_+ and Γ_- represent respectively the number of cations and anions per unit area at the solid surface, a_H is the hydrogen ion activity in the aqueous phase and ψ is the electrical potential at the solid surface.

If the potential determining ions are the protons, and the relevant potential is taken to be the surface potential (see Eq. 5), it follows that Eqs. 44 and 45 can be rewritten as,

$$r_+ = k_+\Gamma_+ (a_H/a_H^o)^{\alpha_+ z_+} \tag{46}$$

$$r_- = k_-\Gamma_+ a_H (a_H/a_H^o)^{\alpha_- z_-} \tag{47}$$

where a_H^o is the proton activity at the pzc. Therefore,

$$\log r_+ = \log (k_+\Gamma_+) + \alpha_+ z_+ \mathrm{pzc} - \alpha_+ z_+ \mathrm{pH} \tag{48}$$

$$\log r_- = \log (k_-\Gamma_-) + \alpha_- z_- \mathrm{pzc} - (1+\alpha_- z_-)\mathrm{pH} \tag{49}$$

It can be seen from Eq. 48 that if the dissolution is controlled by cation removal (z_+ is positive) then the rate is high when the pzc is high, in agreement with the Warren Devuyst adsorption model (see Table 3). On the other hand, according to Eq. 49, a high pzc decreases the dissolution rate if anion removal (z_- is negative) is rate-controlling. The different effects of pzc, illustrated in Fig. 9, can be understood by recalling that a mineral with a high pzc is highly positively charged in acidic solutions. Thus if cation removal is rate-limiting, a high positive double layer potential should enhance the repulsion of cations from the surface thereby promoting dissolution. Similar arguments can be made for the case where the rate is controlled by anion removal.

The fact that the electrical potential at the solid/water interface affects the rate of nonoxidative dissolution has been demonstrated by Engell[35] who found that the dissolution current density of nonstoichiometric FeO decreases with increase in the

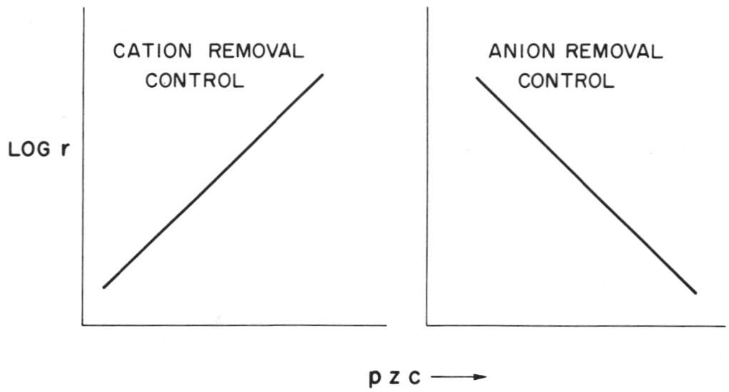

Fig. 9. Relationship between dissolution rate and pzc.

anodic potential, while the opposite effect was obtained with non-stoichiometric CdO. Similar results have been reported by others.[37,39] Theoretical analysis[35-37] based on the ion-transfer model indicates that for a given solid, the relationship between the rate of dissolution and applied potential may be represented as illustrated in Fig. 10.[37] The potential at which the cation removal rate equals the anion removal rate has been termed the freely dissolving potential, E_f. Dissolution maxima have been reported by Engell[35] and Nicol[39] for nonstoichiometric Fe_3O_4. Similar results have been reported by Valverde[40] for $Fe_{0.92}O$, Fe_3O_4, and Fe_2O_3; in these cases, the necessary changes in potential were achieved by introducing a variety of redox couples into the aqueous solution.

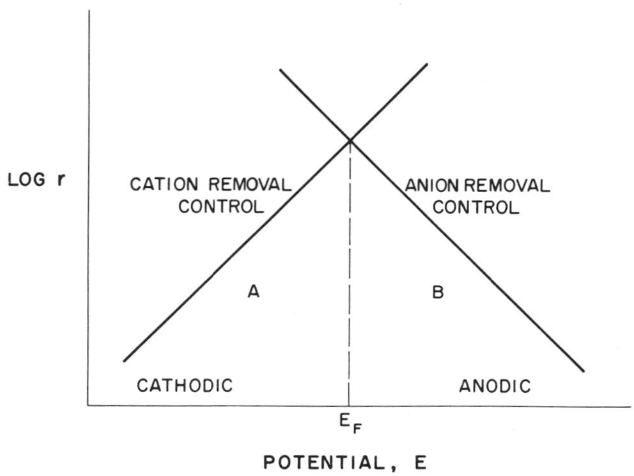

Fig. 10. Relationship between dissolution rate and potential.

Table. 4. Effects of Various Redox Couples on the Dissolution Rates (h^{-1}) of Metal Oxides in Acidic Solutions.[40]

		Dissolution rate (h^{-1})					
Redox couple	E(Ag/AgCl) (Volt)	$Fe_{0.92}O$ (50°C)	Fe_3O_4 (50°C)	Fe_2O_3 (50°C)	CoO (50°C)	NiO (90°C)	CuO (50°C)
Ti(III), Ti(IV)	-0.230	0.014	0.025	0.023	0.024	<0.003	0.280
V(III), V(IV)	0.200	0.45	0.074	0.038	0.013	<0.003	0.120
I^-, I_2	0.364	0.003	0.002	<0.001	0.008	<0.003	0.065
$C_6H_4(OH)_2$, $C_6H_4O_2$	0.414	0.001	0.001	<0.001	0.004	<0.001	0.055
Fe^{2+}, Fe^{3+}	0.520	-	-	-	0.001	0.005	0.065
Br^-, Br_2	0.898	<0.001	<0.001	<0.001	0.0006	0.012	0.055
Ce(III), Ce(IV)	1.270	<0.001	<0.001	<0.001	<0.0006	0.215	0.980

Leaching of Semiconducting Minerals

The effects of potential on dissolution are easiest to view in terms of Eqs. 44 and 45 when the dissolution involves a stoichiometric solid, the variation in potential is achieved by means of changes in pH (Eq. 5) or the cation and anion do not exhibit variable oxidation states. Where two or more oxidation states are possible (e.g. Fe(II), Fe(III); Co(II), Co(III)), application of external potential can lead to redox reactions which can complicate the interpretation of the observed results.[41,42] Table 4 presents a summary of the findings obtained by Valverde.[40] The increase in the dissolution rate of Fe_3O_4 and Fe_2O_3 with decreasing potential (up to 0.2 volt) can be attributed to the increasing reduction of Fe^{3+} in the solid to give the more soluble Fe^{2+} (the Fe(III)-O bond in the solid is much stronger than the corresponding Fe(II)-O bond). Furthermore, where the solids are not completely ionic, application of external potential can have a significant effect on the potential and charge distribution in the space charge region. Thus Diggle[37] has suggested that the trend shown in Fig. 10 is valid only for p-type semiconductors; in the case of n-type semiconductors, the reverse would be observed, i.e. region A would be anion-removal controlled while region B would be cation-removal controlled.

The surface chemistry of a mineral can be altered through the modification of the bulk solid-state properties. Figure 11 shows the effect of the presence of Li^+ and Cr^{3+} as solid-state dopants on the dissolution of NiO in sulfuric acid solution.[43] It can be

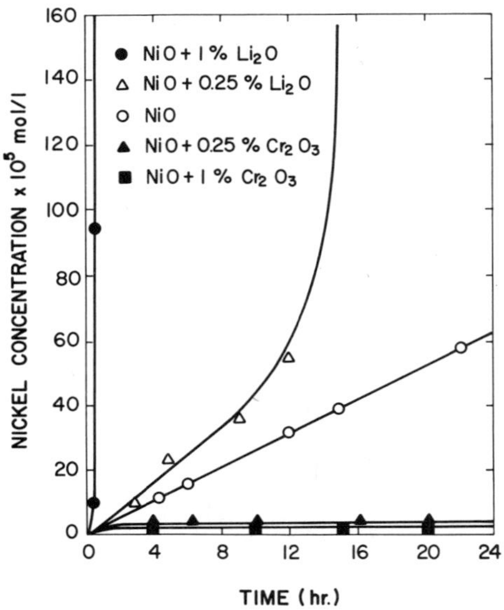

Fig. 11. Dissolution of pure and doped NiO (0.38 kmol m^{-3} H_2SO_4, 60°C).[43]

seen that the presence of Li^+ increases dissolution while Cr^{3+} has the opposite effect. It was found by Lussiez et al[43] that whereas the 1 mol% Li_2O-doped material dissolved completely in 75 minutes, the undoped and the Cr-doped NiO samples attained only 2.5% and 1% dissolution respectively in 48 hours.

These results have been interpreted[43] in terms of the effects of the dopants on the defect structure of NiO, and the subsequent influence of the defect structure on the surface chemistry. Nickel oxide is a p-type semiconductor in which the major defects are holes and cation vacancies. The nonstoichiometry can be described by a chemical reaction involving oxygen and the major defects:

$$1/2\ O_2 = V(Ni)'' + 2\oplus + O(O)^x \tag{50}$$

$$K = [V(Ni)''][\oplus]^2[P_{O_2}]^{-1/2} \tag{51}$$

where the symbols $\oplus$, $V(Ni)''$, $O(O)^x$ represent respectively, an electron hole, a nickel ion vacancy (the double super-script prime indicates the presence of a negative charge of two in comparison to the undisturbed lattice), and an oxygen ion residing at a regular site (the superscript x signifies the absence of excess charge). The presence of an electron hole is equivalent to the replacement of a Ni^{2+} ion with a Ni^{3+} ion.

When Li_2O and Cr_2O_3 are introduced into NiO, the following reactions take place:

$$1/2\ O_2 + Li_2O = 2Li(Ni)' + 2\oplus + 2\ O(O)^x \quad (52)$$

$$Cr_2O_3 = 2Cr(Ni)^{\bullet} + V(Ni)'' + 3\ O(O)^x \quad (53)$$

where Li(Ni)', and $Cr(Ni)^{\bullet}$ represent respectively a Li^+ ion residing at a Ni^{2+} site which has an effective negative charge of unity, and a Cr^{3+} ion sitting at a Ni^{2+} site with a net positive charge of unity. According to Eqs. 52 and 53, the presence of Li^+ increases the number of holes (i.e. Ni^{3+}) whereas the presence of Cr^{3+} increases the number of vacancies (and therefore decreases the number of holes, i.e. Ni^{3+}, according to Eq. 51).

A reasonable explanation of the NiO dissolution behavior can be developed if it is assumed that NiO decomposition requires that SO_4^{2-} ions be adsorbed at the NiO/solution interface, and that these ions interact preferentially with surface Ni^{3+} ions since these cationic sites are more highly charged than Ni^{2+} sites:

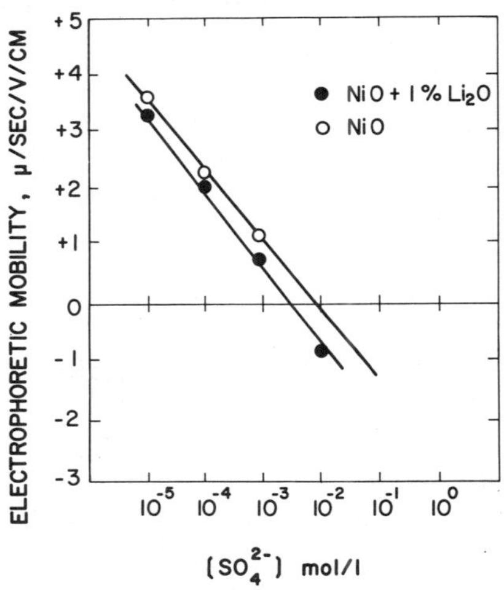

Fig. 12. Electrophoretic mobility of doped and undoped NiO as a function of aqueous sulfate ion concentration; 25°C, pH = 3, ionic strength controlled with 1.0×10^{-2} kmol m^{-3} KNO_3.[43]

$$Ni^{3+}(s) + SO_4{}^{2-}(s) = Ni^{3+}{-}{-}{-}{-}SO_4{}^{2-}(s) \tag{54}$$

$$Ni^{2+}(s) = Ni^{3+}(s) + e(s) \tag{55}$$

$$Ni^{3+}{-}{-}{-}{-}SO_4{}^{2-}(s) + e(s) = Ni^{2+}(aq) + SO_4{}^{2-}(aq) \tag{56}$$

According to this analysis, the greater dissolution of the Li-doped NiO is due to its higher concentration of Ni^{3+} compared with the Cr-doped sample. The fact that variation in the solid-state chemistry alters the surface chemistry is demonstrated by the finding of Aplan et al[44] that whereas the pzc of a pure NiO sample was at pH 7.5, the corresponding Li- and Cr-doped materials had their pzc at pH 6.3 and pH 9.3 respectively. The affinity of the $SO_4{}^{2-}$ ion for the NiO surface is indicated by the electrophoretic mobility data shown in Fig. 12.[43] The solid/liquid interface becomes less and less positive as the sulfate ion concentration in the bulk aqueous phase is increased.

According to recent work by Rogers et al[45] in which Li^+ and Cr^{3+} were used as dopants, the dissolution of ZnO, an n-type semiconductor, can also be modified by means of variations in the solid-state point defect concentrations. Similar results have been reported by Jones et al[46-48] who used variations in the oxygen pressure (see Eqs. 50 and 51) to vary the defect concentrations in NiO. In the case of sulfide minerals, there are also indications that the presence of point defects can change reaction rates in aqueous systems[49-51], although some doubts have been expressed by some investigators.[52] Sulfides are known to possess relatively high electrical conductivities.[53] Thus the presence of point defects in these minerals should affect both the electrical conductivity and the concentration of surface reactive sites. Given the fact that many sulfide dissolution reactions are electrochemical in nature[41,42], it should not be surprising that variation in the solid-state chemistry can alter dissolution rates.

Adsorption Losses in Ammonia Leaching Systems

Oxidized ores are not easily upgraded by conventional mineral processing techniques, such as flotation. Consequently, very often instead of concentrates, large tonnages of low grade material must be treated. Oxidized ores, such as laterites and ocean-floor manganese nodules, occur or are processed as highly divided particulates with specific surface areas of the order of $60m^2/g$ and $200m^2/g$, respectively.[54,55] The presence of large amounts of these colloidal and near colloidal materials in leach residues can lead to

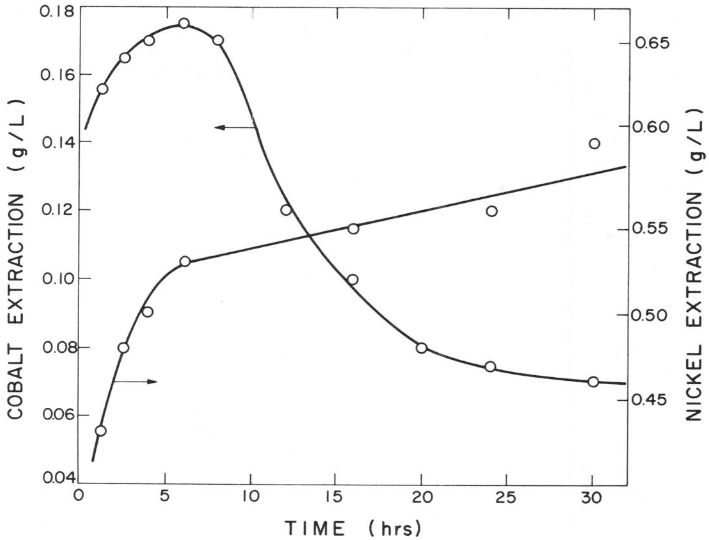

Fig. 13. Ammonia leaching of cobalt and nickel from reduce manganiferous ore.[57]

loss of valuable metal from pregnant liquors through adsorption, ion exchange and coprecipitation. Caron attributed loss of soluble cobalt during ammonia leaching of laterites to the presence of iron hydroxide.[56] Kasherininov found that cobalt leached from a manganese ore with ammonia could be subsequently lost to the manganese dioxide leach residues;[57] see Fig. 13. Osseo-Asare and Fuerstenau[58] have also demonstrated that ammonia-leach residues derived from manganese nodules are capable of interacting with dissolved copper, nickel and cobalt, as shown by Fig. 14. The effect of metal adsorption on the surface charge at the oxide/aqueous ammoniacal solution interface is illustrated in Fig. 3a.[27]

The uptake of hydrolyzable metal ions by oxide adsorbents in aqueous ammoniacal solutions can be interpreted in terms of an adsorption model which combines electrical double layer theory with the aqueous speciation.[26-28] Two types of equilibria may be considered: one in the interfacial region and the other in the bulk solution. The final result obtained for the total adsorption density Γ_T is:[28]

$$\Gamma_T = \frac{\sum_i \Gamma_{m,i} K_i' K_i a_M (a_{OH})^i + \sum_j \Gamma_{m,j} K_j' K_j a_M (a_{NH_3})^j}{1 + \sum_i K_i' K_i a_M (a_{OH})^j + \sum_j K_j' K_j a_M (a_{NH_3})^j} \qquad (57)$$

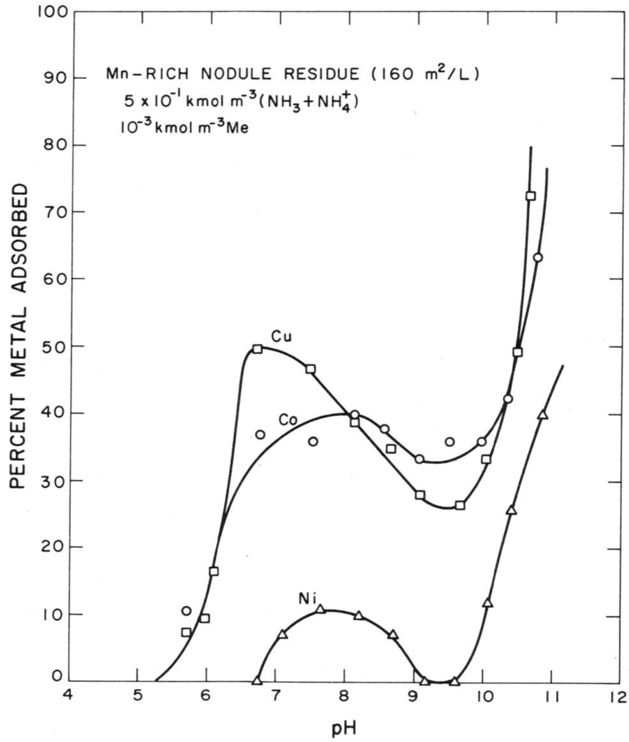

Fig. 14. Uptake of copper, nickel, and cobalt by the manganese-rich fraction of DH-2 nodule as a function of pH at 0.5 kmol m^{-3} total ammonia and 10^{-3} kmol m^{-3} added metal concentration.[58]

where $\Gamma_{m,i}$ is the monolayer coverage of the ith species (mol/m^2), a_M and a_{OH} represent the activity of the uncomplexed metal ion and the hydroxyl ion respectively, K_i and K_j are, respectively, the dissociation constants for the hydroxo and ammine species, while K'_i and K'_j are the corresponding surface adsorption constants.

The free energy of adsorption for each of the solution species, ΔG_{ads}, can be considered as the sum of coulombic, solvation, and chemical contributions (see Eqs. 12, 13 and 15). Figure 15 shows a comparison of the model with experimental results for copper uptake by titania, hematite, alumina and quartz in solutions containing 0.5 kmol m^{-3} (NH_3 + $NH_4{}^+$).[28] The corresponding computed adsorption densities for the various ammine and hydroxo species indicate that the hydroxo species are largely responsible for the metal losses.[28]

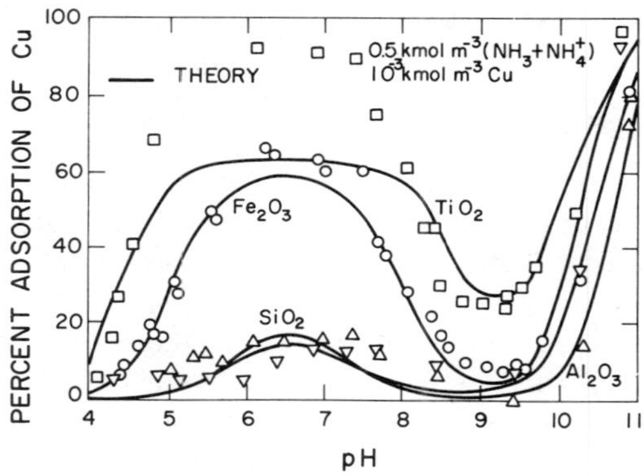

Fig. 15. Comparison of the experimental and theoretical isotherms for the adsorption of copper on various substrates (160 m^2/L) as a function of pH at 10^{-3} kmol m^{-3} added copper and 0.5 kmol m^{-3} $(NH_3 + NH_4^+)$.[28]

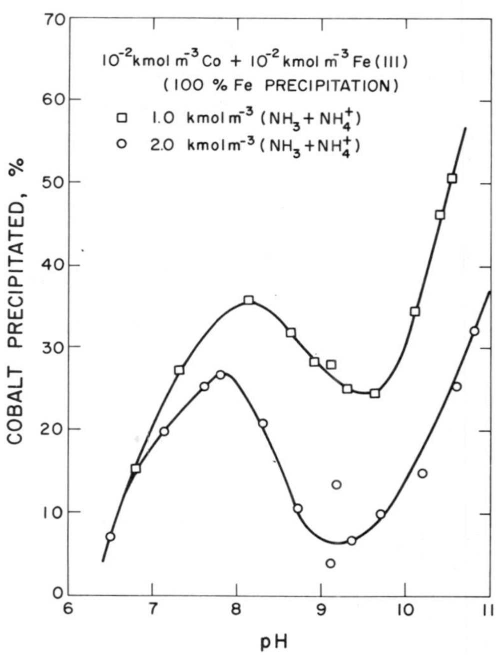

Fig. 16. Effect of pH on cobalt coprecipitation with Fe(III) in an ammoniacal ammonium nitrate solution at 25°C.[59]

The presence of hydroxyl groups on the hydroxo species probably promotes adsorption through hydrogen bonding with the hydroxylated oxide surfaces. The adsorption of the neutral hydroxo species is further favored by the fact that the repulsive solvation term is zero for a species with zero charge (Eq. 15). The model also predicts[28] that metal uptake should increase with increase in the dielectric constant of the solid, in agreement with the experimental results shown in Fig. 15. [ε_s = 80 (TiO_2), 25(Fe_2O_3), 12(Al_2O_3), 4.3 (SiO_2)]. The apparent lack of fit between the experimental and theoretical TiO_2 curves in Fig. 15 has been attributed to the relatively high dielectric constant of this oxide, which promotes multi-layer adsorption or surface precipitation.[28]

As shown in Figs. 14 and 15, metal uptake goes through a maximum, then a minimum, as the pH is increased. A similar behavior, Fig. 16, has been found in the case of coprecipitation experiments in which solutions containing Co(II) and Fe(II) were mixed with ammoniacal solutions.[59] Similar trends have been reported by Han et al[60] who investigated the coprecipitation of Co(II) and Ni(II) with Fe(III), Cr(III) and Al(III), although only the maximum in metal uptake was observed since their experiments were limited to a pH range below 10. Thus it can be concluded that similar processes are at work whether metal removal proceeds via adsorption or coprecipitation; that is, the case of coprecipitation only involves a much larger interfacial area.

Adsorption Phenomena in Gold-Silver Leaching Systems

The carbon/water interface is becoming increasingly important in gold extraction technology. The presence of this interface is a nuisance when processing carbonaceous gold ores since the interactions of the carbonaceous material with gold cyanide ions [$Au(CN)_2^-$] can result in gold loss from solution.[61-70] On the other hand, the adsorption of gold cyanide on activated carbon provides a means of recovering this metal from leach liquors and accounts for the growing commercial interest in the carbon-in-pulp process.[71-76] Since gold uptake by carbonaceous ores has been attributed to the presence in these materials of a natural activated carbon,[61-70] it is clear that detailed knowledge of the surface chemistry of the activated carbon/metal-cyanide-water interface would be extremely useful in the continuing attempts to develop improved gold extraction technologies.

The surface chemistry of activated carbon is controlled by the functional groups present on its surface. The functional groups occupy the edges of the broken graphite plates and can be approxi-

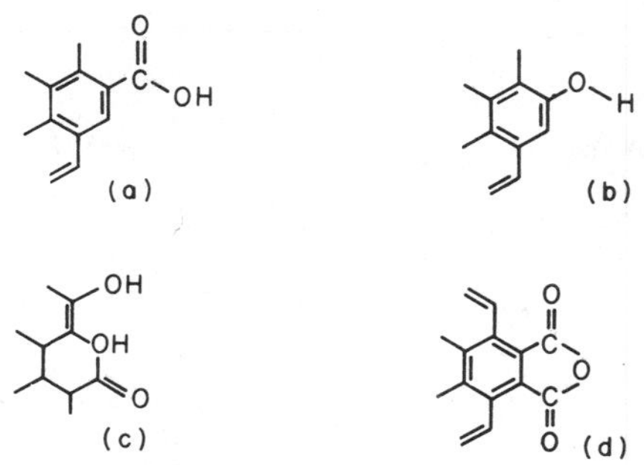

Fig. 17. Structures of some surface oxides present on activated surfaces: (a) carboxylic acid, (b) phenolic hydroxyl, (c) fluorescein-type lactones, and (d) carboxylic anhydrides.

mated by oxygen-containing groups such as those shown in Fig. 17. These functional groups are produced during a surface oxidation or "activation" process, and can undergo acid-base reactions such as those described above for mineral oxides:

$$-COOH(s) \rightleftarrows -COO^- (s) + H^+ (aq) \tag{58}$$

$$-COOH(s) + H^+ (aq) \rightleftarrows -COOH_2^+(s) \tag{59}$$

The acid-dependency of the interfacial charge is demonstrated by the zeta potential results shown in Fig. 18.[70,78] According to these results, the carbon surface is negatively charged over the entire pH range indicated. The effect of pH on gold and silver adsorption in cyanide solutions is also illustrated in Fig. 18.[70,78,81] Both gold and silver adsorption increases with decrease in pH.

The fact that a negatively-charged gold or silver complex adsorbs on a negatively-charged carbon surface indicates the presence of some kind of specific interaction, although the exact nature of this interaction is not yet understood.[70,78-83] Cho and Pitt[78] have suggested that the driving force for adsorption comes from the fact that the large size of the aurocyanide complex makes this ion relatively hydrophobic (poorly hydrated). In support of this argument, they present experimental data which show that the adsorption of anions increases with increase in ionic size, i.e. $Au(CN)_2^- > Ag(CN)_2^- > CN^-$. However, McDougall et al[81] have questioned this analysis since they have found that addition of an excess amount of a large anion such as ClO_4^- has little effect on gold adsorption

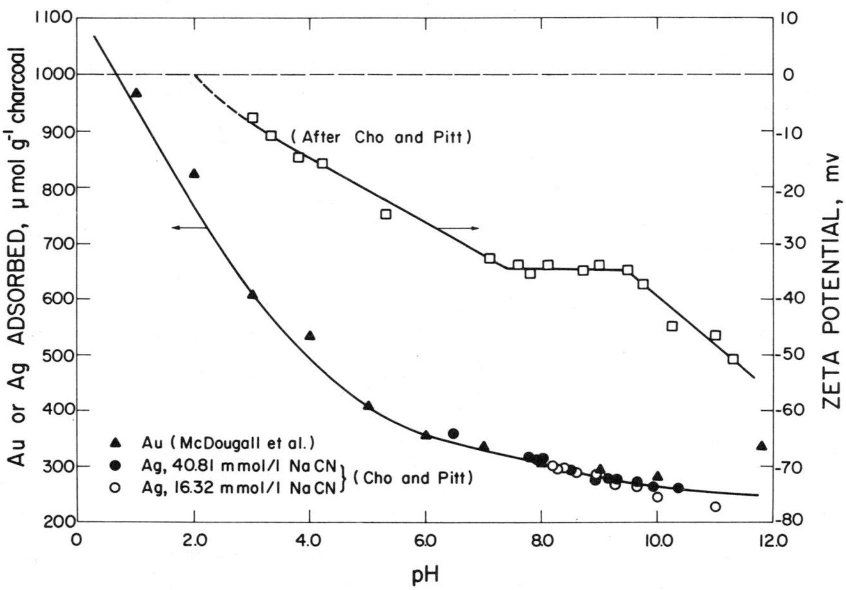

Fig. 18. Effect of pH on (a) the zeta potential at the activated charcoal/water interface,[78] (b) the adsorption of gold and silver cyanides by activated charcoal.[78,81]

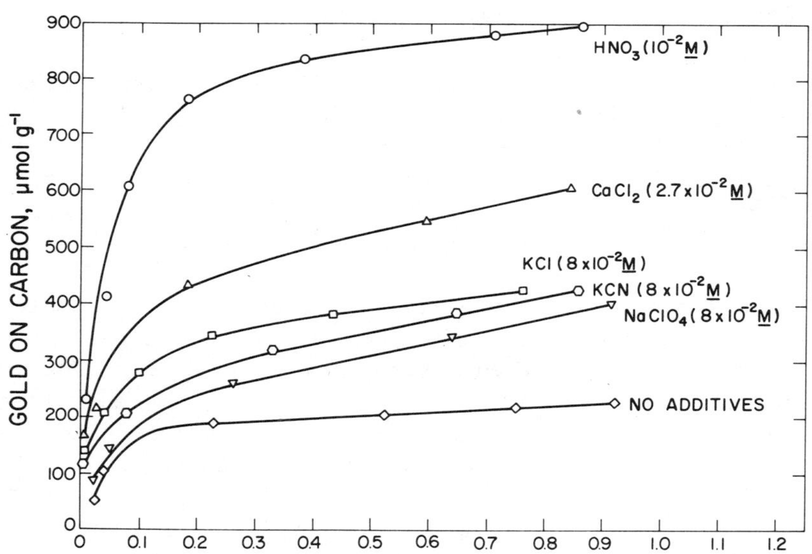

Fig. 19. Effect of various electrolytes on gold cyanide adsorption by activated charcoal.[81]

when compared with Cl^- and CN^- ions (see Fig. 19). An alternative mechanism involving reductive adsorption has been proposed by McDougall et al[81,82] who found with the aid of X-ray photoelectron spectroscopy that the adsorbed gold had an oxidation state of 0.3, i.e. between zero (expected of metallic gold) and unity (expected of $Au(CN)_2^-$ or AuCN). However the nature of this fractional-valent gold species is not known. The role of cations is also not fully understood. As Fig. 19 indicates, there is considerable increase in gold uptake when a cyanide solution containing 8×10^{-2} kmol m^{-3} KCl is replaced with one containing 2.7×10^{-2} kmol m^{-3} $CaCl_2$.[81] The electrophoretic mobility data[70] presented in Fig. 20 indicate that in the presence of calcium ions, the carbonaceous ore/water interface becomes significantly less negatively charged. Thus it is likely that Ca^{2+} enhances gold cyanide uptake partly by neutralizing the negative surface charge. However, since both the KCl and $CaCl_2$ solutions used in Fig. 19 have the same ionic strength (8×10^{-2} kmol m^{-3}), it is clear that the calcium ion participates specifically in the carbon-aurocyanide interaction, although again, details of this interaction are currently not available.

It should be clear from the above discussion that much remains to be understood regarding the adsorption of gold and silver from

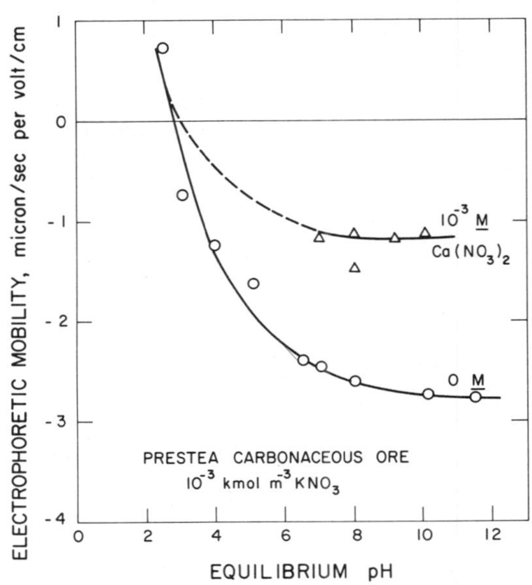

Fig. 20. Effect of calcium on the electrophoretic mobility of Prestea (Ghana) carbonaceous gold ore.[70]

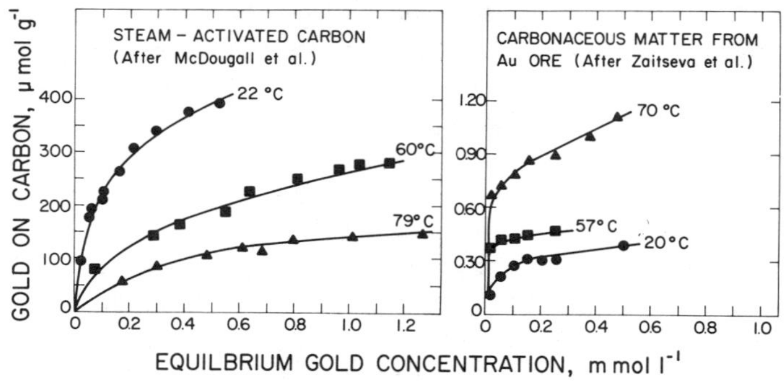

Fig. 21. Effect of temperature on the adsorption of gold cyanide by steam-activated carbon[81] and by carbonaceous matter derived from gold ore.[68]

cyanide solutions. One major obstacle is the lack of adequate characterization of the solid carbonaceous material itself. Commercial activated carbon is prepared from a wide variety of starting materials (e.g. coconut shells, sawdust, coal and petroleum coke, and bone char).[77] The molecular structure of the starting material, the nature and distribution of inorganic matter in this material, the activation atmosphere, the process temperature and reaction time all affect the final activity of carbon. The situation is not any better in the case of carbonaceous gold ores. While it is often asserted that these materials contain carbonaceous matter which is "similar" to activated carbon, the nature and extent of this similarity have been little explored. As can be seen from Fig. 21, it appears that gold adsorption on steam-activated carbon decreases with increase in temperature[81] whereas the opposite effect is reported by Zaitseva et al[68] who used carbonaceous matter extracted from gold ores.

Passive Films in Ammonia Leaching Systems

In the extraction of copper, nickel and cobalt from laterites and manganese nodules, the leaching step is preceded by a reductive roast operation.[84,85] The purpose of this pre-leach step is to convert the oxidic copper, nickel and cobalt to the metallic state and to withdraw these metals out of the crystal lattice of the matrix iron, manganese, and magnesium oxides and silicates. These solid-state changes make the valuable metals more reactive in the aqueous ammoniacal media. The reduction operation does not yield pure

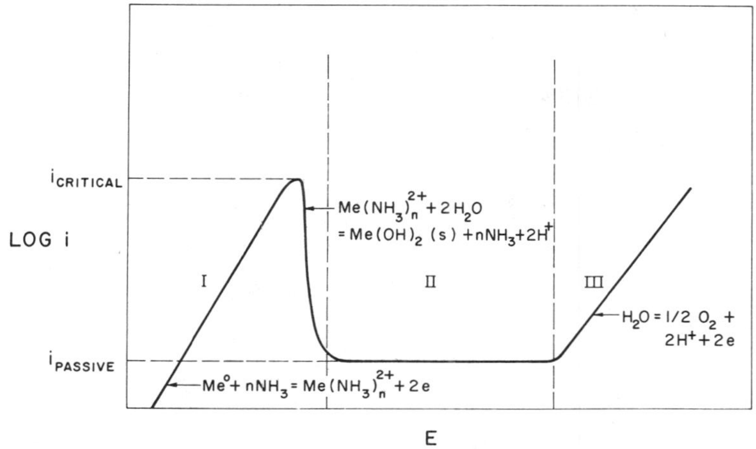

Fig. 22. Schematic illustration of the anodic polarization behavior of a metal electrode in aqueous ammoniacal solution.[90]

metals, rather alloys are formed.[86] Thus leaching of reduced laterites and nodules involves alloy phases. The composition and nature of these alloys are not known. However, given the low concentration of Cu, Ni and Co in the starting ores, it is likely that iron-based alloys are formed.

The poor extraction of cobalt from reduced oxide ores in ammoniacal solutions was noted above (see Fig. 13). There are indications that in addition to the adsorption phenomena discussed previously, electrochemical processes may also be worthy of consideration.[87–90] Figure 22 presents a schematic illustration of the various stages in the electrochemical behavior of a metal electrode in aqueous ammoniacal solution.[90] Figure 23 shows the electrochemical polarization behavior of sintered iron-cobalt powders.[90] According to this figure, iron passivates more readily than does cobalt; as the iron content of the metallic mixture increases, the passivation behavior approaches that of pure iron. Above about 0.6 volt (SHE), the passive currents of the Fe-50% Co and the Fe-10% Co mixtures coincide with the passive current on pure iron. Thus it appears that in the high potential region, iron controls the passive behavior of the Fe-Co mixtures. It has been found[90] that there is practically no cobalt dissolution from a Fe-50% Co sample when the potential exceeds 0.6 volt (SHE). This decrease in cobalt extracti must be attributed to the presence of a passive film of iron oxide on the electrode surface, which serves as a barrier to further meta dissolution.

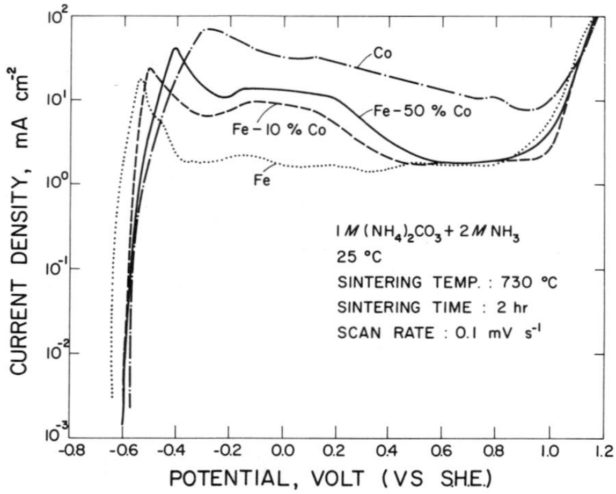

Fig. 23. Polarization behavior of sintered iron-cobalt mixtures in aqueous ammoniacal solution.[90]

Product Layer Modification with Surface Active Agents

Pioneering work by Forward and Veltman[91] revealed that zinc can be extracted directly from zinc sulfide concentrates using oxidative acid pressure leaching to give elemental sulfur:

$$ZnS = Zn^{2+} + S^{o} + 2e \qquad (60)$$

As shown in Table 5, in the experiments of Forward and Veltman[91], zinc extraction increased with temperature up to about 115°C and then declined. This behavior was traced to the presence of molten sulfur (M.p. 119°C) which wets the zinc sulfide surface and thereby causes the ore particles to agglomerate. When Pb concentrates are used instead of Zn concentrates an insoluble $PbSO_4$ product layer is formed. It was found[91] that molten sulfur did not have any adverse affect on PbS decompostion. Apparently molten sulfur did not wet the $PbSO_4$ surface.

If the mineral(S)-water(L)-molten sulfur(V) system is depicted schematically as shown in Fig. 4, then in order for water rather than molten sulfur to wet the mineral surface, the following condition must be satisfied (see Eq. 26):

$$\gamma_{SL} + \gamma_{LV} < \gamma_{SV} \qquad (61)$$

Table 5. Effect of Leaching Temperature on Pressure Leaching of ZnS Concentrates.[91]

Leaching Temp (°C)	90	110	115	120	140
%Zn Extr.	62	88	90	70	73

14% pulp density, -100 mesh, 20 psi O_2 overpressure.

Therefore, the fact that water instead of molten sulfur wets the $PbSO_4$ surface indicates that the interfacial tension at the solid/water interface (γ_{SL}) is smaller for $PbSO_4$ than for ZnS and/or that the interfacial tension (γ_{SV}) at the $S^o/PbSO_4$ interface is greater than that at the S^o/ZnS interface. The hydrophobicity of sulfide minerals is well known[92] and is believed to be caused by the presence of a thin coating of elemental sulfur resulting from the oxidation of the surface layers of the sulfide minerals. On the other hand, $PbSO_4$ is an ionic solid and should therefore present a hydrophilic surface to the aqueous phase. Consequently, the condition that $\gamma_{SL}(PbSO_4) < \gamma_{SL}(ZnS)$ appears likely. Furthermore, since ZnS is more hydrophobic than $PbSO_4$, and molten sulfur is hydrophobic, it follows that the requirement that $\gamma_{SV}(PbSO_4) > \gamma_{SV}(ZnS)$ should be attainable in practice.

It has been discovered by researchers at Sherritt Gordon[93-96] that additives such as tannin compounds (e.g. quebracho), lignosulfonates, and alkylaryl sulfonates at dosages as small as 500-1500 g/ton of concentrate can result in dramatic improvements in the leaching rates of ZnS concentrates. For example, in one test reported by Kawulka et al,[93] the addition of 0.1 g/L calcium lignosulfonate raised zinc extraction at 150°C from 63.3% to 95.9%. Apparently the presence of the additives inhibited the deleterious effect of molten sulfur noted previously by Forward and Veltman.[91]

The mechanism of action of the additives on zinc sulfide dissolution has not yet been studied. However, as discussed above, it is possible to alter the wetting and dispersion characteristics through a modification of the surface chemistry. Lignosulfonates are a byproduct derived from the manufacture of paper, ranging in molecular weight from $1x10^3$ to $20x10^3$.[3] These sulfonates are known to be dispersion agents and emulsion stabilizers.[3] In addition to the sulfonate group, they also contain varying amounts of phenolic, alcoholic and carboxylate groups. In order to prevent the wetting of a mineral by molten sulfur and therefore promote wetting by water

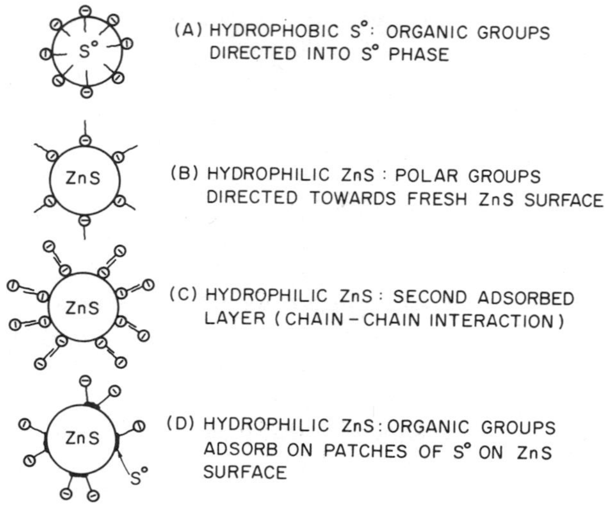

Fig. 24. Adsorption of surfactants on sulfur and zinc sulfide (schematic).

(i.e. 0<<90°, see Fig. 4), the added surfactant must alter the solid/liquid, liquid/vapor, or solid/vapor interfacial tensions. It would also be expected that the surfactants would alter the electrical double layer properties of the sulfide i.e. sulfide/water water and sulfur/water interfaces. Sulfur[20,92] has a point of zero charge (pzc) at about pH2 while the pzc of zinc sulfide[20] is in the neighborhood of pH4. Since sulfur is naturally hydrophobic[92], organic surface active agents will adsorb on molten sulfur droplets, as shown in Fig. 24a, by orienting their nonpolar groups towards the bulk sulfur phase, while the anionic polar groups will be directed towards the aqueous phase. This should produce the desirable effect of reducing γ_{LV} (see Eq. 61). Doyle et al[94] have noted that the presence of the surfactants decreases the particle size of sulfur. This must be another effect of the reduced interfacial tension accompanying adsorption. A fresh ZnS surface should be positively charged in acidic solution (pzc ≅4). Therefore in the case of lignosulfonates, adsorption should occur with the polar groups oriented towards the sulfide surface (Fig. 24b). This mode of adsorption would make the mineral surface hydrophobic and would discourage wetting by water. However, if a higher concentration of the surfactant is added, a second adsorbed layer may form such that the chains of the two layers interact and the hydrophilic groups of the second layer are directed towards the aqueous phase (Fig. 24c). This new situation should produce a negatively charged hydrophilic surface which should promote wetting by water. Furthermore, since under these circumstances, both the ZnS and sulfur particles would be negatively charged, electrostatic repulsion would provide an additional means of separating ZnS from the molten sulfur. On the other hand, in the case where a clean ZnS surface is not available, e.g. where patches of sulfur are present on the mineral surface, the surfactant may adsorb by directing its nonpolar groups towards the surface sulfur layer. This arrangement should result in a hydrophilic mineral surface as shown in Fig. 24d.

The first commercial plant using this direct zinc sulfide pressure leach process commenced operations in 1981. Zinc extraction in excess of 98% and elemental sulfur recovery of higher than 95% have been reported.[97] It is likely that surface active agents may also be used to advantage in other leaching processes where rate-limiting reaction products are formed.

SUMMARY AND CONCLUSIONS

A review has been presented of the basic concepts of interfacial phenomena at the mineral/water interface, particularly as these concepts relate to the adsorption of inorganic and organic surface active species, wetting phenomena and colloid stability. Selected examples from the hydrometallurgical literature have been presented which demonstrate the fact that leaching is inherently heterogeneous in nature and that processes occurring at the solid/water interface can be dominant factors in the observed dissolution behavior. In some cases, interfacial processes are a nuisance, e.g. metal uptake in ammonia leaching systems and gold cyanide interaction with carbonaceous ores. On the other hand, interfacial phenomena can be exploited to enhance and permit metal extraction e.g. use of lignosulfonates in ZnS leaching, and use of activated charcoal in gold recovery from cyanide solutions. Whether the interfacial process is to be avoided or to be encouraged, it should be clear from the above discussion that simultaneous characterization of the solid phase, the interface and the aqueous phase is necessary in order to obtain a complete picture of the fundamental processes which are taking place.

ACKNOWLEDGMENTS

Support of this work by the National Science Foundation and The Pennsylvania Mining and Mineral Resources Research Institute is gratefully acknowledged.

REFERENCES

1. A. W. Adamson: Physical Chemistry of Surfaces, Interscience, New York, NY, 1967.
2. J. T. Davies and E. K. Rideal: Interfacial Phenomena, Academic Press, New York, NY, 1963.
3. M. J. Rosen: Surfactants and Interfacial Phenomena, Wiley, New York, NY, 1978.
4. J. A. Kitchener: Br. Polym. J., 1972, vol. 4, pp. 217-305.
5. J. A. Kitchener: The Scientific Basis of Flocculation, K. J. Ives, ed., Sijthoff and Noordhoff, Alphen aan den Rijn, The Netherlands, 1978, pp. 283-328.
6. J. B. Rosenbaum and J. B. Clemmer: Proc. Intl. Conf. Peaceful Use Atomic Energy, Geneva, 1955, vol. 8, pp. 38-44.
7. D. J. I. Evans and S. C. Lindsay: Trans. AIME, 1960, vol. 217, pp. 41-48.
8. J. Vereecken and R. Winand: Surf. Technol., 1976, vol. 4, pp. 227-235.
9. W. Kunda, D. J. I. Evans and V. N. Mackiw: Modern Developments in Powder Metallurgy, Vol. 1: Fundamentals and Methods, H. H. Hausner and K. H. Roll, eds., Plenum, New York, NY, 1966, pp. 15-29.
10. J. Mizrahi and E. Barnea: Brit. Chem. Eng., 1970, vol. 15, pp. 497-503.
11. G. A. Davies and G. V. Jeffreys: Filtration and Separation, 1969, vol. 6, pp. 349-354.
12. D. M. Hausen: Canad. Met. Quart., 1974, vol. 13, pp. 659-668.
13. D. S. Flett: Acct. Chem. Res., 1977, vol. 10, pp. 99-104.
14. M. Cox and D. S. Flett: ISEC 77, 1977, pp. 63-72.
15. P. R. Danesi and R. Chiarizia: Crit. Rev. Anal. Chem., 1981, vol. 10, pp. 1-126.
16. F. Sebba: Ion Flotation, Elsevier, Amsterdam, 1962.
17. S. A. Zieminski, M. M. Caron and R. B. Blackmore: Ind. Eng. Chem. Fund., 1967, vol. 6, pp. 233-242.
18. G. A. Parks: Adv. Chem. Ser., 1967, vol. 67, pp. 121-160.
19. D. W. Fuerstenau: Pure Appld. Chem., 1970, vol. 24, pp. 135-164.

20. T. W. Healey and M. S. Moignard: Flotation, M. C. Fuerstenau, ed., AIME, New York, NY, 1976, pp. 275-297.
21. K. Osseo-Asare and D. W. Fuerstenau: Croatica Chemica Acta., 1973, vol. 45, pp. 149-161.
22. K. Osseo-Asare, D. W. Fuerstenau and R. H. Ottewill: Adsorption at Interfaces, ACS Symposium Series No. 8, 1975, pp. 63-78.
23. R. O. James and T. W. Healey: J. Colloid Interface Sci., 1972, vol. 40, pp. 42-52.
24. R. O. James and T. W. Healey: J. Colloid Interface Sci., 1972, vol. 40, pp. 53-64.
25. R. O. James and T. W. Healey: J. Colloid Interface Sci., 1972, vol. 40, pp. 65-81.
26. K. Osseo-Asare and D. W. Fuerstenau: Intl. J. Miner. Process., 1979, vol. 6, pp. 85-104.
27. K. Osseo-Asare and D. W. Fuerstenau: Intl. J. Miner. Process., 1980, vol. 7, pp. 117-127.
28. K. Osseo-Asare and D. W. Fuerstenau: Intl. J. Miner. Process., 1980, vol. 7, pp. 219-234.
29. K. Osseo-Asare: NATO Advanced Institute for Hydrometallurgical Process Fundamentals, 1982. (This reference is to other paper by Asare, "Interfacial Phenomena in Hydrometallurgical Liquid-Liquid Extraction Systems.")
30. G. D. Parfitt, Ed.: Dispersion of Powders in Liquids, Elsevier, Amsterdam, 1969.
31. R. H. Ottewill: Nonionic Surfactants, M. J. Schick, ed., Marcel Dekker, New York, NY, 1967, pp. 627-682.
32. I. H. Warren and E. A. Devuyst: I. Chem. E. Symp. Ser. No. 42, London, 1975, pp. 7.1-7.11.
33. I. H. Warren and E. A. Devuyst: International Symposium on Hydrometallurgy, D. J. I. Evans and R. S. Shoemaker, eds., AIME, New York, NY, 1973, pp. 229-264.
34. R. M. Smith and A. E. Martell: Critical Stability Constants, Vol. 4, Plenum, New York, NY, 1976.
35. H. J. Engell: Z. Physik. Chem. NF., 1956, vol. 7, pp. 158-181.
36. D. A. Vermilyea: J. Electrochem. Soc., 1966, vol. 113, pp. 1067-1070.
37. J. W. Diggle: Oxides and Oxide Films, vol. 2, J. W. Diggle, Ed., Marcel Dekker, New York, NY, 1973, pp. 281-386.
38. N. Valverde and C. Wagner: Ber. Bunsenges, Physik. Chem., 1976, vol. 80, pp. 330-333.
39. M. J. Nicol: Hydrometallurgy-Research, Development and Plant Practice, K. Osseo-Assare and J. D. Miller, eds., TMS-AIME, Warrendale, PA, 1982, pp. 177-195.
40. N. Valverde: Ber. Bunsenges. Physik. Chem., 1976, vol. 80, pp. 333-340.

41. E. Peters: Trends in Electrochemistry, J. O'M. Bockris, T. Biegler and B. J. Welch, eds., Plenum, New York, NY, 1977, pp. 267-290.
42. J. B. Hiskey and M. E. Wadsworth: Process and Fundamental Considerations of Selected Hydrometallurgical Systems, M. C. Kuhn, ed., SME-AIME, New York, NY, 1981, pp. 303-325.
43. P. Lussiez, K. Osseo-Asare and G. Simkovich: Metall. Trans. B., 1981, vol. 12, pp. 651-657.
44. F. F. Aplan, E. Y. Spearin and G. Simkovich: Colloids and Surfaces, 1980, vol. 1, pp. 361-371.
45. G. Rogers, G. Simkovich, and K. Osseo-Asare: Hydrometallurgy, 1983, vol. 10, pp. 313-328.
46. C. F. Jones, R. L. Segall, R. S. Smart and P. S. Turner: J. Chem. Soc. Farad. I, 1977, vol. 73, pp. 1710-1720.
47. C. F. Jones, R. L. Segall, R. S. Smart and P. S. Turner: J. Chem. Soc. Farad. I, 1978, vol. 74, pp. 1615-1623.
48. C. F. Jones, R. L. Segall, R. S. Smart and P. S. Turner: J. Chem. Soc. Farad. I, 1978, vol. 74, pp. 1624-1633.
49. G. Simkovich and J. B. Wagner: J. Electrochem. Soc., 1963, vol. 110, pp. 513-516.
50. P. Eadington and A. P. Prosser: Trans. IMM, 1969, vol. 78C, pp. 74-82.
51. A. O. Filmer and M. J. Nicol: J. S. Afri. Inst. Min. Metall., 1980, vol. 80, pp. 415-424.
52. T. Biegler: J. Electroanal. Chem., 1976, vol. 70, pp. 265-275.
53. R. T. Shuey: Semiconducting Ore Minerals, Elsevier, New York, NY, 1975.
54. E. C. Chou, P. B. Queneau and R. S. Rickard: Metall. Trans. B., 1977, vol. 8, pp. 547-554.
55. D. W. Fuerstenau, A. P. Herring and M. Hoover: Trans. SME, 1973, vol. 254, pp. 205-211.
56. M. H. Caron: Trans. AIME, 1950, vol. 188, pp. 67-90.
57. G. O. Kasherininov: J. Appld. Chem USSR, 1961, vol. 34, pp. 1589-1593.
58. K. Osseo-Asare and D. W. Fuerstenau: Complex Metallurgy '78, M. J. Jones, ed., IMM, London, 1978, pp. 43-48.
59. K. Osseo-Asare: COBALT '80, 10th Annual CIM Hydrometallurgical Meeting, CIM, Canada, 1980.
60. K. N. Han, E. Narita and F. Lawson: Hydrometallurgy, 1982, vol. 8, pp. 365-377.
61. W. R. Feldtmann: Trans. IMM, 1914-15, vol. 24, pp. 329-343.
62. H. Fischer: Min. and Scientific Press., 1916, vol. 112, pp. 743-745.
63. A. Dorfman: Bull. CIM, 1922, March, pp. 383-395.
64. E. S. Leaver and J. A. Woolf: USBM Tech. Pap. 481, 1930.

65. R. W. Nice: Canad. Min. J., 1971, pp. 41-49.
66. A. S. Radtke and B. J. Scheiner: Econ. Geol., 1970, vol. 65, pp. 87-102.
67. B. J. Scheiner, R. E. Lindstrom and T. A. Henrie: J. Metal., 1971, pp. 37-40.
68. M. L. Zaitseva, M. D. Ivanovskii and N. K. Larina: Sov. J. Nonferrous Met., 1973, vol. 14(1), pp. 78-80.
69. W. J. Guay: Gold and Silver Leaching, Recovery and Economics, W. J. Schlitt, W. C. Larson and J. B. Hiskey, eds., SME-AIME, New York, NY, 1981, pp. 17-22.
70. K. Osseo-Asare, P. M. Afenya and G. M. K. Abotsi: Precious Metals - Mining, Extraction and Processing, V. Kudryk, D. A. Corrigan and W. W. Liang, eds., TMS-AIME, Warrendale, PA, 1984, pp. 125-144.
71. J. B. Zadra: USBM RI 4672, 1950.
72. R. Sisselman: Eng. Min. J., 1976, May, pp. 90-96.
73. P. A. Laxen, G. S. M. Becker and R. Rubin: J. S. Afric. Inst. Min. Metall., 1979, vol. 79, pp. 315-326.
74. R. J. Davidson, V. Veronese and M. V. Nkosi: J. S. Afric. Inst. Min. Metall., 1979, vol. 79, pp. 281-297.
75. Carbon-In-Pulp Technology for the Extraction of Gold, Austral. Inst. Min. Metall., Parkville, Victoria, Australia, 1982.
76. C. A. Fleming: Hydrometallurgy - Research Development and Plant Practice, K. Osseo-Asare and J. D. Miller, eds., TMS-AIME, Warrendale, PA, 1982, pp. 839-857.
77. J. S. Mattson and H. B. Mark: Activated Carbon: Surface Chemistry and Adsorption from Solution, Marcel Dekker, New York, NY, 1971.
78. E. A. Cho and C. H. Pitt: Metall. Trans. B., 1979, vol. 10, pp. 159-164.
79. E. A. Cho and C. H. Pitt: Metall. Trans. B., 1979, vol. 10, pp. 165-169.
80. E. A. Cho, S. N. Dixon and C. H. Pitt: Metall. Trans. B., 1979, vol. 10, pp. 185-189.
81. G. J. McDougall, R. D. Hancock, M. J. Nicol, O. L. Wellington and R. G. Copperthwaite: J. S. Afric. Inst. Min. Metall., 1980, vol. 80, pp. 344-356.
82. G. J. McDougall and R. D. Hancock: Gold Bull., 1981, vol. 14, pp. 138-153.
83. V. M. Kuz'minykh and N. G. Tyurin: Tsvet. Metall., 1966, No. 12, pp. 21-22.
84. J. H. Canterford: Minerals Sci. Engng., 1975, vol. 7, No. 1, pp. 3-17.
85. D. W. Fuerstenau and K. N. Han: Marine Manganese Deposits, G. P. Glasby, ed., Elsevier, Amsterdam, 1977, pp. 357-390.

86. J. H. Canterford and A. G. Turnbull: Proc. Australas. Inst. Min. Metall., 1980, No. 275, pp. 43-52.
87. G. O. Kashernininov: J. Appld. Chem. USSR, 1960, vol. 33, pp. 1225-1230.
88. C. Y. Chen: Hydrometallurgy - Research, Development and Plant Practice, K. Osseo-Asare and J. D. Miller, eds., TMS-AIME, Warrendale, PA, 1982, pp. 65-85.
89. Q. Gong: Nonferrous Metals Quarterly (China), 1980, vol. 32, No. 2, p. 17.
90. K. Osseo-Asare, J. W. Lee, H. S. Kim and H. W. Pickering: Metall. Trans. B, 1983, vol. 14, pp. 571-576.
91. F. A. Forward and H. Veltman: J. Metals, 1959, vol. 12, pp. 836-40.
92. S. Chander, J. M. Wie and D. W. Fuerstenau: AIChE Symp. Ser., vol. 71(150), pp. 183-188.
93. P. Kawulka, W. J. Haffenden and V. N. Mackiw: U.S. Patent 3,867,268, February 18, 1975.
94. B. N. Doyle, I. M. Masters, I. C. Webster and H. Veltman: Proceed. 11th Commonwealth Min. Metall. Congr., 1978, Hong Kong.
95. G. L. Bolton, N. Zubryckyj and H. Veltman: 13th International Mineral Processing Congress, Warsaw, Poland, June 4-9, 1979.
96. H. Veltman and G. L. Bolton: Erzmetall., 1980, vol. 33, pp. 76-83.
97. E. G. Parker, D. R. McKay and H. Salomon-De-Friedberg: Hydrometallurgy-Research, Development and Plant Practice, K. Osseo-Asare and J. D. Miller, eds., TMS-AIME, Warrendale, PA, 1982, pp. 927-940.

THE INTEGRATED ORGANIC LEACHING - SOLVENT EXTRACTION OPERATION IN HYDROMETALLURGY

G. Thorsen, H. F. Svendsen and A. Grislingås

Department of Chemical Engineering
The Norwegian Institute of Technology
University of Trondheim
7034 Trondheim - NTH, Norway

Solvent extraction of metal ions from leach liquors by liquid cation exchangers such as carboxylic acids is in many situations considered prohibitive for commercial operation because of the need for adding alkali in an amount at least equivalent to the metal extracted. Besides the primary economic burden from the cost of the alkali, the addition of extra chemicals to a process system will at a later stage in the process have to be removed or reclaimed. Besides, the added chemicals will often end up as an environmental deposit problem.

These shortcomings of a solvent extracton operation may be overcome by integrating the leaching and solvent extraction steps, using the concept of organic leaching. The overall process is started by leaching the metal-bearing solid material with the organic acid (liquid cation exchanger) whereby the raw material itself will act as the alkali needed for the further separation process by solvent extraction. The integrated process may thus extend the use and application of liquid cation exchangers such as carboxylic acids and alkyl phosphoric acids.

The principles of the integration of organic leaching and solvent extraction are presented and applications demonstrated for three processes: solvent extraction of iron in zinc hydrometallurgy, recovery of metal values from zinc ash and flue dusts, and separation and recovery of metals from cementation residues from purification of zinc sulphate electrolyte.

INTRODUCTION

In hydrometallurgical processing of metals from solid materials of ores and concentrates or secondary scrap material the first step will in most cases be considered to be a leaching process with an aqueous solution of a solubilizing agent. A pretreatment of the metal-bearing solid material may be included such as the oxidation roast in the roast-leach processes for sulphide concentrates of zinc and copper.

In the leaching process a number of metals considered as impurities will normally enter the resulting aqueous metal salt solutions along with the dissolved valuable metals. The leach solutions will therefore be subjected to purification steps prior to the final separation and reduction processes for recovery of the valuable metals.

Solvent Extraction in Hydrometallurgy

In recent years the technique of solvent extraction has been developed to a great extent for separation, purification and recovery of metal ions from aqueous solutions. Solvent extraction is recognized as a modern tool of considerable impact within hydrometallurgical processing.[1] This is the situation not only for the more costly-metals such as uranium and the rare earths, but indeed also for base metals processing such as in the production of nickel and copper.[2,3]

When solvent extraction is considered to be applied within a hydrometallurgical process, it is generally taken for granted that the metal values appear in an aqueous metal salt solution obtained by leaching a solid metal-bearing material with an aqueous dissolving solution. Starting from the leach solution, thus the technique of solvent extraction may be used either as 1) a recovery process for the valuable metals by selectively extracting these metals into an organic solvent phase, or 2) for purification purposes by extracting the contaminating metals, leaving the valuable metal as a purified aqueous metal salt solution.

As an example of the first kind, the highly successful processes for copper extraction from sulphate solutions by the chelating extractants are to be mentioned.[3] The Falconbridge Matte Leach

Process is an example of the second kind with nickel as the main valuable component in the chloride leach system. Contaminating metals as iron, cobalt and copper are removed from the process solution leaving the purified nickel chloride liquor behind, ready for crystallization of high purity nickel chloride crystals followed by reduction to metallic nickel.[2]

SOLVENT EXTRACTION WITH LIQUID CATION EXCHANGERS

An important type of organic compounds used as solvent extraction reagents are the cationic extractants such as carboxylic acids (i.e. naphtenic acids, Versatic acids) or alkyl phosphoric acids like di(2-ethylhexyl) phosphoric acid (HDEHP).[1,3-7] The mechanism for the extraction of a metal ion M^{n+} with an organic acid RH can be expressed in simplified form by the equation:

$$M^{n+}_{(aq)} + nRH_{(org)} \rightleftarrows R_nM_{(org)} + nH^{+}_{(aq)} \tag{1}$$

In order to transfer the metal into the organic phase the pH of the aqueous phase has to be adjusted, and maintained, within specified limits. During the extraction an equivalent amount of alkali has to be added to neutralize the liberated proton from the organic acid.

Figure 1 shows a typical extraction diagram for the cationic extractant Versatic 10, a mixture of tertiary alkyl carboxylic acids with the formula:

$$\begin{array}{ccccc} & & R_1 & & \\ & & | & & \\ R_2 & - & C & - & COOH \\ & & | & & \\ & & CH_3 & & \end{array}$$

where the average total number of carbon atoms in the molecule is about 10. In short the Versatic acid will be termed RH.

The distribution isotherms are more or less parallel and each metal has a comparatively narrow pH range over which the metal will be transferred from the aqueous phase to the organic phase.

The pH values for extraction of the various cationic species of the metals are related to the complexing tendencies within both the aqueous and the organic phases. Carboxylic acids follow in general

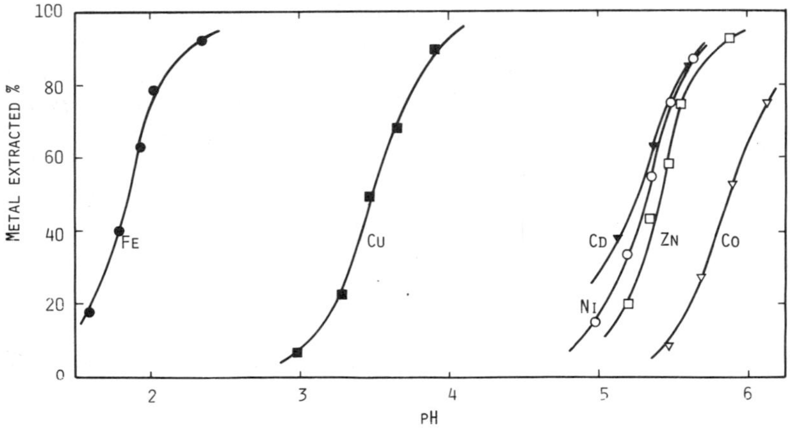

Fig. 1. Equilibrium distribution of metals in 30% Versatic 10-Shellsol system. Temperature 20°C. Extraction from single metal sulphate solutions of 5 g/l metal.

the hydroxide rule, i.e. the extraction takes place just below the pH region for precipitation of the respective hydroxides. There are, however, some deviations from this rule. For instance the isotherms may change with the complexing power of the anions, chloride and sulphate systems, and may thus exhibit different characteristics.[7]

The alkyl phosphoric acids behave in a similar way as the carboxylic acids. Figure 2 shows the extraction diagram for di(2-ethylhexyl) phosphoric acid (HDEHP). The order of extraction of the metals is essentially the same, except for zinc which is extracted at a low pH and has changed position in comparison with copper. The general trend of extraction under more acidic conditions is due to the more acidic nature of the phosphoric acid group compared with the carboxylic acid group.[8]

From Figs. 1 and 2 it is obvious that the extraction may also take place by an exchange reaction between two metal ions, $M1^{n+}$ and $M2^{m+}$ which both may form an organic salt with the cation exchanger in the organic phase. With the valences of the two cations given as n and m respectively, the general exchange reaction will be:

$$nM2^{m+}{}_{(aq)} + mR_nM1_{(org)} \rightleftarrows mM1^{n+}{}_{(aq)} + nR_mM2_{(org)} \qquad (2)$$

where M1 is the more basic of the two metals, i.e. the extraction isotherm for M1 will be at a higher pH value than for M2. However, prior to the exchange reaction the organic phase has to be loaded

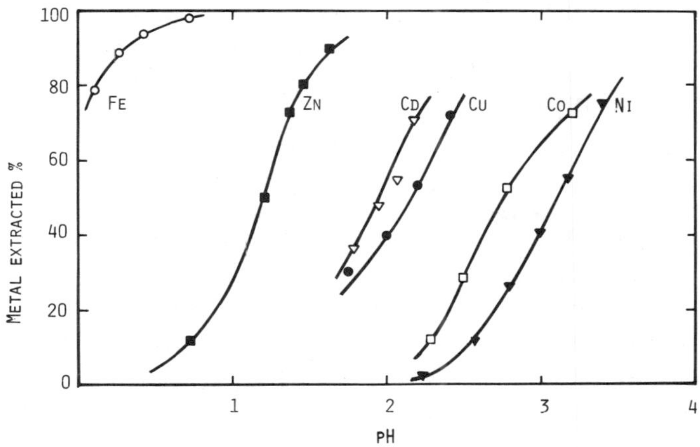

Fig. 2. Equilibrium distribution of metals in 30% HDEHP-Shellsol system. Temperature 20°C. Extraction from single metal sulphate solutions of 5 g/l metal.

with Ml. This loading has usually been connected to Eq. (1) in that the equivalent amount of alkali has been added to the system to neutralize the proton.

Fletcher and Flett[6] studied the extraction of ferric iron from copper and zinc sulphate liquors, and found that iron could be extracted by a single contact with kerosene solutions of copper or zinc naphtenates. Prior to these exchange reactions, however, magnesia was added to the system to prepare the primary loading of the organic phase with copper and zinc. Fletcher and Flett therefore concluded that the major factor in the economics of the use of carboxylic acids for the recovery of metals by solvent extraction is the amount of alkali required which bears a stoichiometric relationship to the amount of metal extracted.

Ashbrook[4] points to the advantage of using exchange reactions for the extraction, thereby eliminating the continuous addition of alkali into the individual stages. Equilibration of the organic acids with ammonia is suggested. On this basis the processing costs for extraction of cobalt or nickel were estimated. Data for three different cation exchangers as extractants has been given. If the total estimated cost for extraction is taken as a basis, then from the data given by Ashbrook the costs for the alkali requirements for pH control amount to approximately 43 per cent of the processing costs for extraction with di(2-ethylhexyl) phosphoric acid (HDEHP), 37 per cent for extraction with naphtenic acid, and 32 per cent for Versatic 911 as extractant.

Undoubtedly, this apparent demand for a separate source o alkali has restricted the interest and use of carboxylic acids and alkyl phosphoric acids as extracting agents for metals.

It seems appropriate at this point to stress the fact that the solvent extraction technique generally is considered as a tool for separation of metals which exist as dissolved species in an aqueous metal salt solution. The recovery of the metals by solvent extraction with carboxylic acids or alkyl phosphoric acids must then unavoidably pay the toll of an equivalent amount of alkali to replace the metal cations removed from the aqueous solution. An alternative and more economical way of making use of the organic acids as cation exchangers will be presented in the following discussions.

INTEGRATION OF LEACHING AND SOLVENT EXTRACTION

The purpose of the present work is to discuss and show how the application of liquid cation exchangers such as carboxylic acids and alkyl phosphoric acids may be extended to include the leaching operation prior to solvent extraction. By this extension the extraction and separation of the metals may be achieved by use of carboxylic acids and alkyl phosphoric acids without adding an external source of alkali to the process streams.

The key to this operation is to start the dissolution process by leaching the solid metal-bearing material with the extracting organic phase itself. A direct reaction between the solid phase metal species and the organic acid will then result in an organic metal salt solution from which the metals may be stripped selectively. In short it can be stated that by this integration of a prior step of organic leaching the resulting overall solvent extraction process utilizes the inherent alkaline property of the solid raw material. The addition of alkali from a separate external source may then only be marginal, serving some scrubbing purposes in the solvent extraction circuits if necessary. Thus the potential uses for commercial application of extractants like the carboxylic acids will be highly increased since the often quoted prohibitive economic factor of the amount of alkali needed will be eliminated.

The conditions and basic principles of the integrated organic leaching-solvent extraction operation will be outlined and illustrated by examples which have been worked out as case studies from bench scale operations.

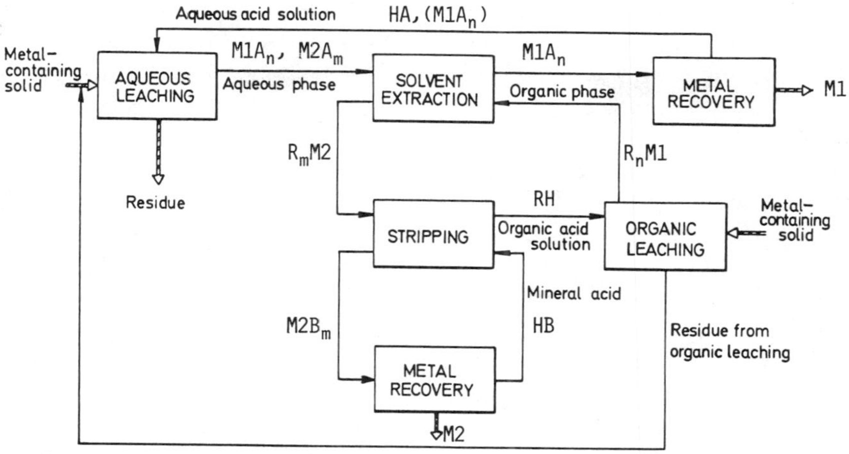

Fig. 3. Integrated leaching and solvent extraction process for solid material containing metals M1 and M2.

The Balanced System: Organic and Aqueous Leaching in Countercurrent Operation

When the solid raw material consists of a mixture of metal species which are partly reactive towards the organic acid, it is often found that the most basic of these metals will preferably react with the organic acid, thus forming a metal-organic salt solution in the organic phase. The metal-loaded organic phase can then in a subsequent step be used for the solvent extraction of a less basic metal from an aqueous salt solution. This aqueous solution may be prepared by a secondary aqueous acid leaching of the solid residue from the organic leaching. Depending on the stoichiometric balance between the leached metals in the organic and the aqueous phases respectively, some of the solid material may go directly to the aqueous acid leaching.

The principle of the integrated leaching and solvent extraction process is shown in Fig. 3. To simplify the illustration it has been assumed that the solid material contains only two soluble metals, M1 and M2, of which M1 is the more basic with reference to Fig. 1. It will be noted that a proper balance within the integrated circuit exists when the amount of M1 dissolved by the organic leaching is equivalent to the amount of M2 in the aqueous leach solution. The solvent extraction process will take place by an exchange reaction according to Eq. 2.

As shown in Fig. 3 the two metal species will appear (after solvent extraction) as separate streams of metal salt solutions, one in the aqueous phase as metal salt $M1A_n$ and the other in the organic phase as metal salt R_mM2. The further processing of these metal salt solutions may follow appropriate recovery methods for the metals. The organic phase may readily be stripped for the metal value by contacting with a mineral acid HB, whereby the organic acid is recovered and thus made ready for another loading with M1 by the organic leaching step. It is to be noted that the mineral acids HA and HB in general may be two different acids.

The integrated leaching-solvent extraction process will in principle comprise a closed-loop system with the only entering materials being the solid feed and the mineral acids equivalent to the metal salt products. The countercurrent solvent extraction process is self-buffering with no need for external pH-adjustment. The mineral acids will also be in closed-loop circuits if the metals are electrolytically deposited in a tankhouse.

Figure 3 shows the simplest possible setup for the integrated leaching-solvent extraction process. For the sake of simplicity it has been assumed that the relative amounts of the two metals present are in stoichiometric balance for the exchange reaction. When several metals are present the process scheme may be adjusted accordingly by adding more circuits. However, the basic principle of the primary leaching of the solid material by the organic phase will remain as the key to the integrated organic leaching - solvent extraction process.

The application of the process principles for a balanced system will be demonstrated by the following case study for the extraction of iron from zinc sulphate leach liquors.

Case I: Solvent Extraction of Iron in Zinc Hydrometallurgy

In the roast-leach-electrowinning process for the production of zinc from sulphide concentrates iron plays a very significant role in the technology and economics of the overall process.

In the roasting step, which converts the sulphides to oxides, most of the present iron combines with zinc forming zinc ferrites. These compounds are complex oxides represented by the formula $ZnFe_2O_4$ and may be considered as a key component in the development of zinc hydrometallurgy.

In the early hydrometallurgical zinc plants the recovery of zinc was deliberately limited to the pure zinc oxide part of the roaster calcine. The oxide part is readily dissolved in dilute sulphuric acid, leaving the zinc ferrites as a residue virtually insoluble by adjusting the leaching conditions at a suitable temperature and acidity. However, this leaching procedure put a limit to the overall zinc recoveries of only 85 to 93% which was a serious drawback of the electrolytic zinc process.

In most modern zinc plants the zinc ferrites are broken down by the attack of hot sulphuric acid. However, the dissolution of the ferritic zinc will be accompanied by the simultaneous dissolution of the iron in the ferrites. This will introduce an iron concentration at the level of 10-25 g/l in the hot acid leach solution.

Over the last 15 to 20 years, different methods have been introduced whereby iron can be precipitated as crystalline, filterable materials, readily separated from the zinc sulphate solution obtained from dissolving zinc ferrite in hot sulphuric acid (spent electrolyte). The iron precipitates are the results of hydrolysis reactions, including the basic iron sulphates known as jarosites, and also iron oxides, both hydrated (goethite) and anhydrous (hematite). An extensive review and discussion of the various processes has been given by Monhemius.[9]

Although the modern technology for removal of iron by precipitation has been successfully adapted by the zinc industry, the main drawback of the precipitated iron materials is that they have to be disposed of by dumping. This is an increasing concern in most places around the world. Obviously a welcome solution to this problem would be to make an iron product suitable at least for iron-making. None of the precipitated materials can be used for this purpose because of their impurity content, in particular of zinc and sulphur.

In this context iron removal by solvent extraction seems to offer a great deal of promise in the right direction. The concept of an integrated organic leaching - solvent extraction operation is particularly well suited for the iron/zinc separation by using the commercial carboxylic acid Versatic 10 as extractant. Referring to Fig. 1 it will be seen that the extraction isotherms for zinc and ferric iron are far apart. Thus a very efficient separation of the two metals is to be expected.

A crucial and very important finding is that Versatic acid is highly reactive in direct reacton with the zinc oxide in the calcines:

$$2RH_{(org)} + ZnO_{(solid)} \rightarrow R_2Zn_{(org)} + H_2O \quad (3)$$

Having formed the Versatic zinc salt of the organic acid, the extraction of ferric iron may take place by the exchange reaction:

$$3R_2Zn_{(org)} + 2Fe^{+3}_{(aq)} \rightarrow 2R_3Fe_{(org)} + 3Zn^{+2}_{(aq)} \quad (4)$$

Thus by reaction (3) and (4) zinc will be transferred from the calcine material into an aqueous sulphate solution via the dissolution of zinc oxide in the organic phase. By the same operation iron has simultaneously been removed from the aqueous solution and extracted into the organic phase as ferric carboxylate salt.

The net result of this combined operation of leaching and solvent extraction is that iron has been extracted without any addition and expense for alkali.

A conceptual flowsheet for the integrated organic leaching - solvent extraction process is shown in Fig. 4. The two steps of sulphuric acid leaching, the neutral leach and the hot acid leach, are basically the same as in the present practice of the iron precipitation processes. By replacing the present precipitation step for jarosite or goethite, the iron extraction process will be brought about by the following steps: the organic leaching of zinc, solvent extraction of iron, and the stripping of iron from the organic phase.

In the organic leaching step enough zinc salt of the organic acid has to be produced in order to 1) neutralize the free sulphuric acid from the hot acid leach and 2) extract iron by the exchange reaction (4) at pH 2-3. Before the exchange reaction between zinc and ferric iron can take place, the free sulphuric acid will have to be neutralized by the organic zinc salt:

$$R_2Zn_{(org)} + H_2SO_{4(aq)} \rightarrow 2RH_{(org)} + ZnSO_{4(aq)} \quad (5)$$

Thus the pH of the aqueous phase will increase until the exchange reaction takes place between pH 2-3.

The iron concentration in the hot acid leach solution will be reduced from some 10-25 g/l Fe to 0.5-1 g/l Fe. The iron may be removed quantitatively if desired, but it is more advantageous to leave a small amount for precipitation as ferric hydroxide due to the scavenging effect of the precipitate on trace metals impuritities.

The final outlet for iron will be by a process called hydrolytic stripping whereby iron is directly precipitated from the organic

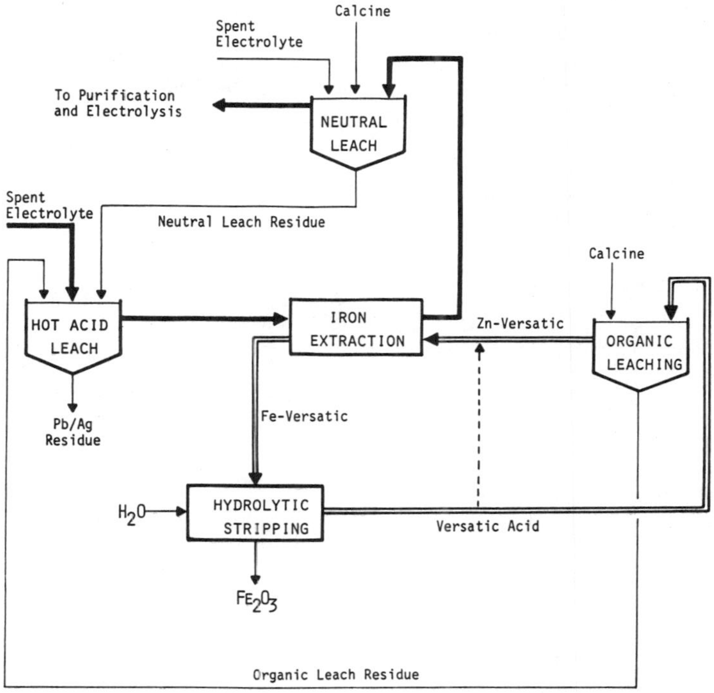

Fig. 4. Integration of organic leaching and solvent extraction of iron in zinc hydrometallurgy.[11]

phase as iron oxide by the reaction:

$$2R_3Fe_{(org)} + 3H_2O_{(aq)} \rightarrow Fe_2O_{3(s)} + 6RH_{(org)} \qquad (6)$$

The hydrolytic stripping process is discussed in another chapter in this proceedings.[10]

The organic leaching may be operated with or without an aqueous phase present. Under optimum conditions an aqueous sulphate solution will be present in order to minimize the contamination of the solid phase by the organic solution.

The leaching rate will be, as expected, highly dependent on the temperature. In Fig. 5 are presented the results from batchwise leaching of zinc calcine at various temperatures. The leaching has in this case been carried out without any aqueous phase present. 100% leaching of zinc corresponds to about 30 g/l zinc in the organic phase. This loading would be only about half the capacity of the organic phase.

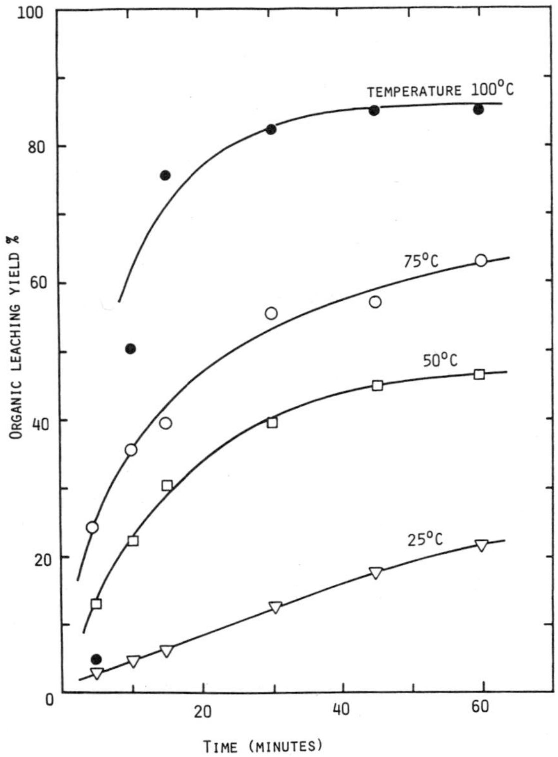

Fig. 5. Leaching of zinc calcine in 30% Versatic.[11]

From Fig. 5 it is seen that the maximum dissolution of zinc from the calcine is as high as about 85%. As expected, zinc ferrites are virtually insoluble in the organic leaching and only traces of iron will be dissolved.

A simplified process flowsheet without the neutral leach is shown in Fig. 6 indicating typical concentrations for process streams in a balanced system on the basis of 60% organic leaching of zinc.[11]

Organic leaching combined with selective stripping

The simplest way to demonstrate the advantage of the organic leaching operation will be to consider a solid material containing a mixture of reactive metal oxides. On attacking this material by an organic phase containing a carboxylic acid, the organic leaching may for an ideal system result in complete dissolution of the oxides. The metals will then appear in the organic phase as a mixture of metal-organic salts in solution.

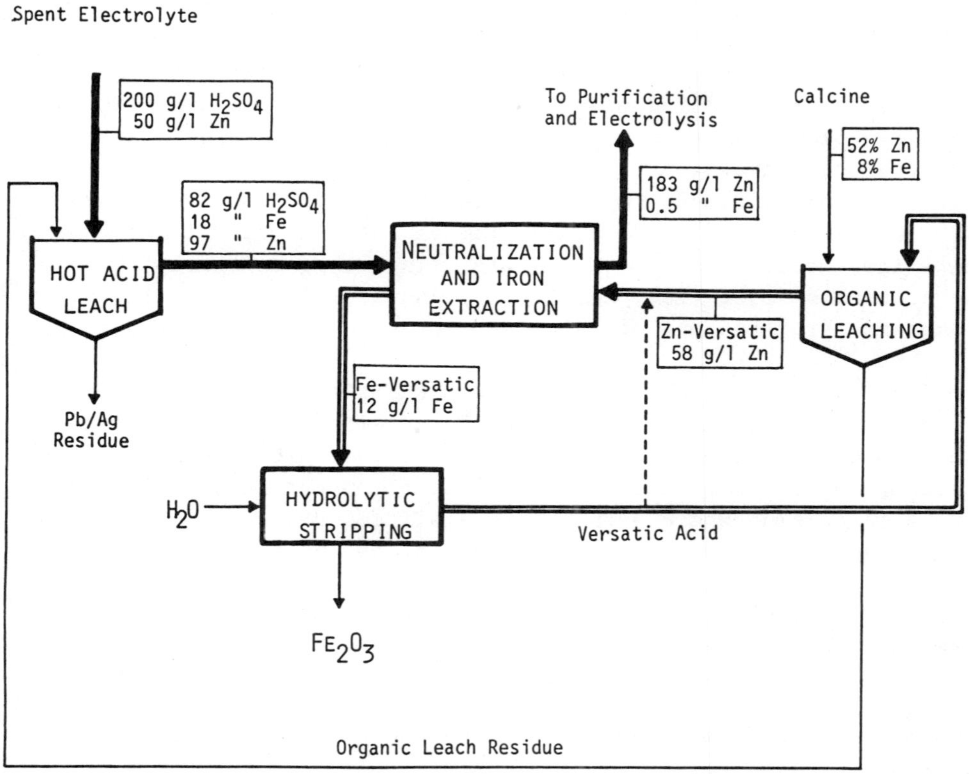

Fig. 6. Solvent extraction of iron in zinc hydrometallurgy. Typical concentrations for process streams without neutral leach.[11]

From Fig. 1 it will be seen that the metals may be stripped selectively from the organic phase, one by one, by an aqueous solution of a mineral acid. Depending on the relative position of the extraction isotherms, the degree of separation, i.e. the purity of the metal products in the resulting separate aqueous streams, will be a function of the number of stages in a counter-current operation within the stripping circuits.

Thus the net result of this combined operation is that the metals will end up as separated aqueous metal salt solutions. This separation will be accomplished without any addition of chemicals apart from the equivalent acid to form the metal salts by the selective stripping.

Case II: Recovery of Metal Values from Zinc Ash and Flue Dusts

An interesting source for organic leaching is to be found in waste oxide materials such as metal-containing ashes and flue dusts. Thus a process has been worked out for the recovery of zinc from zinc ash from galvanizing baths.[12] The oxidized material floats on top of the bath and is removed by mechanical skimming. The resulting zinc ash is a mixture of crude zinc oxide and varying quantities of entrained metallic zinc. There will also be contaminants of some other metals (i.e. iron, lead, etc.) and especially chloride compounds from the use of chloride fluxes in the galvanizing process.

Because of the chloride content, the zinc ash will not be suitable for treatment by a direct leach with sulphuric acid and a subsequent electrowinning of zinc, since the chloride will contaminate the sulphate solution. Chlorides are not tolerated in the electrolysis of zinc where even a very small amount of chloride in the electrolyte is extremely detrimental to the system.

Recovery of zinc by organic leaching with Versatic acid solves the chloride problem quite simply. The zinc oxide part of the zinc ash dissolves readily into the organic phase forming the carboxylate organic zinc salt. Any chloride extracted will be removed by washing the organic phase with water prior to the stripping of zinc into an aqueous sulphate solution.

A simplified flow diagram for processing chloride-containing zinc waste materials is shown in Fig. 7. Versatic 10 is used as a 30% solution by volume in an aliphatic diluent. Recovery of entrained and adsorbed organic phase is accomplished by a hot water and soda wash of the leach residue. Zinc is recovered from the combined aqueous wash solutions by a minor circulating stream of the organic extracting phase.

Zinc is stripped from the loaded organic phase by contacting with sulphuric acid according to Eq. 5. The resulting aqueous zinc sulphate solution may be adjusted at the optimum concentration for electrolytic production of zinc.

Alternatively, by recycling the stripping aqueous phase, solid crystals of $ZnSO_4 \cdot 7H_2O$ may be salted out from the saturated zinc sulphate solution. Thus in this case three coexisting phases will be present in the crystallizer: organic, aqueous, and solid, as illustrated in Fig. 8.

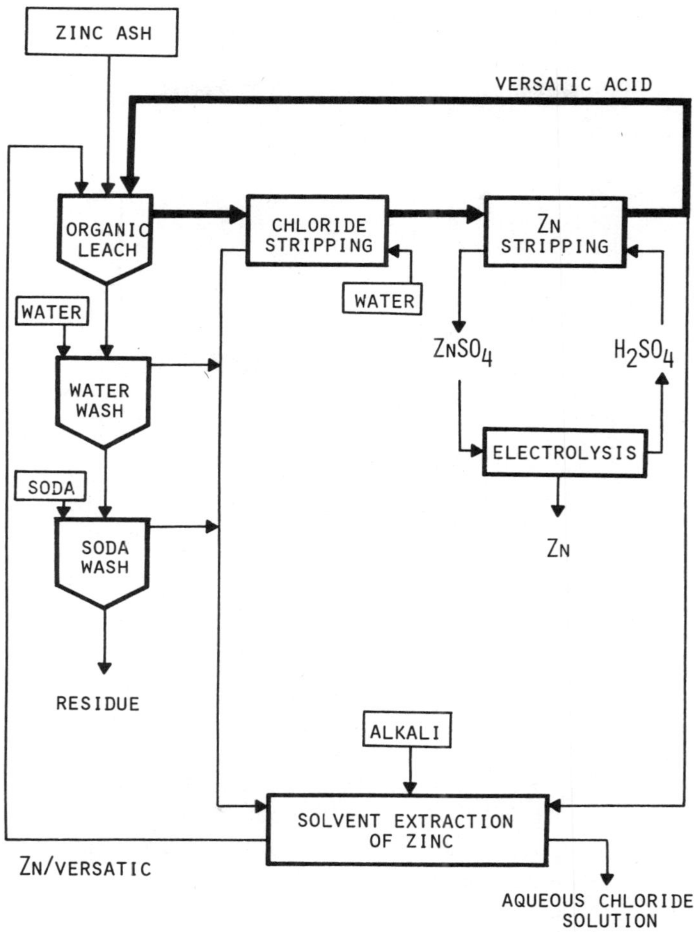

Fig. 7. Process flow diagram for treatment of chloride-containing zinc waste material.[12]

Other zinc salts or zinc salt solutions may optionally be produced by using various aqueous acid solutions for stripping. Alternatively, the Zn-Versatic salt may be used for the extraction of iron and copper from zinc leach liquors obtained in the hot acid leach of zinc ferrites as illustrated in Case 1.

The organic leaching, as presented for the zinc ash material, has also been applied for the treatment of flue dusts from zinc smelting and from a Waelz-process.[12]

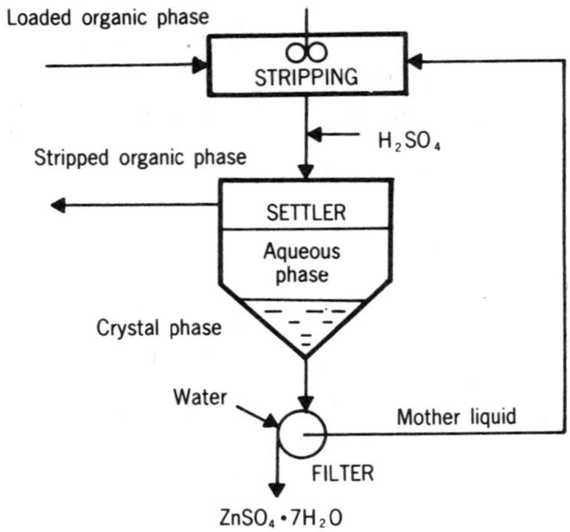

Fig. 8. Stripping of zinc with sulfuric acid from an organic solution of zinc-Versatic.[12]

Case III: Recovery of Metal Values from Zinc Purification Residues

Solution purification in the electrolytic production of zinc is accomplished by zinc dust cementation. Finely divided zinc dust metal is added to the neutral leach zinc liquor whereby secondary and contaminating metals like Cu, Cd, Sb, As, Ni, and Co will precipitate while an equivalent amount of zinc metal dissolves into the zinc sulphate solution. Thus the cementation metallic product will contain the contaminating metals together with the remaining unreacted zinc dust. After filtration some electrolyte will remain in the residue, giving a certain content of zinc sulphate.

The metal contents of the purification residue will vary depending on the composition of the raw material input and the operation of the purification process. A typical set of analytical data in per cent of dry weight is as follows: Zn: 31.3, Cu: 31.3, Cd: 13.7, As: 0.037, Sb: 0.063, Co: 0.138, Ni: 0.113 and sulphate and some fibrous materials account for the rest. From the data it will be seen that the main metal values are zinc, copper, and cadmium. Thus the primary goal for a recovery process will be to obtain the three main metals as separate and purified product streams. In addition there must be an outlet for the minor amounts of the other contaminating metals.

A process based upon an integrated organic leaching-solvent extraction operation has been worked out for both a di(2-ethylhexyl) phosphoric acid system[13] and a Versatic acid system. In the present paper some of the main characteristic features of the Versatic acid system will be presented.

By comparing the diagrams in Figs. 1 and 2 the HDEHP-system may appear to be more suitable than Versatic because of the apparently better separation of cobalt and nickel from the three main metals. However, from the experience obtained by working with both systems, it may be stated that the distribution curves as shown in the diagrams will serve only as a guideline at the start of a process development program. The real process parameters will always have to be worked out on the actual process streams.

This may be further demonstrated by the Versatic acid system. From Fig. 1 it appears that zinc is more basic than cadmium. It has to be noted, however, that the curves are obtained from single metal solutions at 20°C. It turned out that on working in practice with the multicomponent system resulting from the purification residue, cadmium behaved as a more basic metal than zinc. The relative position of zinc and cadmium varies in the literature.[4,7] By following the hydroxide rule, i.e. the order in which the metals are precipitated as hydroxides, cadmium is the more basic metal.[8]

To make the metallic residue suitable for dissolution by the organic acid it has to be oxidized. This may be done by a separate pre-oxidation step in an oven at about 300°C or alternatively, by a combination of the oxidation and the dissolution process. In the latter case the metallic residue is suspended in water and the aqueous slurry contacted with the organic phase, sparging oxygen (air) through the dispersion as an oxidizing agent. The reactions taking place for both kinds of oxidation can be represented by:

$$M^{o} + \frac{1}{2}O_2 + 2RH_{(org)} \rightarrow R_2M_{(org)} + H_2O \qquad (7)$$

where M^{o} represents the various divalent metals in the residue and RH is Versatic acid in the organic phase.

The loading capacity of the organic 30% Versatic acid solution will vary for the various metals. By the stoichiometric relationship of Eq. (7) the maximum solubility of a divalent metal in the organic phase will be about 0.8 mol/l which is about 50 g/l for zinc and copper. In practice it appears that about 35 g/l is near an upper limit for copper. For zinc, however, a concentration of 73 g/l has been achieved[11,12] which indicates that zinc may dissolve as a monovalent hydroxyl compound, $RZn(OH)_{(org)}$.

A conceptual flowsheet of the Versatic acid process is shown in Fig. 9. Both in the pre-oxidation procedure and in the combined oxidation-dissolution process the multicomponent leaching is accomplished by an organic-aqueous dispersion. Pure water is added to the system whereby the dissolved metals will be distributed between the two phases during the leaching operation. The metals will appear in the aqueous phase as sulphate salts because of the zinc sulphate content in the residue from the zinc leach liquor. The distribution of the equivalent amount of metal cations in the aqueous phase is determined by the relative acidic/basic character of the metals present under the prevailing conditions.

A very important feature of the Versatic system is that a separation effect may already be achieved in the leaching step by adjusting the loading of the organic phase at an appropriate level. This may be illustrated by the results obtained from simultaneous oxidation and batch leaching at 50°C for low and high loadings respectively. In Fig. 10 the kinetic curves are given for leaching and extraction of copper, zinc, and cadmium into the organic phase. These results are obtained at relatively low metal loadings (Zn and Cu: 7-8 g/l, Cd: 3 g/l). Thus the main fraction of the three metals is dissolved into the organic phase. About 25% of nickel and cobalt are also extracted into the organic phase but the main bulk of these metals appear in the aqueous phase. The leaching of arsenic and antimony is low and these elements will virtually remain in the residue.

From Fig. 10 it will be noted that the curves for zinc and cadmium are both levelling off at the upper limits, with about 76 and 81% extraction into the organic phase respectively, while copper is approaching quantitative extraction into the organic phase. This is in accordance with the relative positions of the extraction curves in Fig. 1 and the acidic character of copper. The sulphate ions of the aqueous phase have to be balanced by an equivalent amount of cations. Due to their close positions on the diagram, both zinc and cadmium will appear in the aqueous phase.

By increasing the solid/organic phase ratio, higher loading of copper and zinc will be obtained in the organic phase while the organic phase extraction of cadmium, cobalt and nickel will be depressed. In Fig. 11 the leaching and extraction of zinc, cadmium,

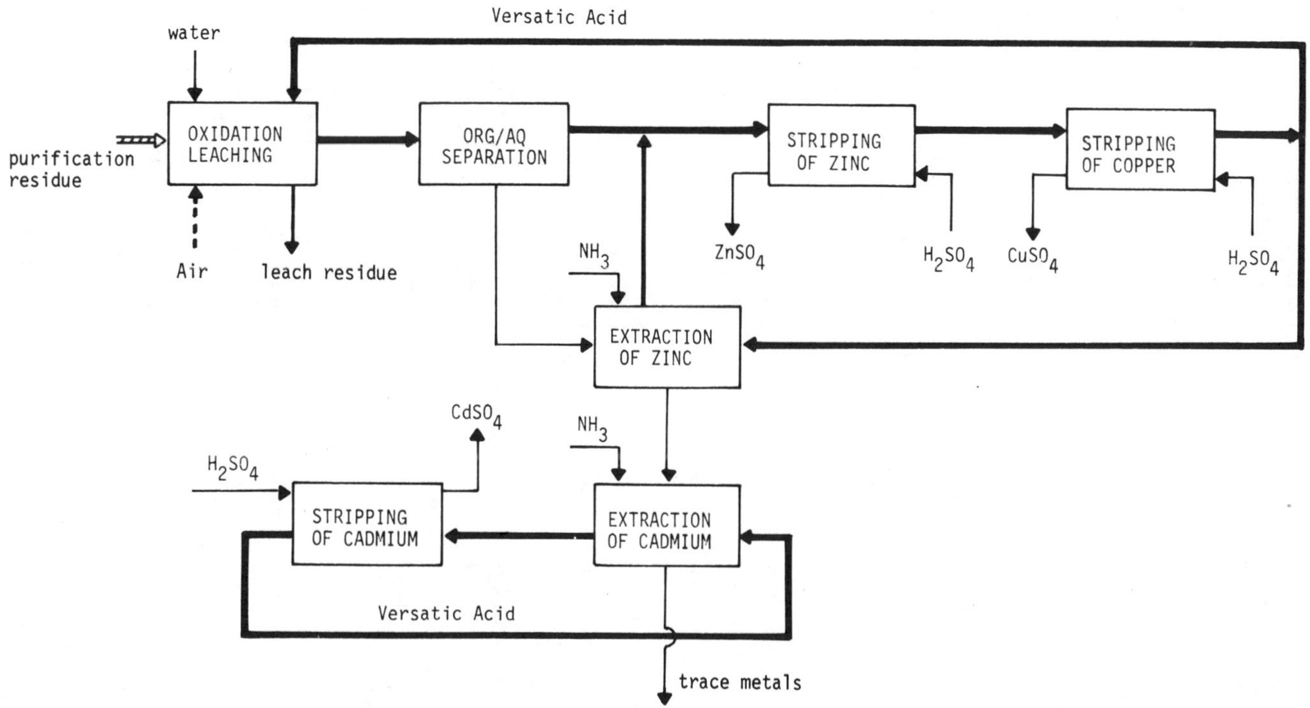

Fig. 9. Recovery of metal values from zinc purification residues.

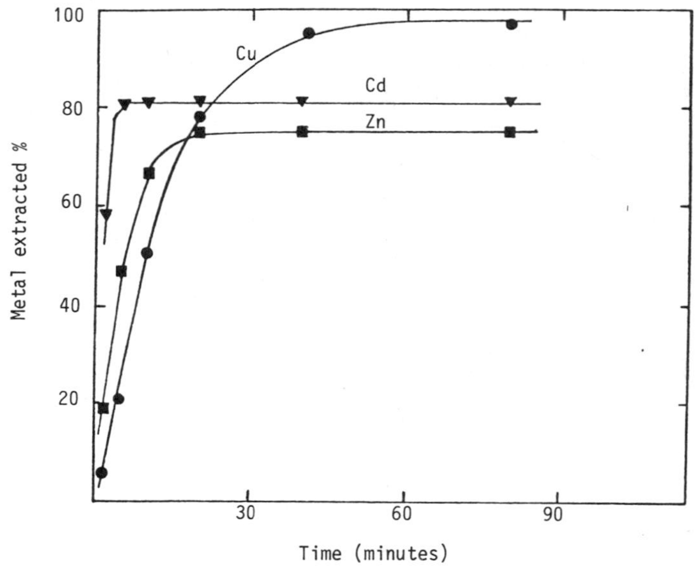

Fig. 10. Oxidizing leaching of zinc purification residues with 30% Versatic acid at 50°C. Metals extracted into organic phase at low loading.

cobalt, and nickel into the organic-aqueous dispersion is given as a function of the increasing loading of the organic phase with copper (and zinc).

In particular the zinc/cadmium relationship is interesting to follow. In view of the relative positions of the extraction curves in Fig. 1 one might be inclined to expect that an opposite effect between zinc and cadmium would appear. However, as already pointed out, the curves in Fig. 1 are obtained by starting from single metal sulphate solutions, all at 5 g/l. Obviously in the present multicomponent system the "crowding effect" from copper and zinc will result in the organic-aqueous phase distributions shown in Fig. 11.

With reference to Fig. 9 it will be seen that there are two main Versatic acid circuits in the process. One of them comes from the leaching step where the organic phase has been primarily loaded with copper and zinc. The other one, the cadmium circuit, is established by complete extraction of cadmium from the aqueous leach solution. Prior to the cadmium extraction, however, the main part of the aqueous zinc content is removed by extraction into the upper Zn-Cu circuit. The zinc extraction will be accomplished at high zinc loading of the organic phase to obtain a satisfactory separation from cadmium.

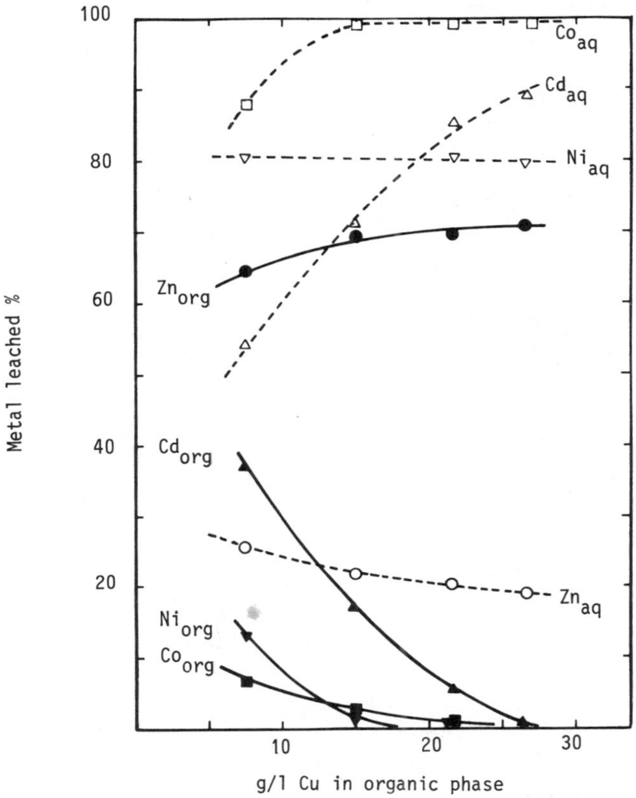

Fig. 11. Oxidizing leaching of zinc purification residues with 30% Versatic acid at 50°C. Metal distribution in organic-aqueous dispersion as a function of copper loading in organic phase.

It has to be noted that the extractions of zinc and cadmium, respectively, from the aqueous leach solution has to be operated by the equivalent addition of a base (ammonia) into the system. On the other hand it was found that the main part of the metal values of copper and zinc dissolves directly into the organic phase during the leaching. This main fraction will therefore represent the net gain in the integrated organic leaching - solvent extraction operation for the purification residue.

Within both extraction circuits there are scrubbing operations prior to the stripping of the organic phase to keep the product purities at specified levels. In this context the separation of cobalt and nickel from the product solutions is essential. Although

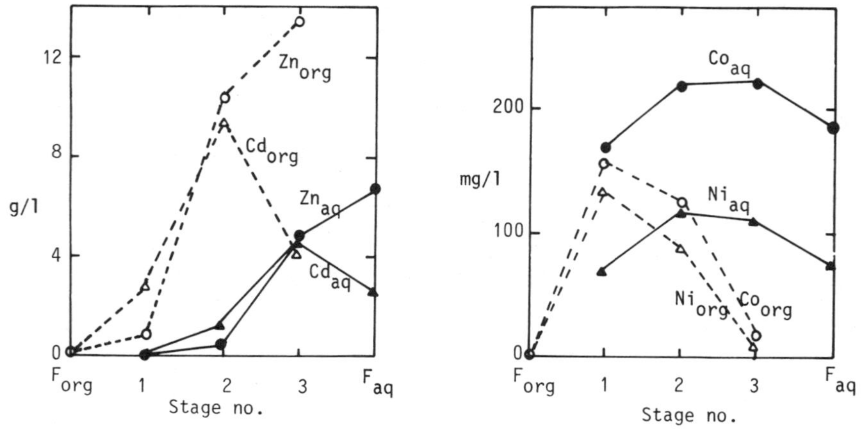

Fig. 12. Extraction of zinc and cadmium in 3 stages.

the "crowding effect" may keep the main part of cobalt and nickel away from zinc it is important for the further processing of both the zinc and cadmium solutions that cobalt and nickel are kept very low.

The efficiency of the Versatic acid system in this respect may be demonstrated by the steady state profiles of the various metal concentrations which were established within extraction and scrubbing circuits containing zinc, cadmium, nickel, and cobalt.

Figure 12 gives the metal profiles for extraction of zinc and cadmium from cobalt and nickel. In countercurrent extraction in three stages, the aqueous concentrations of zinc and cadmium are brought down from 6.8 and 2.4 g/l to 0.05 and 0.1 g/l respectively. The coextraction of cobalt and nickel resulted in organic phase concentrations of 14 and 7 mg/l respectively. The steady state bench scale operations were run at optimum pH, adjusted by addition of ammonia.

Cobalt and nickel concentrations may further be decreased by scrubbing with an aqueous solution at controlled optimum pH. Metal profiles obtained in a scrubbing system are shown in Fig. 13. Although the entering concentrations are as low as 3.7 mg/l Co and 2.1 mg/l Ni, the scrubbing further brings the concentrations down to 0.05 and 0.08 mg/l respectively.

In practice the extraction and scrubbing operations are run in an integrated cascade of stages. Thus the zinc and cadmium content of the aqueous flow will be re-extracted in the extraction section.

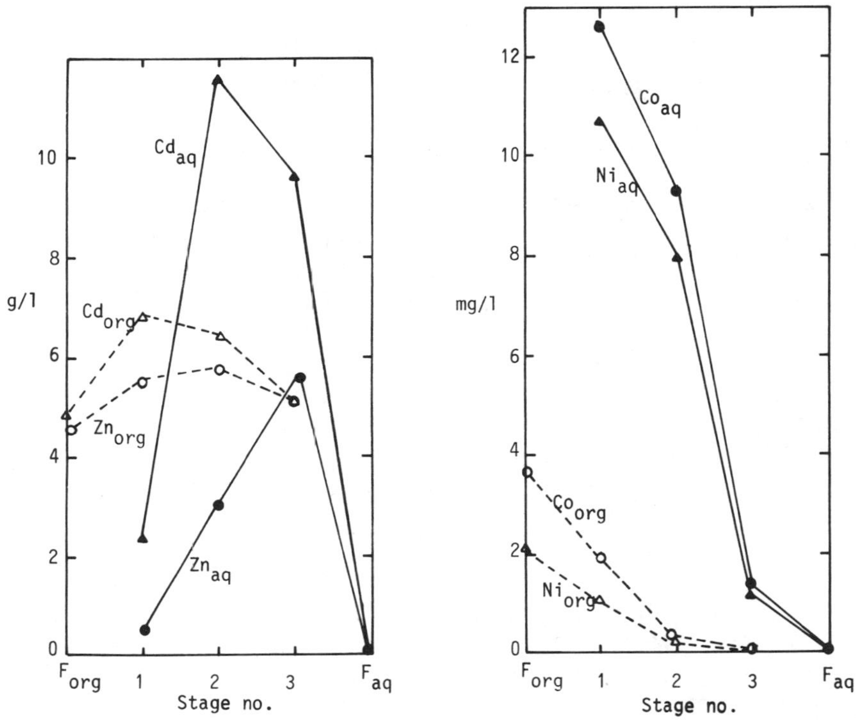

Fig. 13. Scrubbing of cobalt and nickel in 3 stages.

CONCLUDING REMARKS

In this paper three case studies have been presented showing how Versatic acid can be used for extraction and separation of metals from a solid raw material. The three cases have in common that the starting point in the processing is the direct dissolution of metals into the organic Versatic acid phase. By this procedure the alkaline property of the solid material is utilized for the subsequent processing steps of solvent extraction.

Although the processes also have in common that they are all connected to the zinc industry, the general concept of the integrated organic leaching - solvent extraction operation should be applicable whenever a direct dissolution of metals from a solid raw material into an organic acid cation exchanger is possible. The main advantage of this concept will be the possibility of working in closed-loop circuits as demonstrated in Fig. 3. During process developments, however, problems will always arise, and in the end any process has to be proven on technical as well as commercial basis over a long period of time.

In the present situation there are obviously some very important parameters which have to be examined in connection with the possible losses and degradation of the organic phase. Although these parameters are difficult to evaluate from bench scale operations, the estimates made so far look promising.

The processes presented in this paper are presently under technical and commercial exploitation in collaboration with NORZINK.

REFERENCES

1. D. S. Flett: Chem. Ind. (London), 1977 September, pp. 706-712.
2. E. Wigstøl and K. Frøyland: Solvent Extr. Metall. Processes, Proc. Int. Symp., Antwerp, 1972, May, pp. 71-81.
3. D. S. Flett: Inst. Min. Metall., Trans., Sect. C., 1974, vol. 83, pp. C30-C38.
4. A. W. Ashbrook: Miner. Sci. Eng., 1973, vol. 5, no. 3, pp. 169-180.
5. N. M. Rice: Hydrometallurgy, 1978, vol. 3, no. 2, pp. 111-113.
6. A. W. Fletcher and D. S. Flett: Solvent Extraction Chemistry of Metals, McKay, H.A.C. et al., eds. MacMillan, London 1965, pp. 359-375.
7. E.L.T.M. Spitzer: Solvent Extr. Metall. Processes, Proc. Int. Symp., Antwerp, 1972, May, pp. 14-18.
8. John Monhemius: Chem. Ind. (London), 1981 June, pp. 410-420.
9. A. J. Monhemius: Topics in Non-ferrous Extractive Metallurgy (Critical Reports on Applied Chemistry Vol. 1), A. R. Burkin, ed., Blackwell Scientific Publications, London, 1980, pp. 104-130.
10. A. J. Monhemius, L.A.C. Teixeira, and G. Thorsen: Proceedings of NATO Advanced Research Institute on Hydrometallurgical Process Fundamentals, Cambridge, Britain, July 25-31, 1982. Plenum, N.Y. 1984, pp. 279.
11. G. Thorsen and A. Grislingås: "Solvent Extraction of Iron in Zinc Hydrometallurgy." Paper presented at the 109th AIME Annual Meeting, Las Vegas, Feb. 1980.
12. G. Thorsen, A. Grislingås, and G. Steintveit: J. Metals, 1981, vol. 33, no. 1, pp. 24-29.
13. G. Thorsen: "Zinc and Cadmium" in "Handbook of Solvent Extraction," T.C. Lo, M.H.I. Baird, and C. Hanson, eds., John Wiley & Sons, New York, 1983, pp. 709-716.

THERMODYNAMICS OF CONCENTRATED ELECTROLYTE SOLUTIONS APPLIED TO LIQUID-LIQUID EXTRACTION OF METALS AND SOLID-LIQUID EQUILIBRIA

H. Renon, W. Fürst, H. Planche and F. X. Ball

Centre Réacteurs et Processus
Ecole Nationale Supérieure des Mines de Paris
60, Boulevard Saint-Michel, 75006 Paris, FRANCE

A review of the advances in the thermodynamic methods of representing concentrated electrolyte solutions and the understanding of systems of phases in equilibrium of interest to hydrometallurgical processing are considered in this paper.

INTRODUCTION

The Centre "Réacteurs et Processus" of the Ecole des Mines de Paris started a few years ago a program to better understand the properties of systems of phases in equilibrium of interest in hydrometallurgy. The main features of these equilibria are:

1) the presence of an aqueous electrolyte solution, where salts are present at high ionic strength.

2) the importance of chemical reactions in solution: complexation, solvation.

3) the variety of other phases present at equilibrium: organic phase in liquid extraction, unsoluble salt, vapor phase, ion exchange resin.

4) the possible characterization by other experimental methods: electrochemistry, spectroscopy.

This paper is presented in two parts: first, a review of the advances in the method of representation of concentrated electrolyte solutions; and second, a report on the activity of this laboratory on applications to hydrometallurgical systems.

The topic of this paper does not include the study of high temperature equilibria and application of methods where deviations from ideality are not fully taken into account. An earlier review by Barner and Kust[1] summarizes the state of the art in this broader field.

REPRESENTATION OF THERMODYNAMIC PROPERTIES OF CONCENTRATED ELECTROLYTE SOLUTIONS

Description of Models

The general equations and notations for definition of excess properties were recently reviewed by Renon.[2] The infinite reference state is selected for ions. Three scales are commonly used for definition of acivity coefficients, y_i, γ_i, f_i:

$$\begin{aligned} \mu_i &= \mu^o_{i,c} + RT \ln C_i y_i && \text{molarity scale} \\ \mu_i &= \mu^o_{i,m} + RT \ln m_i \gamma_i && \text{molality scale} \\ \mu_i &= \mu^o_{i,X} + RT \ln x_i f_i && \text{mole fraction scale} \end{aligned} \tag{1}$$

where μ_i is the chemical potential of ion i. The osmotic coefficient ϕ describes the deviations from ideality for the solvent w:

$$\phi = \ln a_w \; / \; (- \frac{M_w}{1000} \sum_j \nu_{B_j} m_{B_j}) \tag{2}$$

in a mixture of salts B_j dissociating into ν_{B_j} ions. These coefficients can be related to the excess Gibbs energy.

The excess properties of electrolyte solutions depend on the interaction between charged species in the presence of ionizing solvent and ion-solvent and solvent-solvent interactions and, if dissociation is not complete or solvation is important, on the presence of additional chemical species. The most comprehensive model will incorporate all these interactions.

The early model of Debye-Hückel[3] remains valuable because it predicts the limiting law valid at very low concentration of ions ($< 10^{-3}$ M). In practice, the equation

$$\ln f^2_{\pm} = - \frac{A \; |Z_1 Z_2| \sqrt{\Gamma}}{1 + Ba \sqrt{\Gamma}} + b\Gamma \tag{3}$$

can be used up to 2 M by adjusting the two parameters a and b, where $$B = \left(\frac{8\ \pi\ N\ e^2}{1000\ D\ k\ T}\right)^{1/2} \tag{4}$$

$$\Gamma = 1/2\ \Sigma_i\ C_i\ Z_i^{\ 2} \tag{5}$$

$$A = \left(\frac{2\ \pi\ N}{1000}\right)^{1/2}\ \frac{e^3}{(kDT)^{3/2}} \tag{6}$$

N is Avogradro's number, K, Debye reciprocal length, a is the diameter of the solvated ions, D is the dielectric constant of the solvent, e is the charge of the electron, and k is Boltzmann's constant.

Usually the value taken for D is the dielectric constant of pure solvent. But Scatchard[4,5] incorporated an additional corrective term proposed by Debye and MacAulay[6] to take into account the change of energy of ions brought about by the change of dielectric constant of the solution to obtain a good representation with two adjustable parameters.

Bromley[7,8] proposed an analytical representation which corresponds to Meissner and Kusik's[9] graphical observation of behavior of activity coefficients $\gamma_\pm$:

$$\ln\ \gamma_\pm = \frac{-\ 0.511\ Z_1\ Z_2\ I^{1/2}}{1 + I^{1/2}} + \frac{(0.06 + 0.6\ B)I}{(1 + \frac{1.5}{Z_1\ Z_2}\ I)^2} + \frac{BI}{Z_1\ Z_2} \tag{7}$$

The parameter B can be correlated according to the equation

$$B = B_{cation} + B_{anion} + \delta_{cation}\ \ \delta_{anion} \tag{8}$$

where the last term can sometimes be neglected and the parameters B correlated with standard entropies of aqueous ions.

a) Pitzer's model. In a long series of publications, Pitzer and coworkers[10-18] proposed a consistent model of representation of deviations from ideality and gave values of empirical parameters obtained from extensive data treatment. The excess Gibbs energy is developed in a simple empirical extended Debye-Hückel expression. The model is valid for mixtures of electrolytes and even non-electrolyte.[19] If water remains the main component of the solution, it can be extended to weak electrolytes.

$$G_m^E = RT\ \frac{N_w\ M_w}{1000}\ \left(-\ A_\phi\ \frac{4I}{1.2}\ \ln\ (1 + 1.2\ I^{1/2})\right)$$ (Continues)

$$+ 2\sum_c \sum_a m_c m_a \left(B_{ca} + (\Sigma mZ)C_{ca}\right) + \sum_c \sum_{c'} m_c m_{c'} (\theta_{cc'} + \sum_a m_a \psi_{cc'a}/2)$$

$$+ \sum_a \sum_{a'} m_a m_{a'} (\theta_{aa'} + \sum_c m_c \psi_{caa'}/2)$$

$$\text{where } A_\phi = \frac{1}{3} \left(\frac{2 \pi N d_w}{1000}\right)^{1/2} \left(\frac{e^2}{k D T}\right)^{3/2} \tag{9}$$

The difficulty of using Pitzer's model for hydrometallurgical application is the large number of parameters which are listed only for strong electrolytes at 25°C, and at times with their temperature derivatives. When solvated species are present, there is the formidable task of parameter evaluation, sometimes impossible because they are correlated. We found that the model truncated to two parameters was sufficient for the representation of osmotic coefficients. The C, θ, and ψ, parameters are neglected. The two parameters used are $\beta_{MX}^{(o)}$ and $\beta_{MX}^{(1)}$.

$$B_{ca} = \beta_{ca}^{(o)} + \frac{2\ \beta_{ca}^{(1)}}{\alpha^2\ I} \left(1 -(1 + \alpha I^{1/2}) \exp (- \alpha I^{1/2})\right) \tag{10}$$

As will be shown later, the model is capable of predicting osmotic coefficients in ternary solutions of two salts in water from binary parameters.

However, Fürst and Renon[20] showed that the application to solid-liquid equilibrium of two 1-1 electrolytes with one common ion ($NaCl - NaNO_3$, $KCl - KNO_3$, $NaCl - KCl$, $NaNO_3 - NaNO_2$) in water requires at least 5 of the 8 parameters of the untruncated model. At least one ternary parameter calculated from ternary solid-liquid data is needed. The need of higher order parameter may be explained by the very high ionic strength reached in solution ($I > 10$ M).

b) NRTL models. They are cited here, because they are of some interest, but a recent study by Ball, Furst, and Renon[21] showed their general inferiority to Pitzer's equation. It was not applied to hydrometallurgical systems.

Cruz and Renon's[22] equation represents with high accuracy excess properties of binary mixtures in the whole range of concentration, even for weak electrolytes.[23] The equation is an extended Debye-Hückel form with two additional Debye-MacAulay terms to take into account variation of the dielectric constant and NRTL terms for ion-solvent and solvent-undissociated electrolyte interactions. Its modification and extension to multicomponent mixtures were reported by Ball, Furst, and Renon.[21] The modified form has only two NRTL parameters per binary mixture because the dielectric constant of the solution is calculated according to Pottel's equation.[24]

Chen[25] applied the NRTL model to electrolyte solutions in a different way. No Debye-MacAulay term is used. Instead, additional NRTL terms for interactions between ions of different signs are included. The number of parameters increases rapidly in multicomponent systems.

c) Planche, M.S.A. model. Three problems arise in the application of statistical mechanics to a molecular model of interactions. These are the choice of the potential of interaction between particles, the calculation of radial distribution functions, and the derivation of the thermodynamic properties.

The radial distribution functions were obtained by computer simulation at first for the primitive model using radii of solvated ions much larger than the crystallographic radii along with the dielectric constant of the solvent. The same distribution functions can also be derived from the resolution of the equations of statistical mechanics using approximations and by comparison to the results of simulation. The hypernetted chain approximation gives the best results, and the mean spherical approximation (M.S.A) yield analytical results.

Three equations relate thermodynamic properties to the radial distribution function, the energy equation, the pressure equation and the compressibility equation. The discrepancies between resulting properties indicate weakness in the calculated radial distribution functions.

The new equations derived by Planche and Renon[26] are based on a non-primitive (taking into account other than coulombian interactions between ions) model of interaction. The potential of interaction between two particles takes a general form including hard sphere repulsion, coulombian interaction, and short range interactions:

$$u_{ij}\,(r) = \infty \qquad r < \sigma_{ij} \tag{11}$$

$$N_{av}\, u_{ij}\,(r) = \frac{Z_i Z_j e^2 N_{av}}{\varepsilon_o D\, r} + \frac{W_{ij}\ RT}{2\,\pi\, r\, N_{av}}\left(\delta'\,(r - \sigma_{ij})\right) \qquad r < \sigma_{ij}$$

where δ' is the first derivative of the Dirac function, D is the dielectric constant of the solution and W_{ij} is an energy parameter which characterizes the short range solvent-solvent, ion-solvent and ion-ion interactions.

An advantage of this potential is that an analytical solution of the M.S.A. exists using the method of Baxter[27] and the formalism of Blum.[28] The thermodynamic properties are obtained from that solution using the energy equation. The Helmholtz energy, obtained by integration, is composed of four additive terms, one for ionic interactions, one for short range interactions, one for hard sphere repulsions, and one for contribution of solvation.

$$(a - a_T^o)/RT = a_{ionic} + a_{SR} + a_{HS} + a_{Solv.}$$

$$a_{ionic} = -\frac{\alpha^2}{4\pi} \Sigma\, x_j\, z_j \frac{Z_j\, \Gamma_j + F_j\, P}{1 + \Gamma_j\, \sigma_j} + \frac{\Gamma^3 v}{3\pi N_{av}}$$

$$a_{SR} = -\sum_i \sum_j \frac{x_i\, z_j}{v\Delta} A^L_{ij} \left(1 + \frac{3}{\Delta} \zeta_2\, \sigma_j\right) \qquad (12)$$

$$a_{HS} = \sum_i x_i \ln \frac{x_i\, RT}{P_o\, v\Delta} + 3 \sum_i x_i \frac{\sigma_i^2}{\Delta} \left(\zeta_1 + \frac{\zeta_2^2}{2\Delta}\right)$$

$$a_{Solv.} = \Sigma \frac{x_i}{RT} \left[h_i - (s_i + R)\, T - \frac{3}{2} C_i\, RT\, (\ln T/T_o + \ln 298)\right]$$

where

$$\zeta_n = \frac{\pi}{6} N_{av} \sum_i \frac{x_i}{v} \sigma_i^n$$

$$\Delta = 1 - \zeta_3$$

$$\alpha^2 = \frac{e^2\, N}{\varepsilon_o\, D\, R\, T}$$

$$p = N \sum_i \frac{z_i\, \sigma_i\, x_i}{v(1 + \Gamma_i\, \sigma_i)}$$

$$A^L_{ij} = \frac{A^L_{ijo}}{T} \exp\left(\alpha_{ij}\, T\, (1 - \gamma T)\right) \quad \text{related to } W_{ij}$$

$\Gamma_\alpha = \Gamma$ is the unique scaling inverse length calculated by an easy-to-solve implicit analytic equation. Expressions for chemical potentials can be derived from Eq. 12.

The same equation can be applied to pure solvent, in liquid or vapor state. The model was first applied to polar substances, the solvents of electrolytes, especially water. All parameters are obtained from saturated vapor and liquid volumes and vapor pressure curve. Parameters σ_{ii}, A^L_{iio}, and α_{ii} result from properties of pure solvent. Using the pure solvent parameters, the model is then

applied to aqueous solutions of strong electrolytes. The parameters are: ionic parameters - ionic diameters σ_i of ions
- short range interaction parameters between ions and solvent W: A^L_{iw};

ion-ion interaction parameter - short range interaction parameters between cations M and anion X:A^L_{MX}.

The ion-ion parameter turns out to be a corrective term allowing predictions when no data are available. This is the only model which can be generalized to muticomponent mixture, with only one binary parameter. Moreover Pauling diameter can be used as σ_i for anions but not for cations. The σ_i for cations are correlated with Stokes diameters and the A^L_{iw} parameters related to the strength of hydration. Comparison to the next section shows that the model gives as good representation and prediction of deviations from ideality in strong electrolytes up to 6 M as Pitzer's model. It also has the advantage of a more precise physical meaning.

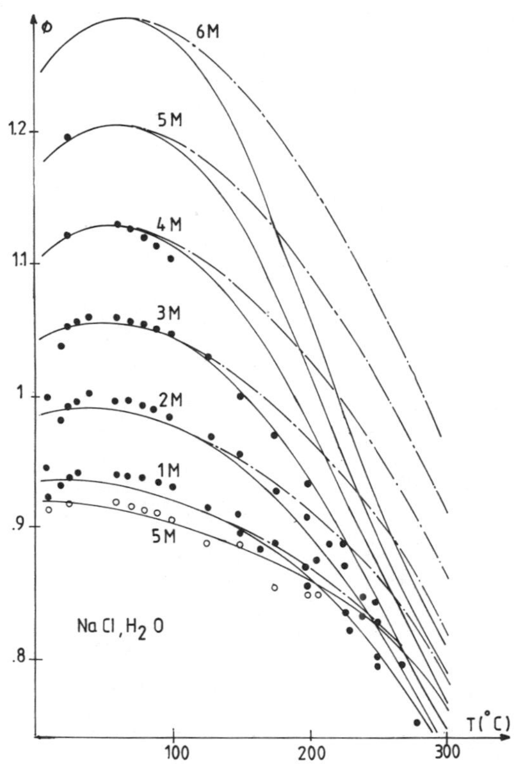

Fig. 1. Application of the Pitzer Model to the NaCl-H_2O mixture in the temperature range 0-300°C.

An application of the model to the NaCl-H_2O mixture in the range of temperature 0 - 300°C is shown in Fig. 1. Qualitative prediction is improved if formation of ion pairs is taken into account.

The M.S.A. model of Planche was also applied with success to the representation of liquid-liquid-vapor equilibria for the system HCl-$CaCl_2$-H_2O-phenol between 25 and 180°C.[29]

Comparison of Models of Deviation to Ideality for Representation and Prediction of Deviations to Ideality in Strong Electrolytes

Ball et al.[21,29] evaluated the NRTL models, and Planche and Renon[26] model by comparison with Pitzer's model.

Table 1 gives a summary of results of osmotic coefficients for the binary mixtures water-salt up to molalities of 6 M.

All models represent to within one percent the 1-1 salts solution but they differ much more for the other salts with a disadvantage for Chen's model.[19,25]

Table 1. Osmotic Coefficients for Water-Salt Binary Mixtures up to 6 Molal.

		100 x r.m.s average deviation in ϕ of water-salt mixture			
		PITZER	NRTL modified CRUZ	NRTL CHEN	M.S.A. PLANCHE
Number of binary parameters		2	2	2	1 + correlated ionic parameters
1-1 salts	27 systems	0.5	0.6	0.9	1.0
other salt	13 systems	1.8	2.2	5.6	2.1

Table 2. Prediction of Osmotic Coefficients in Water - 2 Salts Mixtures without Ternary Parameter.

	100 x r.m.s average deviation in ϕ			
	PITZER	NRTL modified CRUZ	NRTL CHEN	M.S.A. PLANCHE
21 systems	1.5	1.9	1.8	2.1

In Table 2 comparison, the advantage of Pitzer's model truncated after the B terms are illustrated, but it should be noted that the M.S.A. model requires only one adjustable parameter.

APPLICATIONS TO HYDROMETALLURGICAL SYSTEMS

The models of solution of electrolytes were applied at the centre "Réacteurs et Processus" to liquid-liquid, liquid-solid and liquid-vapor systems of interest in hydrometallurgy.

The main problem which makes difficult the application of models of solutions of electrolytes to hydrometallurgy is the presence in solution of many species resulting from association, complexation, and solvation of ions. When these species are present at high concentration, they should be incorporated into the model of deviations from ideality, introducing physical interaction parameters between all species and chemical equilibria constants. Two systems of importance in hydrometallurgy were considered: cupric ions in chloride solutions in relation with liquid extraction of Cu(II), and $PbCl_2$ solution in the presence of NaCl, LiCl or $CaCl_2$ in relation with solid-liquid equilibria. These problems are important in the development of processes for complex sulfide ores. The leaching of galena with concentrated solution of chloride are limited in leaching capacity by the solubility of $PbCl_2$, which is increased by the addition of chlorides which form complexes with Pb(II). Extraction of Cu(II) from chalcopyrite leaching solution is achieved in a three phase reactor in the presence of air (to regenerate the oxidant) and organic phase [to extract Cu(II)]. In both problems, the three kinds of parameters to be found are the parameters in the deviation from ideality model, the constants of complex formation, and the solubility product or ion exchange reaction constant. In the main development, Pitzer's model was used for deviations from ideality with the two parameters $\beta_{MX}^{(o)}$ and $\beta_{MX}^{(1)}$ for each cation-anion interaction. There still remains 8 solutes in $CuCl_2$ solutions (Na^+, H^+, Cl^-, Cu^{2+}, $CuCl^+$, $CuCl_2$, $CuCl_3^-$, $CuCl_4^{2-}$) and 7 in $PbCl_2$ solutions (M^+, Cl^-, Pb^{2+}, $PbCl^+$, $PbCl_2$, $PbCl_3$, $PbCl_4^{2-}$). The interactions parameters involving no complex ions are obtained from binary activity coefficients data. But as it is impossible to obtain directly binary parameters involving mixtures with complexes, we either correlated or optimized them using ternary mixtures data. We make use of a correlation between $\beta^{(o)}$ and $\beta^{(1)}$:

$$\beta_{ij}^{(1)} = 1.419\ \beta_{ij}^{(o)} + 0.0104 \qquad \text{for 1-1 electrolytes}$$
$$\beta_{ij}^{(1)} = 1.491\ \beta_{ij}^{(o)} + 1.461 \qquad \text{for 1-2 salts and 2-1 electrolytes} \tag{13}$$

Figure 2 shows the validity of that correlation for 1-1 electrolyte. In order to further reduce the number of parameters, it is possible to correlate the parameters $\beta_{ij}^{(o)}$ to standard entropies of ions S_j in a manner similar to that proposed by Bromley.[7]

Figure 3 gives an example of such correlation. Table 3 and 4 give values of parameters $\beta^{(o)}$ and $B_{ij}^{(1)}$ obtained in this way from entropies of ions taken from Helgeson.[30]

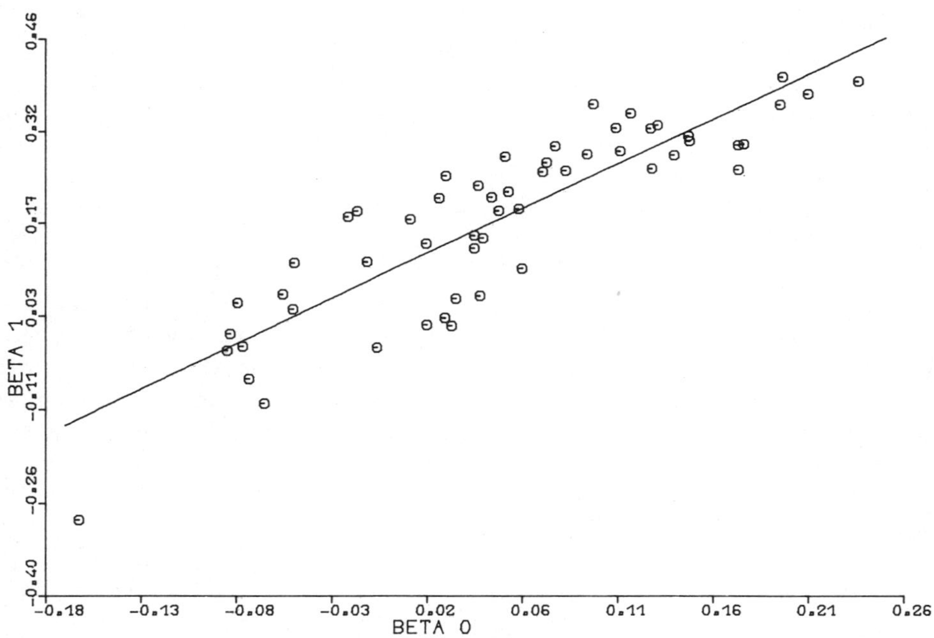

Fig. 2. Correlation of $\beta_{ij}^{(1)}$ - $\beta_{ij}^{(o)}$ for 1-1 electrolytes.

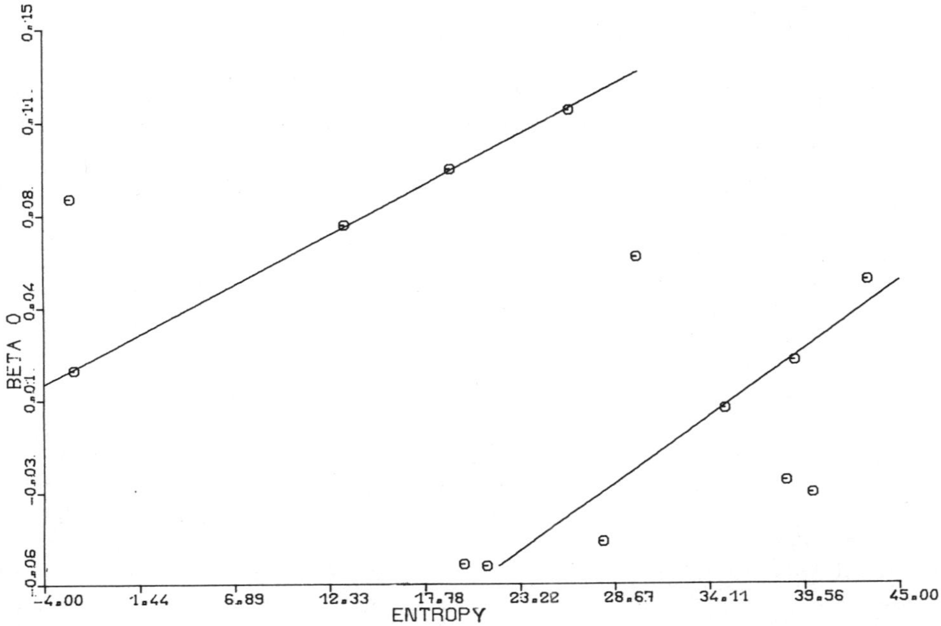

Fig. 3. Correlation between $\beta^{(o)}_{Na^+y}$ and $S^{(o)}_{y^-}$

Table 3. Parameters $\beta^{(o)}_{ij}$.

	Cl^-	$PbCl_3^-$	$PbCl_4^{2-}$	$CuCl_3^-$	$CuCl_4^{2-}$
Na^+	0.825	0.065	1.305	0.091	0
H^+	0.1807	0.394	0	0.243	0
Li^+	0.167	0.22	0	–	–
Ca^{2+}	0.3147	0.948	0	–	–
Pb^{2+}	0.222	0	0	–	–
$PbCl^+$	0.054	-0.16	0	–	–
Cu^{2+}	0.341	–	–	0.405	0
$CuCl^+$	0.078	–	–	0.045	0

Table 4. Parameters $\beta^{(1)}_{ij}$-

	Cl^-	$PbCl_3^-$	$PbCl_4^{2-}$	$CuCl_3^-$	$CuCl_4^{2-}$
Na^+	0.2492	0.231	10.331	0.2311	0
H^+	0.2818	0.619	0	0.4491	0
Li^+	0.2273	0.413	0	-	-
Ca^{2+}	1.6566	1.397	0	-	-
Pb^{2+}	1.993	0	0	-	-
$PbCl^+$	0.218	-0.035	0	-	-
Cu^{2+}	1.9761	-	-	2.068	0
$CuCl^+$	0.212	-	-	0.165	0

The manner of obtaining the parameters are somewhat arbitrary because the ions are ranged in various classes making it a rather imprecise correlation. Sometimes, we have to neglect parameters when data for correlation are not sufficiently precise or just not available. The parameters in Tables 3 and 4 will be used at least in the beginning of the calculations as starting values. Similarly we used some values of the complexation formation constants listed in Table 5 taken from Smith and Martell[31] as initial values in some new adjustment program of these constants by treatment of experimental data. These values are generally not satisfactory for the present work because they were obtained taking a limited number of complexes, using high ionic strength data with simple and not accurate hypothesis on deviations from ideality (that they depend only on ionic strength or calculating them in a way inconsistent with the model used in this work). They were obtained from various types of data including electrochemical potentials, solubilities, electric conductivities, ion exchange equilibria, and spectrophotometric data.

Table 5. Decimal Logarithms of Complexation Formation Constants.

M	Log_{10} (Complex formation constants at 25°C)			
	MCl^+	MCl_2	MCl^-	MCl_4^{2-}
Pb^{2+}	1.59	1.8	1.7	1.4
Cu^{2+}	-0.01	-0.69	-2.29	-4.59

Liquid-Liquid Systems

Cognet and Renon[32] applied Bromley's correlation of activity coefficients in concentrated electrolyte solutions to predict the influence of concentration of major ions in the aqueous phase on the distribution curves of Cu(II) between an organic solution of complexing extractant LIX 64N and aqueous phase at low loading of organic phase. It was possible in this case to predict the relative position of the line representing the logarithm of the distribution coefficient versus concentration of H^+ because Bromley's correlation yielded the needed parameters for the main species (Na^+, NO_3^-, SO_4H^-) and the influence of species in low concentration in the aqueous phase (H^+, Cu^{++}, chlorocomplexes) was negligible. The predicted values were always within the confidence limit of Bromley's correlation (5%).

In the main study now under way Pitzer's model is applied to the extraction of Cu(II) ions from a chloride solution by an organic phase of Escaid 100 loaded with the extractant LIX 65N. The extraction constant can be obtained independently by measuring the distribution of copper between a non-complexing aqueous solution (sulfate) and the organic phase. With dilute solution (10^{-2} M $CuSO_4$ contacting a 0.5% by volume solution of LIX 65N in Escaid 100), solutions can be considered ideal. The liquid extraction data are the measurements made in this laboratory, 65 data points for initial concentration in Cu(II) from 0.0008 to 0.8 mol/l, Cl^- from 2 to 5 mol/l and (H^+) from 2 x 10^{-4} to 0.5 mol/l and concentration of extractant in the aqueous phase 5 and 30% by volume with one to one volumes of aqueous and organic phases.

At first, we tried to calculate the distribution using initial values of parameters reported in Tables 1, 2, and 3 with an extraction constant equal to 1.6. Deviations between calculated and experimental results were very large. Changing the equilibrium constants reduced the deviations from 8 to 0.8% for the average relative deviation in Cu(II) in the aqueous phase and from 31 to 7% for the copper in the organic phase. A more extensive adjustment of parameters still needs to be done.

Solid-Liquid Systems

Data for systems where $PbCl_2$ precipitates from solutions with another concentrated chloride solutions were taken from the literature. Discrepancies between sources of data are apparent in Fig. 4. The data of Kendall and Sloan[33] were used in following treatment

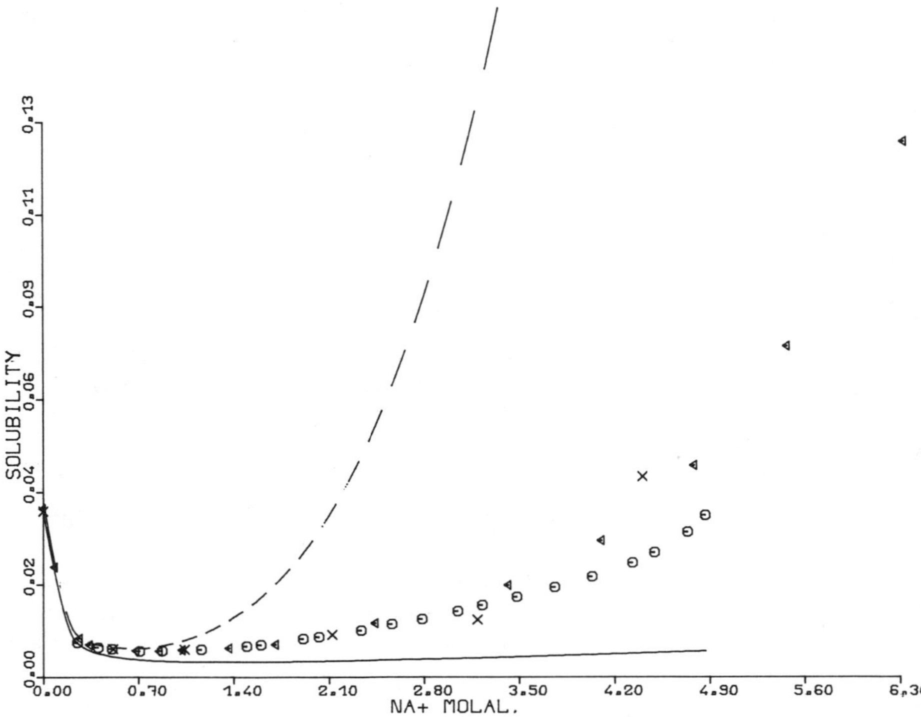

Fig. 4. Representation of $PbCl_2$ solubility in system $NaCl-PbCl_2-H_2O$.
Kendall and Sloan[33]...... O
Herz and Hellebrandt[34]... X
Deacon[35]................ Δ
Calculated with all initial parameters ————
Calculated with parameters for interactions of complexes neglected - - - -

assuming that the solid phase is $PbCl_2$. The initial values of parameters are reported in Tables 1, 2, and 3. The solubility product K_S is taken equal to 1.66×10^{-5} ($\log K_S = -4.78$). With the initial values of parameters the representation of solubility data is very poor as shown in Fig. 4 for the system $PbCl_2$ - NaCl. The representation is also very poor if all physical parameters except $\beta^{(o)}_{NaCl}$ and $\beta^{(1)}_{NaCl}$ parameters are taken equal to zero. Osmotic coefficients of pure $PbCl_2$ solution are also poorly represented in both cases.

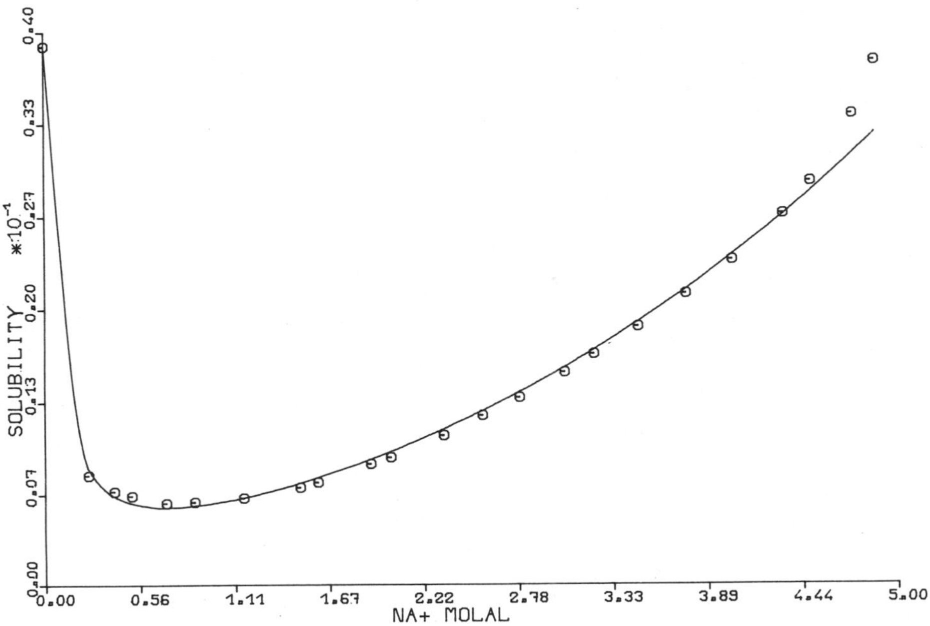

Fig. 5. System NaCl - $PbCl_2$ - H_2O
o : experimental data
—— : calculated with optimized complex formation constants

Several attempts to improve the adjustment will be presented next. The results presented here should be considered as preliminary. First, the chemical constants were adjusted independently for each of the systems $PbCl_2$ - LiCl - H_2O and $PbCl_2$ - NaCl - H_2O using the tabulated values of $\beta^{(o)}$ and $\beta^{(1)}$ but neglecting all $\beta^{(o)}$, $\beta^{(1)}$ parameters for interactions with complex ions (because concentration of complexes are much lower than those of Na^+,Li^+ due to low solubility of $PbCl_2$). Figures 5 and 7 show that a fair representation of data is obtained for both systems. Figures 6 and 8 give the repartition of complex ions. Table 6 gives the values of chemical constants resulting from the adjustment. The values are different for each system. Results for system with $CaCl_2$ are poor, probably because $CaCl^+$ ions are neglected. Then, we tried to fit at the same time data for the two systems with NaCl and LiCl with a unique set of chemical constants, adjusting the values of parameters $\beta^{(o)}$, $\beta^{(1)}$. Results are reported also in Table 6. Values of $\beta^{(o)}$ and $\beta^{(1)}$ for the best adjustment are reported in Table 7. With the exception

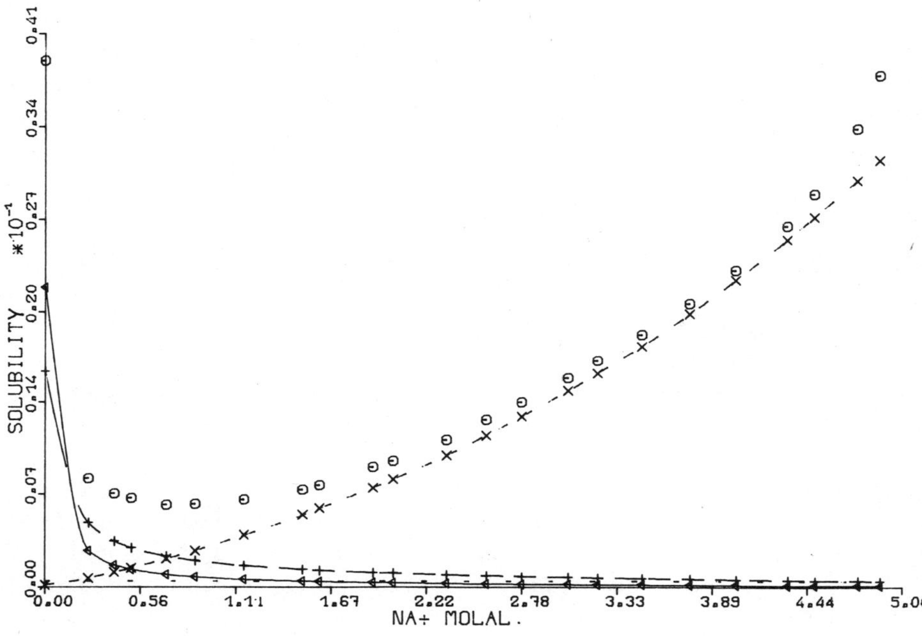

Fig. 6. System NaCl - $PbCl_2$ - H_2O.
o - experimental data;
▲ - mPb^{++}; + $mPbCl^+$; x m $PbCl_3^-$

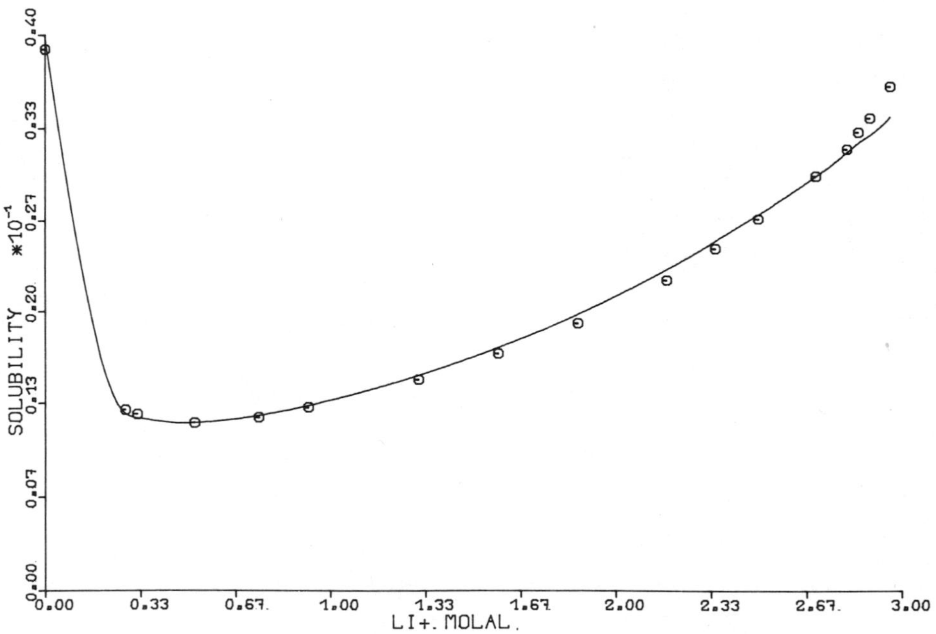

Fig. 7. System LiCl - $PbCl_2$ - H_2O (O...data; —...calculated with optimized complex formation constants).

Table 6. Chemical Constants

SYSTEM	Adjusted	log K_1	log K_2	log K_3	log K_4	log K_S	average relative deviation x 100
$NaCl-PbCl_2-H_2O$	K_1 to K_4, K_S	1.56	1.44	2.18	-13.6	-4.81	4.9
$LiCl-PbCl_2-H_2O$	K_1 to K_4, K_S	0.429	2.66	2.15	-13.4	-4.78	2.5
SMITH and MARTELL	-	1.59	1.8	1.7	1.4	-4.78	-
$NaCl-PbCl_2-H_2O$ and $LiCl-PbCl_2-H_2O$	K_S and $\beta^{(o)}$, $\beta^{(1)}$	1.59	1.8	1.7	1.4	-4.64	16.4
	K_1 to K_4, K_S and $\beta^{(o)}$, $\beta^{(1)}$	1.61	1.77	2.45	1.50	-4.71	13.7

$$K_1 = \frac{a_{PbCl^+}}{a_{Pb^{++}} + a_{Cl^-}} \qquad K_2 = \frac{a_{PbCl_2}}{a_{PbCl^-}\, a_{Cl^-}} \qquad K_3 = \frac{a_{PbCl_3^-}}{a_{PbCl_3^-}\, a_{Cl^-}}$$

$$K_4 = \frac{a_{PbCl_4^{2-}}}{a_{PbCl_3^-}\, a_{Cl^-}} \qquad K_S = a_{Pb^{++}}(a_{Cl^-})^2$$

Table 7. Physical Constants

$\beta^{(o)}$	Cl^-	$PbCl_3^-$	$PbCl_4^{2-}$
Na^+	0.0825	0.1274	2.0348
Li^+	0.167	0.2014	0.8302
Pb^{2+}	-0.0864	0.5958	0
$PbCl^+$	-0.5157	0.958	0.7153

$\beta^{(1)}$	Cl^-	$PbCl_3^-$	$PbCl_4^{2-}$
Na^+	0.2492	1.4127	11.335
Li^+	0.2273	1.2725	-2.3034
Pb^{2+}	2.8127	0.3411	0
$PbCl^+$	1.0715	0.6348	1.001

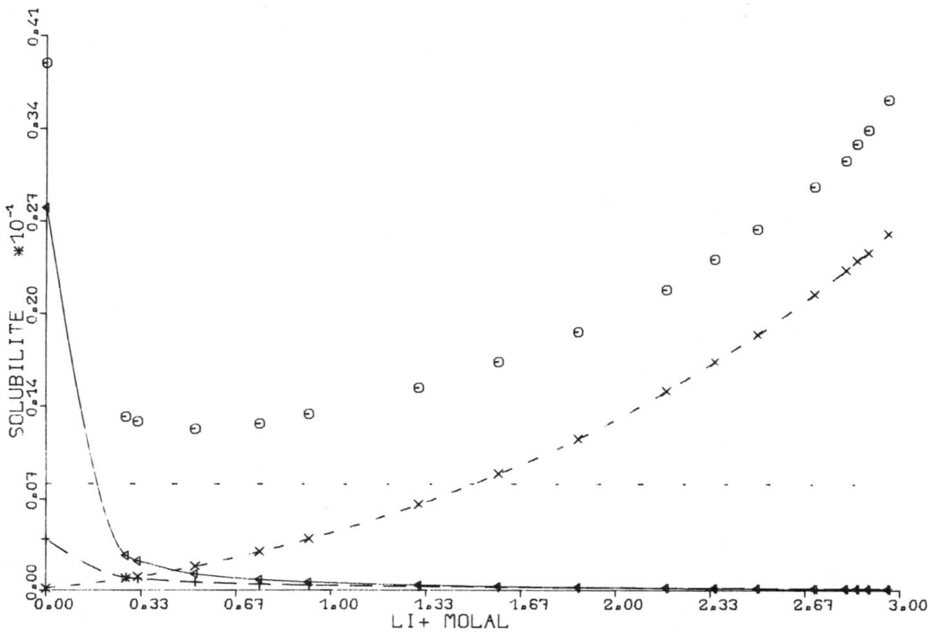

Fig. 8. System LiCl - $PbCl_2$ - H_2O (o - experimental data; Δ - $m_{Pb^{2+}}$; + - m_{PbCl^+}; --- - m_{PbCl_2}; x - $m_{PbCl_3^-}$).

of $\beta^{(1)}_{NaPbCl_4}$ Δ, all parameters having a high value were not very significant parameters. It was also possible to represent the solubility in the system $PbCl_2$ - $CaCl_2$ - H_2O by adjusting the additional parameters, the stability constant of $CaCl^{+2}$ and the parameters $\beta^{(o)}$ and $\beta^{(1)}$ with ions Ca^{++} and $CaCl^+$, obtaining an average deviation of 9%. J. J. Fritz[36,37] gives results using a similar approach with Pitzer model (with three parameters $\beta^{(o)}$, $\beta^{(1)}$, and C) and chemical constants for the system CuCl - NaCl - H_2O, CuCl - KCl - H_2O, and CuCl -NH_4Cl - H_2O.

CONCLUSION

The preliminary results just described should be improved to yield reliable sets of physical and chemical constants able to describe deviations from ideality in complex forming solution. All types of data should be treated simultaneously: solid-liquid and liquid-liquid; spectrophotometric, ion exchange and conductivity data should be reinterpreted in a large range of concentration.

From our experience this will be possible only with a sufficiently simple physical model of interactions using only one parameter for each binary anion-cation interaction. We feel that the model developed by Planche and Renon[26] might be suitable for such application.

REFERENCES

1. H. E. Barner and R. N. Kust: Application of Thermodynamics in Hydrometallurgy, Thermodynamics of Aqueous Systems with Industrial Applications, S. A. Newman Editor, ACS Symp. Ser. 133, pp. 625-641, Washington, D.C., 1980.
2. H. Renon: Deviations from Ideality in Electrolyte Solutions in Foundations of Computer-aided Chemical Process Design, R. S. Meh and W. D. Seider, Engineering Foundation, New York, 1981.
3. P. Debye and E. Hückel: Phys. Z., 1923, vol. 24, p. 185-206.
4. S. Scatchard: Phys. Z, (1932), vol. 33, p. 22.
5. S. Scatchard: Chem. Rev. (1939), vol. 19, p. 309-327.
6. P. Debye and H. MacAulay: Phys. Z, 1925, vol. 26, p. 22-29.
7. L. A. Bromley: J. Chem. Thermodynamics 1972, vol. 4, p. 669-673.
8. L. A. Bromley: AIChE J. 1973, vol. 19, p. 313-320.
9. H. P. Meissner and C. L. Kusik: AIChE J., 1972, vol. 18, p. 294-299.
10. K. S. Pitzer: J. Phys. Chem., 1973, vol. 77, pp. 268-277.
11. K. S. Pitzer and G. Mayorga: J. Phys. Chem., 1973, vol. 77, pp. 2300-2308.
12. K. S. Pitzer and G. Mayorga: J. Solution Chem., 1974, vol. 3, pp. 539-546.
13. K. S. Pitzer and J. J. Kim: J. Amer. Chem. Soc., 1974, vol. 96, pp. 5701-5707.
14. K. S. Pitzer and L. F. Silvester: J. Solution Chem., 1976, vol. 5, pp. 269-278.
15. K. S. Pitzer, R. N. Roy, and L. F. Silvester: J. Amer. Chem. Soc., vol. 99, pp. 4930-4936.
16. K. S. Pitzer: Accounts Chem. Res., 1977, vol. 10, 371-377.
17. L. F. Silvester and K. S. Pitzer: J. Solution Chem., 1978, vol. 7, pp. 327-337.
18. K. S. Pitzer: Activity Coefficients in Electrolyte Solutions, Vol. I, R. Pytkovics, ed., CRC Press Inc., Boca Raton, Florida, 1979, pp. 157-208.
19. C. C. Chen, H. E. Britt, J. F. Boston, and L. A. Evans: AIChE J. 1979, vol. 25, p. 820.

20. W. Fürst, and H. Renon: I.E.C., 1982, vol. 21, p. 396-400.
21. F. X. Ball, W. Fürst, and H. Renon: to be published, 1984.
22. J. Cruz and H. Renon: AIChE J. 1978, vol. 24, p. 817-830.
23. J. Cruz and H. Renon: I.E.C. Fund; 1979, vol. 18, p. 168.
24. R. Pottel: Water, Vol. 3, Plenum Press, New York, NY, 1973, p. 401-431.
25. C. C. Chen: Ph.D. Thesis, MIT, Cambridge, MA, 1980.
26. H. Planche and H. Renon: J. Phys. Chem., 1981, vol. 85, p. 3924-3929.
27. R. J. Baxter: J. Chem. Phys., 1970, vol. 52, p. 4559-4562.
28. L. Blum: Mol. Phys., 1975, vol. 30, p. 1529-1535.
29. F. X. Ball, W. Fürst, H. Planche and H. Renon: to be published 1984.
30. H. L. Helgeson: Am. J. Sci., 1969, vol. 267, p. 724.
31. R. F. Smith and A. E. Martell: Critical stability constant, Vol. 4, Pergamon Press, New York, NY, 1981.
32. M. L. Cognet and H. Renon: Hydrometallurgy, 1977, vol. 2, p. 305-316.
33. J. Kendall and C. H. Sloan: J. Am. Chem. Sci., 1925, vol. 47, p. 2306-2317.
34. W. Herz and M. Hellebrandt: S. Anorg. Allg. Chem, 1923, vol. 177, p. 188-189.
35. G. E. R. Deacon: J. Chem. Soc., 1927, vol. 130, p. 2063-2065.
36. J. J. Fritz: J. Phys. Chem., 1980, vol. 84, p. 2241-2246.
37. J. J. Fritz: J. Phys. Chem., 1981, vol. 85, p. 890.

RATE PROCESSES IN SOLVENT EXTRACTION SYSTEMS

Carl Hanson

Schools of Chemical Engineering
University of Bradford
Bradford, West Yorkshire BD7 1DP United Kingdom

INTRODUCTION

Most processes for the solvent extraction of metals involve chemical complex formation between a metal containing ion and an extractant. This is essential both for transfer of the metal to the organic phase and also, in some cases, for the specificity of the extraction. Since a chemical reaction is involved, the rate of transfer between the two phases will depend upon both the rates of the actual transport processes in the two phases and the kinetics of the chemical reaction(s). A knowledge of the overall rate of transfer is obviously important in plant design. However, interest in a full understanding of the rate determining processes has only developed this last decade. Prior to that, most data for design purposes were generated by relatively simple laboratory or pilot plant experiments, usually interpreted in terms of mass transfer alone as with the simpler and already established industrial extraction processes, thus ignoring any possible contribution from a chemical kinetic resistance.

The last decade has seen a great deal of work reported on both uranium and copper, with lesser amounts on nickel and other metals. The picture may not yet be fully elucidated but the overall complexity of the processes has become clear, giving insight into the many factors which must be taken into account. For an extraction process these may include:

(i) Diffusion of the metal ion in the aqueous phase to the region of the interface.

(ii) Diffusion of the counter-ion in the aqueous phase away from the interface.
(iii) Diffusion of the extractant molecule in the organic phase to the region of the interface.
(iv) Possible dissociation of an extractant aggregate in the region of the interface.
(v) Possible adsorption of extractant molecules on the interface.
(vi) Possible ionisation of extractant molecules on or near the interface.
(vii) Possible distribution of extractant molecules or ions into the aqueous phase followed by their diffusion into that phase.
(viii) Reaction between metal ions and extractant molecules or ions at or near the interface in one or more steps.
(ix) Possible desorption of metal-extractant complex from the interface.
(x) Diffusion of metal-extractant complex in the organic phase away from the interface.

This is not an exhaustive list, nor will all the foregoing occur in every case, but it illustrates some of the steps in the process which must be considered.

It is clear that the overall rate will be dominated by the rates of only one or two of the overall sequence. It is also common to assume that a rate-limiting step can be identified which will be a characteristic of a particular process. This may well be erroneous. The relative rates of the different steps can be expected to vary with (a) the method of study adopted, (b) the degree of approach to equilibrium, and (c) the extent of loading.

The method of study can be expected to have a major influence on the absolute rates of the diffusional steps. To take an extreme example, the Hahn static cell involves only molecular diffusion, whereas in a mechanically agitated system a considerable eddy diffusion component can be expected. A high rate of mass transfer due to a high driving force, e.g., in the early stages of a process, may itself promote interfacial instabilities of the Marangoni type. Any overall picture must also explain the effect of diluent composition on rate and the possible influence of interfacial films. While it is known that temperature has a considerable influence on extraction rate, most fundamental research into the rate processes has been carried out at or near ambient temperature. The temperature coefficients of the different steps will not be the same and so their relative contributions to the overall rate may change significantly with temperature.

In a number of industrially important processes a catalyst is employed to improve the kinetics. Perhaps the best-known example is of LIX 63 in LIX 64N. Elucidation of the mode of operation of such catalysts is a part of the challenge of gaining a full understanding of the mechanism which determines the rate of transfer in such systems.

In early work on mass transfer with chemical reaction in liquid-liquid systems,[1] it was assumed that these could be divided simply into those which were kinetically controlled and those involving a contribution from diffusional resistances. Following from this approach, experimental methods were suggested to allow differentiation between the two groups. These included:

Effect of agitation speed
Effect of phase ratio
Effect of space time
Effect of temperature
Effect of phase inversion

All were to be applied to a polydispersed situation. With the benefit of hindsight, it is clear that such an approach is over-simplistic. Thus an influence of interfacial area on rate (inferred from an influence of phase inversion) was taken as proof of the involvement of a diffusional resistance since it would not arise with a kinetically controlled process taking place homogeneously throughout one phase. However, it ignored the possibility of the overall rate depending on the kinetics of an interfacial reaction.

METHODS OF EXPERIMENTAL STUDY

Numerous techniques have been developed, some of the more significant of which are:

(a) A stirred tank, mostly operated batchwise but occasionally on a continuous basis. This has the attractions of simplicity and a superficial similarity with a plant mixer. However, interpretation of the results on a fundamental basis is impossible since interfacial area is extremely difficult to measure and the hydrodynamic conditions cannot be controlled.

(b) The AKUFVE apparatus was used in much early work and is capable of rapid production of data for conditions close to equilibrium. However, similar limitations apply as in (a).

(c) Constant interfacial area stirred cells of the type originally developed by Lewis. These overcome many of the limitations of methods (a) and (b), although there can still be some uncertainty as to hydrodynamic conditions in the immediate vicinity of the interface.

(d) Laminar jets, again aimed at providing a constant interfacial area. Problems include uncertainty as to hydrodynamic conditions and analytical errors in relation to the amount of transfer taking place in the limited time of contact.

(e) Single drops of dispersed phase rising or falling through a column of continuous phase. This is a simple method and gives rate for a known interfacial area. However, it is subject to end-effects, is difficult to use for study at significant levels of loading, and involves uncertainty as to the hydrodynamic conditions, especially in the dispersed phase.

(f) Static diffusion cells such as the Hahn type. The hydrodynamic conditons are unrealistic in relation to plant practice and can force one step to dominate the overall rate.

(g) A recently-developed technique using a rotating diffusion cell. This gives a known interfacial area and good control of the hydrodynamic conditions.

Examples of the use of each of these methods are given in references 2-8.

INTERPRETATION OF RESULTS

Two major reviews on the kinetics of metal solvent extraction and on the analysis of interphase reactions in liquid-liquid dispersions have appeared recently[9,10] and provide valuable compilations of earlier work. Many attempts have been made to develop interpretative models both for general cases and for specific processes.

Hanson and co-workers[11] suggested that many processes for the solvent extraction of metals probably involved chemical complex formation in a reaction zone within the aqueous phase adjacent to the interface. They modified Astarita's general model for a fast, diffusion controlled reaction to accommodate the fact that the reaction in the case of metals extraction must be reversible to permit both forward extraction and stripping. Relatively complex equations were derived. If it can be assumed that the bulk concentration of the metal-extractant complex in the aqueous phase is negligible and the back reaction is also ignored during forward extraction (and vice versa), then the equations can be simplified and become for a divalent metal:

$$\text{Rate} = -\left[\frac{[\text{HR}]\text{org}}{\phi_{\text{HR}}}\right]^{3/2}\left[\frac{1}{6}D_{\text{HR}}k_1[M^2]\right]^{1/2} \quad \text{for extraction (1)}$$

$$\text{Rate} = \frac{[\text{MR}_2]_{\text{org}}[\text{H}^+]}{\phi_{\text{MR}_2}}\left[D_{\text{MR}_2}k_2\right]^{1/2} \quad \text{for stripping (2)}$$

In each case the rate is per unit interfacial area. This simplified approach suggests that the distribution coefficient of the extractant, ϕ_{HR}, which will be related to its aqueous phase solubility, should be an important parameter in determination of extraction rate, whereas the rate of stripping would depend upon the distribution coefficient of the metal-extractant complex, ϕ_{MR_2}. This appeared to be in general agreement with trends observed in practice. Rod[12] has recently developed this approach in more detail for the case of metals extraction by chelate formation and has achieved good agreement with experimental data (see below).

The above could be described as the classic chemical engineering approach to mass transfer with simultaneous chemical reaction in a multiphase fluid system. The alternative approach, favoured by many workers, has developed from a more chemical standpoint and has postulated reaction either at the interface or involving species adsorbed on the interface. Both include diffusional steps which can contribute to the overall rate and it is not easy to differentiate between them. The key difference in the picture is whether extractant transfers to the aqueous phase prior to reaction or this takes place actually at the interface.

Much work has centered on the extraction of copper from acid sulfate media by hydroxyoxime-type extractants and this illustrates both the complexity of some metals extraction systems and the variations in interpretation described above.

Copper Extraction by Hydroxyoximes

The field has recently been reviewed by the author.[13] A great deal of kinetic data are available, obtained using all the techniques listed above. Early work concentrated on determining the dependence of rate on various parameters, mainly the concentrations of the various species present. Thus there is general agreement on a first-order dependence of rate on aqueous copper concentration. Dependence on the aqueous proton concentration varies with pH but is inverse first-order at low pH values. The nature of the counter-ion has some effect on rate, as does the presence of "inert" salts such as aluminium or magnesium sulfate (partly but not entirely due to their effect on the free proton concentration). Turning to the organic phase, considerable variation can be seen in the reported effect of extractant concentration. It seems likely that this is due to variation in aggregation of the extractant. Attention has been drawn[14] to the fact that all the hydroxyoximes used in the current range of commercial extractants aggregate to some extent and that it is the concentration of monomer which is critical in determining the rate of extraction rather than the gross concentration of oxime. This will also explain, to some extent, the effect of diluent type on rate. Nonyl phenol, present to some extent in most commercial extractants, also plays a part and is a kinetic depressant. The effect of temperature has been studied but is not easy to interpret as it involves a change in aggregation as well as in kinetic rate. Most of the work has centered on the kinetics of forward extraction; stripping has not received such comprehensive attention.

Attempts at quantitative interpretation or modelling of the results have followed the two approaches described above. Thus the author and his co-workers[14] have envisaged reaction taking place in a thin zone in the aqueous phase adjacent to the interface, having previously shown that oximes can exist at finite concentrations in aqueous phases. Three possible rate-determining reaction steps which could take place in the aqueous phase zone were considered, viz.:

$$Cu^{2+} + R^- \rightleftarrows Cu\,R^+ \tag{3}$$

$$Cu^{2+} + HR \rightleftarrows Cu\,R^+ + H^+ \tag{4}$$

$$Cu\,R^+ + HR \rightleftarrows Cu\,R_2 + H^+ \tag{5}$$

After consideration of the kinetic equations involved and the relation of the resultant parameters to known values, it was concluded that step (5) is the rate-determining reaction step. This agrees with Fleming.[15] Attention was also drawn to the possible importance of rate of loss of water from any complex species which is hydrated. Application of equation (1) gave reasonable agreement with experiment. Results for the same system determined by different methods illustrate the influence of experimental method and show the potential importance of diffusional resistances in the organic phase.

In his recent papers, Rod[12] has developed the above approach further and has obtained good agreement between his model and experimental data. He considers a similar reaction sequence to that envisaged by Whewell et al[14] but concludes reaction (3) is the rate-determining step. He also concludes the reaction rate is influenced significantly by the rate of diffusion of the extractant in the organic phase and that this can become the dominant factor determining rate. It appears that the diffusional resistance increases with decreasing extractant concentration and increasing copper concentration in the aqueous phase.

The alternative approach is typified by Cox and co-workers[16], who have postulated a true interfacial reaction and stressed adsorption of key reactants at the interface as a vital parameter. They have proposed the following process sequence:

$$\overline{RH} \rightleftarrows RH_{int} \rightleftarrows RH_{ad} \tag{6}$$

$$Cu^{2+}_{int} + RH_{ad} \rightleftarrows Cu\ R^{+}_{ad} + H^{+}_{int} \tag{7}$$

$$Cu\ R^{+}_{ad} + RH_{int} \rightleftarrows Cu\ R_{2,int} + H^{+}_{int} \tag{8}$$

$$Cu\ R_{2,int} \rightleftarrows \overline{Cu\ R_2} \tag{9}$$

where a bar over a formula denotes bulk species, "ad" identifies species adsorbed at the interface and "int" denotes species adjacent to the interface. Reaction (8) is considered to be rate-controlling. Assuming reaction (7) to be at equilibrium, the rate of extraction will depend on the rate of disappearance of Cu R^{+}_{ad}. Based on this approach, it is possible to predict the rate dependence:

$$\text{Rate} = K[\overline{Cu}^{2+}]\ [\overline{RH}]\ [\overline{H}^{+}]^{-1} \tag{10}$$

which agrees with experimental observation. A similar explanation and model has also been proposed by Komasawa and co-workers.[17]

It is interesting to note that Fisk[8] has recently supported the interfacial reaction model following work with a rotating diffusion cell.

While there are important differences in the two models in terms of visualization of the locale of the rate-determining reaction, it is important to recognize that both approaches encompass the possibility of diffusion making a contribution to the overall resistance and, in fact, there has been general agreement since 1977 that diffusion does play an important part. While some contribution from diffusion in the aqueous phase cannot be ruled out, the organic phase will certainly pose the bulk of the diffusional resistance. From a technological point of view, the existence of a major diffusional resistance is probably of more importance than final elucidation of the locale of the reaction, which could well depend upon the experimental conditions.

Several mechanisms have been proposed to explain the catalytic action of compounds such as LIX 63 and these have been reviewed elsewhere.[13] The more important suggestions are (a) the α-hydroxyoxime assists deprotonation of the β-hydroxyoxime, (b) the α-hydroxyoxime inhibits adduct formation between the oxime and the copper complex at the interface, and (c) that the catalyst complexes copper (probably forming a mixed complex with the β-hydroxyoxime) and transfers it to the bulk organic phase where it re-equilibrates, liberating the catalyst. The third seems the most probable based on present evidence.

Other Factors

With some systems (e.g. nickel extraction with D2EHPA), formation of pseudo-crystalline interfacial compounds has been observed.[18] This will certainly influence rate and it is possible that such complexes do exist in other systems even though they are not visible. Similarly, formation of highly viscous interfacial films, possibly associated with co-extraction of water, has been suggested to explain a progressive reduction in rate of extraction in the Ni/D2EHPA system.[19] The impact of condensed interfacial films in metals extraction systems and methods by which their rheological properties can be studied have been described by Yagodin and co-workers.[20]

TECHNOLOGICAL IMPLICATIONS

Contactor Design

Rate of interphase transfer is one of the factors which should be taken into account when selecting the most appropriate contactor to employ for a particular process. If the rate is slow due to the presence of a significant chemical kinetic resistance, then it is usually necessary to employ mixer-settlers. If kinetics are high, on the other hand, there should also be the possibility of employing column contactors. In most cases the latter offer the prospect of a smaller overall plant size and solvent inventory (the relative difference depends on scale and phase continuity) plus the attraction of an enclosed system.

In addition, whichever type of contactor is fianlly selected, the kinetics of interphase transfer have a direct relation to plant size (residence-time required in the mixer of a mixer-settler to give a certain stage efficiency or height of a transfer unit in a column). Consideration of the kinetic aspects has led to certain developments in mixer-settler design aimed at a reduction in size of the mixer compartment, e.g. multicompartment mixers and the use of static mixers (see Chapter 499).

While use of a more compact contactor will make some contribution to reducing capital costs, a more significant reduction is likely if it is possible to reduce the number of equilibrium stages required. There does seem to be some correlation, even if only superficial, between kinetics and extractant strength. Thus extractants exhibiting fast kinetics usually also have steep equilibrium isotherms under extraction conditions, e.g. Acorga P5100 for copper extraction from acid sulfate systems,[21] making it possible to consider a circuit with only two extraction stages. The economic advantage of eliminating a stage is much greater than that coming simply from a marginal reduction in the size of each stage. The only note of caution to sound is the possible deleterious effect on stripping performance.

Contactor Performance

All workers are agreed that extraction rate for most metals is proportional to interfacial area, giving incentive for generation of a high interfacial area per unit volume through a small drop size provided this does not lead to the need for excessive sized settlers or premature flooding of columns. Of perhaps more importance is the recognition of the likely contribution of a diffusional resistance, most probably located in the organic phase. One then needs to ensure an adequate level of turbulence in the phase concerned. Depending upon the type of contactor selected, there may be a significant difference in rate according to which phase is dispersed. Once again, the effect of phase continuity on extraction rate has to be considered in association with its influence on settling rate. If viscous interfacial films are involved, then it is desirable to facilitate interface renewal by promotion of drop coalescence-redispersion.[19]

Diluent Choice

The choice of diluent is important in respect of equilibrium performance, kinetics, phase disengagement and safety (flash point). Provision of special diluents can be a problem for plants in remote locations and use of extractants with sufficiently fast kinetics for the diluent not to have a significant effect on stage efficiency may be an advantage in allowing locally produced material to be used.

Reagent Testing

Any large-scale user of commercial extractants and diluents will require a series of acceptance tests to apply to newly-delivered supplies. An appreciation of the rate-controlling mechanism is useful in developing tests which will be sensitive to the key parameters. Similar tests are also needed for reagents in use in a circuit, particularly when investigating any unsatisfactory plant performance. It would be desirable for there to be some degree of international standardization in such tests to facilitate comparison of data.

CONCLUSIONS

The study of rate processes in the solvent extraction of metals has important implications both for the performance of industrial plants and an understanding of the mechanism involved. Much has been learned over the last decade but there are still many uncertainties to justify continued research.

REFERENCES

1. C. Hanson: Chapter 12 in Recent Advances in Liquid-Liquid Extraction, Pergamon, Oxford, 1971.
2. M. J. Slater, et al:Proceedings International Solvent Extraction Conference ISEC '74, p. 107, (S.C.I. London, 1974).

3. D. S. Flett, et al.: J. Inorg. Nucl. Chem., 35, 2471 (1973).
4. C. A. Fleming, et al.: Proceedings International Solvent Extraction Conference ISEC '77, p. 193 (C.I.M., Montreal, 1979).
5. R. W. Freeman and L. L. Tavlarides: Chem. Eng. Sci., 35, 559 (1980).
6. R. J. Whewell, et al.: J. Inorg. Nucl. Chem., 37, 2303 (1975).
7. M. A. Hughes and P. D. Middlebrook: Int. J. Miner. Process, 5, 229 (1978).
8. P. R. Fisk: PhD Thesis, University of London, 1981.
9. P. R. Danesi: CRC Critical Reviews in Analytical Chemistry, Nov. 1980.
10. L. L. Tavlarides and M. Stamtoudis: Advances in Chem. Eng., 11, 199 (1981).
11. C. Hanson, M. A. Hughes, and J. G. Marsland: Proceedings International Solvent Extraction Conference ISEC '74, p. 2401 (S.C.I. London, 1974).
12. V. Rod: Chem. Eng. Jl., 20, 131 (1980) and 21, 187 (1981).
13. R. J. Whewell and C. Hanson: Chapter 1 in "Ion Exchange and Solvent Extraction", Ed. J. A. Marinsky and Y. Marcus, vol. 8 (Dekker, New York, 1981).
14. R. J. Whewell, M. A. Hughes, and C. Hanson: Proceedings International Solvent Extraction Conference ISEC '77, p. 185 (C.I.M., Montreal, 1979).
15. C. A. Fleming: NIM Report No. 1793, Johannesburg (1976).
16. M. Cox, C. G. Hirons, and D. S. Flett: Proceedings International Solvent Extraction Conference ISEC '80, paper 118, Liege (1980).
17. I. Komasawa, T. Otake, and T. Muraoka: J. Chem. Eng. Japan, 13, 204 (1980).
18. M. A. Hughes: Hydrometallurgy, 3, 85 (1978).
19. K. Durrani, C. Hanson, and M. A. Hughes: Metallurgical Transactions, 8B, 169 (1977).
20. G. A. Yagodin, V. V. Tarasov, and S. Yu Ivakhno: Hydrometallurgy, 8, 293 (1982).
21. J. A. Tumilty, G. W. Seward, and J. P. Massam: Proceedings International Solvent Extraction Conference ISEC '77, p. 542 (C.I.M., Montreal, 1979).

KINETICS OF EXTRACTION OF IRON (III) FROM CHLORIDE SOLUTIONS BY TRIOCTYLAMINE

J. C. Silva Cassa and A. J. Monhemius

Department of Metallurgy and Materials Science
Imperial College of Science and Technology
London SW7 2BP UNITED KINGDOM

The kinetics of solvent extraction of iron (III) from chloride solutions (HCl/LiCl) by tri-n-octylamine hydrochloride in toluene have been studied using a planar interface stirred cell.

The reaction was studied under conditions of chemical reaction control. It was found that when the viscosity of the aqueous phase increased, due to increased chloride salt content, the aqueous phase Reynolds number was a reasonable criterion of hydrodynamic similarity of mixing conditions. The reaction was shown to be an interfacial process. The extraction rate was first order with respect to the aqueous iron concentration and monomer TOAHCl concentration, whereas the stripping rate was first order with respect to the organic iron concentration and inverse first order with respect to the total TOA concentration. The extraction rate was dependent on the aqueous chloride concentration, whereas the stripping rate was independent of this parameter. Transfer of both neutral and anionic iron chlorocomplexes during extraction was indicated. A mechanism for the reactions involved is presented.

INTRODUCTION

A great deal of information is available on extraction equilibria in amine systems, but little work has been carried out on the kinetics of amine extraction and the fundamental chemistry of the extraction process. There is uncertainty concerning the nature of the extractable metal complexes and to date it is not clear whether amines are neutral or anionic extractants, or both.

The present work is a systematic investigation aimed towards elucidation of the mechanism of amine extraction. The extraction of iron (III) choride by tri-n-octylamine in toluene was selected for study, not on account of its industrial importance amongst amine systems, but rather as a model system which offered certain experimental advantages. Foremost amongst these was the fact that trioctylamine of good purity was commercially available. It is known that purity of reagents is of the utmost importance in obtaining meaningful kinetic data in solvent extraction systems. Another reason for selecting this system was that the iron - TOA complex is strongly coloured. This enabled the use of continuous spectrophotometry for monitoring the rate of iron transfer, a method which was found to give results of very good reproducibility. There is, unfortunately, no standardized method of studying the kinetics of liquid-liquid extraction systems. For this work, a Lewis-type, planar interface, stirred cell was constructed and this was found to give consistent and, hopefully, reliable results.

Apparatus

A fully-baffled, planar interface, stirred cell was used, as illustrated in Fig. 1. All wetted parts of the cell were made of either glass, PTFE or viton rubber.

The cell consisted of a precision bore glass tube (125mm diameter), clamped between two PTFE flanges which were backed by stainless steel plates. The cell was divided at the mid point into two chambers by two concentric horizontal baffles - the central and the ring

Fig. 1. The Stirred Cell.
Key to (a): 1. Glass cell; 2. Top plate (St. steel); 3. Top flange; 4. Bottom plate (St. steel); 5. Bottom flange; 6. Upper stirrer; 7. Lower stirrer; 8. Central baffle; 9. Support tube (glass); 10. Vertical baffles; 11. Ring baffle; 17. Upper stirrer bearing; 18. Lower stirrer bearing. All parts made of PTFE except where specified.

(a)

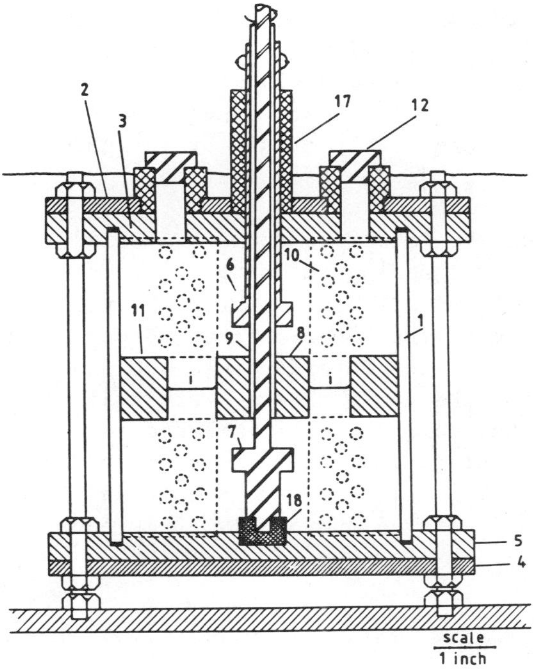

(b)

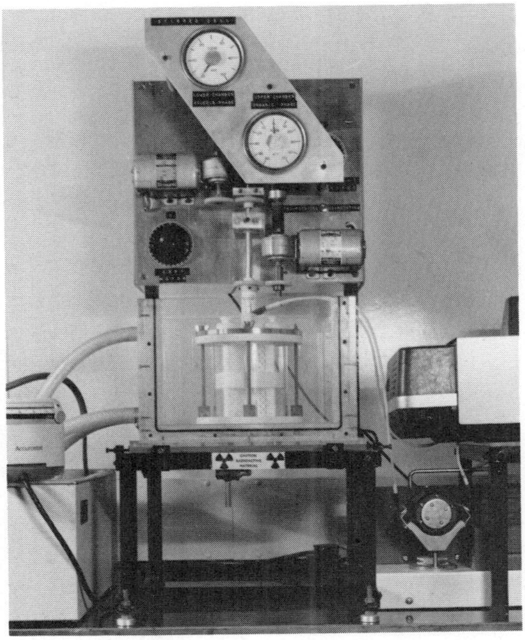

baffle. The interface between the upper (organic) and lower (aqueous) phases was formed in the annular space between these two baffles by completely filling the cell with equal volumes of each phase (480ml of each). Each chamber was equipped with four vertical, perforated, PTFE baffles, set at 90° to each other, and a PTFE agitator. The two agitators were independently driven by variable speed motors and could be used in co- or counter-rotation.

The area of the interface could be varied from 20 to 40cm^2, in 5cm^2 intervals, by changing the size of the central baffle. The top plate of the cell had six access ports, four to the upper chamber and two giving access to the lower chamber through holes in the outer ring baffle, so that the interface was not disturbed. Two further access ports into the lower chamber were available in the base plate of the cell. The whole cell was immersed in a thermostatted water jacket, which enabled temperature control of the cell contents to within ± 0.1°C. The cell was demountable and it was always completely dismantled and cleaned thoroughly after each experimental run.

Materials

Tri-n-octylamine (TOA). Technical grade TOA (Fluka A.G.) was found to have a molecular weight of 352.0 ± 1.6 by potentiometric titration and 353 ± 5 by vapor pressure osmometry (theoretical 353.68). Attempts to further purify the amine were unsuccessful and it was used as received.

Trioctylamine hydrochloride (TOAHCl). Recrystallized TOAHCl was prepared from Fluka TOA by shaking 100ml 0.5M TOA in petroleum ether (40-60° b.p.) with 3 separate 100ml volumes of 1M LiCl and then cooled to -20°C to precipitate TOAHCl. The solid was filtered off and redissolved in fresh petroleum ether and the solution was cooled again to recrystallize TOAHCl. The recrystallization procedure was carried out twenty times. The final product was a waxy white solid with a melting point of 72-73°C. The molecular weight by potentiometric titration was 390.0 ± 2.1 (theoretical 390.23). The melting point of the recrystallized TOAHCl was identical to a sample of pure TOAHCl obtained from Eastman-Kodak.

Stock solutions of TOAHCl for the kinetic experiments were prepared directly from Fluka TOA dissolved in A.R. toluene. The TOA solutions were first equilibrated with an equal volume of 0.1M NaOH and then converted to TOAHCl by equilibration with three equal volumes of 1M HCl. Finally the TOAHCl solution was washed with an equal volume of 1M LiCl. The aggregation behavior, measured by vapor pressure osmometry, and the kinetic behavior of these stock solutions of TOAHCl were indistinguishable from the behavior of solutions prepared directly from recrystallized TOAHCl or from Eastman Kodak TOAHCl.

Lithium chloride. Concentrated stock solutions of lithium chloride (10M LiCl) were prepared by dissolving GPR lithium carbonate in 10M HCl until the reaction was complete. Undissolved solids were filtered off and then the LiCl solution was contacted with TOAHCl solution to remove trace metals. The concentrations of the LiCl stock solutions were determined by chloride analysis and by density measurements.

Other reagents. All other reagents used were analar grade.

Kinetic Experiments

The rates of extraction of iron into the organic phase, or stripping from the organic phase, were obtained by continuous spectrophotometric analysis of the iron concentration in the phase into which transfer was taking place. This technique ensured that the initial rates of reaction were measured with good precision.

In extraction experiments, the aqueous phases consisted of the required concentrations of ferric chloride in solutions containing 1M HCl plus variable concentrations of LiCl to give the required total chloride concentration. The organic phases were prepared by dilution of stock TOAHCl solutions with the appropriate amount of toluene. Each organic phase was pre-equilibrated by shaking with an aqueous phase of the same composition as that to be used in the experiment, but containing no iron.

Before the start of an experiment, the aqueous and organic phases were separately immersed in the thermostatted water bath to equilibrate to the temperature of the cell. When transferring the phases into the cell, care was taken to ensure that no mixing occurred during filling and also that no air bubbles were trapped in the cell. After filling, the continuous analysis system was started and then the experiment was begun by switching on the stirrer motors.

At the end of an experiment, the solutions were drained out of the cell, which was then dismantled for thorough cleaning. Only distilled water, dilute HCl and acetone were used for cleaning and great care was taken at all times to avoid any contamination of the cell or flow system.

Chemical Analysis

Continuous analysis of iron was carried out by circulating solution from the stirred cell by means of a peristatic pump through a flow cell in a spectrophotometer equipped with a recorder.

For extraction experiments, analysis of iron in the organic phase was carried out by monitoring the optical density of the peak at 365 nm. At this wavelength Beers law is obeyed up to 0.02 g/l Fe for a 5mm path length. In the aqueous phase, the position of this peak varies with the chloride concentration over the range 325 to 365 nm. For continuous analysis of the aqueous phase, used during stripping experiments, the wavelength selected was that which gave maximum absorption at the particular chloride concentration of the aqueous phase.

The flow system used for circulating solution from the stirred cell into the spectrophotometer gave a delay time of about 10 seconds, which was negligible for the great majority of the experiments.

In some experiments, the rate of transfer of chloride ions from one phase to the other was measured by the use of ^{36}Cl isotope. The beta emission of the isotope was determined by liquid scintillation coincidence counting. Samples of the organic or aqueous phase (for extraction or stripping, respectively) were taken during these kinetic experiments and mixed with scintillation cocktail and absolute alcohol prior to beta counting. It was found that, in determining ^{36}Cl in the organic phase, colour quenching of the count rate occurred when iron was present, due to the strong yellow colour of the Fe - TOAHCl complex. However there was a linear relationship between counting efficiency (count rate with Fe/count rate without Fe) and organic iron concentration up to 0.2 g/l Fe. This relationship was used to correct the measured count rates for the effect of iron in the organic phase.

Computation of Mass Transfer Rates

The MINITAB statistical data handling package was used to fit polynomials to the measured data, namely the profiles recorded by the spectrophotometric flow system of organic or aqueous iron concentrations versus time. In all cases, the statistically best fit was found with a polynomial of the type:

$$[Fe]/t = a_1 + a_2 t \tag{1}$$

where [Fe] is the iron concentration at time t and the coefficient a_1 is equal to the initial mass transfer rate. This polynomial could be used to fit the kinetic data up to 50% of maximum extraction capacity. Statistical tests such as plots of fitted values against experimental values, residuals against experimental values, and normal scores of residuals against residuals, confirmed the adequacy of this polynomial for representing the experimental concentration-time profiles.

The kinetic parameter used to analyze the experimental data was the initial specific extraction rate R, where R is the initial rate of change of the bulk concentration per unit surface area per unit phase volume, and is given by:

$$R = V_a\ a_1/A = V_o\ a_1/A\ (mol/cm^2 \cdot min) \tag{2}$$

where V_a and V_o are the volumes (l) of the aqueous and organic phases in the cell and A is the interfacial area (cm^2). As the concentration of iron in the phase into which transfer was occurring was measured, a high degree of accuracy was obtained by the spectrophotometric flow system when measuring concentrations close to time zero. The data fitting procedure also gave reliable estimates of the initial mass transfer rate and so this parameter can be used with a good degree of confidence to interpret the kinetic behavior of the system.

The reproducibility of the experimental results is illustrated by the data given in Table 1 in which the mean initial mass transfer rates and the deviation about the mean for pairs of replicated experiments under a variety of conditions are reported.

Table 1. Experimental Reproducibility

TOAHCl ($Mx10^2$)	Fe (III) ($Mx10^3$)	Cl (M)	$R \times 10^3$ (g/l/min) Mean	±%
1·0	1·0	1	0·058	1·9
1·0	2·2	1	0·131	2·0
1·0	2·5	1	0·149	0·6
2·5	2·5	1	0·271	1·5
1·0	1·0	4	0·836	2·8
1·0	1·0	7	1·022	3·1
1·0*	1·74	1	0·526	1·2

* Stripping experiment - iron transferred from organic to aqueous.

Hydrodynamics of the Cell

The effects of stirring conditions on the hydrodynamic behaviour of the stirred cell were investigated qualitatively by means of dye injection and quantitatively by measuring the rate of transfer of chloride, using ^{36}Cl isotope as a tracer, across the interface between TOAHCl in toluene and aqueous HCl-LiCl solutions. Chloride transfer was chosen because it was expected to be a diffusion controlled reaction and it enabled the use of solutions identical to those used for the iron transfer experiments, thus achieving chemical and hydrodynamic similarity between the two systems.

Visual observation of the behaviour of dye injected into each phase showed that good bulk mixing in each phase, with no stagnant volumes, occurred with stirring speeds ranging from 150 to 350 rpm. Talc particles were used to observe the behavior of the interface. At low stirring speeds, this was completely stagnant. As the stirring speed was increased, random movements on the interface appeared and these persisted until, at about 380 rpm, rippling of the interface was observed.

The homogeneity of the organic phase was checked by carrying out a kinetic run in which the aqueous phase was 1M HCl containing 2.5×10^{-3} M Fe^{3+} and the organic phase was 10^{-2}M TOAHCl in toluene. The optical density of the organic phase was continuously monitored by the flow-through spectrophotometric system. Contra-rotational stirring at 150 rpm was used and the organic phase was sampled at several positions within the upper chamber of the cell. The recorded optical density-time profile was a smooth curve, indicating that the concentration of iron in the organic phase was uniform at all positions sampled.

The rate of transfer of chloride from HCl-LiCl solutions into TOAHCl - toluene solutions was measured by spiking the aqueous phase with ^{36}Cl tracer and taking samples of the organic phase for liquid scintillation beta counting. It was assumed that the count rate in the organic phase was proportional to the concentration of chloride transferred from the aqueous phase, at least in the early stages of the experiment, where the back-reaction was negligible. The data were fitted by polynomials similar in form to Eq. 1:

$$[Cl]/t = b_1 + b_2 t \qquad (3)$$

where [Cl] is the concentration of chloride transferred in time t and the coefficient b_1 is equal to the initial mass transfer rate of chloride. The overall mass transfer coefficient, K, with respect to the organic phase was obtained from:

$$K = V_o \, b_1/A[Cl]_o \text{ (cm/sec)} \tag{4}$$

where V_o is the volume (l), and $[Cl]_o$ the total chloride concentration of the organic phase and A is the interfacial area (cm^2).

The effect of stirring speed (contra-rotating impellers) on the rate of transfer of chloride was investigated and it was found that the overall mass transfer coefficient increased continuously over the usable range of stirring speeds, i.e. 150 to 350 rpm. This behaviour is indicative of a diffusion controlled process. In Fig. 2, these results are compared with those of Austin[1] for other

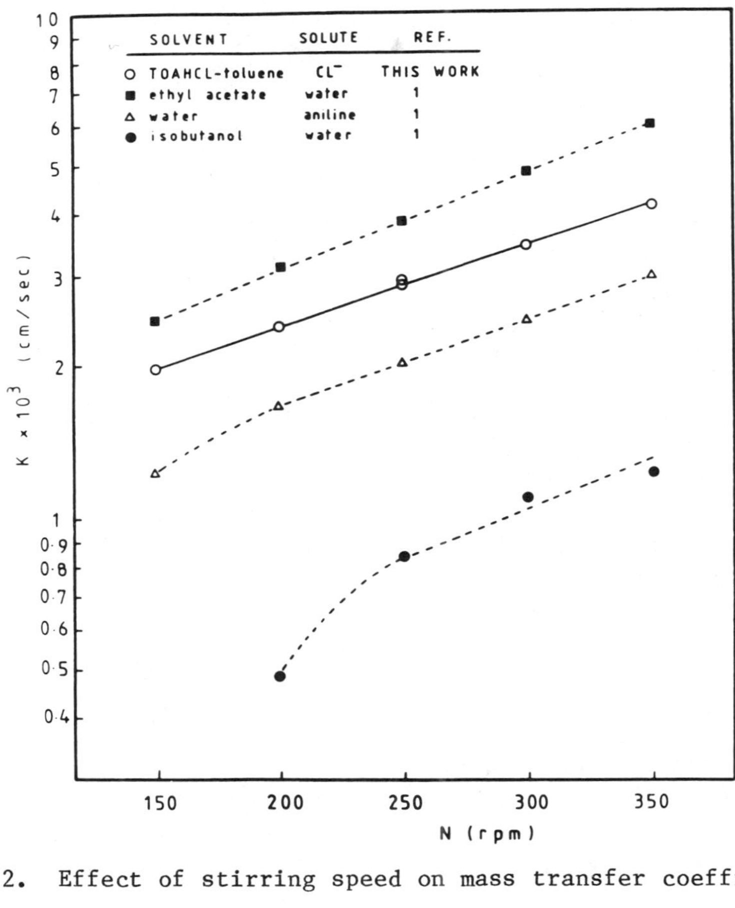

Fig. 2. Effect of stirring speed on mass transfer coefficient

diffusion controlled systems, obtained in a stirred cell of similar design to the one used in the present work. The slopes of the lines plotted in Fig. 2 reflect the mixing conditions in the cells, whereas the vertical positions are due to the effect of the diffusion coefficients of the transferred species on the mass transfer rates. It may be seen that there is good correlation between the behaviour of the various systems. The chloride transfer results could be described by the equation:

$$\log K = -2.97 + 1.66 \times 10^{-3}(N)$$

or (5)

$$K = 1.072 \times 10^{-3} \times 10^{N/602}$$

where N is the stirring speed (rpm). This relationship corresponds to the form of the equation given by Austin for a renewed interface.

The effect of temperature on chloride transfer was investigated and it was found that the overall mass transfer coefficients changed very little with temperature, over the range 15-35°C, the apparent activation energy being 1.7 kJ/mol. This behaviour again suggests a diffusion controlled reaction.

The initial mass transfer rate was found to be directly proportional to the concentration of TOAHCl in the organic phase. Thus the overall mass transfer coefficient, calculated from Eq. 4, was independent of TOAHCl concentration, as would be expected for a diffusion controlled reaction.

Iron-TOAHCl Extraction

The primary objective of this work was to investigate the kinetics of the extraction of Fe^{3+} from aqueous chloride solutions into TOAHCl in toluene. The kinetic parameter studied was the initial specific mass transfer rate of iron which was determined from the measured concentration - time profiles. The first series of experiments were designed to select a range of conditions where the rate of extraction of Fe^{3+} into TOAHCl was under chemical reaction control.

Effect of stirring speed. The change in the initial mass transfer rate was studied over a range of stirring speeds from 100 to 600 rpm. Various modes of stirring were used, e.g. co- and contra-rotating impellers, with equal stirring speeds in each phase; variable stirring speed in the organic phase, with a fixed stirring speed in the aqueous phase; and vice versa. The results obtained are illustrated in Fig. 3. With equal stirring speeds in each phase and contra-rotating impellers, the initial mass transfer rates increased up to about 200 rpm. A plateau region, where the mass transfer rate was independent of stirring speed, occurred in the range 200 to 350 rpm. As the speed was increased further, waves appeared on the interface, causing an increase in interfacial area and in the general turbulence, and this is reflected in Fig. 3 by the increase in mass transfer rates above 350 rpm. At about 500 rpm. the interface was disrupted and mixing of the phases started to occur. The behaviour with co-rotating impellers was virtually identical, but it was decided to carry out all work with contra-rotating impellers, since in principle this should achieve more efficient stirring at the interface. Under conditions where the stirring rate was held constant in one phase, while in the other it was varied, it was found that changing the stirring speed in the organic phase had little or no effect on the mass transfer rate, whereas the results obtained when varying the aqueous phase stirring only were very similar to those obtained when varying both stirrers simultaneously.

Comparing these results with those obtained for the diffusion controlled chloride transfer, shown in Fig. 2, it was concluded that in the plateau region, the transfer of iron from the aqueous to the organic phase was controlled by a chemical reaction and was independent of diffusion processes. Furthermore, the resistance to transfer appeared to be almost entirely located in the aqueous phase, since the mass transfer rate was virtually unaffected by variations of stirring speed in the organic phase.

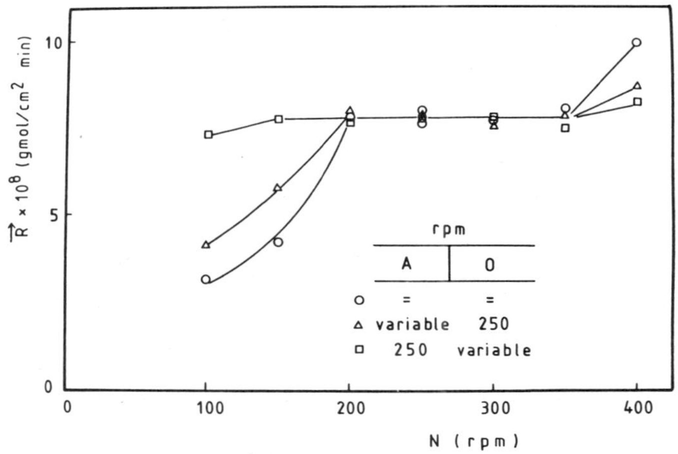

Fig. 3. Effect of stirring speed on extraction rate. (Conditions: Fe = 0.0025 M, TOAHCl = 0.025 M, HCl = 1M.)

The effect of stirring speed on the rate of stripping (i.e. where iron was initially present in the organic phase only) was also investigated and behaviour similar to that shown in Fig. 3 was found, with a plateau region again occurring in the range 200-350 rpm.

The attainment of a plateau region in the stirring curve, which is indicative of chemical reaction control, is dependent not only on the stirring speeed, but also on the concentration of reagents. If the concentrations used are too high, the chemical reactions may be so fast that it is not possible to eliminate diffusion control by increasing the stirring speed, while at the same time maintaining a planar interface between the phases. The effects of varying reagent concentrations on the stirring curve are illustrated in Fig. 4. It

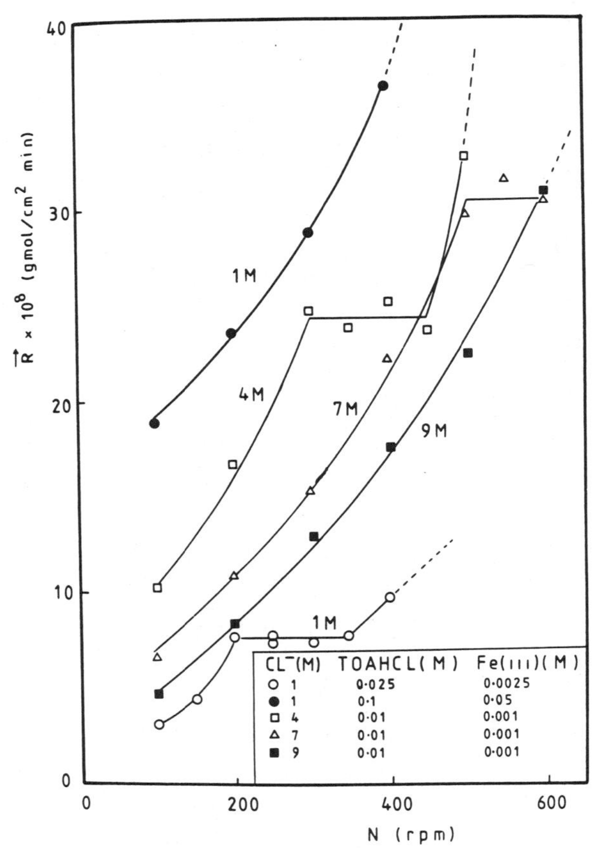

Fig. 4. Effect of aqueous chloride concentration and stirring speed on extraction rate.

may be seen that no plateau was observed at 1M Cl when high concentrations of TOAHCl and Fe were used; the rate increased continuously with stirring speed until the interface was disrupted.

At low TOAHCl and Fe concentrations, it was found that as the aqueous chloride concentration was increased, the plateau region diminished and shifted and then vanished at 9M Cl. These effects were attributed to the increasing viscosity of the aqueous phase as the chloride concentration was increased.[2] Changes in viscosity will affect the mixing conditions in the cell and the diffusivity of the transferred species. In addition, the interfacial tension is likely to increase with the concentration of ionic solutes and this will tend to decrease the stability of the interface.

It was decided that a criterion of hydrodynamic similarity was necessary to correlate results at different chloride concentrations. As it was not possible to calculate the mixing power dissipated with any accuracy, the correlation was based on the Reynolds number in the aqueous phase:

$$Re = NL^2/\nu \tag{6}$$

where N = stirring speed (rps), L = tip-to-tip impeller length (cm), and ν = viscosity (Stokes). It was thought that if the Reynolds number is a reasonable indicator of mixing conditions, the plateau region should start at a constant Reynolds number and the length

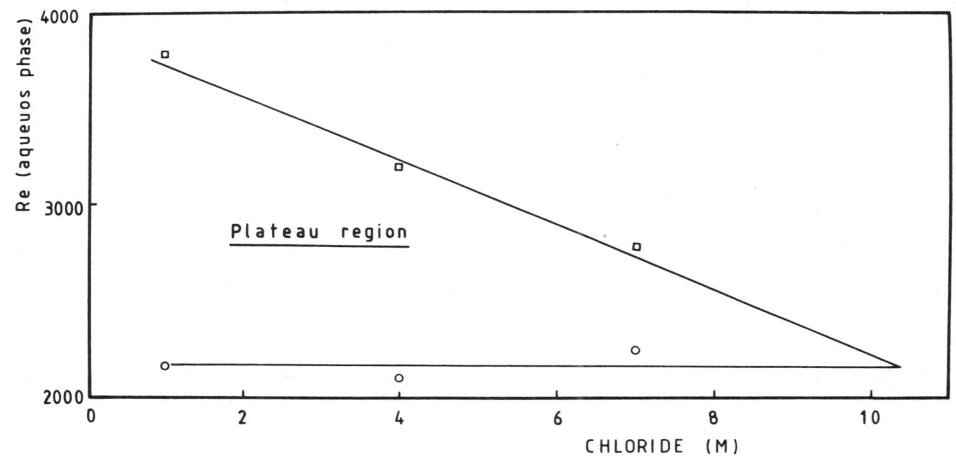

Fig. 5. Effect of aqueous chloride concentration and stirring conditions (Reynolds number) on the plateau region.

of the plateau should be related in some way to the interfacial tension. A plot of the Renolds number at the start and at the end of the plateau region versus aqueous chloride concentration is shown in Fig. 5. It may be seen that the plateau region decreases as the concentration of chloride in the aqueous phase, hence also the viscosity and (probably) the interfacial tension, increase. The results suggest that above 10M chloride there will be no plateau and in the 8 to 10M region, it will be difficult to detect.

It was concluded that up to about 7M chloride, the aqueous Reynolds number was a reasonable criterion of hydrodynamic similarity and it was decided to carry out further experiments at Re = 2680, which is the Reynolds number obtained at 1M chloride and 250 rpm, conditions which lay well within the plateau region shown in Fig. 3. Results presented subsequently will add justification to this approach.

Effect of interfacial area. In order to establish the location of the chemical reaction controlling the mass transfer rate, experiments were carried out at different interfacial areas by varying the annular space between the horizontal baffles. Typical results are illustrated in Fig. 6, where the mass transfer rates at 1M chloride are plotted as a function of specific interfacial area. It may be seen that there is a direct proportionality and the line passes through the origin, demonstrating that the whole area of

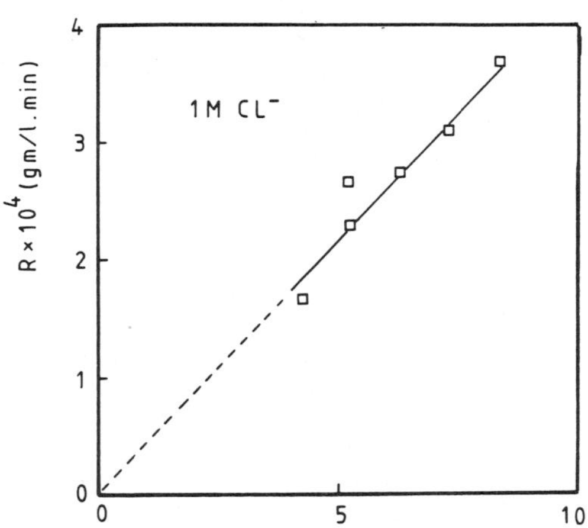

Fig. 6. Effect of interfacial area on extraction rate. (x axis = 0.25 x [interfacial area] [cm^2]) Conditons: Fe = 0.0025M, TOAHCl = 0.025M, HCl = 1M.

the annulus was available for contact between the phases. Similar results were obtained for extraction from 4M and 7M solutions and for stripping into 1M HCl solutions.

It was thus concluded that the mass transfer rate was controlled by a slow chemical reaction located at or near the interface.

Effect of iron concentration. The effect of the aqueous iron concentration on the rate of extraction was investigated over the range 5×10^{-4} to 5×10^{-3}M Fe^{3+}. It was found that the initial mass transfer rate was directly proportional to the initial aqueous iron concentration over this range for the three aqueous chloride concentrations used, namely 1, 4 and 7M. In order to obtain comparable mixing conditions, the stirring rate in the aqueous phase was increased as the chloride concentration was increased to compensate for the viscosity effects discussed previously. The aqueous stirring rates used were 250, 375 and 550 rpm for solutions of 1, 4 and 7M chloride respectively. In each case, these stirring rates gave an aqueous Reynolds number of about 2680, which is in the plateau region. Typical results are presented in Fig. 7 for extraction from 1M chloride solutions. The slope of the corresponding line of initial mass transfer rate against iron concentration at 4M chloride was higher than at 1M chloride, and at 7M chloride it was higher still, indicating that the concentrations of the extractable iron species increase over this range of chloride concentrations. The effect of the concentration of iron in the organic phase on the rate

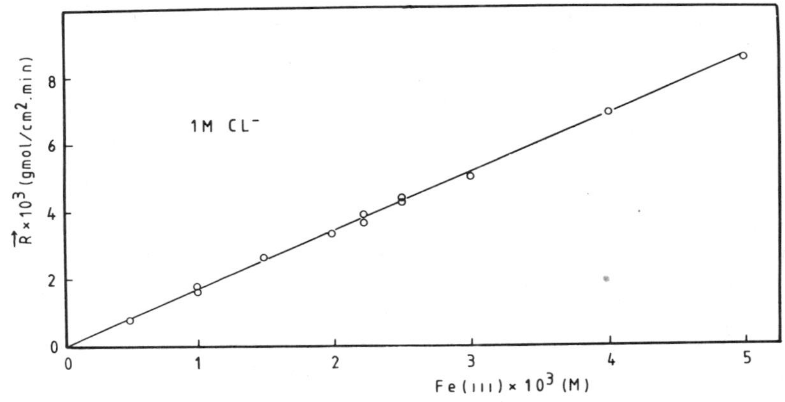

Fig. 7. Effect of aqueous iron concentration on extraction rate. Conditions: TOAHCl = 0.01M, HCl = 1M.

of stripping into 1M chloride solutions was also investigated. Again, a first order, direct proportionality, was obtained.

Effect of TOAHCl concentration. The effect of TOAHCl concentration on the initial mass transfer rate of iron from 1M chloride solutions is shown in Fig. 8. Over the range of TOAHCl concentrations investigated, 2×10^{-3} to 10^{-1}M, the extraction rate increased in parabolic fashion with the total TOAHCl concentration. It is known however that solutions of amine salts in non-polar solvents undergo aggregation. Solutions of TOAHCl in toluene contain monomers and dimers, up to a concentration of about 0.1M TOAHCl and the dimerization constant K_2 is 45 M^{-1} at 25°C.[3] In Fig. 8, the extraction rates are also plotted against the calculated concentrations of monomer and dimer TOAHCl, and it may be seen that there is a linear dependence on the concentration of monomer TOAHCl. Linear dependencies on the monomer TOAHCl concentration were also obtained for extractions from aqueous phases containing 4M and 7M chloride, indicating that under all conditions studied, the extraction of iron is first order with respect to the concentration of monomeric TOAHCl in the organic phase.

The effect of TOAHCl concentration on the rate of stripping into 1M chloride solutions was also studied and it was found that the rates decreased in an inverse parabolic fashion with increasing TOAHCl. A plot of the rate of stripping against the reciprocals of the monomer, dimer and total TOAHCl concentrations is shown in Fig. 9, and it may be seen that a linear relationship with reciprocal

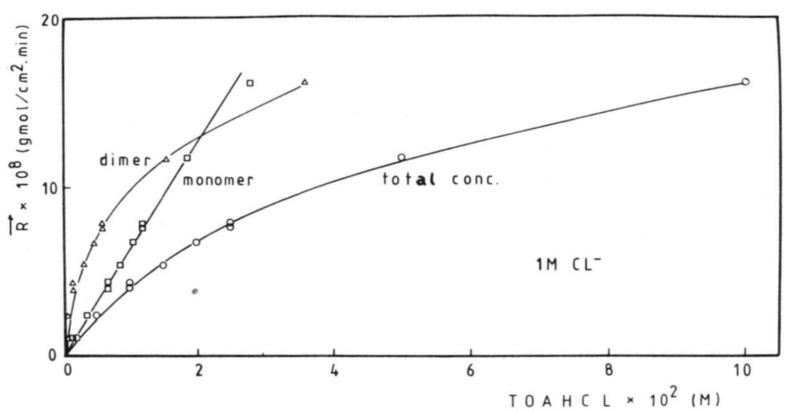

Fig. 8. Effect of TOAHCl concentration on extraction rate. Conditions: Fe = 0.0025M, HCl = 1M.

dimer TOAHCl concentration is obtained. This suggests that there is an inverse first order of the stripping rate with the concentration of dimer TOAHCl. However a linear relationship with the total TOAHCl concentration is also possible and more detailed analysis of the data, discussed subsequently, indicated that this is more probable.

Effect of aqueous chloride concentration. The aqueous chloride concentration was varied in the range 1 to 9M Cl^- at constant Fe and TOAHCl concentrations. Two series of experiments were carried out. In the first, the stirring speed was maintained constant in both phases at 250 rpm. Under these conditions, it was observed that the rate of iron extraction increased up to 4M Cl^- and then steadily decreased, as shown in Fig. 10. However, as already discussed, the viscosity of the aqueous phase increases with chloride concentration and therefore hydrodynamic similarity of the mixing conditions was not maintained, nor were the stirring conditions in the plateau region throughout this series of experiments. Thus it is likely that the conditions changed from chemical control to mixed or diffusion control, which could account for the observed behaviour of the extraction rate.

A second series of experiments was carried out in which the aqueous Reynolds number was kept constant at Re = 2680, by varying the aqueous stirring speed. Under these conditions, the extraction rate increased up to 5M Cl^- and then remained more or less constant, as shown in Fig. 10.

The effect of aqueous chloride concentration in the range 1 to 4M Cl^- on the rate of stripping of iron from the organic phase was investigated. Again hydrodynamics similarity was maintained by varying the aqueous stirring speed to obtain a constant aqueous Reynolds number. It was found that the stripping rate was independent of the aqueous chloride concentration.

These chloride dependencies, combined with previous results, suggest that the rate controlling step for the extraction reaction is located in the aqueous phase, near the interface, whereas for the stripping reaction, the slow step is located in the organic phase, at or near the interface.

Chloride transfer experiments. In order to try to elucidate which iron chlorocomplex species were being transferred across the interface, a number of extraction experiments were carried out in which the aqueous phase was spiked with ^{36}Cl isotope. The rate of

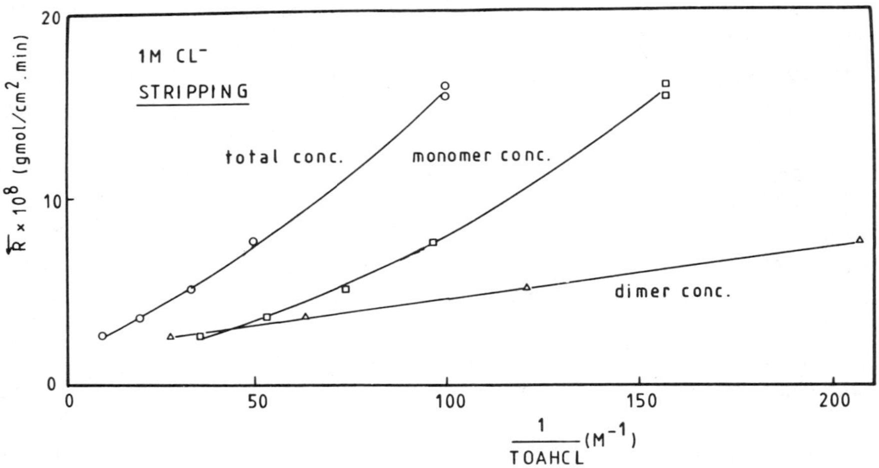

Fig. 9. Effect of TOAHCl concentration on stripping rate. Conditions: $\overline{Fe}$ = 0.00174M, HCl = 1M.

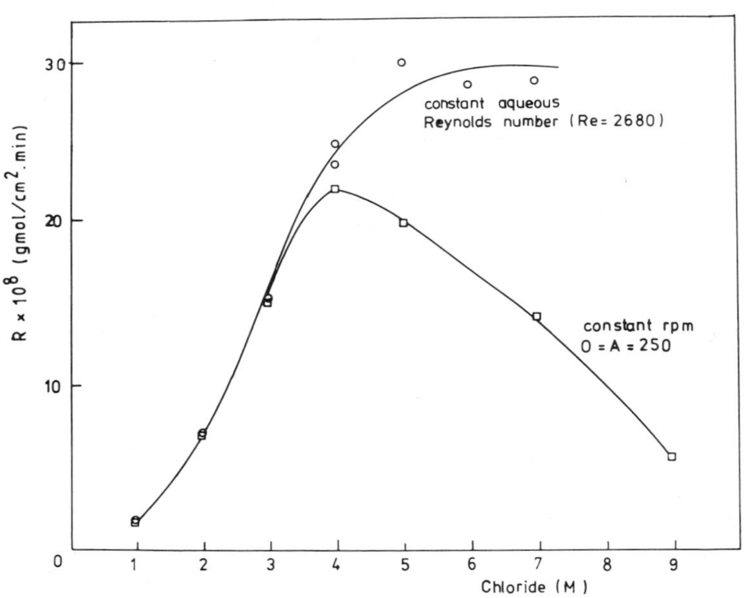

Fig. 10. Effect of aqueous chloride concentration on extraction rate. Conditions: Fe = 0.001M, TOAHCl = 0.01M.

Table 2. Molar Ratios of Chloride Transferred to Iron Transferred into the Organic Phase during Extraction.

[TOAHCl] = 0·025M [Fe] = 0·025M ^{36}Cl = 5μC

$[Cl]_{aq}$ (M)	Cl/Fe transferred
1	3·04 ± 0·09
3	3·15 ± 0·21
5	3·16 ± 0·23
7	3·52 ± 0·34

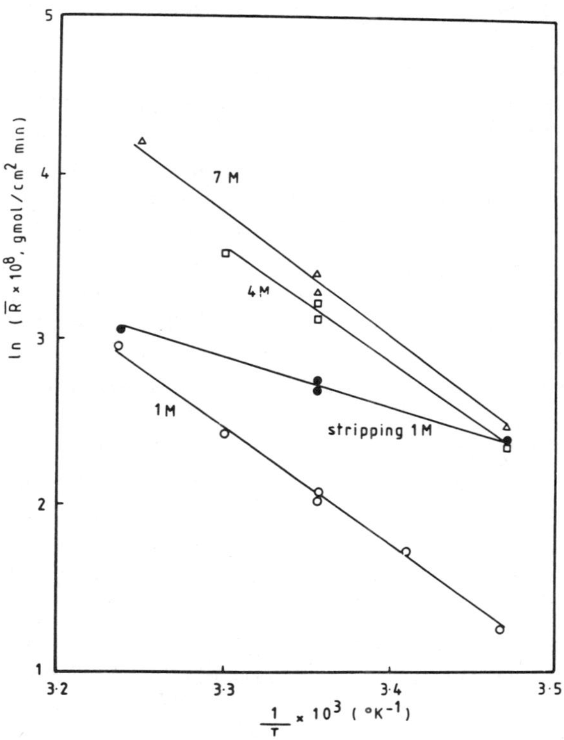

Fig. 11. Effect of temperature on initial mass transfer rate. Conditions: Extraction: Fe = 0.0025M, TOAHCl = 0.02M, HCl = 1M; Stripping: $\overline{Fe}$ = 0.00174M, TOAHCl = 0.02M, HCl = 1M.

Table 3. Apparent Activation Energies.

Cl (M)	Fe ($M \times 10^3$)	TOAHCl ($M \times 10^2$)	E_A (kJ/mol)
1	2·5	2·5	59·2 ± 2·6
4	1·0	1·0	58·8 ± 8·2
7	1·0	1·0	59·9 ±11·0
1 *	1·74	1·0	24·5 ± 2·5

* stripping experiments.

chloride transfer to the organic phase was obtained by sampling the organic phase and measuring the beta activity by liquid scintillation counting. The initial rates of chloride transfer in the presence and absence of iron in the aqueous phase were measured and the chloride associated with extracted iron was obtained from the difference of these two rates. The molar ratios of chloride transferred to iron transferred as a function of aqueous chloride concentration are given in Table 2. It may be seen that the measured ratios are all greater than 3 and increase with increasing aqueous chloride concentration. So also does the error associated with the measured ratio, this being due to the lower count rate in the organic phase as the aqueous chloride concentration is increased. However there is a clear trend in the Cl/Fe ratios and this is interpreted as indicating that both the neutral, $FeCl_3$, and the anionic, $FeCl_4^-$, chlorocomplexes are being transferred into the organic phase. Further evidence for this interpretation is presented subsequently.

Effect of temperature. The effect of temperature on the initial mass transfer rates was studied over the range 15 to 36°C, for extraction at various chloride concentrations and for stipping at 1M chloride. Arrhenius plots of the data are shown in Fig. 11 and the calculated activation energies are given in Table 3.

It may be seen that the apparent activation energies of the extraction (forward reaction) experiments were reasonably independent of the aqueous chloride concentration, suggesting that the mechanism is unchanging over the range of conditions studied.

Reaction Mechanism

The reaction mechanism which was found to best represent the experimental results was as follows:

$$\overline{\mathrm{TOAHCl}} \overset{K_1}{\rightleftarrows} (\mathrm{TOAHCl})_i \qquad \text{(fast)} \qquad (7)$$

$$(\mathrm{TOAHCl})_i \overset{K_2}{\rightleftarrows} (\mathrm{TOAH^+})_i + \mathrm{Cl^-} \qquad \text{(fast)} \qquad (8)$$

$$\begin{matrix} \mathrm{FeCl_3} + (\mathrm{TOAHCl})_i \\ \text{or} \\ \mathrm{FeCl_4^-} + (\mathrm{TOAH^+})_i \end{matrix} \quad \underset{k_2}{\overset{k_1}{\rightleftarrows}} \quad (\mathrm{TOAHFeCl_4})_i \qquad \text{(slow)} \qquad (9)$$

$$(\mathrm{TOAHFeCl_4})_i + \overline{\mathrm{TOAHCl}} \underset{k_4}{\overset{k_3}{\rightleftarrows}} (\mathrm{TOAHCl})_i + \overline{\mathrm{TOAHFeCl_4}} \qquad \text{(slow)} \qquad (10)$$

$$\overline{\mathrm{TOAHFeCl_4}} + \overline{\mathrm{TOAHCl}} \overset{K_3}{\rightleftarrows} \overline{\mathrm{TOAHFECl_4 \cdot TOAHCl}} \qquad \text{(fast)} \qquad (11)$$

where $(\quad)_i$ are the interfacial species and the overline indicates the bulk organic phase species.

In this reaction sequence, the adsorption of the surface-active TOAHCl on to the interface, Eq. 7, and its dissociation, Eq. 8, are both likely to be fast processes, which under conditions of chemical reaction control will be at equilibrium. The rate controlling steps are considered to be the reactions represented by Eqs. 9 and 10, namely: reaction between either the neutral or the anionic iron chlorocomplex with interfacial TOAHCl or $\mathrm{TOAH^+}$ to form the intermediate organic iron complex at the interface; followed by displacement of the intermediate complex into the bulk organic phase by adsorption of fresh TOAHCl on to the interface. Finally an equilibrium reaction in the organic phase, Eq. 11, results in the formation of the complex containing two moles of TOA per mole of iron, which is the composition of the organic complex under equilibrium conditions, as shown by distribution and spectroscopic studies.[4-7]

A rate equation can be derived from this reaction sequence on the assumption that there is a steady state concentration of the intermediate organic iron complex at the interface. The derivation of the rate equation is given in the Appendix, the final result being:

$$\frac{V}{A}\frac{d(\overline{\mathrm{Fe}})}{dt} = k_5^* \, \alpha_n(\mathrm{Fe})(\overline{\mathrm{TOAHCl}}) - k_6^* \frac{(\overline{\mathrm{Fe}})}{(\overline{\mathrm{TOAHCl}})} \qquad (12)$$

where Fe and $\overline{\mathrm{Fe}}$ are the bulk concentrations of iron in the aqueous and organic phases, respectively; α_n is the fraction of iron present in the aqueous phase as the extractable chlorocomplex(es); $\overline{TOAHCl}$ is the bulk concentration of TOAHCl in the organic phase; and k_5^* and k_6^* are combinations of the rate and equilibrium constants governing the reactions described by Eqs. 9 to 11 and which are defined in the Appendix.

The first term on the right-hand side of Eq. 12 governs the initial rate of extraction and predicts that this should be first order with respect to both the aqueous iron concentration and the organic TOAHCl concentration, and to α_n, which in turn is related in a complex fashion to the aqueous chloride concentration. The second term on the right-hand side of Eq. 12 relates to the initial stripping rate and shows that this should be first order with respect to the organic iron concentration, inverse first order with respect to the organic TOAHCl concentration and that it should be independent of the aqueous chloride concentration.

These predictions are in agreement with the experimental observations and thus the proposed mechanism is consistent with the kinetic data.

There is no "a priori" reason or need to assume that either monomer, or dimer, or total TOAHCl, is the reactive species in the organic phase. However, the experimental results suggest that, for extraction, it is monomeric TOAHCl which takes part in the reaction, whereas for stripping, either the dimer only, or the total TOAHCl concentration may be involved.

Rate equation evaluation. Equation 12 was tested by fitting all the experimental extraction and stripping rates obtained at 1M chloride. Extensive statistical evaluation of the adequacy of the rate equation was carried out.

The extraction data were fitted to the expression

$$\frac{d\overline{(Fe)}}{dt}\,\frac{V}{A} = k_5^*\,\alpha_n(Fe)(\overline{TOAHCl}) \tag{13}$$

The three possibilities for $\overline{TOAHCl}$, namely monomer, dimer, or total concentration were considered and, in every case, the concentration of monomer gave the best fit, indicating that this is the most likel organic species involved in the extraction reaction. For the data at 1M chloride a value of

$$k_5^* \alpha_n = (2.59 \pm 0.03) \times 10^{-9} \text{ cm}^4/\text{mol.min} \quad (25°C)$$

was found. This of course is not a true constant since it includes α_n, the fraction of extractable chlorocomplex(es). α_n will vary with the chloride concentration of the aqueous phase and in order to compare data obtained at different chloride concentrations, it is necesssary to estimate values for α_n by the use of stability constant data for the iron chlorocomplexes. Unfortunately, there is little or no agreement amongst the published sets of stability constant data for this system.

The experiments reported previously in which labelled chloride was used in conjunction with iron extraction showed that the most likely species being transferred were the neutral ($FeCl_3$) and the anionic ($FeCl_4^-$) chlorocomplexes. Values of α_3 and α_4 were

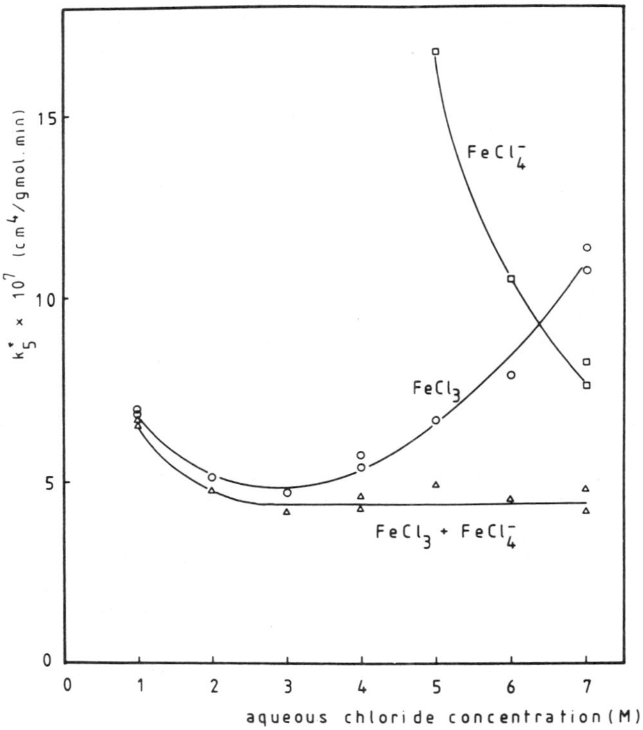

Fig. 12. Calculated k_5^* value as a function of aqueous chloride concentration for the various possible active chlorocomplexes.

calculated from the various sets of stability constant data and were tested in Eq. 13 with rate data obtained at various aqueous chloride concentrations to determine whether a constant value for k_5^* resulted. Results obtained using stability constants from reference 8 are shown in Fig. 12 and it may be seen that the assumption that both $FeCl_3$ and $FeCl_4^-$ are transferring species (i.e. $\alpha_n = \alpha_3 + \alpha_4$), gives a reasonably constant value for k_5^* except at low chloride concentrations. A second set of stability constant data[9] gave results very similar to those shown in Fig. 12, another set[10] agreed at chloride concentrations above 4M, whereas a fourth set[11] showed no agreement. However the fact that good agreement using two different sets of stability constant data, and partial agreement using a third, was achieved is perhaps indicative that the assumption that both neutral and anionic chloro-complexes are transferred during extraction is reasonable. On the basis of this assumption, the

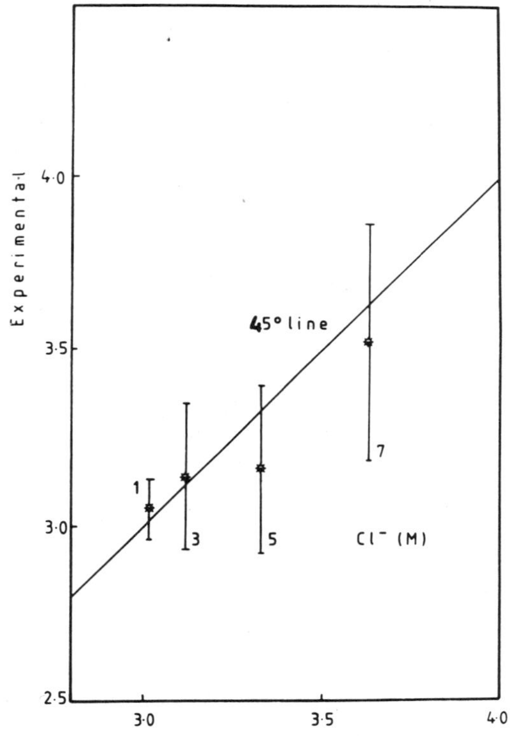

Fig. 13. Experimental versus predicted Cl/Fe molar ratios for the active chlorocomplexes.

theoretical average Cl/Fe molar ratios for the transferring species were calculated from the stability constant data and these are compared with the experimental data, reported in Table 2, in Fig. 13. Taking account of the experimental errors in the measured values, they are in reasonable agreement with the predicted values.

The stripping rate data were fitted to the expression:

$$\frac{d(Fe)}{dt}\frac{V}{A} = k_6^* \frac{(\overline{Fe})}{(\overline{TOAHCl})} \tag{14}$$

Again all three possibilities for TOAHCl (monomer, dimer, or total) were considered, but in this case it was not possible to distinguish

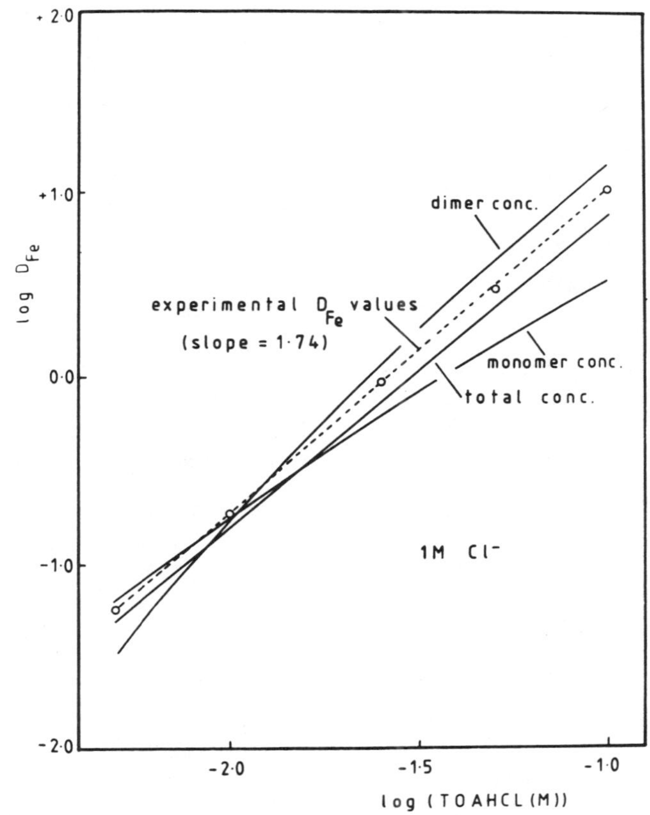

Fig. 14. Experimental and predicted distribution coefficients as a function of TOAHCl concentration.

by statistical tests which organic species was involved in the reaction. In order to resolve this difficulty, the kinetic data were compared with equilibrium distribution data. At equilibrium the extraction and stripping rates will be equal and thus:

$$k_5^* \alpha_n(\text{Fe}) \, (\overline{\text{TOAHCl}})_1 = k_6^* \, (\overline{\text{Fe}})/(\overline{\text{TOAHCl}})_{1,\ 2 \text{ or } T} \qquad (15)$$

where subscripts 1, 2 and T, indicate monomer, dimer and total TOAHCl, respectively. Hence,

$$D_{Fe} = \frac{(\overline{\text{Fe}})}{(\text{Fe})} = (k_5^* \alpha_n/k_6^* \quad (\overline{\text{TOAHCl}})_1 \, (\overline{\text{TOAHCl}})_{1,\ 2 \text{ or } T} \qquad (16)$$

Experimentally determined distribution coefficients are compared with those predicted from the kinetic data in Fig. 14 and it may be seen that the assumption that the total concentration of TOAHCl is involved in the stripping rate equation gives predicted D_{Fe} values most consistent with the experimental results.

The final version of the rate equation was thus established as:

$$\frac{d(\overline{\text{Fe}})}{dt} \frac{V}{A} = k_5^* \, (\alpha_3 + \alpha_4) \, (\text{Fe}) \, (\overline{\text{TOAHCl}})_1 - k_6^* \frac{(\overline{\text{Fe}})}{(\overline{\text{TOAHCl}})_T} \qquad (17)$$

where $k_5^* = (4.5 \pm 0.3) \times 10^{-7}$ cm^4/mol min,
and $k_6^* = (9.2 \pm 0.3) \times 10^{-7}$ mol/min cm^2, at 25°C.

Equation 17 was integrated and the resulting function was examined to determine whether it would predict concentration - time profiles of the form observed experimentally. It was found that the predicted profiles for both extraction and stripping were in close agreement with the experimental profiles over a wide range of solution compositions.

CONCLUSIONS

The initial extraction and stripping rates of iron with TOAHCl under the experimental conditions were controlled by the rates of chemical reactions taking place at or near the interface between the two phases. A mechanism was proposed for the reactions involved (Eqs. 7 to 11) from which a rate equation was derived (Eq. 12). Evaluation of this equation indicated that the initial extraction rate was first order with respect to the total aqueous iron concentration and to the concentration of monomer TOAHCl in the organic phase. The initial stripping rate was first order with respect to the organic iron concentration and inverse first order with respect to the total TOAHCl concentration. At 1M chloride, this model was able to predict the initial extraction and stripping rates and the equilibrium distribution coefficients, and in the integrated form it was used to predict concentration - time profiles of the form observed experimentally. These results indicated that the proposed mechanism was an adequate representation of the kinetic data. The rate equation could also be used to account for the dependence of the rates on the aqueous chloride concentration. The initial extraction rates were related to the concentrations of the neutral ($FeCl_3$) and anionic ($FeCl_4^-$) iron chlorocomplexes in the aqueous phase, whereas the initial stripping rates were independent of the aqueous chloride concentration.

It was concluded that the main kinetic resistance to the extraction of iron into the organic phase was in the reaction between the aqueous iron chlorocomplexes and monomeric TOAHCl at the interface. For stripping iron from the organic phase, the difficulty of displacing monomer TOAHCl at the interface by $TOAHFeCl_4$ appeared to be the rate controlling step.

REFERENCES

1. P. J. Austin: Ph.D. Thesis, University of London, 1966.
2. R. C. Weast (ed.): CRC Handbook of Chemistry and Physics, 49th ed., CRC Press, 1968.
3. J.C.S. Cassa: Ph.D. Thesis, University of London, 1981.
4. S. Lindebaum and G. Boyd: J. Phys. Chem., 1963, vol. 67, p. 1238.
5. L. Kuca: Solvent Extraction Chemistry, Eds. D. Dyrssen, J. O. Liljenzin and J. Rydberg, p. 454, North Holland, Amsterdam, 1967.
6. M. Good: J. Inorg. Nucl. Chem, 1965, vol. 27, p. 2429.
7. G. Michel, A. Artoisenet and G. Duyckaerts: Anal. Chim. Acta, 1973, vol. 63, p. 53.
8. J. Dousma, T. J. Hoven and P. L. Bruyn: J. Inorg. Nucl. Chem, 1978, vol. 40, p. 1089.
9. Y. Marcus: J. Inorg. Nucl. Chem, 1960, vol. 12, p. 287.
10. H. C. Hegelson: Am. J. Sci., 1967, vol. 267, p. 729.
11. G. A. Gamlem and D. O. Jordan: J. Chem. Soc, 1953, p. 1435.

ACKNOWLEDGEMENTS

J. C. S. Cassa wishes to acknowledge the financial support he received from the British Council and the Federal University of Bahia, Brazil.

APPENDIX

Derivation of the rate equation

The mechanism given in Eqs. 9 to 11 may be written in simplified form as:

$$A + B_i \underset{k_2}{\overset{k_1}{\rightleftarrows}} AB_i \qquad \text{(slow)} \qquad (A1)$$

$$AB_i + \overline{B} \underset{k_4}{\overset{k_3}{\rightleftarrows}} \overline{AB} + B_i \qquad \text{(slow)} \qquad (A2)$$

$$\overline{AB} + \overline{B} \overset{K_3}{\rightleftarrows} \overline{AB.B} \qquad \text{(fast)} \qquad (A3)$$

where $A = FeCl_3$ or $FeCl_4^-$; $\overline{B} = \overline{TOAHCl}$; $B_i = (TOAHCl)_i$ or $(TOAH^+)_i$; and $(\ )_i$ = interfacial species.
From Eq. A1, the extraction rate is:

$$\frac{V}{A}\frac{d(AB)_i}{dt} = k_1\ (A)\ (B)_i - k_2\ (AB)_i \qquad (A4)$$

From Eq. 7, $(TOAHCl)_i\ /\ \overline{TOAHCl} = B_i/\overline{B} = K_1$.

At time close to zero, $\overline{B}$ will be constant, thus

$$B_i = \overline{B}\ K_1 = \text{constant} \qquad (A5)$$

therefore

$$k_1\ (B)_i = k_1^* \qquad (A6)$$

Substituting Eq. A6 into A4:

$$\frac{V}{A}\frac{d(AB)_i}{dt} = k_1^*\ (A) - k_2(AB)_i \qquad (A7)$$

The extraction rate derived from Eq. A2 is:

$$\frac{V}{A}\frac{d(AB)_i}{dt} = k_3(AB)_i\ (\overline{B}) - k_4(\overline{AB})(B)_i \qquad (A8)$$

Using Eq. A5 to eliminate $(B)_i$

$$k_4(B)_i = k_4^*$$

therefore

$$\frac{V}{A}\frac{d(AB)_i}{dt} = k_3(AB)_i\ (\overline{B}) - k_4^*\ (\overline{AB}) \qquad (A9)$$

Equating (A7) and (A9):

$$k_1^*(A) - k_2(AB)_i = k_3(AB)_i\ (\overline{B}) - k_4^*(\overline{AB}) \qquad (A10)$$

therefore:

$$(AB)_i = \frac{k_1^*(A) + k_4^*\ (\overline{AB})}{k_2 + k_3(\overline{B})} \qquad (A11)$$

Assuming a steady state concentration of the interfacial complex $(AB)_i$, the rate of transfer of iron into the organic phase will be equal to the rate of formation of $(AB)_i$, thus substituting Eq. A11 into Eq. A9:

$$\frac{V}{A}\frac{d\overline{Fe}}{dt} = \frac{(k_1^*k_3/k_2)\ (A)(\overline{B}) - k_4^*\ (\overline{AB})}{1 + (k_3/k_2)(\overline{B})} \qquad (A12)$$

From Eq. A3:

$$(\overline{AB}) = \frac{(\overline{AB.B})}{K_3(\overline{B})} \qquad (A13)$$

Writing $\dfrac{k_1^*\ k_3}{k_2} = k_5^*$ and $\dfrac{k_4^*}{K_3^*} = k_6^*$

and substituting Eq. A13 into A12:

$$\frac{V}{A}\frac{d(\overline{Fe})}{dt} = \frac{k_5^*\ (A)(\overline{B}) - k_6^*(\overline{AB.B})/(\overline{B})}{1 + (k_3/k_2)(\overline{B})} \qquad (A14)$$

(A) is the concentration of the extractable iron chlorocomplex(es) which, for a constant aqueous chloride concentration, may be written as α_n(Fe), where Fe is the total aqueous iron concentration and α_n is the fraction of iron present as the extractable complex(es).

$\overline{AB.B}$ is equal to the concentration of iron in the organic phase. Thus writing Eq. A14 in terms of the species involved, gives:

$$\frac{V}{A}\frac{d(\overline{Fe})}{dt} = \frac{k_5^* \alpha_n (Fe)(\overline{TOAHCl}) - k_6^* (\overline{Fe})/(\overline{TOAHCl})}{1 + (k_3/k_2)(\overline{TOAHCl})} \qquad (A15)$$

It may be seen from Eqs. A1 and A2 that k_2 is the rate of dissociation of the interfacial complex and k_3 is the rate of desorption of the interfacial complex into the organic phase. It is likely that $k_3 < k_2$ and since, under all conditions studied, $\overline{TOAHCl}$ was less than 0.1M, the approximation that $(k_3/k_2)(\overline{TOAHCl}) << 1$ can be made. Thus Eq. A15 becomes:

$$\frac{V}{A}\frac{d(\overline{Fe})}{dt} = k_5^* \alpha_n (Fe)(\overline{TOAHCl}) - \frac{k_6^*(\overline{Fe})}{(\overline{TOAHCl})} \qquad (A16)$$

INTERFACIAL PHENOMENA IN HYDROMETALLURGICAL LIQUID-LIQUID EXTRACTION SYSTEMS

K. Osseo-Asare

Department of Materials Science and Engineering
The Pennsylvania State University
University Park, PA 16802 U.S.A.

A review is presented of interfacial chemical concepts such as adsorption and interfacial tension, monolayers, and the surface charge on oil drops. The interfacial activity of solvent extraction reagents is discussed in terms of two parameters: the efficiency of adsorption, i.e. the minimum bulk concentration needed to produce a saturated interface, and the effectiveness of adsorption, i.e., the adsorption density of a saturated interface. The effects of the structure and composition of the extractant molecules, the nature of the organic diluent and the nature and composition of the aqueous phase on the efficiency and effectiveness parameters are considered. The role of interfacial activity in metal extraction processes is illustrated with the copper-hydroxyoxime system, and mixed extractant systems containing carboxylic or sulfonic acids in combination with chelating extractants.

INTRODUCTION

It is now well established that hydrometallurgical liquid-liquid extraction reagents can be usefully considered as surface active agents.[1-42] The interfacial activity of these extractants is a result of their amphipathic nature, i.e. the presence of both hydrophobic and hydrophilic groups in the same molecule. The hydrophobicity is contributed by the hydrocarbon radicals, which may be straight or branched alkyl chains, or alkylaryl groups involving benzene or naphthalene rings. The polar functional groups such as

Table 1. Functional Groups of Selected Solvent Extraction Reagents

FUNCTIONAL GROUP	COMPOUND
$-C(OH)=O$	CARBOXYLIC ACIDS
$=P(OH)=O$	PHOSPHORIC ACIDS
$-SO_2-OH$	SULFONIC ACIDS
$-CH(OH)-C(=N-OH)-$	HYDROXYOXIMES
$=C(OH)-C(=)-N=$	8-HYDROXYQUINOLINES (OXINES)
$-NH_2, =NH, \equiv N$	1°, 2°, 3° AMINES
$\equiv NR^+X^-$	QUATERNARY AMMONIUM SALTS

those shown in Table 1, are responsible for the hydrophilic nature. The amphipathic characteristic is necessitated by the fact that an extractant must be able to interact with aqueous phase species while still maintaining a preferential organic phase solubility. The hydrophilic groups provide the functional groups which form the metal-ligand bonds while the organic phase radicals ensure the desired organic phase solubility.

In spite of the growing recognition of the surfactant properties of solvent extraction reagents, it cannot be said that there is complete agreement concerning the significance of interfacial phenomena in extraction processes. In a plenary lecture presented during ISEC'80, Freiser[43] was insistent that interfacial effects are of little importance in metal extraction. On the other hand, in another plenary lecture at the same conference, Hanson[44] suggested that reaction occurs in "a narrow zone in the aqueous phase immediately adjacent to the interfaces but with a significant (and possibly sometimes dominating) diffusional resistance in a zone (likely to be of greater thickness) on the organic side of the interface."

Liquid-liquid extraction involves ligand exchange processes whereby metal-organic ligand bonds substitute for the metal-oxygen bonds present in the hydrated aqueous species:

$$Me(H_2O)_n^{z+}\ (a) + zHR(o) = Me(H_2O)_{n-z}R_z(o) + 2H^+(a) + zH_2O \qquad (1)$$

where the symbols (a) and (o) designate aqueous phase and organic phase species respectively. In general, extractant molecules possess extremely low aqueous solubilities. For example, according to Foakes et al[45], the aqueous solubility of commercial hydroxyoxime copper extractants can be as low as 10^{-6}kmol m^{-3} (anti-LIX65N, pH 1.78, 25°C). Thus, it is not unreasonable to expect the locale of the extraction reaction to be near the two phase boundary. However, the tendency in the literature to appeal to indirect evidence (e.g. mathematical models of metal extraction rates, measurements of bulk phase metal distribution) as support for the significance or insignificance of interfacial phenomena in metal extraction processes is clearly unacceptable. More direct experimental evidence derived from interfacial and bulk phase physico-chemical investigations is needed.

In 1967, two pioneering studies appeared in which interfacial tension measurements were used to elucidate metal extraction mechanisms: those by McDowell and Coleman[1,2] on the uranium-amine sulfate system, and Scibona et al.[3] on the kinetics of the reaction of tetraheptyl ammonium chloride with bromide ions. However, since then, with the exception of a few investigators such as Scibona, Chiarizia, Danesi et al.[4-9], van Dalen et al.[11-15], Flett, Cox et al.[16-20], Hughes et al.[23-27], Vandegrift et al.[28-30], Yagodin et al.[31-33], and Osseo-Asare and Keeney[35-39], there has been little attempt to undertake systematic and extensive investigations which focus on the interfacial region.

In this paper, a review is presented of the basic concepts of interfacial chemical phenomena of relevance to metallurgical liquid-liquid extraction.[46-56] This is followed with a discussion of selected interfacial studies conducted in the context of hydrometallurgical liquid-liquid extraction research. An attempt is made to point out the gaps in the literature in the hope that a greater awareness of the current deficiencies would encourage the solvent extraction community to take a more serious interest in interfacial studies. While the present discussion focuses on the chemical aspects of interfacial phenomena, it must be noted that the physical aspects (e.g., the role of interfacial-tension-driven instabilities) are equally important.[57,58] Unfortunately, previous research in the field of metallurgical solvent extraction has paid little attention to the hydrodynamic aspects of interfacial phenomena.

BASIC CONCEPTS OF INTERFACIAL CHEMISTRY

Interfacial Tension

Below the boiling point, the molecules present in a liquid do not fly off because they are held together by intermolecular attractive forces (e.g. London dispersion, dipole-dipole, dipole-induced dipole, hydrogen-bond) which are large enough to oppose the thermal forces which tend to destroy the liquid state. When an aqueous solution is brought into contact with an immiscible organic liquid, the molecules present at the phase boundary are in contact with unlike molecules on adjacent sides of the interface. In general, the interfacial molecules will not experience the same degree of attraction from the foreign molecules as they would if they were present in their respective bulk regions. Consider a system consisting of a layer of oil in contact with water. We can imagine three possibilities at the phase boundary according to whether the attraction of the water molecules for the oil phase molecules is (a) greater than the attraction of H_2O for H_2O; (b) equal to the attraction of H_2O for H_2O; (c) less than the attraction of H_2O for H_2O. In the case of (a) and (b), we would expect the oil and aqueous phases to be completely miscible. On the other hand, in the case of (c), the water molecules present at the phase boundary would not enjoy as much attractive interaction as those molecules in the bulk phase. Thus the interfacial molecules would possess a higher interfacial energy. The interfacial energy or interfacial tension is the most common property which is used to characterize the interfacial state of two-phase systems.

In view of the above discussion, it would be expected that the interfacial tension (γ) would decrease with increase in the strength of the attractive forces experienced by the dissimilar molecules. Table 2 indicates that the interfacial tensions of n-octane and

Table 2. Surface and Interfacial Tensions (20°C, mN m^{-1})

	γ(air)	γ(water)
water	72.75	--
benzene	28.85	35.0
toluene	28.5	--
n-octanol	27.53	8.5
carbon tetrachloride	26.95	45.0
n-octane	21.8	50.8

n-octanol are 50.8 and 8.5 mN m^{-1} respectively against water. It should be concluded therefore that the attractive forces between water and octanol molecules are greater than those between water and octane molecules. The interaction of water with alkanes such as n-octane is derived mainly from dispersion forces. On the other hand, in the case of the alkanols, hydrogen bonding provides an additional means of interaction with water. This hydrogen bonding is made possible by the presence of the polar hydroxyl group in the alkanol structure. In fact, the presence of both polar and non-polar properties in the same molecule allows the octanol molecules at the oil/water interface to act as a bridge between the two phases, with the polar group dipping into the water while the non-polar organic portion is kept immersed in the organic phase as shown in Fig. 1a. Thus, at the oil/water interface there is a certain amount of interfacial packing and orientation of octanol molecules.

Adsorption

Suppose now we consider a three component system e.g. water in contact with a benzene solution of n-octanol. The octanol molecules, through their hydroxyl groups, will again interact strongly with the water molecules at the oil/water interface. However, the presence of the water-insoluble hydrocarbon chains will discourage complete immersion of octanol molecules in water. Thus the polar hydroxyl and nonpolar organic portions will be immersed in the opposite sides of the interface as shown in Fig. 1b. The possibility of interaction with water draws octanol molecules from the bulk organic phase to the interfacial region. This population of octanol molecules at the phase boundary represents adsorption, since it corresponds to an increased concentration of solute molecules in the interface relative to the bulk organic phase.

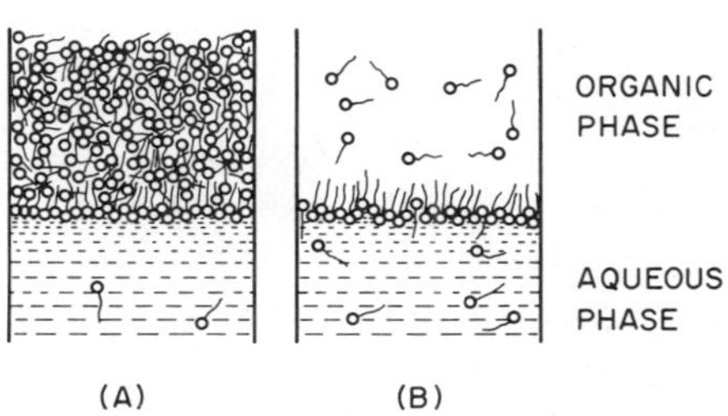

Fig. 1. Schematic illustration of adsorption and orientation of molecules at the organic/water interface[47]; (a) octanol/H_2O (b) octanol-benzene/H_2O.

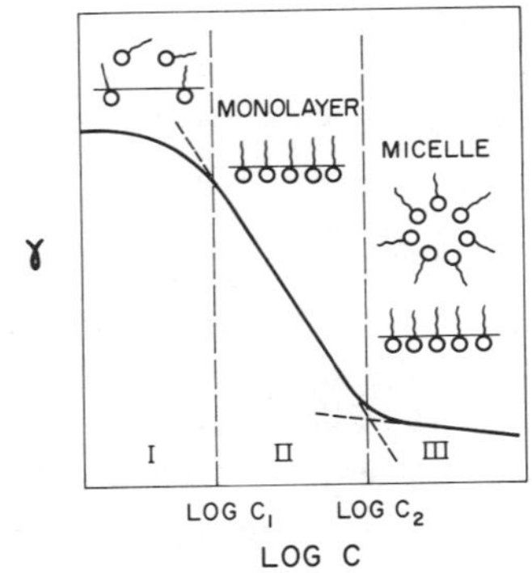

Fig. 2. Schematic illustration of the effect of surfactant concentration on interfacial tension.

The adsorption of a solute at the oil/water interface leads to a decrease in interfacial tension, as shown schematically in Fig. 2. The γ versus log C curve can be divided into three parts. In region I, $C < C_1$, the liquid/liquid interface is sparsely populated by surfactant molecules and there is very little effect of concentration on interfacial tension. In region II, $C_1 \leq C \leq C_2$, the interfacial tension varies linearly with log C and the interface is characterized by a monolayer of adsorbed molecules. In region III, $C > C_2$, there is a change in slope of the γ versus log C curve, increase in surfactant concentration now has a less dramatic effect on interfacial tension. This region corresponds to the situation where the surfactant undergoes aggregation or micellization in the bulk organic phase, and C_2 is termed the critical micelle concentration or CMC.

The Gibbs Equation

When two homogeneous immiscible phases are in contact, the phase boundary may be viewed as an interfacial phase or "interphase" of finite thickness, the material within which can assume the three states of matter: gas, liquid and solid. The Gibbs adsorption equation provides a means of deriving the number of adsorbed molecules per unit area, or the interfacial molecular area, A, from interfacial tension versus concentration data. For a given temperature (T), the general form of the Gibbs equation is:

$$(d\gamma)_T = -\Sigma_j \Gamma_j (d\mu_{ji})_T \tag{2}$$

where μ_{ji} refers to the chemical potential of the jth species in the interfacial region (i). Consider a three component system consisting of a solute (j = 3) as well as oil (j = 1) and aqueous (j = 2) phases. Application of Eq. 2 gives

$$d\gamma = -\Gamma_3 d\mu_{3i} - \Gamma_2 d\mu_{2i} - \Gamma_1 d\mu_{1i} \tag{3}$$

The Gibbs convention postulates the presence of a hypothetical plane within the phase boundary region such that either $\Gamma_2 = 0$ or $\Gamma_1 = 0$. If $\Gamma_2 = 0$ is chosen, Eq. 3 becomes

$$d\gamma = -\Gamma_3 d\mu_{3i} - \Gamma_1 d\mu_{1i} \tag{4}$$

Since for dilute solutions, $d\mu_{3i} \gg d\mu_{1i}$ (i.e. as the concentration of solute is varied, the interfacial chemical potential of the solute changes much more than the chemical potential of the organic diluent), it follows that

$$d\gamma = -\Gamma_3 d\mu_{3i} \tag{5}$$

Furthermore, using the fact that at equilibrium, $\mu_{3i} = \mu_{31}$ (i.e. the interfacial chemical potential of the solute equals the corresponding value in the bulk organic phase), and that

$$\mu_{31} = \mu^{\circ}_{31} + RT\ln C_{31} \tag{6}$$

where C_{31} is the bulk organic phase concentration of the solute, it follows that

$$d\gamma = -\Gamma_3 d\mu_{31} \tag{7}$$

Therefore,

$$\Gamma_3 = 1/A = -(1/RT)d\gamma/d\ln C_{31} \tag{8}$$

where A is the interfacial moelcular area of the solute.

According to Eq. 8, a small or zero value of $d\gamma/d\ln C$ implies a small or zero adsorption density. Referring to Fig. 2, this would imply that in the region $C > C_2$, there is negligible adsorption. However, obviously this cannot be the case since the low interfacial tension in region III indicates significant adsorption.

In order to explain this apparent paradox, the aggregation of the solute molecules must be considered.

Consider an organic phase solute HR which aggregates according to Eq. 9:

$$mHR = (HR)_m \qquad K_m \tag{9}$$

where K_m is the equilibrium constant for the aggregation reaction. Let C_R be the total analytical concentration of R in the organic phase. Then

$$C_R = C_{HR} + mC_{(HR)_m} \tag{10}$$

Therefore, when $C_{HR} \gg C_{(HR)m}$, i.e. at low concentrations of R where aggregation is not significant, $C_R \cong C_{HR}$ and

$$d\gamma/d\ln C_R = d\gamma/d\ln C_{HR} \tag{11}$$

At higher concentrations of R, $C_R \cong mC_{(HR)_m}$ and therefore,

$$d\gamma/d\ln C_R = d\gamma/d\ln C_{(HR)_m} \tag{12}$$

Furthermore, according to Eq. 9,

$$C_{(HR)_m} = K_m C_{HR}^m \tag{13}$$

Thus it follows from Eqs. 12 and 13 that in region III of Fig. 2,

$$d\gamma/d\ln C_R = d\gamma/md\ln C_{HR} \tag{14}$$

The right hand side of Eq. 14 approaches zero as the aggregation number, m, becomes infinitely large.

Monolayers

It was shown above that the adsorption of a solute at the organic/aqueous interface may lead to the formation of an interfacial film or monolayer of adsorbed molecules. The presence of the adsorbed molecules gives rise to a two-dimensional interfacial pressure, or film pressure, π. The interfacial pressure is given by

$$\pi = \gamma_o - \gamma \tag{15}$$

where γ_0 and γ represent respectively the interfacial tension in the absence and presence of the film. The relationship between interfacial pressure and the interfacial area per molecule, presented as a π-A or force-area curve, provides a useful way of studying the nature of interfacial films.[39,46-52,59-68] A distinction can be made between adsorbed monolayers (where the surfactant molecules are soluble in at least one of the two adjacent phases) and insoluble monolayers (where the adsorbed molecules are insoluble in both phases). The force-area curves for adsorbed monolayers can be derived from interfacial tension data. According to Eq. 15, interfacial tension data can be converted to interfacial pressures once the interfacial tension of the surfactant-free system is known. The corresponding A values can be determined with the Gibbs equation (see Eq. 8). In the case of insoluble monolayers, the π-A curve can be determined more directly. A small amount of the surfactant molecules is introduced at the organic/aqueous or gas/aqueous interface and allowed to spread in a confined area. By systematically varying the interfacial area available for the interfacial molecules while at the same time measuring the corresponding interfacial pressure, the force-area curve can be generated.[39,46-52,59-68] Figure 3 presents a schematic illustration of various kinds of π-A curves.

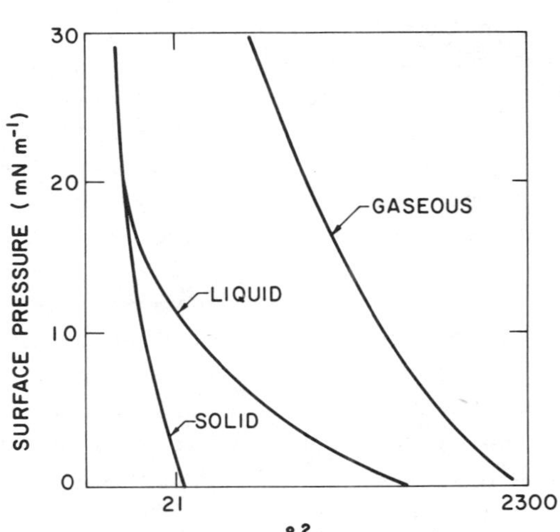

Fig. 3. Schematic surface pressure-area isotherms.

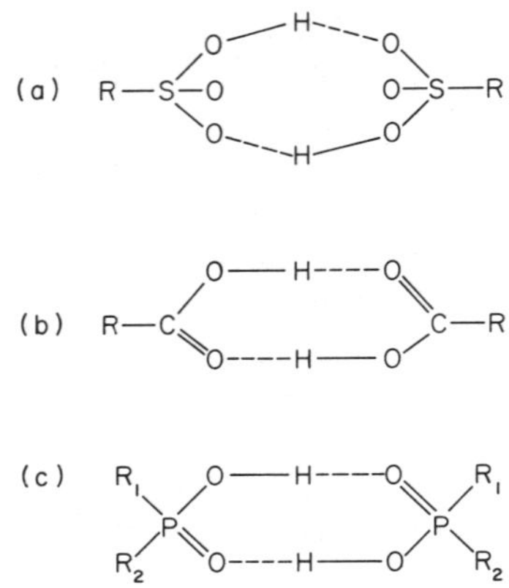

Fig. 4. Aggregation and hydrogen bonding.

Aggregation and Micellization

The aggregation of surface active agents, such as the sulfonic, carboxylic and organophosphorous acids, first gives rise to dimers. As shown in Fig. 4, the driving force for aggregation comes from intermolecular hydrogen bonding. Aggregation of surfactants in aqueous solutions has received much more attention in the literature than has aggregation in organic solvents. In organic solvents, aggregation is believed to be stepwise so that in a given surfactant solution, a dynamic equilibrium exists among several species of varying degrees of aggregation.[55] On the other hand, in aqueous systems, aggregation takes place over a very narrow range of surfactant concentrations and the result is a sharply defined CMC. In many organic solutions, aggregation may start at very low concentrations e.g. in the range of 10^{-7} to 10^{-6} kmol m^{-3} (cf>10^{-4} kmol m^{-3} for aqueous micelles). The average aggregation numbers tend to be smaller (~10) in organic solvents (cf > 100 in aqueous solutions). In view of the stepwise nature of organic phase aggregation, Kertes[55] has questioned the validity of the CMC values reported for surfactants in organic solvents. Perhaps it would be more correct to view the reported CMC values as the surfactant concentrations which represent the average degree of aggregation. It cannot be denied however, that there are numerous property-versus-concentration plots which exhibit more or less abrupt slope changes attributable to changes in aggregation. The properties which have been investigated include oil/water interfacial tensions, dye solubilization, fluorescence depolarization, ultra-violet spectroscopy, and NMR spectroscopy.[55]

The principal driving force for the organic phase aggregation of surfactants comes from the dipoles of the surfactant molecules. There is a decrease in free energy when these dipoles interact with each other. Therefore, there is a tendency for the dipoles (i.e. the polar heads) to get together to form a central core shielded from the organic solvent by the hydrocarbon groups. Thus in contrast to aqueous micelles, organic phase micelles have their polar groups located away from the solvent, hence the term "reversed" or inverse micelles. The ultimate micelle size reflects a balance between the dipole-dipole interactions, and the entropy decrease accompanying the aggregation.

The major factors controlling aggregation are temperature, the nature of the surfactant (molar volume, geometry, dipole moment, concentration), and the nature of the solvent (molecular geometry, dipole moment, dielectric constant, solubility parameter). The solvent affects aggregation by modifying the dipole-dipole interactions of the surfactant molecules with solute-solvent interactions.

Electrostatics and Electrokinetics

When a polar organic phase is brought into contact with an aqueous phase with which it is immiscible, a potential difference is found to exist between the interior of the two liquids.[47-56] This potential difference is a result of two factors: (a) the unequal distribution of positive and negative ions between the two phases; and (b) the presence of an adsorbed film of dipolar or charged molecules at the organic/aqueous interface. The ions and adsorbed species may be present as impurities (e.g. organic acids in the organic phase, and ions such as NH_4^+, H^+, HCO_3^- in the aqueous phase) or they may be deliberately added. At the organic/aqueous interface, the net charge is of the same magnitude on the adjacent sides, but with different signs.

In general, in the absence of specific adsorption, dispersions of oil droplets in water tend to acquire a negative charge on their surfaces. To see how this charge arises, let us consider the electrochemical potential of an ion j in the organic phase (phase 1). Let ϕ_1 and ϕ_2 be the bulk phase potentials in the organic and aqueous phases respectively. Then,

$$\bar{\mu}_{j1} = \mu_{j1} + z_j e \phi_1 \qquad (16)$$

where $\mu_{j1}(=\mu_{j1}^{o} + RT\ln C_{j1})$ is the chemical potential and z_j is the charge (including sign) on the ion. At equilibrium the electrochemical potential should be the same in both phases:

$$\bar{\mu}_{j1} = \bar{\mu}_{j2} \tag{17}$$

For a 1-1 electrolyte of concentration C_1 and C_2 respectively in the two phases, the chemical potentials of the constituent ions may be written as:

$$\mu_{\pm k} = \mu^{\circ}_{\pm k} + RT\ln C_{\pm k} \tag{18}$$

where the plus and minus subscripts refer to the cations and anions respectively, and k = 1 or 2 depending on whether the organic or aqueous phase is under consideration.

At equilibrium, the electrochemical potential of each ion should be identical in both phases. That is,

$$\bar{\mu}_{\pm 1} = \bar{\mu}_{\pm 2} \tag{19}$$

Using Eqs. 16-19, it can be shown[47] that

$$2RT\ln(C_1/C_2) = (\mu^{\circ}_{+2} - \mu^{\circ}_{+1}) + (\mu^{\circ}_{-2} - \mu^{\circ}_{-1}) \tag{20}$$

and

$$2e(\phi_1 - \phi_2) = (\mu^{\circ}_{+2} - \mu^{\circ}_{+1}) - (\mu^{\circ}_{-2} - \mu^{\circ}_{-1}) \tag{21}$$

As a first approximation, the standard chemical potentials $\mu^{\circ}_{\pm k}$ can be estimated by the energy required to move an ion of radius $r_{\pm}$ from vacuum to the interior of the respective liquid of dielectric constant ε_k, which according to the Born equation is given by

$$\mu^{\circ}_{\pm k} = \frac{-e^2}{2r_{\pm}}(1 - 1/\varepsilon_k) \tag{22}$$

It follows from Eqs. 20-22 that

$$\ln(C_1/C_2) = \frac{e^2}{4kT} \cdot \left(\frac{r_- + r_+}{r_- r_+}\right)\left(\frac{\varepsilon_1 - \varepsilon_2}{\varepsilon_1 \varepsilon_2}\right) \tag{23}$$

and

$$\phi_1 - \phi_2 = \left(\frac{e}{4}\right)\left(\frac{r_- - r_+}{r_- r_+}\right)\left(\frac{\varepsilon_1 - \varepsilon_2}{\varepsilon_1 \varepsilon_2}\right) \tag{24}$$

According to Eq. 24, for $r_- > r_+$ and $\varepsilon_1 < \varepsilon_2$, the value of $(\phi_1 - \phi_2)$ is negative. Thus since anions are generally larger than cations, and the dielectric constants of organic phases tend to be lower than that of water, it should be expected that the organic phase would be negative relative to the aqueous phase. In the extreme case of nonpolar organic phases where there is negligible solubility of ions in the interior of the organic phase, the above analysis implies that anions would be closer to the organic/aqueous interface. In other words, the negative charge observed on nonpolar oil droplets in water is a result of the desorption of aqueous phase cations from the interfacial region.

In the presence of specific adsorption, the total potential difference $(\phi_1 - \phi_2)$ across the organic/aqueous interface can be treated as the sum of ψ, the distribution potential resulting from the unequal distribution of positive and negative ions between the two phases, and V, the Volta potential, due to the dipoles or charges on species adsorbed at the organic/aqueous interface. That is,

$$\phi_1 - \phi_2 = \psi + V \tag{25}$$

The change in Volta potential (ΔV) which is caused by the presence of an adsorbed layer at a previously clean interface provides a useful means of characterizing the electrostatics of the interfacial region. This potential, which is often referred to as the adsorption or interfacial potential has the general form,[47]

$$\Delta V = 4\pi \Gamma \mu_D + \psi_o \tag{26}$$

where ψ_o is the interfacial potential due to the presence of polar ionic groups on the adsorbed species, and μ_D represents the dipole moment of the polar groups. As discussed elsewhere in this book for the solid/water system,[69] the presence of a potential difference between the nonpolar oil/water interface gives rise to an electrical double layer and therefore electrophoretic mobility experiments can be used to determine zeta potentials in liquid/liquid systems.

Interfacial Viscosity

Interfacial viscosity changes can be related to changes in film structure in a monolayer and to variations in film-substrate interactions. Closer molecular packing, for instance, enhances film viscosity. With increased compression, the solid film state is approached and the viscosity rises and finally becomes nonNewtonian. Measurements of surface viscosity have been successfully used as a means of detecting phase transitions. Several techniques are available, including the canal or slit viscometer--a two dimensional analog of the conventional Ostwald viscometer,[46-48] and laser light-scattering.[70]

INTERFACIAL ACTIVITY OF SOLVENT EXTRACTION REAGENTS

General Considerations

The interfacial activity of an extractant is a function of the structure and properties of its hydrophobic and hydrophilic groups. In addition, the nature of the organic diluent and the aqueous phase is also important. To fully discuss the contributions of these various factors to extractant interfacial activity would require many more systematic studies than are presently available. However, the available data do provide some indication of the general trends. Table 3 presents a summary of the interfacial activity data for a variety of solvent extraction reagents. The data, covering a wide spectrum of reagent types, clearly show that solvent extraction reagents should be treated as surface active compounds.

Efficiency of Adsorption (C_1)

The value of C_1 (see Fig. 2) can be used as a measure of the efficiency of adsorption since it represents the minimum bulk phase concentration required to produce a saturated interface.[49] If C_1 is small (or $pC_1 = -\log C_1$ is large), it means that interfacial saturation occurs at a relatively low extractant concentration, thus indicating a strong interfacial activity. Examination of Table 3 reveals that the most interfacially active acidic extractant is HDNNS with a C_1 value of 3×10^{-6} kmol m^{-3}. At the other extreme are the commercial oxime reagents, whose C_1 values occur at about 10^{-4} kmol m^{-3} and 10^{-2} kmol m^{-3} respectively in hexane and toluene.

Effect of Hydrophobic Group. The efficiency of adsorption is dependent on the structure of the extractant and the nature of the aqueous phase. In general pC_1 increases with the molecular weight of a surfactant, a reflection of the well known linear relationship between the efficiency of adsorption and the number of carbon atoms present in a straight chain hydrophobic residue.[46-51] This chain length effect is due to the presence of attractive chain-chain van der Waals interactions which tend to stabilize the adsorbed molecules. The presence of a branched chain in the hydrophobic group, interferes with the approach of the chains and therefore leads to a decrease in the chain-chain interactions. The effect of alkyl chain structure on the efficiency of adsorption can be illustrated with the pC_1 data for HDOP, HDEHP, HDNOP and HD(DIBM)P (4.7, 3.6, 3.7, 1.6 respectively). The most highly branched of these alkyl phosphoric acids, HD(DIBM)P is the least interfacially active.[29]

Effect of Organic Diluent. If the extractant interacts strongly with diluent molecules, there will be competition between extractant-diluent interactions and extractant-aqueous solution interactions. If the extractant-diluent interactions are sufficiently strong, there would be a withdrawal of extractant molecules from the interfacial region i.e. a decrease in adsorption would occur. Figure 5 presents a comparison of the γ-logC behavior of purified anti LIX65N in hexane and toluene.[23] It can be seen that while with hexane a pC_1 value of about 4 is obtained, the corresponding value in toluene is much lower, at about 2. Figure 5 also shows that a similar, though less pronounced trend, is found in the case of LIX 63. These results suggest that toluene solvates the oxime more strongly than does hexane. Aromatic diluents, on account of the presence of the delocalized cloud of π electrons, are more polar than aliphatic diluents. Consequently, compared with hexane, there will be a smaller tendency for toluene to reject the polar groups of extractant molecules. A similar trend is found in the behavior of HDHOEP: at 25°C, pC_1 = 5.8 and 4.7 with dodecane and xylene respectively.[29]

Effect of Hydrophilic Group. Any factors which increase the aqueous phase reactivity will tend to enhance the interfacial activity. It can be seen from Table 3 that the pC_1 values increase in the order sulfonic acid > phosphoric acid > hydroxyoxime. This trend follows that of the acid strengths i.e. sulfonic acid > phosphoric acid > carboxylic acid.[71] The acidity of a compound is related to the relative stability of the acid and its anion. Any factors which stabilize the anion will increase the acid strength. For example, the acidity of carboxylic acids, i.e. the ability to deprotonate the -OH group comes from the fact that the anion is stabilized by the formation of a resonance hybrid; that is, the negative charge is smeared over the two oxygen atoms.[71] Thus, sulfonic acids are stronger acids than carboxylic acids because in this case, the negative charge is spread over three oxygen atoms instead

Table 3. Interfacial Parameters for Selected Solvent Extraction Reagents.

Reagent Class	Compound		M.W. g/mol	Concentration Range, M	Organic Diluent	Aqueous Phase
	Name (Abbr.)	Structure				
Acidic	Versatic 911	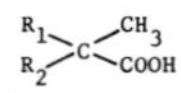		0–0.5	Toluene	pH 5.0 $[Na^+]$=0.0025M with Na_2SO_4
	Dodecanoic acid (Lauric Acid)	$[CH_3(CH_2)_{10}]COOH$	200	10^{-5}–10^{-1}	Hexane	pH 2.5 $[Na^+]$=0.04M with $NaNO_3$
	Dinonylnaphthalene Sulfonic Acid (HDNNS)	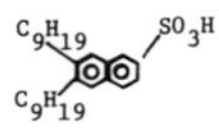	460	$3x10^{-8}$–$1.5x10^{-2}$	Hexane	pH 2.5 0.05M KNO_3
	Dinonylnaphthalene Sulfonic Acid (HDNNS)	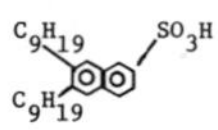	460	$2x10^{-8}$–$3x10^{-3}$	Toluene	0.102M $HClO_4$
	Di(n-Butyl) Phosphoric Acid (HDBP)	$(C_4H_9)_2HPO_4$	210	10^{-2} ~2	Toluene	10^{-1}M HCl
	Bis(2-Ethylhexyl) Phosphoric Acid (HDEHP)	CH_2CH_3 $[CH_3(CH_2)_3CHCH_2]_2HPO_4$	322	10^{-2} ~2	Toluene	10^{-1}M HCl
	*Bis(2-Ethylhexyl) Phosphoric Acid (HDEHP)	CH_2CH_3 $[CH_3(CH_2)_3CHCH_2]_2HPO_4$	322	10^{-7}–1	Dodecane	10^{-3}M HNO_3
	**Bis(2-Ethylhexyl) Phosphoric Acid (HDEHP)	CH_2CH_3 $[CH_3(CH_2)_3CHCH_2]_2HPO_4$	322	10^{-8}–1	o-Xylene	10^{-3}M HNO_3
	*Bis(2-Ethylhexyl) Phosphoric Acid (HDEHP)	CH_2CH_3 $[CH_3(CH_2)_3CHCH_2]_2HPO_4$	322	10^{-8}–10^{-1}	Dodecane	1M HNO_3
	*Di(n-Octyl) Phosphoric Acid (HDOP)	$[CH_3(CH_2)_7]_2HPO_4$	322	10^{-10}–1	Dodecane	10^{-3}M HNO_3
	*Di(n-Octyl) Phosphoric Acid (HDOP)	$[CH_3(CH_2)_7]_2HPO_4$	322	10^{-8}–10^{-1}	Dodecane	1M HNO_3
	*Bis(2-Ethylhexyl) Phosphinic Acid (H[DEHP])	CH_2CH_3 $[CH_3(CH_2)_3CHCH_2)_2HPO_2$	290	10^{-8}–10^{-1}	Dodecane	1M HNO_3
	*Bis(2-Ethylhexyl) Phosphonic Acid (HEH[EHP])	CH_2CH_3 $[CH_3(CH_2)_3CHCH_2)_2HPO_3$	306	10^{-8}–$3x10^{-1}$	Dodecane	1M HNO_3
	*Bis(hexoxyethyl) Phosphoric Acid (HDHOEP)	$(C_6H_{13}$-O-$C_2H_4)_2HPO_4$	354	10^{-9}–10^{-1}	Dodecane	1M HNO_3
	*Bis(hexoxyethyl) Phosphoric Acid (HDHOEP)	$(C_6H_{13}$-O-$C_2H_4)_2HPO_4$	354	10^{-9}–10^{-1}	o-Xylene	1M HNO_3
	Bis(diisobutylmethyl) Phosphoric Acid HD(DIBM)P	$(C_9H_{19})_2HPO_4$	350		Dodecane	1M HNO_3
	Bis(2,2-dimethylhexyl) Phosphoric Acid (HDNOP)	$(C_8H_{17})_2HPO_4$	322		Dodecane	1M HNO_3
	2-Hydroxy-5-Nonyl-Benzophenone Oxime (LIX65N)	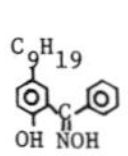	339	10^{-5}–10^{-1}	Hexane	pH 5.0 0.025M Na_2SO_4
	2-Hydroxy-5-Nonyl-Benzophenone Oxime (LIX65N)	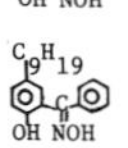	339	10^{-5}–$6.4x10^{-2}$	Toluene	pH 5.0 0.025M Na_2SO_4
	2-Hydroxy-5-Nonyl-Benzophenone Oxime (LIX65N)	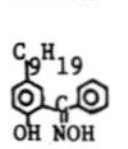	339	0–10^{-1}	Toluene	I=0.5M (0.167M Na_2SO_4) pH 2.0
	2-Hydroxy-5-Nonyl Benzophenone Oxime (LIX65N)	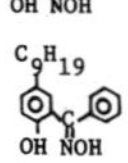	339	0–10^{-1}	Hexane	I=0.5M (0.167M Na_2SO_4) pH 2.0

** Interface never saturated; +For dimeric areas; N.G., not given
* Aggregation occurs at high solute concentrations

Table 3 (continued)

Method	Temperature (°C)	log C_1	log C_2	$A_\circ$ ($\mathring{A}^2$/molecule)	$\Gamma_\circ$ (molecule/cm^2)	References
Du Nouy Ring	N.G.			~115	~8.7 x 10^{13}	(18)
Du Nouy Ring	N.G.					(16)
Du Nouy Ring	25	-6.5	-5.3	38	26.3x10^{13}	(35)
Wilhelmy Plate	25			40	25x10^{13}	(5)
Du Nouy Ring	N.G.			158$^+$	6.3x10^{13+}	(94)
Du Nouy Ring	N.G.			213$^+$	4.7x10^{13+}	(94)
Du Nouy Ring	25	-2.1		88	11.4x10^{13}	(28)
Du Nouy Ring	25					(28)
Du Nouy Ring	25	-3.6		86	11.6x10^{13}	(29)
Du Nouy Ring	25	-3.1		60	16.7x10^{13}	(28)
Du Nouy Ring	25	-4.7		78	12.8x10^{13}	(29)
Du Nouy Ring	25	-5.1		54	18.5x10^{13}	(29)
Du Nouy Ring	25	-3.6		87	11.5x10^{13}	(29)
Du Nouy Ring	25	-5.8		46	21.7x10^{13}	(29)
Du Nouy Ring	25	-4.7		100.5	10.0x10^{13}	(29)
Du Nouy Ring	25	-1.6		80	12.5x10^{13}	(29)
Du Nouy Ring	25	-3.7		100	10x10^{13}	(29)
N.G.	N.G.					(17)
N.G.	N.G.					(17)
Drop Weight	28			136	7.4x10^{13}	(23)
Drop Weight	28			85	11.8x10^{13}	(23)

(Continued)

Table 3 (continued)

Reagent Class	Compound: Name (Abbr.)	Compound: Structure	M.W. g/mol	Concentration Range, M	Organic Diluent	Aqueous Phase
Acidic	2-Hydroxy-5-Nonyl-Benzophenone Oxime (LIX65N)	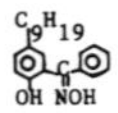	339	$6.3x10^{-5}$-$7x10^{-1}$	Hexane	pH 5.0
	5,8-Diethyl-7-Hydroxy-6-Dodecanone Oxime (LIX63)	C_2H_5 C_2H_5 $CH_3(CH_2)_3CHCCHCH(CH_2)_3CH_3$ HON OH	272	0-10^{-1}	Toluene	I=0.5M (0.167M Na_2SO_4) pH 2.0
	5,8-Diethyl-7-Hydroxy-6-Dodecanone Oxime (LIX63)	C_2H_5 C_2H_5 $CH_3(CH_2)_3CHCCHCH(CH_2)_3CH_3$ HON OH	272	0-10^{-1}	Hexane	I=0.5M (0.167M Na_2SO_4) pH 2.0
	5,8-Diethyl-7-Hydroxy-6-Dodecanone Oxime (LIX63)	C_2H_5 C_2H_5 $CH_3(CH_2)_3CHCCHCH(CH_2)_3CH_3$ HON OH	272	$8.4x10^{-5}$-10^{-1}	Hexane	pH 5.0
	5,8-Diethyl-7-Hydroxy-6-Dodecanone Oxime (LIX63)	C_2H_5 C_2H_5 $CH_3(CH_2)_3CHCCHCH(CH_2)_3CH_3$ HON OH	272	$5.7x10^{-6}$-$8x10^{-2}$	Hexane	pH 2.5 0.5M KNO_3
Basic	Didecylamine	$[CH_3(CH_2)_9]_2NH$	297	10^{-7}-$6x10^{-3}$	Benzene	$5x10^{-3}$M H_2SO_4
	*Tetraheptylammonium Chloride	$[CH_3(CH_2)_6]_4N^+Cl^-$	445.5	$5x10^{-7}$-$8x10^{-3}$	Benzene	1M LiCl
	*Tetraheptylammonium Nitrate	$[CH_3(CH_2)_6]_4N^+NO_3^-$	472	$4x10^{-6}$-$8x10^{-3}$	Benzene	1M $LiNO_3$
	*Tetraheptylammonium Zinc Tetrachloride	$[(CH_3(CH_2)_6)_4N^+]_2ZnCl_4^{2-}$	1027	$7x17^{-7}$-$8x10^{-3}$	Benzene	0.4M $ZnCl_2$ + 0.2M HCl
	*Tridodecylamine (Trilaurylamine; TLA)	$[CH_3(CH_2)_{11}]_3N$	521	10^{-7}-$3.8x10^{-1}$	Toluene	1M NaCl
	*Tridodecylammonium Chloride (Trilaurylammonium Chloride; TLAHCL)	$[CH_3(CH_2)_{11}]_3\overset{+}{N}H(\bar{Cl})$	557.5	10^{-7}-$9x10^{-1}$	Toluene	1M HCl
	*Tridodecylammonium Nitrate (Trilaurylammonium Nitrate) ($TLAHNO_3$)	$[CH_3(CH_2)_{11}]_3\overset{+}{N}H(\overset{-}{NO_3})$	584	$4x10^{-7}$-$2x10^{-1}$	o-Xylene	4M HNO_3
	Tridodecylmethylammonium Chloride (TDMAC)	$[CH_3(CH_2)_{11}]_3\overset{+}{N}CH_3(\bar{Cl})$	571.5	10^{-8}-10^{-2}	o-Xylene	2M HCl
	Trioctylmethylammonium Chloride (TOMAC)	$[CH_3(CH_2)_7]_3\overset{+}{N}CH_3(\bar{Cl})$	403.5	10^{-7}-$3x10^{-4}$	o-Xylene	2M HCl
	*Trioctylmethylammonium Nitrate (TOMAN)	$[CH_3(CH_2)_7]_3\overset{+}{N}CH_3(\overset{-}{NO_3})$	430	10^{-9}-10^{-1}	o-Xylene	2M HNO_3
	*Tridodecylmethylammonium Nitrate (TDMAN)	$[CH_3(CH_2)_{11}]_3\overset{+}{N}CH_3(\overset{-}{NO_3})$	598	10^{-9}-10^{-1}	o-Xylene	2M HNO_3
	*Tridodecylmethylammonium Acetate (TDMAAC)	$[CH_3(CH_2)_{11}]_3\overset{+}{N}CH_3(CH_3COO^-)$	595	10^{-8}-10^{-1}	o-Xylene	2M CH_3COOH
	Tridodecylmethylammonium Perchlorate (TDMAP)	$[CH_3(CH_2)_{11}]_3\overset{+}{N}CH_3(ClO_4^-)$	635.5	10^{-9}-$5.6x10^{-3}$	o-Xylene	2M $HClO_4$
	Aliquat 336	$(R_3\overset{+}{N}\ CH_3)Cl^-$ R = C_8 or C_{10}	~425	10^{-9}-10^{-3}	o-Xylene	1M HCl
Neutral	Nonylphenol	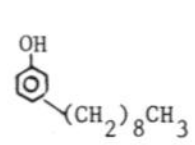	220	0-10^{-1}	Hexane	I=0.5M (0.167M Na_2SO_4) pH 2.0
	Nonylphenol	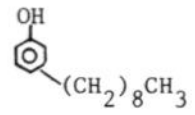	220	0-$5x10^{-1}$	Toluene	I=0.5M (0.167M Na_2SO_4) pH 2.0

*Aggregation occurs at high solute concentrations

Table 3 (continued)

Method	Temperature (°C)	log C_1	log C_2	$A_\circ$ ($\AA^2$/molecule)	$\Gamma_\circ$ (molecule/cm^2)	References
Du Nouy Ring	Room Temp.					(22)
Drop Weight	28	~-2.0		176	5.7×10^{13}	(23)
Drop Weight	28	~-3.7		101	9.9×10^{13}	(23)
Du Nouy Ring	Room Temp.					(22)
Du Nouy Ring	25	-4.7		108	9.2×10^{13}	(35)
Du Nouy Ring	N.G.	-5.2	-4.0	78	12.8×10^{13}	
Wilhelmy Plate	N.G.	-4.5		48	20.8×10^{13}	(4)
Wilhelmy Plate	N.G.	-4.1		41.2	24.3×10^{13}	(4)
Wilhelmy Plate	N.G.			98.0	10.2×10^{13}	(4)
Wilhelmy Plate	25	-2.0		133	7.52×10^{13}	(6)
Wilhelmy Plate	25	-3.7		180	5.5×10^{13}	(6)
Wilhelmy Plate	25	-3.3,	~-0.63	322	3.1×10^{13}	(6)
Du Nouy Ring	25	-6.4	~-4.5	51	19.6×10^{13}	(30)
Du Nouy Ring	25	-6.4	~-4.5	48	20.8×10^{13}	(30)
Du Nouy Ring	25	-6.0	~1.5	223	4.5×10^{13}	(30)
Du Nouy Ring	25	-5.3		225	4.4×10^{13}	(30)
Du Nouy Ring	25	-4.7		257	3.9×10^{13}	(30)
Du Nouy Ring	25	-5.00		422	2.4×10^{13}	(30)
Du Nouy Ring	25	-6.0	~-4.0	60	16.7×10^{13}	(28)
Drop Weight	28	~-2.7		73	13.7×10^{13}	(23)
Drop Weight	28	~-1.3		106	9.4×10^{13}	(23)

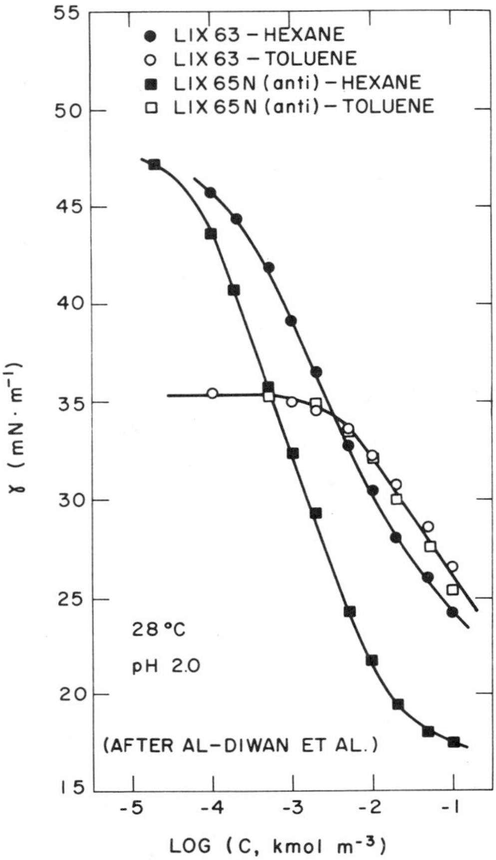

Fig. 5. Effect of purified LIX65N and LIX63 on the interfacial tension at the hexane/aqueous and toluene/aqueous interfaces.[23]

of two. The presence of a relationship between acid strength and interfacial activity has been demonstrated by Vandegrift and Horwitz[29] who obtained a linear plot of pC_1 versus pKa (acidity constants determined in 75% ethyl alcohol) for a series of dialkylphosphoric acids (see Table 3). The most acidic reagent HDHOEP (pKa = 2.71) is also the most interfacially active (pC_1 = 5.8). On the other hand, the least acidic reagent, HD(DIBM)P (pKa = 4.74), has the least interfacial activity (pC_1 = 1.6). The difference in the acidity of the two extreme reagents can be explained in terms of the effect of the structure of the hydrophobic group on the polarity of the hydrophilic group. The presence of oxygen, a strongly electronegative atom, in the hexoxyethyl group of HDHOEP increases the electron withdrawal ability of the hydrocarbon chain, and thereby increases the tendency of the molecule to release the proton of the OH group. On the other hand, in the case of the highly branched HD(DIBM)P, the electron releasing alkyl groups are brought much closer to the polar group and thereby have a greater tendency to destabilize the anions which may result from deprotonation.

Factors which decrease organic phase reactivity also enhance interfacial activity. According to Table 3 the phosphoric and phosphonic acids [HDEHP, HEH(EHP)] have the same pC_1 value (i.e. 2.6) compared with a pC_1 value of 5.1 for the phosphinic acid H(DEHP).[29] According to the electronic theory of acidity[71], the acidity should be expected to increase in the order, phosphoric acid > phosphonic acid > phosphinic acid since the number of oxygen atoms available to share the negative charge on the anions is highest for the phosphoric acid: 4, followed by the phosphonic acid: 3, and the phosphinic acid: 2. In fact, the pKa values of 3.49, 4.8, and 5.88 respectively for HDEHP, HEH(EHP) and H(DEHP) are consistent with the expected trend.[29] However, contrary to the correlation suggested above between acid strength and interfacial activity, the weakest acid is now found to give the highest pC_1 value. Vandegrift and Horwitz[29] have explained these results in terms of steric hindrance effects. According to these investigators, molecular models of HDEHP, HEH(EHP) and H(DEHP) indicate that as C-O-P bonds are replaced with C-P bonds, there is a tendency for the ethyl side chains to crowd the oxygens on phosphorous. Since the oxygens are needed for hydrogen bonding during aggregation, the net result is a decrease in the tendency towards aggregation on going from C-O-P to C-P bonds. Vandegrift and Horwitz provide interfacial tension data which show that the dimerization of H(DEHP) does not occur until the concentration exceeds 3×10^{-4} kmol m^{-3}. Since aggregation results in the formation of polymeric species whose outer surface is hydrophobic, it would be expected that monomers would be the most interfacially active extractant species. Thus factors which inhibit aggregation are likely to enhance interfacial activity.

Effect of Aqueous Species. The efficiency of adsorption can also be affected by counterions. The simplest case is found in the effect of pH on interfacial tension. Figure 6, based on the work of Flett et al.[16-18], illustrates the pH dependency of the interfacial tension for a number of acidic reagents. The results

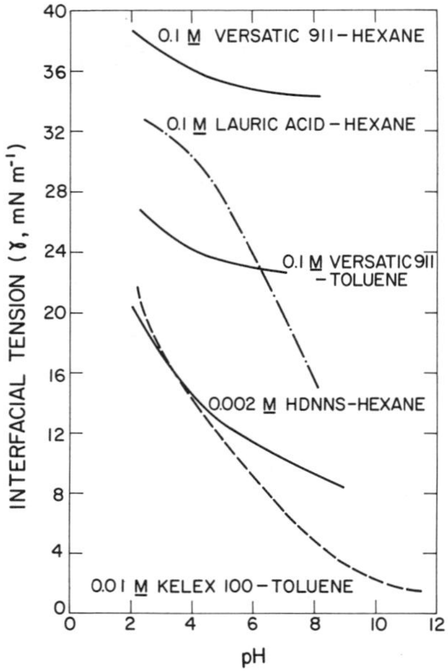

Fig. 6. Effect of pH on the interfacial tension of selected solvent extraction systems.[16-18]

indicate a decrease in interfacial tension with increase in pH. According to the work of Al-Diwan et al.[23], the same general trend is followed by commercial hydroxyoxime reagents, with the exception that at very low pH, the interfacial tension goes through a maximum and begins to decrease as the pH is lowered further. See Fig. 7.

This pH effect on interfacial tension is well known in the surface chemistry literature and has been attributed to the ionization of the polar groups of the surfactant molecules. Figure 8 shows the effect of pH on the interfacial tension at the benzene/water interface for hexadecanoic acid (palmitic acid) and hexadecylamine.[72] The reagents have the same chain length (C_{16}) but differ in the nature of the polar head group, $-COOH$ and $-NH_2$ respectively. The drop in γ at pH5 and pH9 respectively for palmitic acid and hexadecylamine signify the onset of ionization:

$$-COOH = -COO^- + H^+ \qquad K_a \qquad (27)$$

$$-NH_2 + H^+ = -NH_3^+ \qquad K_b \qquad (28)$$

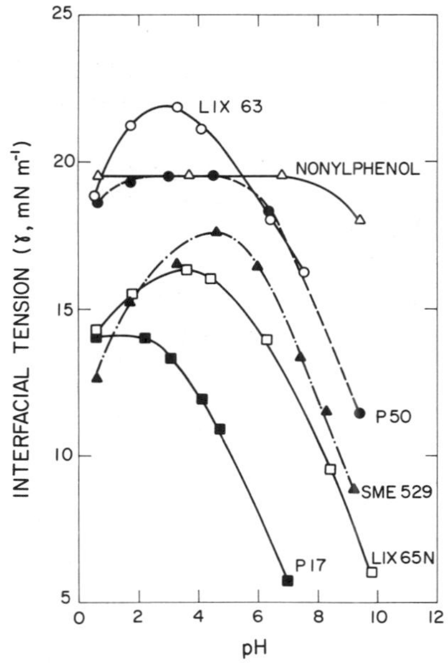

Fig. 7. Effect of pH on the interfacial tension at the Escaid 100/aqueous (0.167 kmol m^{-3} Na_2SO_4) interface in the presence of various unpurified reagents: nonyl phenol (~0.42 kmol m^{-3}), LIX63 (0.05 kmol m^{-3}), LIX65N, P17, P50, SME 529 (~0.18 kmol m^{-3}) 28°C.[23]

Ionization increases the polarity of the hydrophilic groups. The increased polarity makes the surfactant species less compatible with the nonpolar organic phase and therefore leads to increased adsorption. As can be seen from Eqs. 27 and 28 the ionization of the carboxylic acid and the amine occurs in opposite directions, in agreement with Fig. 8 which shows a decrease in γ at high and low pH respectively for palmitic acid and hexadecylamine. If the data in Fig. 8 are considered as titration curves, then the inflection point (pH7) in the hexadecylamine data may be taken as representing the pH at which the ionic and neutral amine species occur at equal concentrations. Thus, a value of 7.0 is obtained for pK_b. Similarly pK_a can be estimated to be about 7.5 for carboxylic acids at the benzene/aqueous interface.[72] In general, it is found that the pH at which a long chain carboxylic acid is half ionized at the benzene/water interface is about 3 pH units above that observed for the corresponding ionization in bulk aqueous solution. Similarly, amines ionize at the benzene/water interface at 3-4 pH units below that observed in the bulk solution.

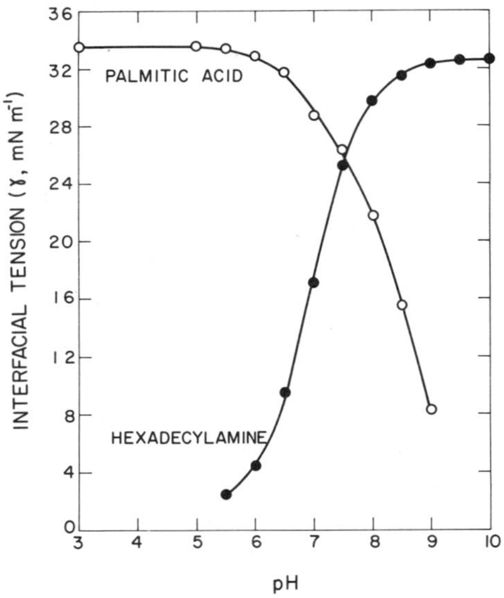

Fig. 8. Effect of pH on the interfacial tension at the benzene/water interface in the presence of hexadecanoic (palmitic) acid and hexadecylamine.[72]

The apparent shift in the pH at which ionization occurs can be viewed as a reflection of the fact that the interfacial pH differs from the bulk pH. In other words, the interfacial pK value is the same as the bulk value, it is rather the H^+ concentrations that differ. The hydrogen ion concentration in the interfacial region ($C_{H,i}$) is related to the bulk concentration (C_H) by Eq. 29:

$$C_{H,i} = C_H \exp{-e\psi/kT} \tag{29}$$

Thus,

$$(pH)_i = pH + e\psi/2.3\ kT \tag{30}$$

and

$$(pH)_i > pH \text{ if } \psi > 0 \tag{31}$$

$$< pH \text{ if } \psi < 0 \tag{32}$$

The effects of aqueous species become more complicated when counter ions other than the proton are involved. Pizzichini et al.[6] studied the interfacial tension behavior of the systems trilaurylamine (TLA) - toluene/1 kmol m^{-3} $NaCl-H_2O$. It was found that the C_1 values were 10^{-2}, 3.7×10^{-4} and 8×10^{-4} kmol m^{-3} respectively for TLA, TLAC and TLAN. In their study of quaternary ammonium salts, Vandegrift et al.[30] obtained the following values for C_1: 4.5×10^{-7} kmol m^{-3} for tridodecylmethylammonium nitrate (TDMAN), 1×10^{-5} kmol m^{-3} for tridodecylmethylammonium perchlorate (TDMAP), and 2×10^{-5} kmol m^{-3} for tridodecylmethylammonium acetate (TDMAAc). The results of Pizzichini et al.[6] and Vandegrift et al[30] suggest that C_1 increases in the order $Cl < NO_3 < ClO_4 < CH_3CO_2$.

Effectiveness of Adsorption (A_o)

The adsorption density of a saturated interface, Γ_o, is the highest interfacial concentration attainable by an extractant. The value of Γ_o therefore provides a convenient means of ascertaining the effectiveness of adsorption at the liquid/liquid interface.[49] As previously discussed (see Eq. 8) the interfacial area, A_o, is related to Γ_o ($A_o = 1/\Gamma_o$). Values of Γ_o and A_o for several types of extractants are included in Table 3. The A_o data presented in this table were derived indirectly by application of the Gibbs equation (Eq. 8). Table 4 presents interfacial area derived by Hunt using spread monolayer techniques.[59] Although Hunt's experiments were conducted at the air/water interface, they are nevertheless of interest since they represent the first deliberate attempt to apply the spread monolayer technique to solvent extraction research. The A_o values in Table 3 range from as low as 38 $Å^2$/molecule for HDNNS to as high as 422$Å^2$ for TDMAAc, although it must be noted that in the absence of aggregation data, it is not clear whether the TDMAAc interfacial area refers to a monomer or polymer basis (see Eq. 14).[30] As was found above in the case of C_1, the variations in A_o are also attributable to physicochemical factors related to the structure and properties of the organic and aqueous phase species.

A comparison of the experimental A_o data with theoretical values derived from molecular models gives useful information concerning the orientation of the interfacial molecules. When an aliphatic chain is oriented perpendicular to the interface it gives a cross-sectional area of about 10$Å^2$; the corresponding value for a benzene ring is 25$Å^2$. In the case of a $-CH_2-$ group oriented parallel to the interface, the cross-sectional area is about 7 $Å^2$.[49] Molecular models give cross-sectional area of 16 $Å^2$ and 24$Å^2$ respectively for the carboxylate and sulfonate anions.[73]

Table 4. Effect of Multicharged Cations on the Limiting Areas of Dialkyl Phosphate Monolayers.[59]

Cation	Limiting area (Å^2/molecule) Dimyristyl phosphate	Dicetyl phosphate	Distearyl phosphate
H^+/K^+	42.5	41.2	41.0
La^{III}	42.0	40.5	41.4
Ce^{III}	42.6		
Pr^{III}	41.6		
Nd^{III}	43.9		
Sm^{III}	43.9		
Gd^{III}	43.5	39.9	
Tb^{III}	34.3	33.0	43.0
Dy^{III}	24.7	28.7	42.1
Ho^{III}			35.0
Er^{III}	25.7	28.1	35.2
Yb^{III}		27.4	30.6
Y^{III}		30.8	36.7
Se^{III}		45.0	

Effect of Hydrophobic Group. The effect of alkyl chain structure on interfacial activity can be illustrated by comparing the data for C_8 straight chain versus branched chain phosphoric acids: HDOP, HDEHP, HDNOP. The presence of branched chains tends to increase interfacial area/molecule. Thus HDNOP being the most branched, has the greatest A_o value. According to Vandegrift and Horwitz[29], molecular models for HDEHP give A_o values of 58Å^2 and 103Å^2 respectively depending on whether the alkyl groups are arranged perpendicular or parallel to the interface. The average of these two theoretical A_o values is 81Å^2 which when compared with the experimental value of 91Å^2/monomer suggests that in the HDEHP - dodecane/1 kmol m^{-3} HNO_3 system, the orientation of the alkyl groups is between the perpendicular and parallel modes.

Another example of the effect of the hydrophobic group is provided by the different monolayer properties of 2-(n-tetradecyl)-naphthalene sulfonate (TDNS), 6-(n-tetradecyl)-tetralin sulfonate (TDTS), and nonadecylbenzene-p-sulfonate (NDBS) at the air/water interface.[60,61] As can be seen from Fig. 9, TDNS and TDTS give an

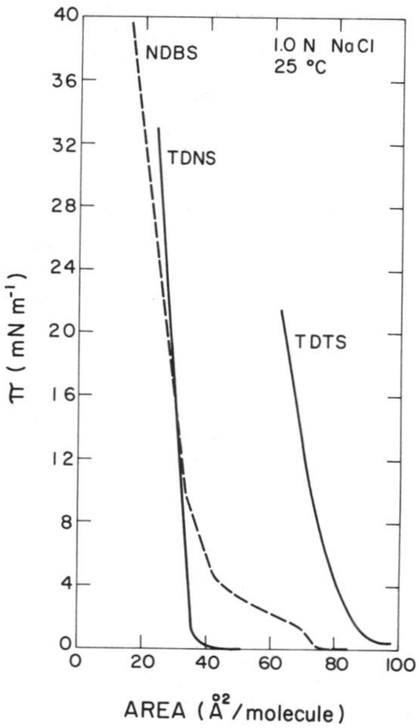

Fig. 9. π-A isotherms of TDNS, TDTS, and NDBS at 25°C on 1.0 kmol m^{-3} NaCl subsolutions.[60,61]

incompressible monolayer and liquid condensed monolayer respectively. The NDBS monolayer shows an intermediate behavior, giving a liquid condensed film at high areas and a solid phase at low areas. The difference in behavior of these three sulfonates can be attributed to the difference in the hydrophobic interactions of the naphathalene, tetralin, and benzene rings. The bonds in the naphathalene ring are more saturated than those in the tetraline ring and therefore the naphathalene compound possesses a greater concentration of delocalized electrons which promote greater interfacial activity. The TDNS molecule readily arranges itself at the interface with the napthalene rings oriented vertically. On the other hand, apparently the tetradecyl chain presents a steric hindrance to close packing of TDTS. In the case of NDBS, the transition from a liquid to a solid film is probably caused by the change from a parallel orientation of the benzene ring to a vertical arrangement.[60,61]

Effect of Hydrophilic Group. In general, interfacial areas decrease in the order $R\text{-}SO_3^- \cong R\text{-}SO_4^- > R\text{-}COO^-$. For example, van Voorst Vader[73] obtained the following A_o values at the

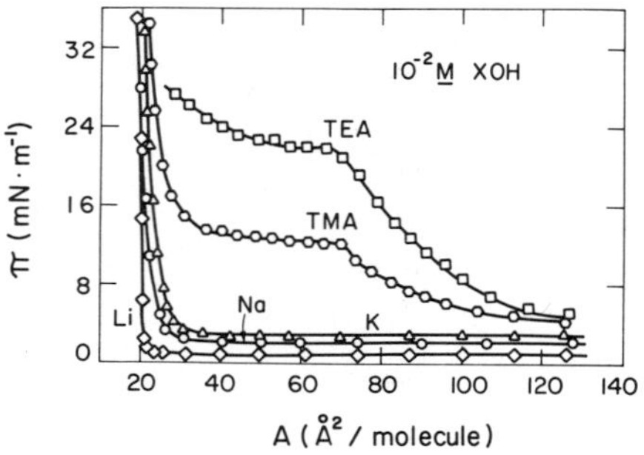

Fig. 10. Effect of cations (10^{-2} kmol m^{-3}) on the π-A isotherms of docosanoic (behenic) acid at 26°C.[62]

heptane/ water interface: 53Å^2 for sodium dodecyl sulfate, 54Å^2 for sodium undecylsulfonate, and 45Å^2 for sodium dodecanoate (laurate). Apparently, the cross section area of the hydrated polar group determines the interfacial area per molecule. Thus the sulfonate and sulfate groups are more hydrated than the carboxylate group.

Effect of Aqueous Phase Species. The ability of the polar groups of monolayers to interact with aqueous phase species can cause pronounced variations in the observed interfacial areas. For example, in carboxylate systems, the interfacial area has been found to increase in the order of increasing size of the unhydrated ion, i.e. Li < Na < K < TMA as illustrated in Fig. 10.[62] On the other hand, as shown in Fig. 11, in the case of sulfate monolayers, the order is reversed, i.e. A_o increases with increasing size of the hydrated ion: K < Na < Li < TMA.[63] Similarly the size of the hydrated ion seems to control the interaction of anions with the $-(CH_3)_3N^+$ polar group. As shown in Fig. 12, Goddard et al.[63] found the following order for A_o: SCN < NO_3 < Cl < F. According to Hunt[59], the interaction of alkyl phosphate monolayers with rare-earth ions increases with the charge density of the unhydrolyzed ions.

Monolayer-ion interactions are controlled by (a) ion-water interactions and (b) polar head-group interactions. The first kind of interaction, termed water structure-enforced ion pairing by Diamond[74] refers to the fact that large ions have a lesser tendency to become hydrated, i.e. they are relatively hydrophobic. Thus, large ions disturb the water structure and are therefore rejected

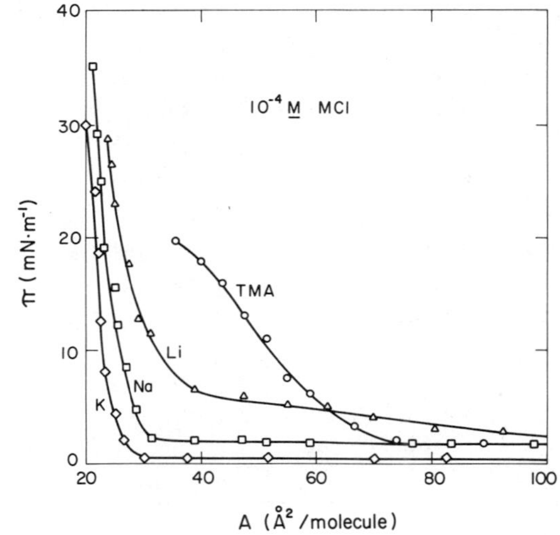

Fig. 11. Effect of cations (10^{-4} kmol m^{-3}) on the π-A isotherms of sodium docosyl sulfate at 25°C.[63]

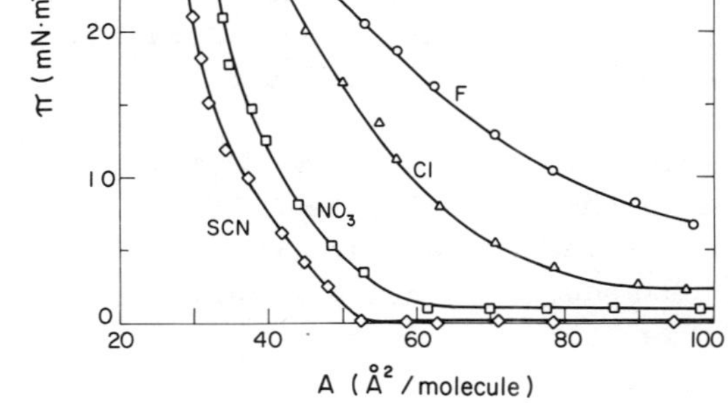

Fig. 12. Effect of anions (10^{-2} kmol m^{-3}) on the π-A isotherms of docosyltrimethyl ammonium bromide at 25°C.[63]

from the aqueous phase and subsequently interact with the interfacial hydrophilic groups. According to this scenario, large ions should promote the adsorption of surfactants. The second type of interaction relates to the fact that the polar groups in a charged monolayer experience mutual repulsion. However, the penetration of counterions neutralizes the interfacial charge and thereby permits further adsorption. Since small highly charged ions favor electrostatic interactions, it follows that small ions would promote adsorption. The net effect of counterions on the observed interfacial area in a given system is a result of the contributions from these two opposing tendencies.

INTERFACIAL ACTIVITY AND METAL EXTRACTION

Interfacial Reactions in Hydroxyoxime Systems

The first attempt to develop a reaction mechanism for copper extraction with LIX65N(HR) was presented in 1973 by Flett et al.[75] who proposed that the rate-controlling step is an interfacial reaction involving the formation of the 1:2 complex CuR_2 (Eq. 35):

$$HR(o) = HR(i) \qquad \text{fast} \qquad (33)$$

$$Cu^{2+}(a) + HR(i) = CuR^{+}(a) + H^{+}(a) \qquad \text{fast} \qquad (34)$$

$$CuR^{+}(a) + HR(i) = CuR_2(i) + H^{+}(a) \qquad \text{slow} \qquad (35)$$

$$CuR_2(i) = CuR_2(o) \qquad \text{fast} \qquad (36)$$

Theoretically, Eq. 35 should give a zero or second order dependency on R depending on whether the organic/aqueous interface is saturated or unsaturated with LIX65N molecules. The actual order observed was lower, about 1.0. Flett et al. sought to explain this by postulating that the liquid/liquid interface was saturated with LIX65N molecules which, therefore, gave HR(i) in Eq. 34 a zeroth order contribution to the overall rate. This argument is however not satisfactory since the presence of surface saturation would mean that HR(i) in Eq. 35 should also give a zeroth order contribution. Thus the overall order of R should be zero instead of the observed value of unity.

The effect of LIX63 on copper extraction with LIX65N, was also investigated by Flett et al.[75] According to these authors, the role of LIX63 (HL) is to accelerate the formation of the CuR_2 complex through the substitution of Eq. 35 with a more rapid formation of a transitional mixed Cu-LIX65N-LIX63 complex:

$HL(o) = HL(i)$	fast	(37)
$CuR^+(a) + HL(i) = CuRL(i) + H^+(a)$	slow	(38)
$CuRL(i) = CuRL(o)$	fast	(39)
$CuRL(o) + HR(o) = CuR_2(o) + HL(o)$	fast	(40)

In this scheme, the rate controlling step was considered to be Eq. 38. According to this mechanism, the reaction should be zero or first order with respect to LIX63 depending on whether or not the surface is saturated with this compound. The observed fractional order of 1/2 was attributed to partial coverage of the organic/aqueous interface with LIX63 molecules. Noting that Flett et al. do not provide any reasons to justify the initial formation of a Cu-LIX65N complex (Eq. 34), Ashbrook[76] has proposed an alternative which involves the initial formation of a Cu-LIX63 complex:

$Cu^{2+}(a) + HL(i) = CuL^+(i) + H^+(a)$	fast	(41)
$CuL^+(i) + HR(i) = CuRL(i) + H^+(a)$	slow	(42)
$CuRL(i) + HR(i) = CuR_2(i) + HL(o)$	fast	(43)
$CuR_2(i) = CuR_2(o)$	fast	(44)

He argues that Eq. 41 is more likely than Eq. 34, since the hydroxyl proton of LIX63 is labile and α-acyloinoximes are capable of forming 1:1 complexes with Cu(II).

Atwood et al.[77] also proposed an interfacial reaction mechanism with Eq. 47 as the rate determining step.

$HR(i) + HL(i) = H_2L^+(i) + R^-(i)$	fast	(45)
$R^-(i) + Cu^{2+}(a) = CuR^+(i)$	slow	(46)
$CuR^+(i) + HR(i) = CuR_2(o) + H^+(a)$	fast	(47)
$CuR^+(i) + R^-(i) = CuR_2(o)$	fast	(48)
$H_2L^+(i) = HL(i) + H^+(a)$	fast	(49)

According to this mechanism, the role of LIX63 is that of a proton acceptor. The authors justify this protonation by referring to the fact that protonation is the first step in the Beckmann rearrangement reactions of oximes. They do not explain however, why LIX63 should be more easily protonated compared with LIX65N. Nevertheless, a protonation mechanism has also been proposed by Whewell et al.[78]

Interfacial Tension Studies. The mixed ligand models as proposed by Flett et al.[75] and Ashbrook[76], and the protonated oxime model of Atwood et al.[77] all suggest that interfacially adsorbed LIX65N and LIX63 species play key roles in the reaction mechanism. These models were formulatd at a time when there were no data on the interfacial activity of the oxime-type chelating extractants. However interfacial tension data now available in the literature permit one to evaluate the relative interfacial activities of LIX65N and LIX63. Figure 5, taken from Al-Diwan et al.[23] shows that in hexane, LIX65N is more interfacially active than LIX63. Similar results have also been presented by Flett et al.[79] and Fleming[22] which show that in mixed solutions of LIX63 and LIX65N, the interfacial tension is controlled by LIX65N. On the basis of these results, one might conclude that models for LIX63-LIX65N systems which rely on interfacially adsorbed LIX63 molecules are suspect. However, further examination of Fig. 5 reveals that contrary to the trend found with hexane, in the case of toluene solutions, the γ versus logC curves for LIX65N and LIX63 coincide, suggesting equal interfacial activity. Moreover, earlier, Flett et al.[17] presented interfacial tension data according to which in the pH range 3-10, the hexane/water interfacial tension decreased by about 2 mN m^{-1} when a 10^{-3} kmol m^{-3} LIX65N - hexane solution was replaced with a 10^{-3} kmol m^{-3} LIX65N-10^{-4} kmol m^{-3} LIX63 - hexane solution. It must be noted that the LIX63/LIX65N ratio used in the later study[79] (2×10^{-3} kmol m^{-3} LIX63/10^{-1} kmol m^{-3} LIX65N) was much less than that used in the earlier work[17] (10^{-4} kmol m^{-3} LIX63/10^{-3} kmol m^{-3} LIX63). It would be worthwhile to repeat these experiments over a much wider range of mixture compositions. If the results of ref. 17 can be substantiated, it would mean that there might be some interaction (e.g. the protonation reaction proposed by Atwood et al.[77], see Eq. 45) between LIX63 and LIX65N which yields an interfacial species which is more surface active than the individual extractants.

The above discussion demonstrates that before meaningful interfacial mechanisms can be proposed, it is imperative that the interfacial speciation be well characterized. Fleming[22] included interfacial tension and zeta potential measurements in his extraction mechanism studies with LIX63 and LIX65N. He noted that even though HR (=LIX65N) was more interfacially active than CuR_2, the interfacial tension obtained in the presence of CuR_2 and HR were lower than that in the presence of only HR. On the basis of these results, Fleming has proposed that the slow rate of metal extraction with LIX65N alone is due to the formation of a highly interfacially active species $2CuR_2.HR$. When this species occupies the liquid/liquid interface at the expense of HR molecules, the reaction rate as given by Eq. 35 is inhibited. The role of LIX63 then is to prevent the formation of the $2CuR_2.HR$ complex. Fleming bases this conclusion on the fact that addition of a small amount of LIX63 (~5 volume %) to the CuR_2 + HR system gives interfacial tension results which are essentially the same as those obtained for HR alone. The mode of action of LIX63 as conceived by Fleming[22] is presented by Eqs. 50-53:

$$Cu^{2+}(a) + HL(i,a) = CuL^{+}(i,a) + H^{+}(a) \tag{50}$$

$$CuL^{+}(i,a) + HR(i) = CuRL(i) + H^{+}(a) \tag{51}$$

$$CuRL(i) = CuRL(o) \tag{52}$$

$$CuRL(o) + HR(o) = CuR_2(o) + HL(o) \tag{53}$$

The rate-limiting step is taken to be Eq. 51. Fleming argues that Eq. 51 does not involve an interfacial LIX63 molecule and is therefore consistent with the interfacial tension results which exclude LIX63 from the liquid/liquid interface in LIX63-LIX65N mixtures. The first complexation reaction step in Fleming's model is the formation of the 1:1 Cu-LIX63 complex (Eq. 50). Equations 50 and 51 are based on preliminary studies which supposedly indicate that the aqueous solubility of CuL^+ exceeds that of CuR^+ and that furthermore, the aqueous phase reaction to form the 1:1 copper-ligand complex proceeds much more rapidly for LIX63 than for LIX65N. Unfortunately no details of the preliminary studies were presented. Therefore, it is difficult to evaluate the merits of this argument.

Additional evidence for the possible role of interfacial chemical reactions in hydroxyoxime systems is given by the effects of diluents on extraction rates. Table 5 provides a summary of the results of Dalton et al.[34] who investigated the effects of several diluents on the rate of copper extraction with o-hydroxy aryl oximes. Their results show that the extraction rates are higher for the aliphatic diluents. At the same time the interfacial

Table 5. Effect of the Diluent on Interfacial Tension and Copper Extraction Rate with o-hydroxyaryloximes.[34]

Diluent	% Aromatic Content	Solubility Parameter	Acorga P17 Interfacial Tension (mNm^{-1})	Acorga P17 % Approach to Equil. in 15 s.	Acorga P50 Interfacial Tension (mNm^{-1})	Acorga P50 % Approach to Equil. in 15 s.
Isopar L	0	8.0	18.3	63	16.7	100
Hexane	0	7.3	18.8	50	16.8	93
Escaid 100	20	9.2	19.9	45	18.9	97
Toluene	100	8.9	22.2	15	25.2	45
Aromasol H	100	12.0	22.3	13	25.3	65

tension was found to be highest for the aromatic diluents. The interfacial tension represents the free energy of a unit area of liquid/liquid interface; the smaller the value of the interfacial tension, the easier it is to create interfacial area, i.e. the smaller the droplets. Thus the correlation of enhanced rates with low interfacial tension may be simply a result of the associated increase in the area available for interfacial reaction. On the other hand, the interfacial tension also provides a measure of the interfacial population density (through the Gibbs equation). Flett[19] and Whewell et al.[24] have shown that the changes in metal extraction rates obtained by varying diluent composition, are directly related to changes in the interfacial area per molecule (see Fig. 13). The adsorption density of an extractant will decrease with increase in the solvating power of the extractant (see Table 3). Thus Fig. 13 suggests that the rate-determining reaction must involve an interfacial process.

The above discussion shows that interfacial chemical reactions have been invoked by several investigators in their attempts to explain metal extraction with hydroxyoximes. Unfortunately, the available experimental data are too sketchy to permit unambiguous conclusions to be drawn. For example, additional work needs to be done to verify Flett et al.'s postulate of an interfacial LIX63-LIX65N mixed complex, and Fleming's suggestion of an interfacial 2 $Cu(LIX65N)_2$•LIX65N complex. Detailed aggregation studies are also indispensable. As has been demonstrated by Fleming[22], the fractional reaction order first reported by Flett et al.[75] for LIX63, can be explained by dimerization of this extractant. In addition, in support of the homogeneous reaction mechanism of Freiser et al.[43,80-84], Freiser and Akiba[83] have presented a correlation plot similar to Fig. 13 in which reaction rate is plotted against extractant solubility in the aqueous phase. Thus,

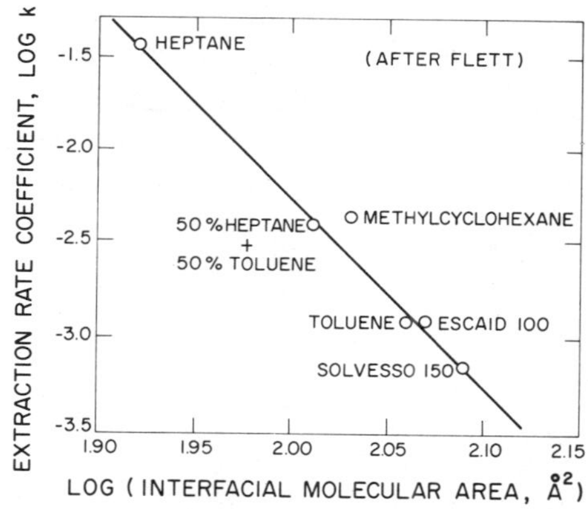

Fig. 13. Cu extraction by LIX65N: effect of interfacial molecular area on extraction rate.[19]

detailed physicochemical measurements are needed for the bulk aqueous phase, the interfacial region, as well as the bulk organic phase.

Interfacial Reactions in Mixed Extractant Systems Containing Carboxylic Acids

There has been much interest in exploring the possibility of synergic interactions between organic acids (e.g. carboxylic, dialkphosphoric, and sulfonic acids) and the commercial oxime-and oxine-type reagents. Much of the observed synergism is undoubtedly related to the increased hydrophobicity of the mixed complexes which are formed when more than one extractant is utilized. There are indications, however, that interfacial effects may also be significant actors as far as the extraction rates are concerned. The first attempt to relate interfacial chemical measurements to extraction characteristics in mixed oxime-carboxylic acid systems was made by Flett et al.[16] They were interested in finding an explanation for the slow extraction of nickel observed previously by Flett and West.[85] They investigated the nickel-LIX63-lauric acid system and on the basis of spectrophotometric measurements concluded that the extracted species was a 1:2:2 nickel:oxime: carboxylic acid complex. The organic/water interfacial tension measurements with oxime-lauric acid mixtures showed that the interfacial activity of lauric acid far exceed that of LIX63. With LIX63 alone, the presence of nickel increased the interfacial tension thus signifying that the extracted Ni-LIX63 complex was less surface active than LIX63. However, with LIX63-lauric acid mixtures extraction of Cu(II), Ni(II) and Co(II) was accompanied by a decrease in interfacial tension indicating that there was some interfacial interaction between the metal ions and the adsorbed carboxylic acid groups. The interfacial tension lowering in the case of nickel was reported to be much more dramatic compared with that for copper and cobalt. This led Flett et al.[16] to propose that the anomalously slow extraction of nickel in the mixed extractant system was a result of the formation of a nickel carboxylate interfacial film which served as a barrier to mass transport.

Flett[86] and Cox and Flett[19] have reported that subsequent work has shown the previously reported drastic interfacial tension lowering in the nickel-oxime-carboxylate system to be in error and that nickel behavior is similar to that of copper and cobalt. This has reopened the question regarding the slow nickel extraction in

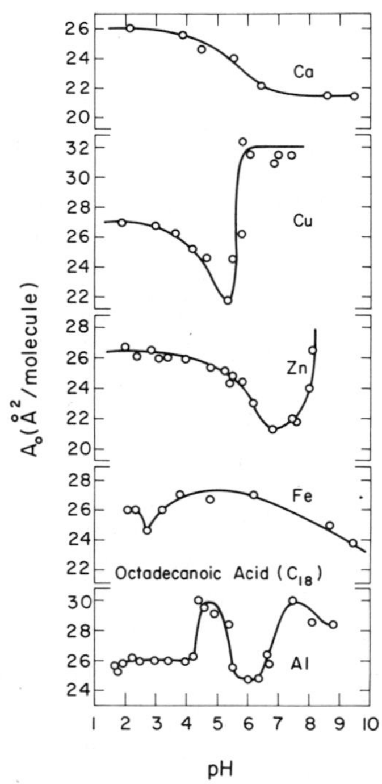

Fig. 14. Effect of pH on the interfacial molecular area of stearic acid on 5 x 10^{-4} kmol m^{-3} Ca(II), Cu(II), Zn(II), Fe(III) chlorides, and $Al_2(SO_4)_3$; 21 ± 5°C.[64]

the mixed oxime-carboxylate system. The earlier suggestion by Flett et al.[16] that interfacial carboxylate films may be present deserves further attention. As Fig. 14 clearly demonstrates[64], the interfacial packing of carboxylates is a strong function of both the pH and the nature of the metal ions in the aqueous substrate. The interfacial viscosity of carboxylate monolayers is also a function of the aqueous phase composition.[65]

The rate of nickel extraction by LIX63-lauric acid mixtures has been investigated by Flett et al.[17] They obtained the following rate law for the forward reaction:

$$-dC_{Ni}/dt = k_f C_{Ni(a)}\, C_{HL(o)}\, C_{HA(o)}/C_{H(a)} \tag{54}$$

They proposed the following interfacial reactions (LIX63 = HL; lauric acid = HA) with reaction 56 as the rate limiting step:

$$Ni^{2+}(i) + HA(i) = NiA^+(i) + H^+(i) \quad \text{fast} \tag{55}$$

$$NiA^+(i) + HL(i) = NiAHL^+(i) \quad \text{slow} \tag{56}$$

$$NiAHL^+(i) + HL(i) = NiL_2HA(i) + H^+(i) \quad \text{fast} \tag{57}$$

$$NiL_2HA(i) + HA(i) = NiL_22HA(i) \quad \text{fast} \tag{58}$$

However, these investigators themselves recognized that the suggested interfacial reaction (Eq. 56) could not be easily reconciled with available nonkinetic data. It is not clear why nickel would react with a carboxylic acid at a bulk aqueous phase pH of about 2.0 when previous equilibrium results indicate that $pH_{0.5}$ has values of 5.50 and 5.79 respectively for 0.1 kmol m^{-3} hexane solutions of lauric acid and the α-hydroxy oxime.[16] Further, how does the interfacial HL species react with NiA^+ (Eq. 56) if (as interfacial tension data[16] would seem to indicate) the interface is preferentially occupied by carboxylic acid molecules?

Mixed extractant systems involving carboxylic acids and hydroxy-quinoline-type reagents have also been investigated. Lakshmanan and Lawson[87] studied the extraction of cobalt by Kelex 100-Versatic 911 mixtures. They concluded that the extracted complex has the stoichiometry: $CoL_2 \cdot 2HA$ (HA = carboxylic acid). These investigators found that the extracted cobalt in the mixed system was easier to strip than the corresponding complex in the Kelex-alone system, however, they did not find any evidence for the presence of synergism. Contrary to the findings of Lakshmanan and Lawson, Hummelstedt et al.[88] have reported that addition of Versatic 911 to Kelex 100 leads to the synergistic extraction of cobalt as well as nickel. These authors attributed the poor cobalt stripping in the Co-Kelex system to air oxidation of Co(II) to Co(III) in the organic phase. They did not explain however, why the presence of carboxylic acid molecules in the mixed complex should prevent cobalt oxidation. Flett et al.[18] re-examined the Co-Kelex-Versatic 911 system and demonstrated that the cobalt oxidation has an interfacial origin. They presented interfacial tension data which, as shown in Fig. 15, indicate that even though Versatic 911 is less surface active than Kelex 100, the interfacial tension in the mixed Kelex-Versatic system reflects the higher interfacial tension of Versatic 911. This kind of behavior suggests the presence of a bulk organic phase interaction between the two extractants leading to removal of Kelex molecules from the liquid/liquid interface. Support for such an interaction was provided by NMR spectra of the mixed extractants which indicated a shift in the OH signal of Versatic 911 which varied linearly with Kelex concentration.

Interfacial Reactions in Mixed Extractant Systems Containing Sulfonic Acids

The addition of dinonylnaphthalene sulfonic acid (HDNNS) to LIX63 affects both the extent and rate of nickel and cobalt

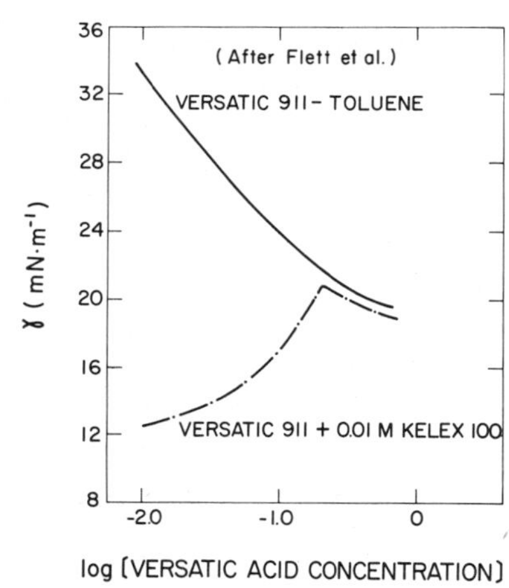

Fig. 15. Effect of versatic acid concentration on the interfacial tension at the toluene/water and toluene-Kelex 100/water interfaces.[18]

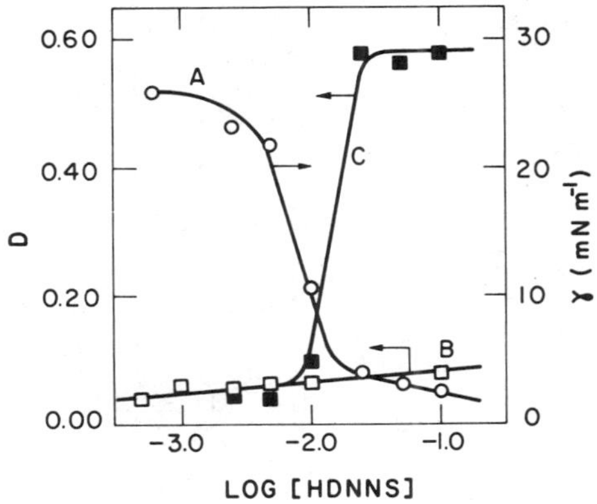

Fig. 16. Comparison of interfacial tension and nickel extraction as a function of HDNNS concentration in the presence and absence of LIX63 at 25°C. Aqueous solutions maintained at 0.5 kmol m^{-3} ionic strength with KNO_3. Initial [Ni(aq)] = 1.0 x 10^{-1} kmol m^{-3}; initial pH = 2.5. (A) Interfacial tension at the hexane (10^{-1} kmol m^{-3} LIX63)/water interface; (B) nickel distribution coefficient for HDNNS. O/A = 1.1; (C) nickel distribution coefficient for 10^{-1} kmol m^{-3} LIX63/HDNNS mixtures. O/A = 1/1.[35]

extraction.[35,36] Equilibrium distribution data obtained with purified reagents[89] indicate that the predominant extracted complex has the stoichiometry, Ni/LIX63/HDNNS = 1/3/2. Further, as can be seen from Fig. 16, in the case of the mixed extractant, there is an abrupt increase in nickel extraction at a sulfonic acid concentration which is nearly coincident with the CMC. This behavior has been attributed to the ability of HDNNS micelles to solubilize LIX63 molecules.[35-37] Using infrared and HNMR techniques, it has been demonstrated by Keeney and Osseo-Asare[35,38,90] that LIX63 and HDNNS molecules interact through hydrogen bonding between the oxime and sulfonic acid functional groups. The presence of LIX63-HDNNS interaction is also indicated by the fact that in the presence of LIX63, the CMC of HDNNS shifts to higher concentrations.[35,36,38,90]

The presence of HDNNS affects the interfacial charge at the organic/aqueous interface. The electrophoretic mobility data[91] presented in Fig. 17 demonstrate that adsorption of HDNNS at the liquid/liquid interface results in highly negatively charged organic

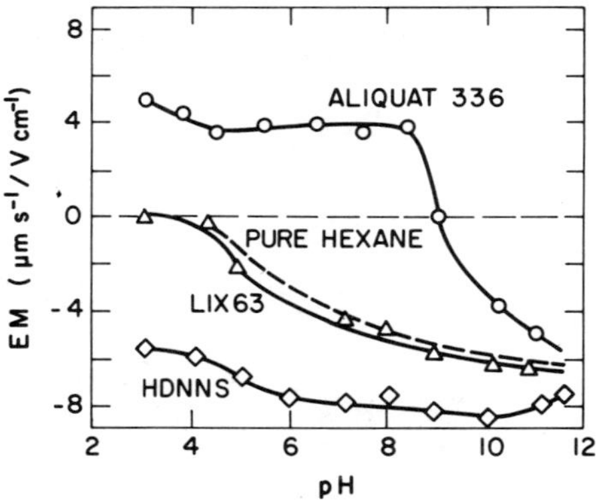

Fig. 17. Effect of solvent extraction reagents on the electrophoretic mobility of hexane drops: [HDNNS] = 1.4 x 10^{-4} kmol m^{-3}, [LIX63] = [Aliquat 336] = 1.5 x 10^{-4} kmol m^{-3}; aqueous phase: 4 x 10^{-3} kmol m^{-3} KNO_3.[91]

droplets. The high negative charge must be attributed to the dissociation of the adsorbed sulfonic acid molecules. It would be expected that the interfacial concentration of extractable cations would increase as a result of electrostatic attraction to the negatively charged interface. According to Fig. 17, organic drops containing Aliquat 336 (a quaternary ammonium compound) are positively charged. Thus, the previous report[37] that the presence of Aliquat 336 decreases nickel extraction by HDNNS must not be attributed only to the replacement of adsorbed HDNNS molecules by the ammonium molecules. An additional factor that must be considered, is the repulsion of the cationic Ni^{2+} species by the positively charged ammonium-saturated organic/aqueous interface. As can be seen from Fig. 18,[91] the interfacial potential can have a dramatic effect on extraction rate, particularly in systems involving multi-valent ionic species.

Metal extraction rate studies have shown that for both rising drop[91] and liquid membrane[92] experiments, there is a tendency for the LIX63-HDNNS mixed extractant behavior to coincide with the HDNNS alone results at high HDNNS concentrations; see Fig. 19. It is possible to explain these results by postulating the existence of a catalytic micelle whose concentration changes with HDNNS concentration. That is, the incorporation of LIX63 molecules into HDNNS micelles leads to an increase in the effective concentration of LIX63 at the micelle/aqueous core interface. Given the

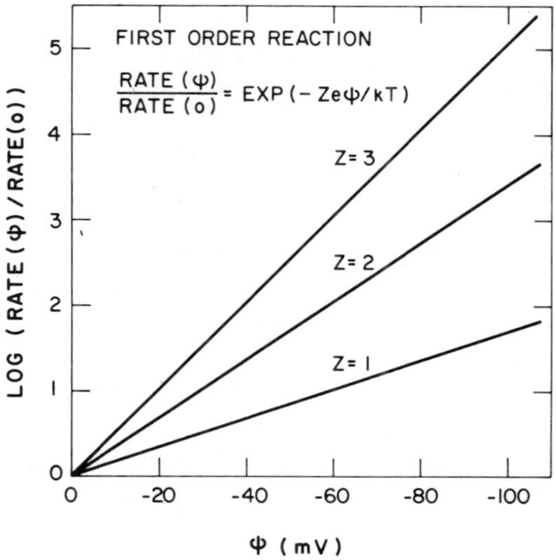

Fig. 18. Theoretical dependency of reaction rate on surface charge.[91]

fact that the stoichiometry of the mixed complex has been determined[89] to be Ni/LIX63/HDNNS = 1/3/2, it seems reasonable to assume that in its simplest form, the proposed catalytic species must contain at least three LIX63 molecules, i.e. $[(HD)_m \cdot 3H_2Ox]$ where H_2Ox and HD represent LIX63 and HDNNS respectively. It appears that metal extraction with HDNNS-LIX63 mixtures involves both phase transfer catalysis and micellar catalysis.[35–37] The phase transfer catalysis arises from the fact that the sulfonic acid is more interfacially active than the oxime and the fact that the presence of the negatively charged sulfonate groups increases the interfacial concentration of the extractable cations:

$$Ni^{2+}(a) + 2(HD)(i) = NiD_2(o) + 2H^+(a) \qquad (59)$$

The presence of the mixed micelle $[(HD)_m \cdot 3H_2Ox]$ with an aqueous core, permits further reaction of the extracted nickel sulfonate with LIX63:

$$NiD_2(o) + (HD)_m \cdot 3H_2Ox(o) = Ni(H_2Ox)_3D_2(o) + (HD)_m(o) \qquad (60)$$

It would be expected that the rate of the reaction represented by Eq. 60 would depend on the concentrations of both the mixed micelle and the pure micelle. As the total HDNNS concentration increases,

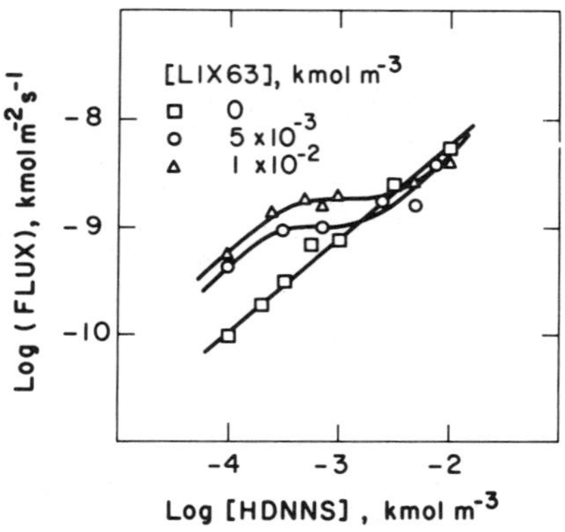

Fig. 19. Effects of HDNNS and LIX63 concentrations on the rate of nickel extraction into hexane drops.[91]

$[(HD)_m \cdot 3H_2Ox]$ will initially predominate leading to an increase in the rate of metal extraction via Eq. 60. However, as the total HDNNS concentration becomes very high, $[(HD)_m]$ becomes the predominant species and the reverse rather than the forward direction of Eq. 60 becomes the favored reaction. Under these circumstances, the predominant path for metal extraction is the reaction with the pure micelle:

$$NiD_2(o) + (HD)_m(o) = NiD_2(HD)_m(o) \qquad (61)$$

Use of HDNNS in combination with other reagents has also been studied by van Dalen et al.[11-15] With the aid of vapor pressure osmometry results, these authors proposed that HDEHP molecules are solubilized within the micelle core of HDNNS. The presence of HDNNS-HDEHP interaction is suggested in the apparent shift in the CMC of HDNNS which accompanies the addition of HDEHP, as shown in Fig. 20.[11] A similar interaction has been demonstrated in the LIX63-HDNNS system[35,36,38,90] on the basis of interfacial tension, infrared and HNMR results; however, apparently in the case of van Dalen's work[11], infrared spectra gave no indication of such an interaction. The fact that the participation of HDNNS micelles is a crucial factor in the extraction of metals in the mixed HDNNS-HDEHP system, is demonstrated by Fig. 21.[11] This figure shows the same trend as that previously observed in the case of the Ni-LIX63-HDNNS system (Fig. 16). According to van Dalen, the observed synergic extraction of metal is due to the solubilization of metal-HDEHP complexes within the HDNNS micelle ore. Addition-

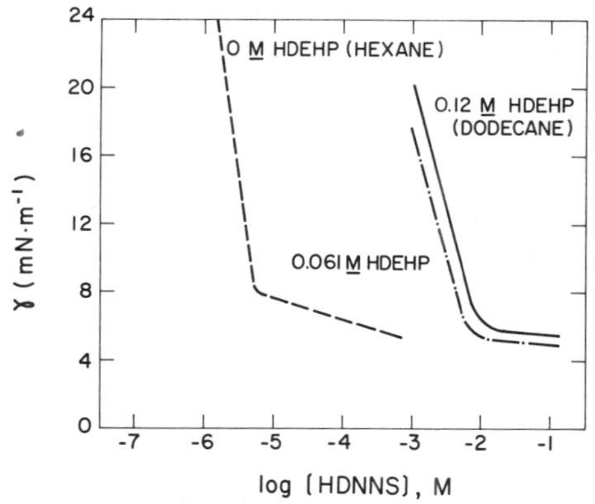

Fig. 20. Effect of HDNNS on the interfacial tension of the hexane/water[35] and the dodecane-HDEHP/water[11] systems.

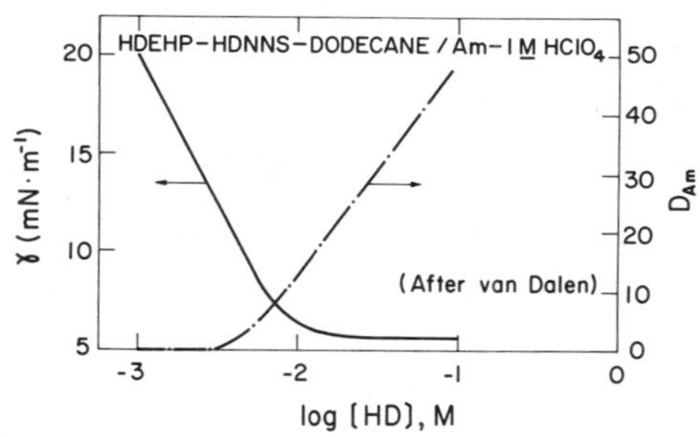

Fig. 21. Comparison of interfacial tension and Americium extraction as a function of HDNNS concentration in the presence of HDEHP.[11]

ally, selectivity is related to the compatibility of the size of the metal-HDEHP complex with the dimensions of the micelle core. If the complex is too big to fit into the micelle core, it is rejected and no synergism is observed. While the role of the HDNNS micelle in the extraction process cannot be disputed, it has not been conclusively determined that the extracted metal resides within the micelle core. An alternative approach would be to consider the participation of a catalytic micelle such as that proposed above for the LIX63-HDNNS system.

The ability of surfactant aggregates to enhance extraction rates through a concentrative effect presents interesting possibilities. Ching-Kwang et al.[40] have demonstrated the presence of water-in-oil microemulsions in certain solvent extraction systems. Bauer et al.[41,42] have shown that microemulsions, which can be considered as "swollen" micelles[93], are capable of increasing gallium extraction with a hydroxy quinoline-type reagent. In this case, the microemulsion was intentionally introduced by adding an alcohol and a long chain carboxylic acid to the organic phase.

SUMMARY AND CONCLUSIONS

In this paper, a review of selected interfacial chemistry concepts relevant to hydrometallurgical solvent extraction has been presented. In addition, the literature on the interfacial chemical aspects of liquid-liquid metal extraction processes has been reviewed. There is now little doubt that solvent extraction reagents are interfacially active. However, there are not sufficient systematic studies in the literature to permit definite conclusions to be made of the role and nature of this interfacial activity in the metal extraction process. Practically all the reported interfacial physicochemical measurements have been confined to interfacial tension determinations. However, even in these cases, not all investigators have taken the trouble to ascertain the nature of the "extractant" being studied (see Table 3). Commercial reagents used without purification give results that are difficult to interpret unambiguously since these reagents contain diluents and modifiers which may be themselves interfacially active or alter the adsorption behavior of the active molecule through bulk phase solvation. In many instances, details of the experimental conditions are omitted e.g. temperature, composition of the aqueous phase. It is not enough to give the ionic strength, the nature of the ions present is also of importance, as demonstrated for example by the different effects of simple cations such as Li^+, Na^+ and K^+ on anionic monolayers (Figs. 10 and 11).

Interfacial studies are still largely conducted as an afterthought. There is a need for more deliberate investigations which focus on the interfacial arena. Furthermore, future studies need to use a variety of techniques: e.g. interfacial viscosity, spread monolayers (Langmuir trough), interfacial potentials (ΔV), and electrophoresis.[46-52,66-68] Studies with extractants of systematically varied molecular structure covering a broad range of organic and aqueous phase compositions would be invaluable. Finally, these interfacial studies must be complemented with studies of bulk phase phenomena which can influence interfacial behavior e.g. aggregation (vapor pressure osmometry), complex chemistry (UV-visible spectroscopy, magnetic susceptibility), extractant-diluent and mixed extractant interactions (infrared, HNMR), and extracant solubility in the aqueous and organic phases. Finally, the largely neglected subject of the effects of interfacial activity on the hydrodynamics of hydrometallurgical solvent extraction processes, deserves serious attention. For an inherently heterogeneous process such as solvent extraction, both chemical and mass transfer processes contribute to the overall rate of metal extraction.

ACKNOWLEDGMENT

Support of this work by the National Science Foundation is gratefully acknowledged.

REFERENCES

1. W. J. McDowell and C. F. Coleman: J. Inorg. Nucl. Chem., 1967, vol. 29, pp. 1325-1343.
2. W. J. McDowell and C. F. Coleman: Solvent Extraction Chemistry, 1967, North-Holland, Amsterdam, pp. 540-546.
3. G. Scibona, P. R. Danesi, F. Orlandini, B. Scuppa and M. Magini: Solvent Extraction Chemistry, 1967, North-Holland, Amsterdam, pp. 547-551.
4. G. Scibona, P. R. Danesi, A. Conte, and B. Scuppa: J. Colloid Interface Sci., 1971, vol. 35, pp. 631-635.
5. R. Chiarizia, P. R. Danesi, G. Alessandro, and B. Scuppa: J. Inorg. Nucl. Chem., 1976, vol. 38, pp. 1367-1369.
6. M. Pizzichini, R. Chiarizia and P. R. Danesi: J. Inorg. Nucl. Chem., 1978, vol. 40, pp. 669-671.
7. P. R. Danesi and R. Chiarizia: J. Appl. Chem. Biotechnol., 1978, vol. 28, pp. 581-598.
8. P. R. Danesi, R. Chiarizia and G. F. Vandegrift: J. Phys. Chem., 1980, vol. 84, pp. 3455-3461.
9. P. R. Danesi and R. Chiarizia: Crit. Rev. Anal. Chem., 1981, vol. 10, pp. 1-126.
10. "Organic Reagents and Mass Transfer - Panel Discussion Part 2 Surface-Active Properties of Extractants," Proc. Int. Solv. Extr. Conf. '77, pp. 268-270, 1977.
11. A. Van Dalen: Solvent Extraction and Inclusion Compound Formation with Dinonylnapthalene Sulfonic Acid, RCN-141, Petten, The Netherlands, 1971.

12. A. Van Dalen, K. W. Gerritsma and J. Wijkstra: Proc. Int. Solv. Extr. Conf. '71, pp. 1096-1100, 1971.
13. A. Van Dalen, K. W. Gerritsma and J. Wijkstra: J. Colloid Interface Sci., 1974, vol. 48, pp. 122-126.
14. A. Van Dalen, K. W. Gerritsma and J. Wijkstra: J. Colloid Interface Sci., 1974, vol. 48, pp. 127-133.
15. A. Van Dalen, J. Wikjstra and K. W. Gerritsma: J. Inorg. Nucl. Chem., 1978, vol. 40, pp. 875-881.
16. D. S. Flett, M. Cox, and J. D. Heels: Proc. Int. Solv. Extr. Conf. '74, pp. 2559-2575, 1974.
17. D. S. Flett, M. Cox, and J. D. Heels: J. Inorg. Nucl. Chem., 1975, vol. 37, pp. 2533-2537.
18. D. S. Flett, M. Cox and J. D. Heels: J. Inorg. Nucl. Chem., 1975, vol. 37, pp. 2197-2201.
19. D. S. Flett: Acct. Chem. Res., 1977, vol. 10, pp. 99-104.
20. M. Cox and D. S. Flett: Proc. Int. Solv. Extr. Conf. '77, pp. 63-72, 1977.
21. E. S. Perez de Ortiz: J. Appl. Chem. Biotechnol., 1978, vol. 28, pp. 149-156.
22. C. A. Fleming: Natl. Inst. Met. Sth. Afric. Rept. No. 1973, 1976.
23. T. A. B. Al-Diwan, M. A. Hughes and R. J. Whewell: J. Inorg. Nucl. Chem. 1977, vol. 39, pp. 1419-1424.
24. R. J. Whewell, M. A. Hughes and C. Hanson: Advances in Extractive Metallurgy, 1977, The Institution of Mining and Metallurgy, pp. 21-25.
25. K. Durrani, C. Hanson and M. A. Hughes: Metall. Trans. B, 1977, vol. 8, pp. 169-174.
26. M. A. Hughes, E. Mistry and R. J. Whewell: J. Inorg. Nucl. Chem., 1978, vol. 40, pp. 1694-1696.
27. M. A. Hughes: Hydrometallurgy, 1978, vol. 3, pp. 85-90.
28. G. F. Vandegrift and E. P. Horwitz: J. Inorg. Nucl. Chem., 1977, vol. 39, pp. 1425-1432.
29. G. F. Vandegrift and E. P. Horwitz: J. Inorg. Nucl. Chem., 1980, vol. 42, pp. 119-125.
30. G. F. Vandegrift, S. M. Lewey, G. R. Dyrkacz and E. P. Horwitz: J. Inorg. Nucl. Chem., 1980, vol. 42, pp. 127-130.
31. G. A. Yagodin, V. V. Tarasov, A. V. Fomin and S. Yu. Ivakhno: Proc. Int. Solv. Extr. Conf. '77, vol. 1, pp. 260-265, 1977.
32. G. A. Yagodin, S. Yu. Ivakhno and V. V. Tarasov: Proc. Int. Solv. Extr. Conf. '80, Paper No. 80-140, 1980.
33. G. A. Yagodin, V. V. Tarasov and S. Y. Ivakhno: Hydrometallurgy, 1982, vol. 8, pp. 293-305.

34. R. F. Dalton, F. Hauxwell and J. A. Tumilty: Chem. Ind. (London) 1978, pp. 181-184.
35. K. Osseo-Asare and M. E. Keeney: Metall. Trans. B, 1980, vol. 11B, pp. 63-67.
36. K. Osseo-Asare and M. E. Keeney: Sep. Sci. Technol., 1980, vol. 15, pp. 999-1011.
37. K. Osseo-Asare and M. E. Keeney: Proc. Int. Solv. Extr. Conf. '80, Paper No. 80-121, 1980.
38. M. E. Keeney and K. Osseo-Asare: Polyhedron, 1982, vol. 1, pp. 453-455.
39. M. E. Keeney and K. Osseo-Asare: Polyhedron, 1982, vol. 1, pp. 541-542.
40. W. Chin-Kwang, K. Hung-Chang, C. Tien, L. Seng-Chung, K. Tien-Chu and H. Kwang-Hsien: Proc. Int. Solv. Extr. Conf. '80, Paper No. 80-23, 1980.
41. D. Bauer, P. Fourre and J. C. Lemerle: C. R. Hebd. Seances Acad. Sci. II, (Paris), 1981, vol. 292, pp. 1019-1022.
42. P. Fourre and D. Bauer: C. R. Acad. Sci. II, (Paris), 1981, vol. 292, pp. 1077-1080.
43. H. Freiser: Proc. Int. Solv. Extr. Conf. '80, Paper No. 80-II, 1980.
44. C. Hanson: Proc. Int. Solv. Extr. Conf. '80, Paper No. 80-I, 1980.
45. H. J. Foakes, J. S. Preston and R. J. Whewell: Anal. Chim. Acta, 1978, vol. 97, pp. 349-356.
46. A. W. Adamson: Physical Chemistry of Surfaces, 3rd Ed., Wiley, New York, 1976.
47. J. T. Davies and E. K. Rideal: Interfacial Phenomena, 2nd Ed., Academic Press, New York, 1963.
48. G. L. Gaines: Insoluble Monolayers at Liquid-Gas Interfaces, Interscience, New York, 1966.
49. M. J. Rosen: Surfactants and Interfacial Phenomena, Wiley, New York, 1978.
50. R. Aveyard and B. Vincent: Prog. Surf. Sci., 1977, vol. 8, pp. 59-102.
51. Lucassen-Reynders: Prog. Surf. Membr. Sci., 1976, vol. 10, pp. 253-360.
52. Ter-Minassian-Saraga: Prog. Surf. Membr. Sci., 1975, vol. 9, pp. 223-256.
53. J. T. Davies: Adv. Catalysis, 1954, vol. 6, pp. 1-65.
54. F. MacRitchie: Interfacial Synthesis, Vol. 1, F. Millich and C. E. Carraher, Jr., eds., 1977, Marcell Dekker, New York, pp. 103-139.
55. A. S. Kertes: Surf. and Colloid Sci., 1976, vol. 8, pp. 193-295.

56. G. Scibona, P. Danesi and C. Fabiani: Ion Exchange and Metal Extraction, 1981, vol. 8, pp. 95-227.
57. H. Sawistowski: Recent Advances in Liquid-Liquid Extraction, C. Hanson, ed., Pergammon, London, 1971, pp. 293-366.
58. J. C. Berg: Recent Developments in Separation Science, vol. 2, N. N. Li, ed., CRC Press, Cincinatti, Ohio, 1975, pp. 1-32.
59. E. C. Hunt: J. Colloid and Interface Sci., 1969, vol. 29, pp. 105-115.
60. K. D. Dreher and J. E. Wilson: J. Colloid Interface Sci., 1970, vol. 32, pp. 248-255.
61. K. D. Dreher and P. S. Ossip: J. Colloid Interface Sci., 1974, vol. 46, pp. 442-447.
62. E. D. Goddard, O. Kao, and H. C. Kung: J. Colloid Interface Sci., 1967, vol. 24, pp. 297-309.
63. E. D. Goddard, O. Kao, and H. C. Kung: J. Colloid Interface Sci., 1968, vol. 27, p. 616-624.
64. J. A. Spink and J. V. Sanders: Trans. Farad. Soc., 1955, vol. 57, pp. 1154-1165.
65. R. D. Neuman: J. Colloid Interface Sci., vol. 53, pp. 161-171.
66. J. H. Brooks, B. A. Pethica: Trans. Farad. Sc., 1964, vol. 60, pp. 208-215.
67. J. A. G. Taylor and J. Mingins: J. Chem. Soc. Faraday I, 1975, vol. 71, pp. 1661-1171.
68. J. Mingins and J. A. G. Taylor: A Manual for the Measurement of Interfacial Tension, Pressure and Potential at Air or non-Polar Oil/Water Interfaces, Unilever Research Laboratory, Port Sunlight, 1974.
69. K. Osseo-Asare: NATO Advanced Research Institute on Hydrometallurgical Process Fundamentals, July 15-31, 1982, Churchill College, Cambridge Univ., U.K.
70. S. Hard and R. D. Neuman: J. Colloid Interface Sci., 1981, vol. 83, pp. 315-334.
71. R. T. Morrison and R. N. Boyd: Organic Chemistry, 1973, Allyn and Bacon, Boston.
72. R. A. Peters: Proc. Roy. Soc., 1931, vol. A133, pp. 140-154.
73. F. Van Voorst Vader: TRans. Farad. Soc., 1960, vol. 56, pp. 1067-1084.
74. R. M. Diamond: J. Phys. Chem., 1963, vol. 67, pp. 2513-2517.
75. D. S. Flett, D. N. Okuhara and D. R. Spink: J. Inorg. Nucl. Chem., 1973, vol. 35, pp. 2471-2487.
76. A. W. Ashbrook: Coord. Chem. Rev., 1975, vol. 16, pp. 285-307.
77. R. L. Atwood, D. N. Thatcher and J. D. Miller: Metall. Trans. B., 1975, vol. 6B, pp. 465-473.

78. R. J. Whewell, M. A. Hughes and C. Hanson: J. Inorg. Nucl. Chem., 1976, vol. 38, pp. 2071-2075.
79. D. S. Flett, J. Melling and D. R. Spink: J. Inorg. Nucl. Chem., 1977, vol. 39, pp. 700-701.
80. S. P. Carter and H. Freiser: Anal. Chem., 1980, vol. 52, pp. 511-514.
81. K. Akiba and H. Freiser: Sep. Sci. Technol., 1982, vol. 17, pp. 745-750.
82. V. Bagreev and H. Freiser: Sep. Sci. Technol., 1982, vol. 17, pp. 751-762.
83. K. Akiba and H. Freiser: Anal. Chim. Acta, 1982, vol. 136, pp. 329-337.
84. H. Watarai, L. Cunningham and H. Freiser: Anal. Chem., 1982, vol. 54, pp. 2390-2392.
85. D. S. Flett and D. W. West: Proc. Int. Solv. Extr. Conf. '71, pp. 214-223, 1971.
86. D. S. Flett: Chem. Ind., 1977, vol. 6, pp. 223-224.
87. V. I. Lakshmanan and G. J. Lawson: J. Inorg. Nucl. Chem., 1973, vol. 35, pp. 4285-4294.
88. L. Hummelstedt, H. E. Sund, J. Karjaluoto, L. O. Berts and B. G. Nyman: Proc. Int. Solv. Extr. Conf. '74, pp. 829-848, 1974.
89. D. R. Renninger and K. Osseo-Asare: unpublished research.
90. M. E. Keeney and K. Osseo-Asare: unpublished research.
91. K. Osseo-Asare and K. L. Lin: Symposium on Mass Transfer-Theories and Applications, ACS National Meeting, New York, August 23-28, 1981.
92. D. J. Chaiko and K. Osseo-Asare: unpublished research.
93. K. Shinoda and H. Kunieda: J. Colloid Interface Sci., 1973, vol. 42, pp. 381-387.
94. S. J. Lyle and D. B. Smith: J. Colloid Interface Sci., 1977, vol. 61, pp. 405-412.

KINETICS OF MASS TRANSFER WITH CHEMICAL REACTION IN GAS-LIQUID AND LIQUID-LIQUID SYSTEMS

H. Renon, A. Gaunand, P.-E. Lévy, A. Kamen, and N. Papassiopi

Centre Réacteurs et Processus, Equipe de Recherche Associée au C.N.R.S. n° 768
Ecole Nationale Supérieure des Mines de Paris
60, Boulevard Saint-Michel, 75006 Paris, FRANCE

The kinetics of oxidation of Cu(I) by oxygen gas in concentrated NaCl solutions was interpreted using the film theory. Extension of the theory of mass transfer with chemical reaction beyond the film model gave inconclusive results. The kinetics of the liquid extraction of Cu(II) by LIX 65N was also studied. Initial results are reported on the interaction between kinetics and hydrodynamics in a three phase gas-liquid-liquid reactor.

INTRODUCTION

The need for basic research studies in the development of the Minemet process[1] initiated two fundamental investigations into the chemical engineering aspect of mass transfer with chemical reaction between fluid phases. The first study is the oxidation by oxygen of Cu(I) to Cu(II) in concentrated NaCl solutions. The second is the liquid extraction of Cu(II) by LIX 65N. Both studies were carried out experimentally in especially designed laboratory equipment. The measurements allow an interpretation of results using a kinetic model in the first case. Finally some efforts are in progress to fully understand the interaction between kinetics and hydrodynamics in a three phase gas-liquid-liquid reactor.

THE KINETICS OF OXIDATION OF CU(I) BY OXYGEN GAS IN CONCENTRATED NaCl SOLUTIONS

The heterogeneous chemical reaction can be written:

$$4Cu(I) + O_2(g) + 4H^+ \rightarrow 4Cu(II) + 2H_2O \qquad (1)$$

The solubility of copper in cuprous state is very low in aqueous solution if no complexing agent like chloride ions is present. The above reaction carried out in concentrated brines is therefore an excellent method of regeneration of leaching solutions and many experimental studies have been reported in the literature.[2-10] The superposition of kinetics of diffusion of oxygen into the aqueous phase, diffusion of reactant and products and chemical kinetics are dependent upon hydrodynamic and concentration conditions. Experimental results were at first obtained around the conditions of the industrial process, in order to determine the influences of the main variables of the process. Then the film theory was considered to interpret the results and new measurements were carried out to ascertain the reaction regime. Finally application of the theory of mass transfer with chemical reaction of Toor-Marcello[15,16] were found to be inconclusive. On the contrary the consideration of transients within the film theory seems necessary for a complete interpretation of measurements.

EXPERIMENTAL METHOD AND RESULTS

The experiments were carried out in the standard glass reactor represented in Fig. 1. The solution is stirred mechanically by a four-bladed radial impeller made of teflon. Four equally spaced vertical baffles prevent vortexing. A water jacket controls the temperature within 0.5°C. The impeller speed is measured by stroboscopy, and remains closely stable during each test. The gas phase is introduced by six calibrated capillaries located under the tips of the impeller blades. The gas is water-saturated before being injected into the reactor in order to keep constant the solution volume, and to prevent any capillary clogging at high NaCl concentrations.

The experimental set-up is schematically presented in Fig. 2. During each experiment, the solution pH is maintained constant by means of an automatic delivery of reagent. Concentrated H_2SO_4 solution is added in doses of 10^{-5} dm^3, by means of a Tacussel type

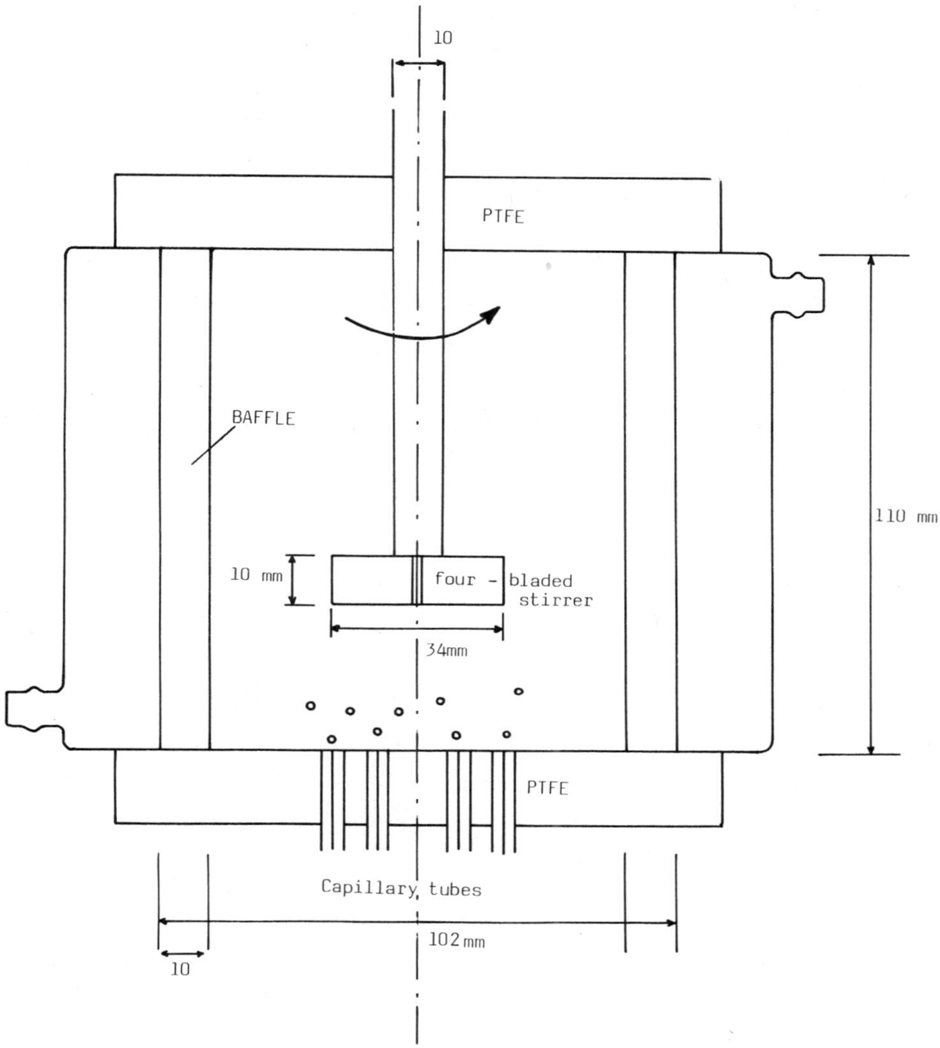

Fig. 1. The standard reactor.

EBX digital syringe burette controlled by a Tacussel type Urectron 6 automatic titrator. The amount of acid required to maintain the solution pH constant leads to the overall oxidation rate. It was verified under these conditions that reaction (1) is stoichiometric and the sulfate ions have a negligible influence on the reaction rate. It is also possible to maintain the solution potential constant by means of another electrode-titrator-burette system.

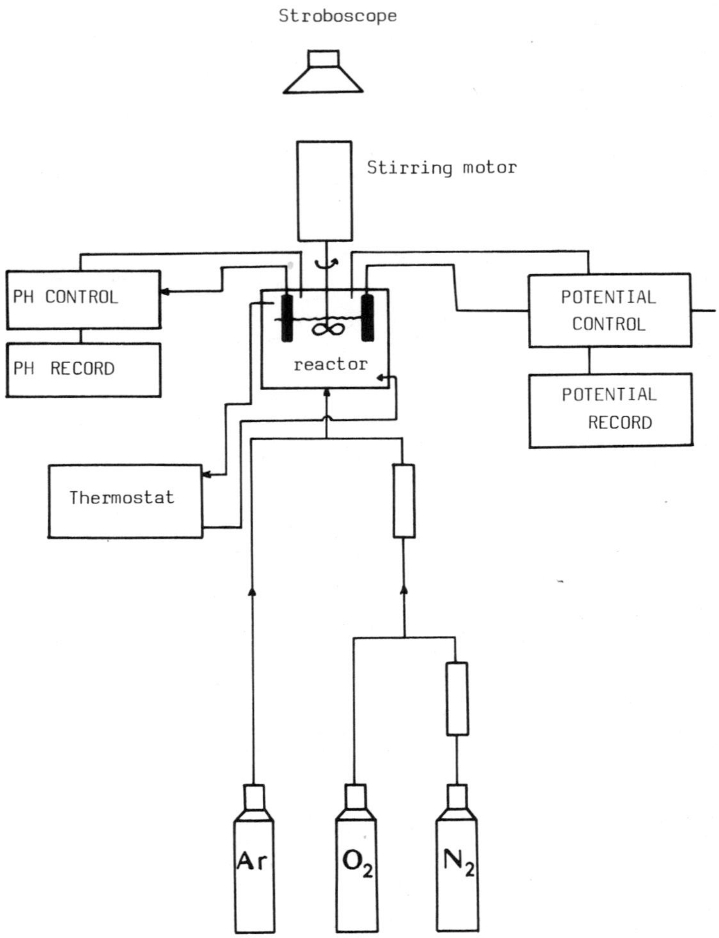

Fig. 2. Experimental set up.

The composition of the reagent added for this purpose is identical to initial solution composition in the reactor, except for the copper ions which are all in the cuprous oxidation state. The amount of reagent added to maintain the solution potential constant provides an alternate means of measuring the overall oxidation rate. This alternate method is more accurate than the former at high potentials, when the reaction rate levels off and the amount of acid needed to control pH decreases drastically.

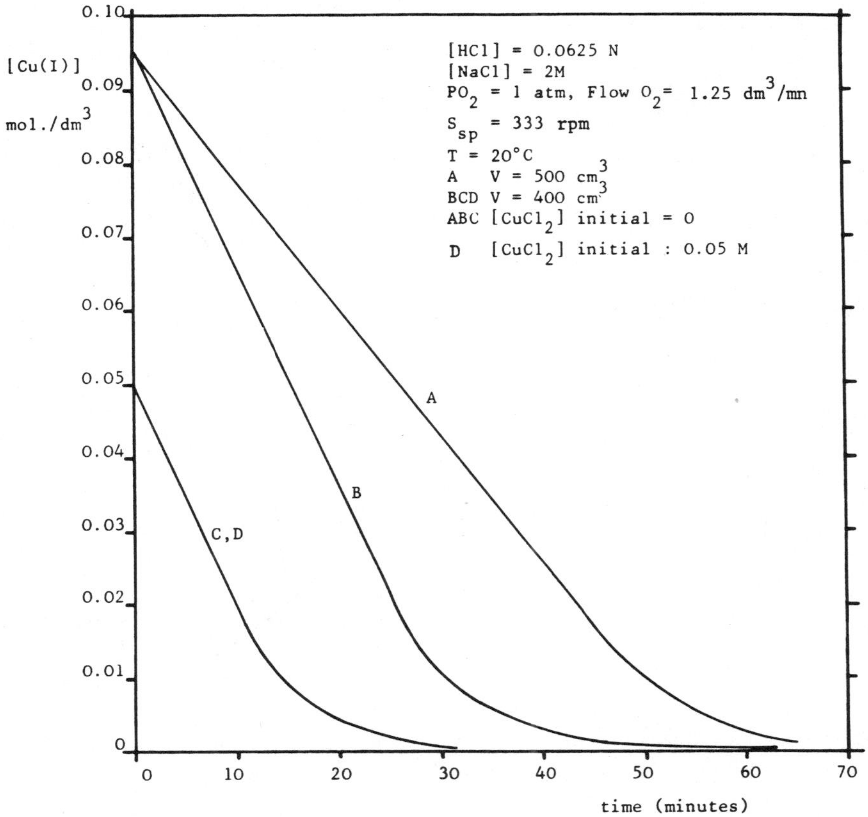

Fig. 3. Observed kinetics of oxidation of Cu(I).

The reaction rate is studied as a function of solution concentration, oxygen partial pressure, temperature, stirring speed and solution volume.

Typical curves of Cu(I) concentration versus time otained at constant pH are presented in Fig. 3. In the range of conditions studied here, the oxidation of Cu(I) by oxygen gas follows linear kinetics up to 85% of the copper ions in the cupric state.

In Fig. 3, curve A and curve B are the results of five and seven identical tests respectively. After each test, the oxidation rate V_{ox} was computed from the linear portion of the curve. The standard deviation represents about 6% of the mean value, which means a fair reproducibility on the value of V_{ox}.

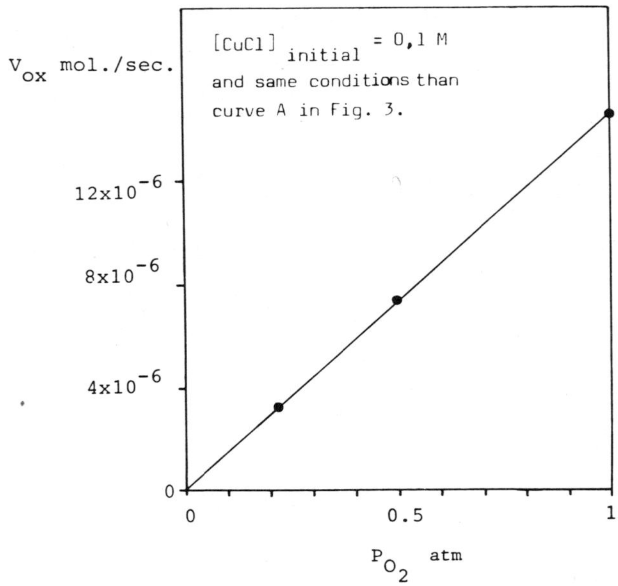

Fig. 4. Effect of oxygen partial pressure on oxidation rate of Cu(I).

Two experiments (C and D) also presented in Fig. 3, were carried out under conditions similar to those of the "B" test, except for the Cu(I) and Cu(II) concentration of the starting solution. A comparison of B and D confirms that V_{ox} does not depend upon the Cu(I) concentration of the solution. A comparison of C and D shows that V_{ox} is also independent of the total copper concentration of the solution. The oxygen partial pressure was made to vary from 21 vol.% to 100 vol.% at constant total gas flow rate. The results depicted in Fig. 4 form a straight line passing through the origin of coordinates.

The apparent dependence of oxidation rate upon cuprous ion concentration and oxygen partial pressure is then found to be zero order and first order respectively.

The effect of temperature was studied at temperatures ranging between 20°C and 50°C. Temperature has a moderate effect on reaction rates.

The effects of chloride ions and solution pH cannot be studied independently since adding NaCl into the solution modifies the ionic strength, then the activity coefficient of the H^+ ions, and hence

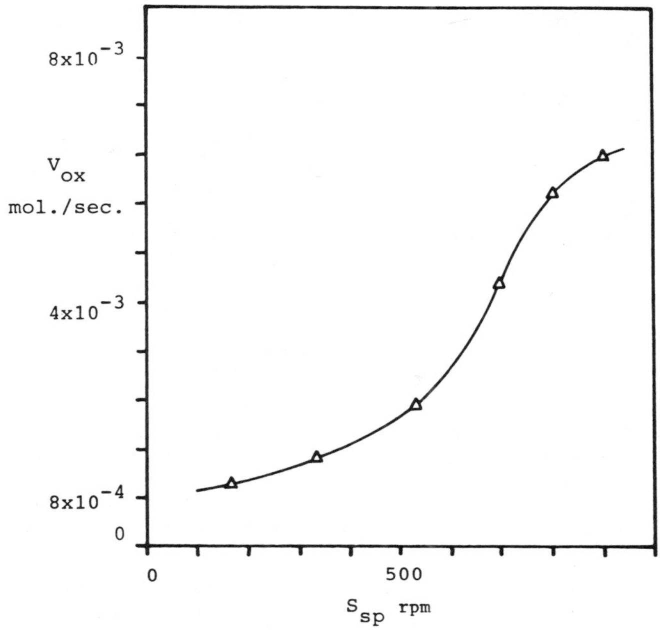

Fig. 5. Dependence of oxidation rate upon stirring rate (same conditions as Fig. 3)

the solution pH. The oxidation of Cu(I) is slightly faster when pH is lowered, but markedly slower when chloride ion concentration increases.

The oxidation rate is very sensitive to the hydrodynamics of the system. The effect of stirring speed in the range 160 to 900 r.p.m. is depicted in Fig. 5. The rate V_{ox} increases mostly between 600 and 800 r.p.m. The volume of solution in the reactor is a critical parameter. The overall oxidation rate drops when the solution volume rises.

Application of the film theory to the interpretation of experimental results

The first attempt to interpret experimental results used a steady-state assumption and the film theory with the usual assumptions of Hatta's theory[11] which includes diffusion film of thickness δ and negligible convection within the film. The chemical reaction is assumed to be first order with respect to oxygen and nth order with respect to Cu(I). It is assumed not fast enough that the Cu(I) concentration within the film varies. Under these conditions the rate of reaction V_{ox} is given by:

$$V_{ox} = 4R_{O_2} = 4Ak_L \frac{Ha}{tahh(Ha)} [(O_2)_i - \frac{(O_2)_o}{cosh(Ha)}] \quad (2)$$

where the Hatta number Ha

$$Ha = \frac{1}{k_L} \sqrt{kD_{O_2}(Cu(I))_o^n} \quad (3)$$

is a measure of the relative velocities of chemical reaction and mass transfer of oxygen, A is the interfacial area, k_L the mass transfer coefficient of oxygen, D_{O_2} is its diffusion coefficient, $(O_2)_i$ is the equilibrium solubility of oxygen, and k the reaction rate constant. R_{O_2} is the interfacial oxygen flow, proportional to V_{ox} according to the steady state assumption.

If Hatta's number is larger than 3 the reaction is "fast", i.e., it is fully achieved within the film of diffusion of oxygen.

$$\frac{V_{ox}}{4} = R_{O_2} = Ak_L \, Ha \, (O_2)_i = A(O_2)_i \sqrt{k(Cu(I))_o^n \, D_{O_2}} \quad (4)$$

The rate does not depend on the mass transfer coefficient k_L, but on the equilibrium solubility of oxygen $(O_2)_i$. The linear behavior of the curves Cu(I) vs. time indicate that Hatta's theory can be applied in the situation of a fast reaction regime with the steady state assumption, only if the order of the homogeneous reaction with respect to Cu(I) is zero (n = 0).

$$R_{O_2} = A(O_2)_i \sqrt{k \, D_{O_2}} \quad (5)$$

In order to apply the theory in this form, the value of Hatta's number should be checked by independent experiments.

Initially, measurements of mass transfer rate at given interfacial areas were achieved at a stirring speed (167 r.p.m.) such that the liquid surface remains horizontal. No gas was injected through the capillaries but pure oxygen was continuously introduced at the top of the vessel. Knowing the diffusion coefficient of oxygen and its interfacial concentration (assuming equilibrium and Henry's law applies), Eq. (5) gives the kinetic constant k since the interfacial area is known if we still assume a fast reaction regime.

This makes the interfacial area the only unknown in Eq. (4) for all experiments carried out at standard conditions of oxygen injection through capillaries and high stirring speed.

However, the validity of Eq. (4) has to be checked by an estimation of Hatta's number. It was decided to measure k_L and $k_L A$ in the same cell under the same hydrodynamic conditions for the absorption of carbon dioxide into water. k_L was measured at low stirring speed with plane interface without injection and $k_L A$ at standard condition. The values of k_L and $k_L A$ were estimated for the $O_2/Cu(I)$ system by using Calderbank's correlation[12-13] for k_L and Westerterp et al[14] correlation for interfacial areas A.

$$\frac{k_L\ (O_2/Cu(I))}{k_L'\ (CO_2/H_2O)} = \left(\frac{\mu}{\mu'}\right)^{-0.33}\left(\frac{\rho}{\rho'}\right)^{0.33}\left(\frac{D}{D'}\right)^{0.67} \tag{6}$$

$$\frac{k_L A(O_2/Cu(I))}{k_L' A'(CO_2/H_2O)} = \left(\frac{\mu}{\mu'}\right)^{0.67}\left(\frac{\rho}{\rho'}\right)^{0.83}\left(\frac{\sigma}{\sigma'}\right)^{-0.5}\left(\frac{D}{D'}\right)^{0.67} \tag{7}$$

The system oxygen/sodium sulfite was studied experimentally using the same equipment under the same conditions. This system is used to measure interfacial areas in chemical reactors because it is known to obey the fast reaction regime in Hatta's theory where Eq. (5) applies. The chemical kinetic constant for the $O_2/SO_3^=$ system was calculated from the plane interface experiments. Knowing kLA from Eq. (7) and A from $O_2/Cu(I)$ system by using Westerterp correlation yields kL. For both plane interface and dispersion reactions, k is derived from Eq. (5), then Ha from Eq. (3) for the k posteriori validation in Eq. (5).

Table 1 summarizes the results. The system $O_2/Cu(I)$ does not satisfy the inequality Ha $>$ 3 but it approaches it. This interpretation of results satisfies the observed linearity of the curves Cu(I) vs. t up to 85% of completion of the oxidation reaction.

In this interpretation, the interfacial area is found to increase from 227 to 308 cm^2 if the reaction volume decreases from 500 to 400 cm^3 because the impeller stirring effect is better. The same relative increase of A is observed for the $O_2/SO_3^=$ system.

Figure 6 illustrates the parallel effects of stirring rate on oxidation in the systems $O_2/Cu(I)$ and $O_2/SO_3^=$, confirming the dominant influence of interfacial area on the oxidation rate.

System	$Cu(I)/O_2$ ■	SO_3^{2-}/O_2 Δ
Initial concentrations mol./dm^3	[CuCl] = 0.095 [HCl] = 0.0625 [NaCl] = 2.00	$[Na_2SO_3]$ = 0.8 $[CoSO_4]$ = 10^{-4}
pH	1.03	8.00
θ (°C)	20	30
V_{cm^3}	500	500

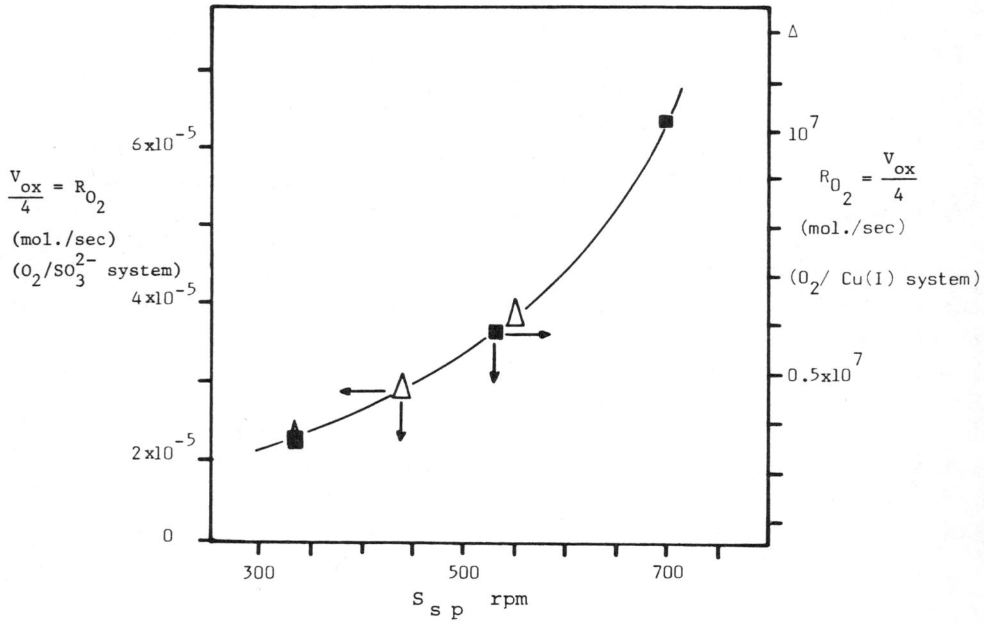

Fig. 6. Comparison of effect of stirring speed for two systems.

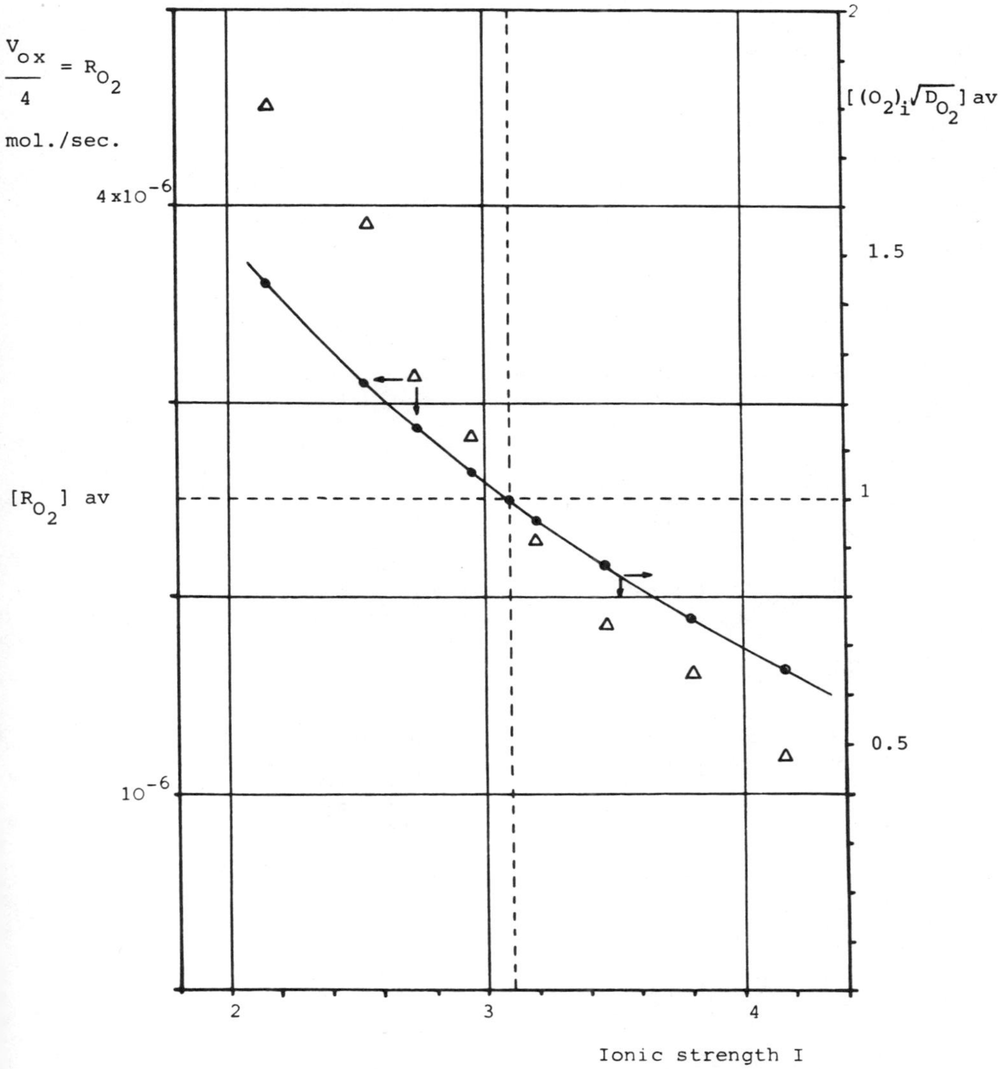

Fig. 7. Influence of ionic strength on oxidation rate.

Figure 7 illustrates the effect of ionic strength (I) on rate of oxidation of Cu(I) in conditions of Fig. 3. The points are experimental values and the solid lines show the variations of the expressions $(O_2)_i\sqrt{D_{O_2}}$ estimated as a function of ionic strength from independent measurements and correlation. The weak residual of the influence of ionic strength would come mainly from the influence of the concentration of NaCl upon the kinetic constant k.

The value of Hatta number close to 3 can explain conflicting conclusions found in the literature[2-10] about the reaction regime.

Table 1. Application of Film Theory (Fast Regime)

Experiments carried out with fixed interfacial area, stirring speed 167 r.p.m., no gas injection through capillaries.

SYSTEM	k	k_L(m/s)	Ha
CO_2/H_2O	-	9.55×10^{-5} (measured)	-
$O_2/SO_3^=$	4.76×10^3 $m^3/mol\text{-}s^{-1}$ (Eq. 5)	8.4×10^{-5} (correlation)	23.4
O_2/Cu(I)	$30.8\ s^{-1}$ (Eq. 5)	8.1×10^{-5} (correlation)	2.87

Experiments carried out with gas injection through capillaries (1.25 dm^3/min.), stirring speed 333 r.p.m., V = 500 cm^3.

SYSTEM	$K_LA/V(s^{-1})$	$A(cm^2)$	K_L(m/s)	Ha
CO_2/H_2O	4.02×10^{-3} (measured)	-	-	-
$O_2/SO_3^=$	5.67×10^{-3} (correlation)	201 (Eq. 5)	14.1×10^{-5}	13.9
O_2/Cu(I)	4.13×10^{-3} (correlation)	227 (Eq. 5)	9.1×10^{-5}	2.55

Experimental conditions

CO_2/H_2O — $(Co_2)_i = 3.82 \times 10^{-2}$ mol/dm^3 followed by pH measurements

$O_2/SO_3^=$ — $(Na_2SO_3) = 0.8$ mol/dm^3, $(CoSO_3) = 10^{-4}$ mol/dm^3, T = 30°C, P_{O_2} = 1 atm., pH = 8 followed by solution sample analysis

O_2/Cu(I) system — (CuCl) = 0.1 mol/dm^3, (NaCl) = 2 mol/dm^3, (HCl) = 0.0625 mol/dm^3, T = 20°C.

Discussion on the applicability of the film theory: non-steady-state calculations

The main shortcoming of the above treatment is the inconsistency between the kinetics of reaction assumed (zero order with respect to Cu(I)) with the kinetics of end of reactions when the reaction becomes homogeneous because the oxygen concentration in the bulk solution (Cu(I) - Cu(II)) is close to its solubility.

Precise kinetics measurements were then taken during the end of reaction.

The kinetics of the homogeneous reaction was assumed to be represented by

$$r_{O_2} = k_2(O_2)^m(Cu(I))^n \tag{8}$$

Data with variable concentration of (Cu(I)) and (O_2) were obtained by using pure oxygen and synthetic air, and by controlling the pH through additions of measured amounts of sulfuric acid until completion of 99.5% of the reaction. Treatment of all these data gave the following parameter values:

$$k_2 = 1.44 \times 10^{-3}(m^3/mol)^2\ s^{-1}, \qquad m = 1 \text{ and } n = 2 \tag{9}$$

These are within 1.3% of the data.

This result is obviously in contradiction with the above treatment of heterogeneous kinetics.

As more sophisticated surface renewal models[15,16] yield similar results, the film theory was still applied but to the system of non-steady state mass transfer with chemical reaction. A suite of quasi-steady-state mass transfer was assumed. The assumption of small oxygen concentration compared to that of Cu(I) was kept. Equations (2) and (3) were used with n = 2 and $k = k_2$. A fast reaction regime was not assumed.

The thickness of the diffusion film is δ, the volume of solution in the film V_I (for distance to interfaces x is smaller than δ), and the volume of solution in the bulk is V_S ($x > \delta$). Oxygen balance in V_S yields:

$$V_S \frac{d(O_2)_o}{dt} = N_\delta\ A - k_2\ (O_2)_o(Cu(I))_o^2 V_S \tag{10}$$

Cu(I) balance over whole volume $V = V_I + V_S$ yields

$$V \frac{d(Cu(I))}{dt} = - 4R^1_{O_2} - 4k_2(O_2)_o(Cu(I))^2_o V_S \qquad (11)$$

with

$$R^1_{O_2} = (N - N_\delta) - \frac{d \int_o^\delta (O_2)_x dx}{dt} \qquad (12)$$

From the solution of Hatta's equation,

$$(O_2)_x = \frac{(O_2)_o \sinh(\alpha\ x) + (O_2)_i \ \sinh(\alpha(\delta - x))}{\sinh(\delta\ \alpha)} \qquad (13)$$

with

$$\alpha = \sqrt{\frac{k_2\ (Cu(I))^2_o}{D_{O_2}}} \qquad (14)$$

from which are derived N, N_δ. and the integral $\int_o^\delta (O_2)_x dx$

$$N = k_L \frac{Ha}{\tanh\ (Ha)} \left[(O_2)_i - \frac{(O_2)_o}{\tanh\ (Ha)}\right] \qquad (15)$$

$$N_\delta = N \frac{(O_2)_o \ \tanh\ (Ha) - O_2)_i}{(O_2)_o - \ (O_2)_i \ \tanh\ (Ha)} \qquad (16)$$

$$\int_o^\delta (O_2)_x \ dx = \frac{[(O_2)_o + (O_2)_i]\ (Sh\ \alpha\ Ha - 1)}{\alpha Sinh(Ha)} \qquad (17)$$

This system of equation was solved by numerical techniques.

Figure 8 shows the results with the following conditions. It was assumed that homogeneous kinetics prevails and that Eqs. (8) and (9) apply.

$D_{O_2} = 1.75 \times 10^{-9}\ m^2\ s^{-1}$

$(O_2)_i = 6.85 \times 10^{-1}\ mol/m^3$, $(O_2)_o = 0$, $(Cu(I))_o = 9.5\ mol/m^3$

$V = 500\ cm^3$ (conditions C and D of Fig. 3)

$A = 227\ cm^2$

The adjusted parameter is $k_L = 20 \times 10^{-5}\ m\ s^{-1}$ which is to be compared to 9.1×10^{-5} obtained from correlation with CO_2/H_2O absorption. The same interfacial area A as before was used in the present treatment.

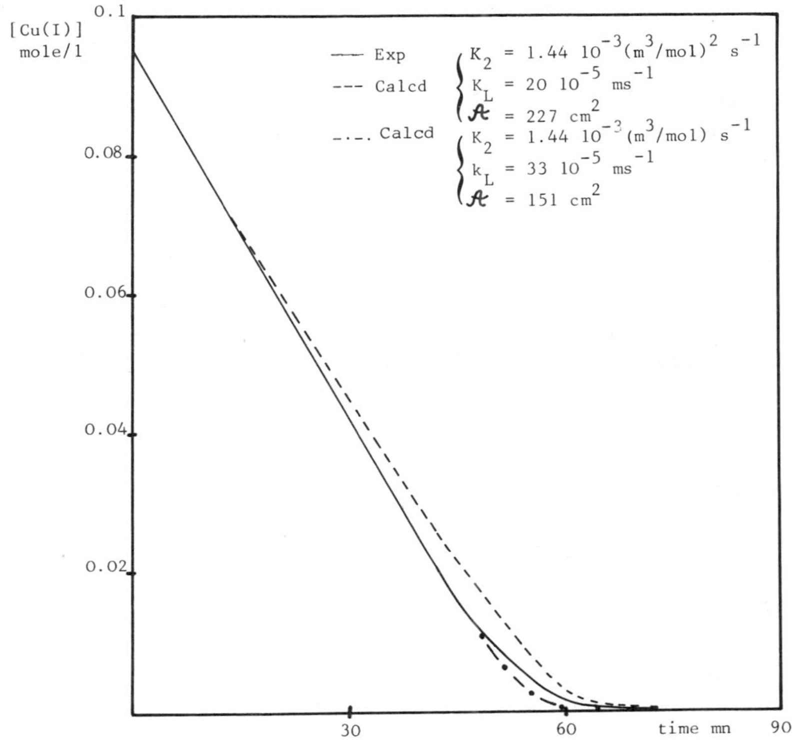

Fig. 8. Representation of oxidation kinetics by a non-steady state film model.

The results are satisfactory, taking into account that the theory is valid only if $(Cu(I))_o >> (O_2)_i$ (Fig. 8).

According to this new interpretation of results, Hatta's number should be much lower. It is not any more justified to keep the value of interfacial area determined as before from Eq. (5) assuming a fast reaction regime and zero order with respect to Cu(I). A reasonable assumption is to take the value of interfacial area from experiments using sulfite oxidation ($A = 201\ cm^2$) and to apply Westerterp[14] correlation to obtain the area in the $Cu(I)/O_2$ system ($A = 151\ cm^2$). Adjusting the mass transfer coefficient to the measured value of the rate of oxidation assuming the same chemical kinetics (2nd order with respect to Cu(I)) obtained from independent measurement, one obtains a value of $k_L = 33.10^{-5}$ m/s which is still higher than before and an initial value of Ha = 0.46. The representation of the oxidation curve with these new parameter values is still good since the product $k_L A$ remains approximately the same. It is also possible to reinterpret the measurements at a fixed area. From initial rate of oxidation, one finds $k_L = 19.10^{-5}$ m/s and Ha = 0.84. There is consistency of the ratio of k_L for fixed horizontal interface and dispersed gas in the $O_2/SO_3^=$ and O_2/Cu(I) oxidation systems, although one is in the fast regime and the other in the intermediate regime of Hatta's theory

$$\frac{k_L(O_2/SO_3^= \text{ dispersed } O_2)}{k_L(O_2/SO_3^= \text{ fixed area})} = \frac{14.1}{8.4} = 1.68 \quad \text{(Table I)}$$

$$\frac{k_L(O_2/\text{Cu(I), dispersed } O_2)}{k_L(O_2/\text{Cu(I), fixed area})} = \frac{32}{19} = 1.68$$

One should also consider that k_L is probably not uniform in space (because there are regions close to the stirrer where turbulence is high, and other quiet regions) and time (Marangoni effect).

THE KINETICS OF EXTRACTION OF Cu(II) BY LIX 65N FROM CHLORIDE SOLUTIONS

The initial experiments were carried out in the standard reactor described earlier. The chemical exchange reaction is:

$$\text{Cu(II)} + 2\ \overline{\text{RH}} \rightleftarrows \overline{\text{CuR}}_2 + 2\ \text{H}^+ \tag{18}$$

Mass transfer rates were measured by batch and relaxation methods. It was possible to measure either the variation of Cu(II) or pH in the water, or to control the pH by recorded additions of sodium hydroxide solution. In the relaxation method, equilibrium is perturbed by introduction of a small quantity (1 to 5 cm^3) of 1 M $CuCl_2$, or LIX 65N and the kinetics of return to equilibrium followed by regulation of pH. The results are obtained with a volume phase ratio of 1 and continuous aqueous phase. The results can be represented by an equation of the form:

$$R = k \frac{[\text{Cu(II)}]\ (\overline{\text{RH}})}{(\text{H}^+)} \tag{19}$$

where Cu(II) is the concentration of Cu(II) in aqueous phase, and $(\overline{\text{RH}})$ is the concentration of LIX 65N in the organic phase.

The kinetic constant k depends on the stirring rate. The interfacial area however depends on all conditions. These results are consistent with those obtained by Flett et al.[17] using the AKUFVE device.

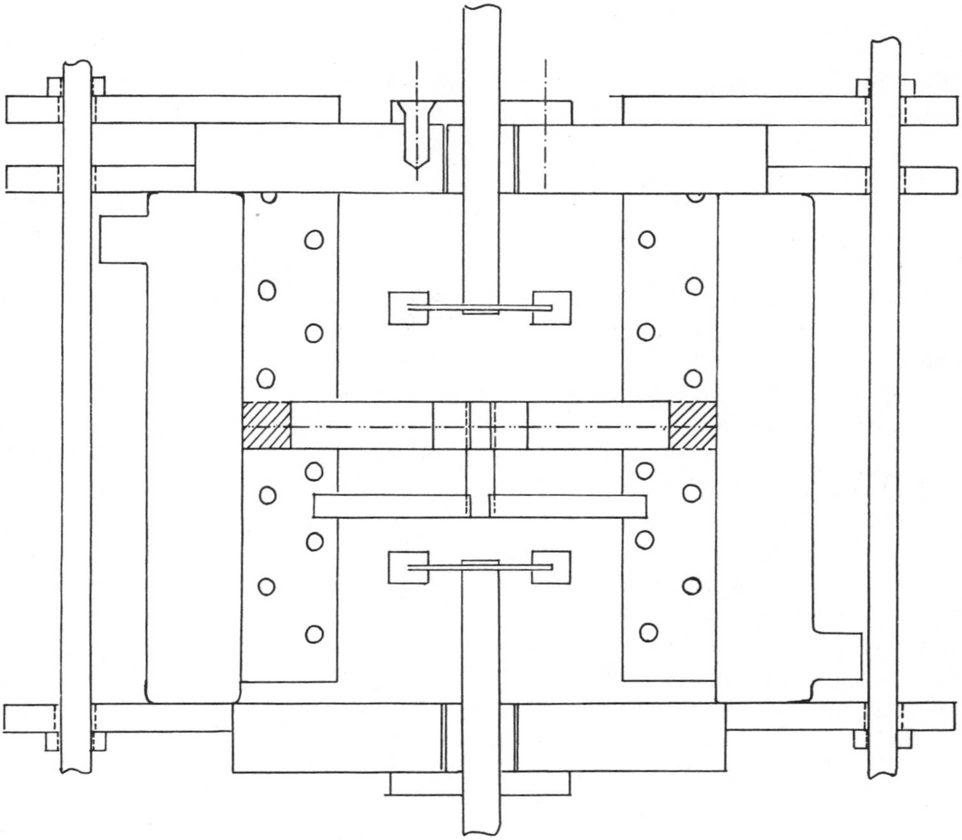

Fig. 9. Lewis Cell for extraction studies.

In order to better understand the rate of reaction, a second series of experiments were carried out in a constant interfacial area Lewis cell (Fig. 9). Aqueous phase pH is controlled by addition of NaOH, and proportional quantities of extractant are added in the organic phase at the same time in order to keep the concentrations constant in all phases. Samples of organic phase are taken to measure the variation of (CuR_2) with time. The main difficulty of this method is the very slow variation of composition because the interfacial area is small and this limits the study to initial kinetics. The results indicate that the rate of copper extraction is proportional to the interfacial area, and the stirring rate has an influence in the organic phase but not in the aqueous phase in the range 50-250 rpm. The influence of the initial concentrations of Cu(II),

RH, and H^+ are consistent with Eq. (19). An increase in the chloride concentration raises the mass transfer rate. Figure 10 shows that the kinetics are much faster in chloride solution than in sulfate solution.

An activation energy of 26 kJ/mole was obtained between 20 and 50°C.

In a third series of experiments the mass transfer rate of extractant RH in the absence of chemical reaction was measured in the same hydrodynamic conditions. It was possible to measure the concentration of LIX 65N in the aqueous phase by UV spectrometry at 203 nm. The mass transfer coefficient was found equal to 6.3×10^{-6} m/s in agreement with the hydrodynamic model of the cell.[18] The acceleration coefficient, ratio of the rate with and without chemical reaction, was found to be very high, of the order of 100 in sulfate solutions and 1000 in chloride solutions.

In a fourth series of measurement, the homogeneous reaction kinetic constant of the reaction of RH with Cu(II) in the aqueous phase was obtained by addition of Cu(II) in an aqueous solution saturated with RH and following the complex formation by UV measurement at 350 nm. The results can be represented by:

$$R' = K \left[\frac{(Cu(II))(RH)}{(H^+)} - \frac{1}{K_E}\frac{(CuR_2)(H^+)}{(RH)}\right] \tag{20}$$

This is consistent with the mechanism:

$$\begin{aligned} &RH \rightleftarrows R^- + H^+ \\ &Cu^{2+} + R^- \rightarrow CuR^+ \qquad \text{rate determining step} \\ &Cu\,R^+ + R^- \rightleftarrows CuR_2 \end{aligned} \tag{21}$$

The constants found are $K = 8 \times 10^{-3}\ s^{-1}$ and $K_E = 1.7$.

It turns out that this homogeneous kinetics is much too slow to explain the enhancement factor observed before in the framework of a simple Hatta's theory taking into account the resistance to mass transfer in the organic phase.

Another study on the same system was recently published by Kojima and Miyauchi.[19]

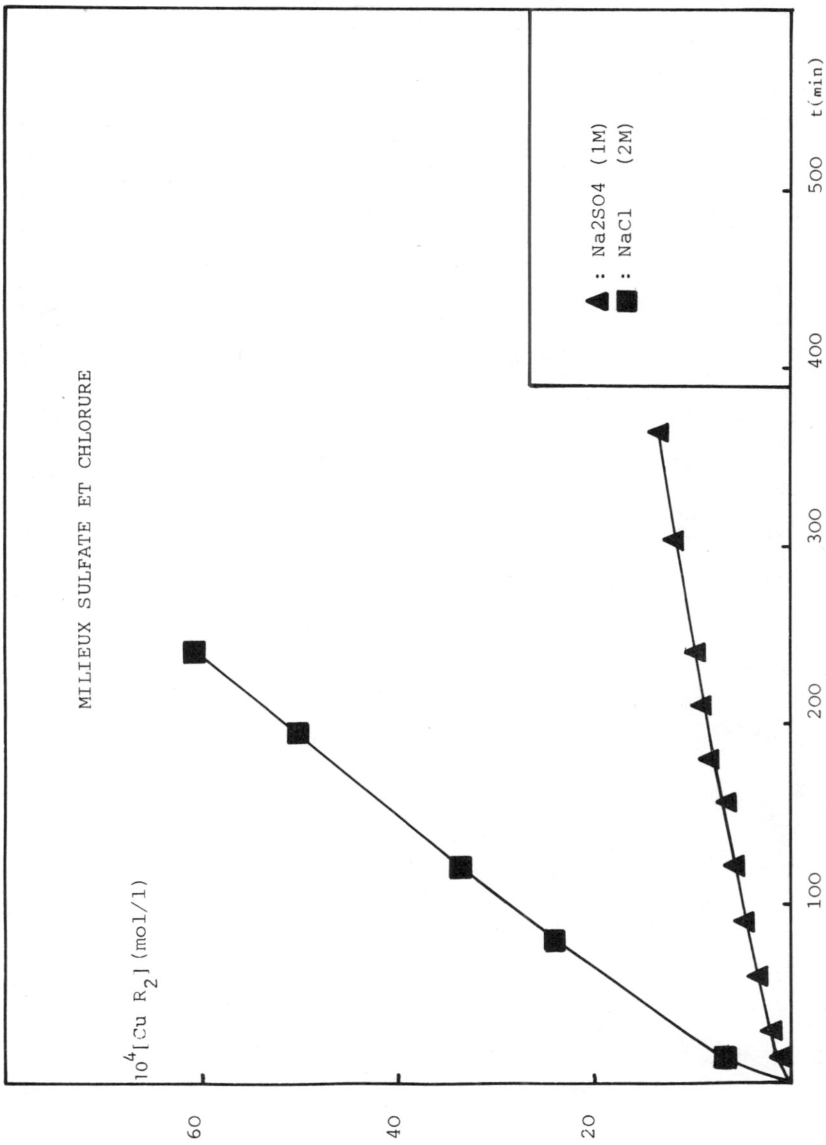

Fig. 10. Influence of chloride ions.

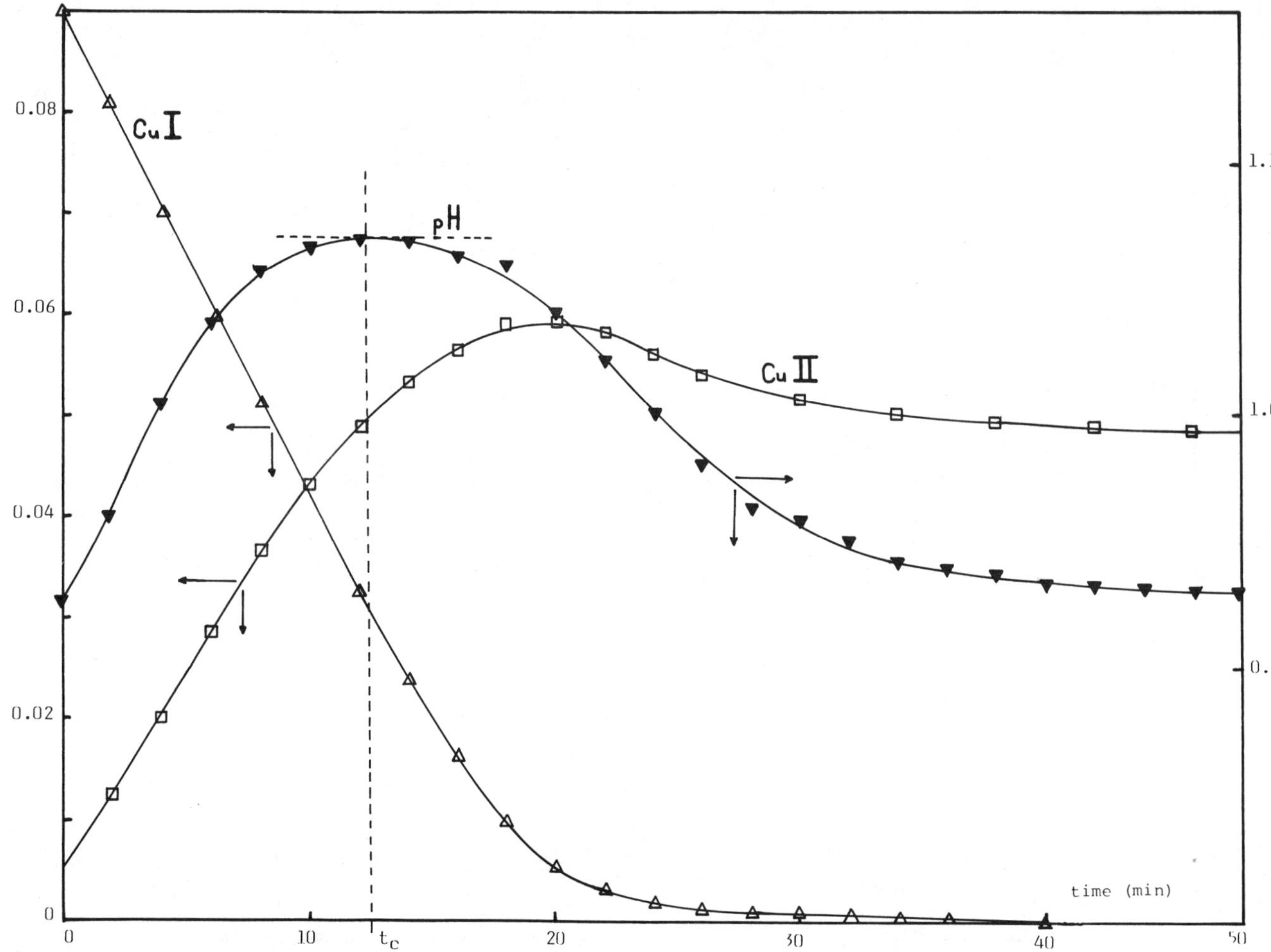

Fig. 11. Simultaneous oxidation and extraction.

THREE PHASE REACTOR

Two types of studies were started to try to understand the synergism observed[1] when oxidation and extraction reactions are carried out in the same reactor with a continuous aqueous phase with bubbles of air and drops of organic phase. It had been shown that even an inert organic phase enhances the oxidation in a stirred

with bubbles of air and drops of organic phase. It had been shown that even an inert organic phase enhances the oxidation in a stirred cell. Therefore, this would indicate a hydrodynamic effect of changing interfacial area and residence time. There are also probable kinetic effects.

Two problems in reactors with known interfacial area (plane interfaces) were examined at the laboratory scale.

First was the transfer of oxygen through the organic phase and its influence on oxidation kinetics. Initial results indicate that the transfer of oxygen is fast through the organic phase. If a stirred thick organic layer is placed between oxygen and a stirred aqueous solution of Cu(I), the oxidation rate was found to be the same as if no organic layer were present.

The second was the effect of carrying out simultaneously oxidation and extraction at an organic aqueous interface. Oxygen was circulated above the organic phase with extractant $\overline{RH}$. The aqueous phase was a solution of Cu(I). The extraction rate was found to be slower than the oxidation rate. The pH of the aqueous phase also increased. The oxidation rate increased slightly when Cu(II) was extracted.

Hydrodynamic studies are presently being carried out in a 330 l continuous reactor where the residence time distributions of each phase, the power required on the stirrer, and the volume of each phase are measured. The reactor is made of glass and some observation of the structure of the interfaces can be made. No reaction is carried out. The hydrodynamic characteristics of the phases of the reactive system are reproduced (viscosities, densities and interfacial tensions). It is observed that drops of organic phase are often surrounding the oxygen bubbles with a ratio of diameters of 3.

The results now available show that the interfacial area organic + aqueous and air + aqueous are both useful for oxidation. The results shown in Fig. 11 were carried out in a one liter three phase batch reactor where oxygen is bubbled through capillaries and illustrate the relative rate of oxidation and extraction as the reactions proceed.

REFERENCES

1. J. M. Demarthe, L. Gandon, and A. Georgeaux: Int. Symp. on Copper Extraction and Refining, J. C. Yannopoulos and J. C. Agarwal, eds., AIME, 1976, Chap. 43, p. 825-848.
2. H. Nord: Acta Chem. Scand., 1955, vol. 9, p. 430-441.
3. E. M. Moldavskaja, T. I. Kunin, and A. A. Shutov: Izvestija Vyshikh Uchonykh Zavedenjj S.S.S.R. "Khimija i Khimicheskaja Tekhnologija," 1968, vol. 11, p. 1211-1214.
4. A. S. Jhaveri and M. M. Sharma: Chem. Engng. Sci., 1968, vol. 23, p. 669.
5. A. G. Sokolovski and B. P. Maceevskij: Latvijas PSR Zinatnu Akademijas Vestis., Kimijas serija, 1973, vol. 2, p. 149-152.
6. A. G. Sokolovski and B. P. Maceevskij: Latvijas PSR Zinatnu Akademijas Vestis., Kimijas serija, 1973, vol. 2, p. 186-190.
7. B. P. Maceevskij and A. G. Sokolovskij: Latvijas PSR Zinatnu Akademijas Vestis. Kimijas serija, 1973, vol. 2, p. 153-156.
8. T. I. Kunin, A. A. Shutov and E. M. Moldavskaja: Isvestija Vuzov Kimija i Khimicheskaja Tekhnologija, 1971, vol. 14, p. 971-974.
9. A. S. Jhaveri and M. M. Sharma: Chem. Engng. Sci., 1967, vol. 22, p. 1-6.
10. Y. Awakura, M. Iwai, K. Nabeoka, and H. Majima: Denki Kagaku, 1980, vol. 48, no. 2, p. 104-110.
11. P. V. Danckwerts: Gas-Liquid Reactions, McGraw Hill, New York, 1970.
12. P. H. Calderbank: Trans. Instn. Chem. Engrs., 1958, vol. 36, p. 443-463.
13. P. H. Calderbank: Trans. Instn. Chem. Engrs., 1959, vol. 37, p. 173.
14. K. R. Westerterp, L. L. Van Dierendonck, and J. A. DeKraa: Chem. Engng. Sci., 1963, vol. 18, p. 157-176.
15. H. L. Toor and J. M. Marchello: A.I.Ch.E. Journal, 1958, vol. 4, p. 97-101.
16. P. V. Dankwerts: Ind. Engr. Chem., 1951, vol. 43, p. 1460-1467.
17. D. S. Flett, D. N. Okuhata, and D. R. Spink: J. Inorg. Nucl. Chem., 1973, vol. 35, p. 2471.
18. D. L. Olander: Chem. Eng. Sci., 1963, vol. 18, p. 123.
19. T. Kojima and T. Miyauchi: Ind. Eng. Chem. Fund., 1981, vol. 20, p. 14.

THE CHEMICAL REACTIONS OF TRIBUTYL PHOSPHATE IN THE SOLVENT EXTRACTION OF METALS

KNona C. Liddell
Department of Chemical Engineering
Washington State University
Pullman, Washington 99164

and

Renato G. Bautista*
Ames Laboratory USDOE and Dept. of Chemical Engineering
Iowa State University
Ames, Iowa 50011

Recent investigations concerning applications of tributyl phosphate to separation and recovery of metals are reviewed. Potential applications that may be commercially developed in the future are discussed along with existing processes that have been tested on a pilot plant scale or put into industrial practice. Emphasis has been placed on metals and process media of particular interest to hydrometallurgists. Developments occurring in the past 10 years are stressed. Although to date the greatest success of TBP in hydrometallurgy has been in multistage separation of rare earths, the great versatility of this reagent, combined with its relatively low cost and desirable physical properties, suggests that it may be much more widely applied to metal recovery in the future. Based on the fundamental and developmental work that has been recently carried out, commercial hydrometallurgical applications of TBP appear most likely for recovery of refractory and rare metals.

---

*now located at Dept. of Chemical and Metallurgical Engineering University of Nevada Reno, Reno, Nevada 89557.

INTRODUCTION

Tributyl phosphate is an extremely versatile reagent. To varying degrees, it extracts most of the elements of the periodic table. Careful selection of process conditions is therefore necessary to effect good separations and many stages are often required. In spite of these difficulties, hydrometallurgical applications of TBP are increasing and new processes continue to be developed. The literature on the physical chemistry of extraction by TBP suggests that the potential of this reagent has not yet been fully exploited.

Since the subject of hydrometallurgical uses of TBP has been reviewed before, we have concentrated our attention on the developments of the past 10 years. In cases where the original papers or patents may be difficult to obtain, we have cited Chemical Abstracts in addition to the original source. The chapter is organized in six sections following the customary terminology of extractive metallurgists. First is a section on alkali metals and alkaline earths. This is followed by a section on the base metals Mn, Fe, Co, Ni, Cu, Zn, In, Sn and Al. Sc, Ti, V, Cr, Nb, Mo, Ta and W are discussed under refractory metals. In the rare metals section, Ga, Tl, As, Sb, Te, Cd and Hg are considered. Ag, Au, Ru, Rh, Pd, Re, Os, Ir and Pt are covered in the section on precious metals and the platinum group. In the final section, rare earths are discussed.

ALKALI METALS AND ALKALINE EARTHS

Alkali Metals

Extraction of alkali metals by TBP has received only limited study and there appear to be no industrial-scale applications aimed at separating these elements. However, alkali metal extraction can be significant in certain systems and concern must sometimes be given to coextraction.

Hasegawa et al.[1] reported that the sodium distribution ratio in the aqueous $NaClO_4$-TBP system had a maximum value of about 0.2 near initial aqueous $NaClO_4$ concentrations of 1 M. The lithium distribution coefficient has been reported to depend strongly on the anion; for extraction from 2.5 M aqueous solutions, the distribution coefficients were 0.003, 0.044 and 0.270 for LiCl, LiBr and LiI, respectively.[2] From perchlorate solutions, the extracted sodium complex[3] is reportedly $NaClO_4 \cdot 5TBP$. $NaNO_3$ is extracted[4] as $NaNO_3 \cdot 3TBP$.

Alkaline Earths

Somewhat more attention has been paid to the extraction of alkaline earths. The solvent extraction behavior of these elements is important in view of the fact that they occur in many minerals along with metallic elements for which extraction processes have been developed. Brunisholz et al.[5] pointed out, for example, that beryllium is present in conjunction with yttrium in the mineral gadolinite.

Brunisholz et al.[5] studied the extraction of Be, Mg, Ca and Sr from HNO_3 by TBP in kerosene. Distribution coefficients were generally low. With initial aqueous metal concentrations below 2 M, the order of extraction was Ca > Be > Sr > Mg; above 2 M this changed to Ca > Be > Mg > Sr. Distribution coefficients ranged from 6×10^{-2} for Ca to 6×10^{-3} for Mg for initial aqueous concentrations of 2 M. More data were reported for Be, indicating that K_D increased with the initial aqueous Be concentration, went through a minimum as the initial acid concentration was increased and increased as the volume fraction of TBP was increased relative to the kerosene diluent.

Kolarik[3] has reported that the strontium species extracted from perchlorate media is $Sr(ClO_4)_2 \cdot 5TBP$.

A number of synergic alkaline earth solvent extractions have been examined. Extraction of beryllium from sulfuric acid solutions by the sodium salt of di-(2-ethylhexyl) phosphoric acid (NaDEHP) and TBP has been reported by El-Yamani et al.[6] In order to find the optimum conditions for extraction, synergistic factors (S) were determined

$$S = \frac{K_{Dmix}}{K_{D1} + K_{D2}} \tag{1}$$

where K_{Dmix}, K_{D1} and K_{D2} are distribution ratios of the mixture and the individual extractants 1 and 2. An S value of 5.11 was found for 0.2 M NaDEHP + 0.74 M TBP; higher NaDEHP concentrations caused formation of emulsions. The organic loading of Be under these conditions was 0.4 M Be for n-hexane diluent. The optimum pH of the aqueous phase was dependent on the diluent; the value was about 2 for kerosene but about 3 for cyclohexane and xylene. Lower pH values led to sharp decreases in K_D and higher values resulted in turbidity. Extractability was highest in diluents of low dielectric constant. Under the optimum extraction conditions, the extracted species was believed to be $Be(NaDEHP)_2(TBP)_2$.

McDowell found[7] that di-(2-ethylhexyl) phosphate and tributyl phosphate synergically extracted strontium when the diluent was an aliphatic hydrocarbon; with benzene as the diluent, there was no synergic effect. When the DEHP was nearly all protonated, $Sr(DEHP)_2{\cdot}2HDEHP{\cdot}TBP$ was believed to be present. If the ratio of sodium di-(2-ethylhexyl) phosphate to total di-(2-ethylhexyl) phosphate was above 0.2, $Sr(DEHP)_2{\cdot}4TBP$ was the predominant species. The synergism was most pronounced when 60% of the total phosphate was TBP and 30% of the DEHP was NaDEHP.

BASE METALS

Manganese

The physical chemistry of manganese extraction by tributyl phosphate apparently has not been studied nor have any industrial-scale processes come to our attention. An analytical separation scheme for Mn has been developed[8] but it is subject to interferences from many other metallic elements, especially mercury, copper, bismuth, iron and vanadium. The extracted species was believed to be $MnCl_2{\cdot}3TBP$ but this is somewhat doubtful because the TBP concentration was varied over a very narrow range and the data extrapolated far beyond this range.

Iron

Trivalent iron species[9] which can be extracted by TBP from perchloric acid solutions include $Fe(ClO_4)_3$, $FeOH(ClO_4)_2$ and $Fe_2(OH)_2(ClO_4)_4$; solvation numbers were not determined in this work. Over the pH range from -1 to 3, the distribution coefficient went through a minimum between pH -1 and 0, then increased sharply to reach a maximum of close to 10 near pH 2.5. The same log K_D vs pH curve was obtained for total Fe(III) concentrations from 2.8×10^{-4} to 8.0×10^{-2} M. Absorption spectra of TBP extracts indicated that different species were extracted at higher and lower acid and Fe concentrations. At high pH, $FeOH(ClO_4)_2$ was believed to be extracted; if the Fe concentration was greater than 10^{-2}, the high pH extract also contained the dimeric species $Fe_2(OH)_2(ClO_4)_4$.

Haggag et al.[10] have compared the extraction of ~ 10^{-3} M Fe from various acid solutions. A multistage extraction was necessary to effect the separation of Fe(II) and Fe(III) using pure TBP; HCl, HBr, H_2SO_4, HCl + H_2SO_4, and HBr + H_2SO_4 were considered. Halide complexes were extracted in all cases except that of H_2SO_4 solutions. Fe(III) was more extractable than Fe(II). Much higher acid concentrations were required for significant extraction (K_D > 10) in the sulfuric acid system than in the others; the authors suggested the use of HCl in analytical applications.

TBP has been suggested for purification of HCl solutions in metallurgical processing[11]; use of 50 vol % TBP in kerosene was suggested over the temperature range 10-60°C.

In the Falconbridge matte leach process, Fe(III) has been separated from HCl, Ni(II), Co(II) and Cu(II) using ~5% TBP in an aromatic hydrocarbon[12,13]; two extraction stages and three stripping stages are used. The pregnant solution contains roughly 120 gpl Ni, 160 gpl HCl and 2 gpl each of Fe, Co and Cu. Following the separation of Fe, cobalt and copper are extracted from the liquor with tri-iso-octylamine.[13]

Cobalt

From acidic chloride solutions, the cobalt-containing species extracted reportedly depend on the acid concentration[14,15] and on the concentration of added salts such as $CaCl_2$[14] but the details of these reports are not in agreement. For HCl concentrations up to 8 M, NaCl up to 5.5 M and $CoCl_2$ between 1.7×10^{-6} and 1.7×10^{-3} M, the Co distribution ratio was independent of the $CoCl_2$ concentration.[15] The species extracted from solutions with less than 4.8 M HCl was $CoCl_2 \cdot 3TBP$; with greater acid concentrations $HCoCl_3 \cdot 3TBP$ was extracted. The TBP concentration had no effect; both benzene and hexane were used as diluents. Extraction of $[TBP \cdot H(H_2O)_4]_2[CoCl_4]$ from 7.8 - 12.4 M HCl and $[Ca(TBP)_2(H_2O)_2][CoCl_4]$ from concentrated $CaCl_2$ have been reported.[14] At low HCl and $CaCl_2$ concentrations, $CoCl_2 \cdot 2TBP$ was the main compound extracted. Fifty percent TBP in toluene has been used to separate Co from HCl and $CaCl_2$ in a pulsating packed column.[16] 99.5% extraction was achieved.

Apelblat[17] has reported that the species extracted by TBP from aqueous cobalt nitrate solutions is $Co(NO_3)_2(TBP)_3(H_2O)_h$; the hydration number h was not determined. This confirms an earlier report by Healy et al.[4] The distribution ratio is fairly low[17] but increases with the $Co(NO_3)_2$ concentration; with the aqueous phase 4 M in Co, the organic phase was about 0.85 M. The activity of TBP in the organic phase was determined to be close to 0.5 for all $Co(NO_3)_2$ concentrations between 0 and 3.5 M.

Cobalt extraction by heated TBP is endothermic with HCl as the aqueous medium.[18]

Copper

Extraction of copper by TBP has scarcely been studied, but it is known that extraction by pure TBP from hydrochloric acid is endothermic.[18]

Zinc

Extraction of zinc by TBP has been much studied. A great deal of the reported work was done in chloride solutions. The species extracted depend on other salts that may be present. From LiCl-$ZnCl_2$ solutions, $ZnCl_2 \cdot 2H_2O \cdot 2TBP$, $[Li \cdot 4H_2O \cdot TBP]$ and $[Li \cdot 4H_2O \cdot TBP][ZnCl_4]$ are extracted. From $CaCl_2$-$ZnCl_2$ solutions, the species extracted by pure TBP are $ZnCl_2 \cdot 2H_2O \cdot 2TBP$, and $[Ca \cdot 6H_2O \cdot 2TBP][ZnCl_4]$. Only $ZnCl_2 \cdot 2H_2O \cdot 2TBP$ was said to be extracted from solutions containing NaCl.[19] Other workers[20] stated that 1-4 wt % TBP in hexane extracted Zn as $HZnCl_3 \cdot 3TBP$ from 1-7 M HCl and as $H_2ZnCl_4 \cdot 3TBP$ when more acid was present; from NaCl-HCl solutions, a small amount of zinc was reported to be extracted in the form of solvated $NaZnCl_3$. Rice and Smith[21] investigated the thermodynamics of zinc extraction from chloride solutions; the extracted species was taken to be $ZnCl_2 \cdot 2TBP$. Tributyl phosphate behaved almost ideally in the H_2O-NaCl-TBP and H_2O-$ZnCl_2$-TBP systems; the H_2O-HCl-TBP system, however, showed nonideal behavior which was tentatively attributed to formation of the two complexes TBP•HCl and TBP•2HCl. Assuming that the NaCl-$ZnCl_2$-H_2O-TBP system was ideal, the equilibrium constant for

$$ZnCl_2(aq) + 2TBP(org) = ZnCl_2 \cdot 2TBP(org) \qquad (2)$$

was determined to be 9.8.

A process to recover zinc from industrial liquors containing NaCl has been developed by Bressa et al.[22] With multistage operation using a kerosene-TBP organic phase, Zn yields exceeded 95%. Iron interfered with the separation.

Ritcey et al.[23] have developed a process for the separation of Zn from chloride liquors also containing Fe and Pb. The extraction of zinc was most selective with 40-50% TBP. The best diluent tested was Solvesso 150; use of other diluents was associated with phase separation problems. A zinc circuit was proposed involving 6 extractions. Three scrubbing and 6 stripping stages were proposed.

Zinc can be extracted synergically by LIX 34 + TBP.[24] The HDEHP-TBP system is antagonistic;[25] the distribution ratio drops sharply when TBP is added to HDEHP. This was ascribed to an interaction taking place in the organic phase between HDEHP and TBP to form $(HDEHP)_2(TBP)_n$.

The separation of zinc and cadmium in chloride solutions has potential industrial applications and has been reported on by Forrest et al.[26], Fletcher et al.[27] and Rice and Smith.[28] Zinc is more readily extracted than cadmium; Zn/Cd separation factors reported by Forrest et al.[26] range from 9 to 200. The separation was enhanced by diluting the TBP in kerosene and by reducing the metal (Cd or Zn) concentration; 2 M $CaCl_2$ gave larger separation factors than 3 M NaCl. The highest separation factor was obtained with 50% TBP in kerosene, 2 M $CaCl_2$ and less than 10^{-2} M Cd or Zn. From neutral solutions, the predominant extracted species were reportedly $ZnCl_2 \cdot 3TBP$ and $CdCl_2 \cdot 2TBP$; in dilute acid, $HZnCl_3$ and $HCdCl_3$ are also extracted and there were indications that H_2CdCl_4 was extracted under some conditions (HCl > 0.5 M, with NaCl and $CaCl_2$). Rice and Smith[28] surveyed various methods of separating Zn and Cd for a liquor from a zinc plant. Of the reagents tested that would preferentially extract zinc, TBP was the best. A separation factor of 2.6 was obtained under simulated plant conditions. Fletcher et al.[27] have presented details of two Zn-Cd separation processes that have been tested on a pilot plant scale. Using kerosene solutions of either TBP or naphthenic acid, it is possible to obtain a raffinate with a Cd:Zn ratio of better than 10000:1 with 99% Cd recovery. The TBP process had significant advantages from a process control point of view; it was much less sensitive to pH or to variations in the Zn content of the feed and crud formation was less of a problem. Using TBP would, however, require more stages than would naphthenic acid and phase separation would be more difficult with TBP.

Alian and El-Kot[29] devised a Zn-Cd separation using H_2SO_4; although their main interest was in analytical applications, the method gave very good separations of macro amounts of these metals and they indicated their belief that it would be adaptable to industrial use.

The extraction of zinc from HCl solutions is exothermic when pure TBP is used.[18]

Cobalt, Nickel, Copper, Zinc and Cadmium

Sato[30] has made some interesting comparisons among Co, Cu, Zn and Cd extraction from hydrochloric acid. For equilibration at 20°C with no diluent, the distribution coefficients of Zn, Cd and Cu had maxima as the initial acid concentration increased. The distribution coefficient of cobalt, however, showed a steady increase. Below 8 M HCl, the order of the distribution coefficients was Zn > Cd > Cu > Co; at greater acid concentrations, it changed to Zn > Cd > Co > Cu. If LiCl was added to the hydrochloric acid, Co and Cu formed tetrasolvates; with LiCl not present, these elements formed disolvates. Zinc and cadmium formed tri- or tetrasolvates in each instance. These results may not be directly comparable since different diluents were used--benzene for Co and Cu and kerosene for Zn and Cd. Distribution coefficients of Cu, Zn and Cd decreased with higher temperatures; for cobalt, the distribution coefficient increased with temperature.

Extraction of Co, Cu, Ni and Zn perchlorates was investigated by Aggett et al.[31] At constant ionic strength, the distribution ratio for Co and Ni increased with $NaClO_4$ concentration; for Zn, the ratio had a broad minimum. Sodium perchlorate itself was extracted to some extent, with distribution ratios near 0.1 between 0.1 and about 5 M $NaClO_4$.

The extraction was represented by

$$M^{2+}(aq) + 2ClO_4^-(aq) + m\,H_2O(aq) + 2TBP(org) = [M^{2+}(H_2O)_m(TBP)_2][ClO_4^-]_2(org) \quad (3)$$

Indium and Tin

Extraction of Sn(II) and Sn(IV) has been compared by Woidich and Pfannhauser.[32] In solutions with diluents of fairly low dielectric constant, Sn(IV) had a higher distribution coefficient than Sn(II); this was true in cyclohexane, petroleum ether, undecane, *n*-hexane and benzene. In xylene, CCl_4 or $CHCl_3$, Sn(II) was more readily extracted. Generally, considerably higher K_D values were observed with the first group of diluents. It would, therefore, be fairly easy to extract Sn(IV) from Sn(II) but more difficult to do the reverse. Two extraction reactions were proposed based on data from the cyclohexane and benzene systems

$$SnCl_6{}^{2-}(aq) + 2TBP(org) = SnCl_4 \cdot 2TBP(aq) + 2Cl^-(aq) \tag{4}$$

$$SnCl_3{}^-(aq) + 2TBP(org) = SnCl_2 \cdot 2TBP(org) + Cl^-(aq) \tag{5}$$

The distribution ratio increases continuously with total Cl^- concentration.

Sn(IV) and In(III) were studied by Golinski.[33,34] Tin was much more readily extracted than In; both elements were better extracted if the diluent was nonpolar. Two reactions were suggested for the extraction of In(III)

$$In^{3+}(aq) + 3H^+(aq) + 3Cl^-(aq) + nTBP(org) = InCl_3 \cdot nTBP(org) + 3H+(aq) \tag{6}$$

$$In^{3+}(aq) + 4H^+(aq) + 4Cl^-(aq) + nTBP(org) = HInCl_4 \cdot nTBP(org) + 3H^+(aq) \tag{7}$$

where the value of n is 2 or 3. The first reaction was important at lower acid concentrations, the second at higher concentrations. Tin was believed to be extracted according to the reactions

$$Sn^{2+}(aq) + 2Cl^-(aq) + nTBP(org) = SnCl_2 \cdot nTBP(aq) \tag{8}$$

$$Sn^{2+}(aq) + 3H^+(aq) + 3Cl^-(aq) + nTBP(org) = HSnCl_3 \cdot nTBP(org) + 2H^+(aq) \tag{9}$$

$$Sn^{4+}(aq) + 4Cl^-(aq) + nTBP(org) = SnCl_4 \cdot nTBP(org) \tag{10}$$

$$Sn^{4+}(aq) + 6H^+(aq) + 6Cl^-(aq) + nTBP(org) = H_2SnCl_6 \cdot nTBP(org) + 4H^+(aq) \quad (11)$$

with n = 2 or 3. Extraction of indium was not influenced by the oxidation state of the tin and only slightly influenced by its concentration. Separation factors as high as 400 were obtained with 0.5 M TBP, 2.379 gpl Sn(IV) and 40 gpl indium.

Aluminum

Aluminum is readily extractable from $NaClO_4$-HCl solutions. Over 90% extraction from 7.41×10^{-3} M Al, 1.6-3.2 M $NaClO_4$ and pH 4 to 0.5 has been reported.[35]

Somewhat surprisingly, hydrolyzed and even polymerized Al has been said to be extracted from perchlorate solutions.[36] Extraction reportedly increased as the solution pH and extent of hydrolysis rose.

REFRACTORY METALS

Scandium

Solvent extraction with tributyl phosphate is an attractive route to the recovery of scandium and has received much attention, particularly in the USSR. Most work has been done on extraction from chloride systems.

When scandium is extracted from HCl solutions by TBP, the solvate in the organic phase is $ScCl_3 \cdot HCl \cdot 2TBP$ according to Samodelov;[37] Morris and Sutton[38] maintained, however, that it is likely that there are several distinct solvates of the general formula $ScCl_3 \cdot xTBP$.

In the extensive process development work carried out in the USSR, the effect of iron has been a major concern. Infrared spectroscopy has indicated that Fe is extracted from HCl as $HFeCl_4 \cdot nTBP$; if the aqueous phase contains $MgCl_2$ and $FeCl_3$, the solvated species is $FeCl_3 \cdot nTBP$.[39,40,41] The presence of iron favors the extraction of scandium since Sc distribution ratios for extraction from $MgCl_2$ solutions by TBP saturated with $FeCl_3$ are much higher than those for extraction by TBP alone.[41] From 5 N $MgCl_2$, Sc extraction dropped from 62.5% if iron were present to 16.6% in its absence.[42] Iron had less effect on extraction from HCl. Scandium and iron can be readily separated by washing and back-extraction. Washing the extract with HCl converts $FeCl_3 \cdot nTBP$ to $HFeCl_4 \cdot nTBP$; the latter species is more extractable.[39] Scandium can then be back-extracted by 2 N HCl.[39,40,41] The salting out ability for Sc of various salts was determined to be[42] KCl < $CaCl_2$ < $AlCl_3$ < $MgCl_2$ < LiCl < $FeCl_3$. Small amounts of Ca and Al are also extracted along with scandium in the system $ScCl_3$-$CaCl_2$-$AlCl_3$-HCl-H_2O-TBP-kerosene.[43] With 70% TBP in kerosene, the maximum Sc extraction capacity was 43.5 gpl. Regardless of the HCl, Ca or Al concentrations, 7-10% of the Ca and 3-7% of the Al were extracted. Gallium forms a binuclear solvate with Sc; with heated TBP extracting Sc and Ga from 0.1-4 M HCl, the extracted species include $ScCl \cdot (GaCl_4)_2 \cdot 3TBP$.[44] Seventy percent TBP in kerosene was used in a process developed by Favorskaya et al.[39] to separate Sc from Ca, Al, Mg, Fe(II) and Fe(III). Two back-extraction steps were employed. First 10 N HCl was used to back-extract Ca, Al, Mg and Fe(II). Scandium was then back-extracted by H_2O or < 2 N HCl. A hydroxide precipitate containing 7.6% Sc_2O_3 was obtained in about 90% yield by adding NH_3.[45] Back-extraction of Sc without first back-extracting the other elements gave a precipitate containing only 0.78% Sc_2O_3. This process has been successfully applied to a leach liquor containing $MgCl_2$, $FeCl_2$, $TiCl_4$, $FeCl_3$, KCl, NaCl, $CaCl_2$, $AlCl_3$, and $ScCl_3$.[45] The hydroxide precipitate initially contained 5.3-6% Sc_2O_3; when dissolved in H_2O and recrystallized the solid contained 20-30% Sc_2O_3.

A synergic effect has been reported for the extraction of scandium from HCl solutions.[46] Addition of petroleum sulfoxides or trialkylbenzylammonium chlorides to TBP caused synergic extraction. However, di-(2-ethylhexyl)-phosphate was antagonistic.

A number of other aqueous media have been investigated. Korovin et al.[47] compared Sc distribution coefficients in hydrochloric, perchloric, nitric and sulfuric acid. Below an acid concentration of 6 M, the distribution coefficients decreased in the order $HClO_4 > HNO_3 > H_2SO_4 > HCl$. At higher acid concentrations the order changed to $HClO_4 > H_2SO_4 > HCl > HNO_3$.

There are several scandium species extracted from aqueous $HClO_4$. From ir and NMR spectroscopic work, Berezhko[48] et al. concluded that the extracted species is $[Sc(TBP)_i(H_2O)_{6-i}](ClO_4)_3$ with i = 1,2,3. Samodelov[49] reported that the species extracted depends on the scandium concentration, being $Sc(ClO_4)_3 \cdot 3TBP$ at "high" concentrations; detailed conditions were not given.

Distribution coefficients were determined for Sc extraction from $HClO_4$ and H_2SO_4 in work by Berezhko et al.[50] and it was found that H_2SO_4 gives a much lower distribution ratio than $HClO_4$. With 50% TBP in xylene, K_D was 7.2 for extraction from 5 M $HClO_4$; K_D dropped to 1.2 for extraction from 4.5 M $HClO_4$ and 0.45 M H_2SO_4 and declined to 0.11 for a system that contained 4.9 M H_2SO_4. The initial aqueous phase Sc concentration in these experiments was 1 gpl Sc_2O_3.

Separation of Sc and Ca in the system $Sc(NO_3)_3$-$Ca(NO_3)_2$-HNO_3-TBP-H_2O has been studied by Radhakrishnan and Owens.[51] The Sc distribution ratio increased with the sum of the aqueous $Sc(NO_3)_3$ and $Ca(NO_3)_2$ equilibrium concentrations but decreased with increasing HNO_3 concentration. The Sc/Ca separation factor increased with both the total equilibrium salt concentration and with the HNO_3 concentration. The ratio of aqueous Sc to Ca had no effect on the separation factor. Radhakrishnan and Owens envisioned that a process based on the extraction of Sc from nitrate solutions could be used for the recovery of Sc from slag resulting from reduction of Sc_2O_3.

During separation from basic nitrate media, scandium was reported to be extracted as $ScOH(NO_3)_2 \cdot nTBP$, $Sc_2(OH)_2(NO_3)_4 \cdot 2nTBP$ and $Sc_2OOH(NO_3)_3 \cdot 2nTBP$.[52] The last complex formed only if the $OH^-/Sc(III)$ ratio was 1.5 - 2.0; the value of n was either 2 or 3.

Distribution ratios have been found to be increased when mixed acids were used. At a total acid concentration of 4 M, extraction was greater from HCl-$HClO_4$ mixtures than from either 4 M HCl or 4 M $HClO_4$; an analogous result was found for HNO_3- $HClO_4$ mixtures compared to pure HNO_3 and $HClO_4$. Sekine et al.[53] attributed this behavior to the extraction of mixed ligand complexes but were unable to relate changes in the distribution ratio to changes in the speciation in the aqueous phase.

Above 6 M acid, the Sc distribution ratio decreased in the order[53] $HClO_4 > HCl > HNO_3$. Below 6 M, the order changed to $HClO_4 > HNO_3 > HCl$. The distribution ratio was fairly insensitive to the H^+ concentration itself but depended strongly on the anion concentration.

Titanium

The solvent extraction of titanium is complicated by hydrolysis. Roland and Gilbert[54] obtained Raman, infrared and ultraviolet spectra for $TiCl_4$-HCl-H_2O-TBP systems; CS_2 and CCl_4 were used as diluents. The initial aqueous Ti(IV) concentrations ranged from 0.1 M to about 0.6 M. At high acid concentrations (~11 M HCl) and low TBP concentrations (< 0.1 M), the extracted compound was $(TBP)_2TiCl_4$. Hydrolysis occurs at low HCl concentrations and it is likely that there are several titanium species in the organic phase; these could not be characterized spectroscopically. Startsev and Krylov[55] reported that the Ti species extracted by pure TBP from HCl solution is $TiCl_4 \cdot 2TBP$ near HCl concentrations of 6 M or without excess TBP; when both the HCl and TBP concentrations were high $H_2TiCl_6 \cdot 3TBP$ was extracted.

Sorokin and Denisova[56] developed a process to recover Ti and several other metals from titanium slag chlorination wastes. The organic phase was 70% TBP in kerosene. The aqueous phase had a pH ⩾ 9 and contained a minimum of 5 equivalents of Cl^- per liter. With these process conditions, 92 -100% of the Nb, 50 - 70% Ti, 60 -100% Ta and 10 - 40% V were extracted.

Recovery of Ti from sand has been shown to be technically feasible by Tedesco and de Rumi;[57] they stated that the process could be economic if Fe were also recovered and if acid were recycled. Titanium-bearing sands were first refluxed with 9 M HCl. The residue was then treated with 7 M H_2SO_4 at 100°C for 4-12 h; additional acid was added until the H_2SO_4 concentration was 13 M. Ti was then extracted at room temperature with TBP and back-extracted with H_2O. The titanium in solution was oxidized with air and precipitated as $TiO_2 \cdot xH_2O$. Upon ignition, solid TiO_2 of 99.9% purity was obtained.

Tedesco and de Rumi[58] have also done fundamental work on Ti extraction from H_2SO_4 solutions. The acid concentration had a strong effect on titanium extraction. Only negligible amounts were extracted at H_2SO_4 concentrations below 8 M. Between 8 and 13 M, the Ti distribution coefficient increased with the acid concentration; extraction with acid concentrations above 13 M was not practical because of a gradual decomposition of the organic phase. Between 10 and 13 M, two extraction reactions were reported:

$$TBP(org) + TiOSO_4(aq) = TBP \cdot TiOSO_4(org) \quad (12)$$

$$2TBP(org) + TiOSO_4(aq) = (TBP)_2 \cdot TiOSO_4(org) \quad (13)$$

The complexes in the organic phase were anhydrous as long as the TBP concentration was above 0.5 M but were hydrated at lower values. CCl_4 was the diluent used. In 13 M acid, the Ti distribution ratio was about 10 for extraction by 1 M TBP in CCl_4. Acid was preferentially extracted when its concentration was greater than 11 M.

Extraction of Ti(IV) from hydrochloric acid has also been investigated.[59,60]

Vanadium

The solvent extraction chemistry of vanadium is complicated and there is no consensus in the literature as to the nature of the species in the extract. Extraction from HCl solutions has been investigated by Vinarov et al.,[61] Tedesco and de Rumi[62] and Sato.[63] Vinarov et al.[61] concluded that the V(V) compound extracted is likely to be $H_2VO_3Cl \cdot nTBP$ with n = 1 or 2. The largest distribution ratio they reported was 17.1 and was obtained with an initial acid concentration of 6 N and an initial aqueous vanadium concentration of 0.1 at an ionic strength of 5.05. Tedesco and de Rumi[62] rejected the hypothesis of Vinarov et al., stating that the extraction reactions of V(V) are

$$O_2Cl(aq) + 2TBP(org) = VO_2Cl(TBP)_2(org) \quad (14)$$

$$O_2Cl(aq) + 3TBP(org) = VO_2Cl(TBP)_3(org) \quad (15)$$

They reported that the distribution coefficient increased with the HCl concentration and with the concentration of TBP in CCl_4; with 8 M HCl and 2 M TBP, K_D was nearly 25. Sato[63] reported on the extraction of V(IV) by 12 M TBP in benzene. For extraction from HCl, the distribution coefficient had a maximum at 7 M HCl but with 0.1 M HCl and varying amounts of LiCl, K_D increased with increasing Cl^- concentration. Under all conditions Sato studied, the value of K_D was less than 1. The extracted species were believed to have been hydrolyzed to $VOOHCl \cdot TBP$ and $VOOHCl \cdot 2TBP$.

Chromium

The chromium species extracted from 1-4 N HCl are $Cr(H_2O)_3Cl_3 \cdot 3TBP$ or $[Cr(H_2O)_2(OH)Cl_3][H(H_2O)TBP]$.[64]

Aggett and Udy[65] studied Cr extraction from $HClO_4$ because they were interested in the role of hydrolysis in solvent extraction and wanted to work with a trivalent cation about the same size as Fe^{3+} but which would not be so prone to hydrolyze. They found that the distribution ratio increased with $NaClO_4$ concentration and Cr(III) concentration; most experiments also showed an increase in K_D with increasing pH although a maximum in K_D was found as the pH was varied at 10^{-4} M Cr(III) and 5 x 10^{-2} M $NaClO_4$. The K_D values reported covered the range from 10^{-2} to 1. Absorption spectra indicated that hydrolyzed species were not extracted; the TBP extracts, however, were thought to contain a mixture of polymerized species.

TBP can be used to remove Cr(VI) from waste streams coming from plating baths.[66] Cuer et al.[66] compared the performance of TBP and amine extractants and concluded that use of TBP was more economical. In 2-4 N acid, the major Cr(VI) species are $Cr_2O_7^{2-}$ and $HCr_2O_7^-$; both are extracted as $H_2Cr_2O_7 \cdot nTBP$. While TBP was not as good an extractant as Alamine 336 or LA2, the TBP loading increased with acid concentration and was 30 gpl Cr in 2N acid. TBP was more stable than the amines under the oxidizing, strongly acidic conditions of interest. Up to 99.5% of the chromium in a waste stream can be recovered using TBP.

Niobium

Niobium can be quite readily extracted from chloride solutions under some conditions. Nb is not extracted from HCl solutions by 100% TBP if the acid concentration is less than 7 M;[67] above this concentration, the distribution ratio increases with the amount of acid. The extracted compound is $HNbOCl_4 \cdot 3TBP$ and/or $H_2NbOCl_5 \cdot 3TBP$ when niobium is extracted by TBP in benzene.[67] Two extraction reactions were reported by Sellier et al.,[68] both giving the same extracted species

$$NbO_2H^{2+}(aq) + 2H^+(aq) + 4Cl^-(aq) + 3TBP(org) = Nb(OH)_2Cl_4H(TBP)_3(org) \qquad (16)$$

$$NbO_2H_2^{3+}(aq) + H^+(aq) + 4Cl^-(aq) + 3TBP(org) = Nb(OH)_2Cl_4H(TBP)_3(org) \quad (17)$$

The distribution ratio increased sharply with HCl concentration; from a value of 10^{-2} at 4 N HCl, it increased to 10 at 6 N. A 1.093 M solution of TBP in dodecane was used. The distribution ratio also increased with TBP concentration up to 2 M; data were not reported for higher concentrations but it appears likely that the K_D value goes through a maximum just above 2 M HCl. Interestingly, for this series of experiments, the distribution coefficient was said to be higher for 4.72 N HCl than for 5.27 N HCl, in seeming contradiction to the data on the effect of variable HCl concentrations at 1.093 M TBP. At 1.093 M TBP and a constant H^+ concentration, the distribution ratio was larger at higher Cl^- concentration; LiCl was added in these runs.

Tantalum and niobium can be separated in fluoride solutions.[69] The niobium species extracted are $[nTBP\cdot H^+\cdot(H_2O)_m]NbOF_4^-$, $[nTBP\cdot H^+\cdot(H_2O)_m]NbF_6^-$ and $[nTBP\cdot H^+\cdot(H_2O)_m]HNbF_4^-$.

TBP has an antagonistic effect on the extraction of niobium(V) from aqueous perchlorate solutions by di-(2-ethylhexyl) phosphoric acid.[70]

Molybdenum

Molybdenum has a complicated aqueous phase chemistry which has made it difficult to understand its solvent extraction behavior.

The species extracted from solutions 1-4 N in HCl is $MoO_2Cl_2\cdot 2TBP$;[71,72] at very high HCl concentrations $HMoO_2Cl_3\cdot 2TBP$ is extracted. With 5.1 M HCl and 10^{-8} to 10^{-2} M Mo, log K_D was 1.3.[72] At slightly higher molybdenum concentrations, 0.01 to 0.1 M, the value of log K_D dropped to about 0.1; this was attributed to dimerization. Maximum extraction was observed at an HCl concentration of 9 M.

The distribution ratio for extraction of Mo(VI) from HCl solutions by TBP increases with acid concentration but decreases with temperature.[73] The extracted species were believed to be less stable at higher temperatures causing the extraction equilibrium to shift in the direction of the aqueous phase.

Extraction of molybdenum from HF solutions can be very high but is quite dependent on the acid concentration. No Mo was extracted from 4 M HF but 65.9% extraction was achieved at 0.25 M.[74] Mixtures of HF and H_2SO_4 gave considerably higher extraction; 88-94% of the Mo was extracted in one step from 1 M HF, 2-4 M H_2SO_4 and 1-50 gpl Mo. The extraction was selective with respect to Fe(III), Sn(IV), Ti(IV) and Mn(II).

The extraction of anions of $H_3PMo_{12}O_{40}$ and $H_4SiMo_{12}O_{40}$ by TBP in benzene or nitrobenzene was studied by Lakshmanan and Haldar;[75] the former compound was extracted more readily. The distribution ratio of $H_3PMo_{12}O_{40}$ increased with TBP concentration and with the concentration of the heteropoly acid; extraction was greater with nitrobenzene-TBP than with benzene-TBP media. Distribution ratios between 10^{-4} and 16.8 were reported for $H_3PMo_{12}O_{40}$.

Urbanski et al.[76] have reported on the extraction of Mo(VI) from H_2SO_4 solutions by HDEHP and TBP. Addition of TBP altered both the kinetics and the thermodynamics of this extraction system. The extraction was considerably faster in the HDEHP-TBP system. The distribution ratio, however, was decreased. The slope of the log-log plot of distribution ratio vs HDEHP concentration decreased with increasing initial Mo concentration in the aqueous phase, suggesting that different complexes were being extracted under different conditions.

Extraction of Mo(VI) from HNO_3 has been reported on by Cruywagen and McKay.[77] Distribution ratio values were 0.10 or lower. A maximum in the distribution ratio was seen between pH 0.5 and 1.0; K_D was greater at higher values of KNO_3 concentration. Slow aqueous phase condensation reactions had a significant effect on the solvent extraction. K_D was determined as a function of the time elapsed between preparation of the aqueous solution and the beginning of the extraction; the distribution ratio gradually increased with time. This was attributed to the gradual formation of additional species extractable from the aqueous phase and could be prevented by using a lower pH.

The species extracted from HNO_3 solutions containing H_2O_2 is reportedly $[H_3O(H_2O)_3 \cdot 3TBP]_2[Mo_2O_{11}(H_2O_2)_2]$.[78]

Hexavalent molybdenum has been found to be synergically extracted by TBP and tri-n-octylammonium chloride;[79] species extracted from 3 M HCl containing 0.004 M Mo(VI) were $MoO_2Cl_2 \cdot 2TBP$, $MoO_2Cl_2 \cdot 2TOA$ and $MoO_2Cl_2 \cdot TOA \cdot TBP$.

Tantalum

Influence of H_2SO_4, HCl, HBr, water-soluble alcohols and acetone on the extraction of trace amounts of tantalum fluoride have been investigated by Sanad et al.[80] The distribution ratio rose steadily with increases in the H_2SO_4 concentration from 2 to 14 M; between 10 and 14 M H_2SO_4, the distribution ratio was about 10. It was postulated that several fluoride complexes are extracted by the generalized reaction

$$H_nTaF_{5+n}(aq) + nTBP(org) = H_nTaF_{5+n} \cdot nTBP(org) \tag{18}$$

Distribution ratios for extraction from HCl or HBr were much lower than for sulfuric acid. It was suggested that in many situations separation of Ta from other elements could be improved by addition of organic additives such as alcohols and acetone.

Salting out agents were found to be necessary for the extraction of Ta from solutions of K_2TaF_7;[81] LiCl, $LiNO_3$ and Li_2SO_4 were effective. Tantalum was extracted as $LiTaF_6$.

Tantalum and tungsten are very effectively separated in a process reported by Favorskaya et al.[82] The aqueous phase contained ~ 0.09 M WO_3, ~ 0.08 M Ta_2O_5, 2-8 M HF and 1-2 M H_2SO_4. With 70% TBP in kerosene, 96-99% of the Ta was extracted but less than 1% of the tungsten. Tungsten extraction, however, increased with sulfuric acid concentration.

Korovin and Kol'tsov[69] investigated the separation of niobium and tantalum in HF solutions. The extracted species for tantalum are $TaOHF_4 \cdot nTBP$ and $TaF_5 \cdot nTBP$. With hydrofluoric acid concentrations less than 1 M, the distribution ratios of Nb and Ta were < 0.1 and ~ 10, respectively. Between 1 and 3 M HF, the Nb distribution coefficient decreased but the coefficient for Ta increased, enhancing the possibility of separating these elements easily.

Recovery of high purity tantalum and niobium oxides from ores is possible with the process developed by Prasad et al.[83] With pure TBP and 0.3 N H_2SO_4, 98% of the Ta can be extracted in three stages; 96% of the Nb was extracted from 6N H_2SO_4 in six stages.

The process was developed specifically for columbite-tantalite ores. The ore composition was 45.6% Nb_2O_5, 21.0% Ta_2O_5, 4.7% TiO_2, 16.95% Fe_2O_3, 6.17% MnO and 1.3% SiO_2; ore was fused with ammonium bifluoride and the fused material leached with cold water to provide the solution on which the solvent extraction experiments were carried out.

Tungsten

Ammonium paratungstate has been obtained from the liquor left after Mo was extracted by TBP.[84] The process has been applied to mother liquors resulting from autoclave leaching of scheelite concentrates with Na_2CO_3 and to solutions resulting from the water leaching of melts obtained by fusing tungsten waste with $NaNO_3$. A pH of 0.4-0.9 is required; H_2O_2 must be added to prevent formation of mixed Mo-W polyanions and coextraction of tungsten. Hydrogen peroxide is added in considerable excess, 1.5-3 mol per mol of W + Mo;[85] nitric, hydrochloric or sulfuric acids may be used.

Tungsten can be extracted from solutions containing fluoride or sulfate; with HF-H_2SO_4 mixtures containing 1-8 M H_2SO_4, 5-30% W extraction was observed.[82] There was no tungsten extraction in the absence of sulfate ion. With multistage countercurrent apparatus, Ta and Nb were extracted from 1-2 M H_2SO_4 and separated from W, which was only slightly extracted.

$H_3PW_{12}O_{40}$ was reportedly 100% extracted from 1-8 N HNO_3[136] by TBP in kerosene. Extraction of this compound by TBP in benzene has also been reported on.[75] The extracted species was solvated by 3 TBP molecules. The distribution ratio increased with TBP concentration and was much higher when nitrobenzene was used as the diluent rather than benzene; in nitrobenzene, the extracted acidic compound was believed to be ionized.

RARE METALS

Gallium

Gallium extraction from chloride solutions has been quite thoroughly investigated and processes developed for recovery of gallium from secondary sources.

Iron can have positive or negative effects on gallium extraction, depending on its oxidation state (ferrous or ferric) and concentration. Fe(II) reportedly causes salting out and increases the gallium distribution ratio;[86] concentrations of iron and gallium were not specified. Fe(III) has little effect on gallium extraction when present in the organic phase at a concentration of 0-26 gpl; at higher iron concentrations, distribution ratios of both Fe and Ga decreased.[87] Aqueous phase Ga concentrations of 0.05-5 gpl were used along with 4.5-7 M HCl; the distribution coefficient increased with the acid concentration.

The salting out efficiency of Li, Na, K, Mg, Ca, Zn and Cd chlorides was tested and found to decrease in the order

Li > K > Na > Mg > Ca > Zn > Cd

All of these elements were coextracted with gallium.[88]

Morris et al.[89] found that two gallium complexes were extracted from HCl. Between 0.2 and 0.6 M HCl, solvated $GaCl_3$ was the extracted species; at greater HCl concentrations, Raman spectra gave evidence of $HGaCl_4 \cdot 2TBP$. Bhat and Sundararajan[90] reported that species extracted from hydrochloric acid were $HGaCl_4$ and $GaCl_3$; if a high concentration of LiCl was also present, $LiGaCl_4$ was extracted. All these species were said to be solvated with 3 TBP molecules. The distribution ratio increased with total chloride concentration for LiCl + HCl, LiCl and $CaCl_2$ but passed through a maximum in HCl solutions. This was attributed to competition between HCl and Ga extraction. The maximum occurred at 4 N HCl. Extraction increased with TBP concentration and was 98% complete with 20% TBP in benzene. The distribution ratio decreased with increases in the gallium concentration. K_D was 8.1 with 0.019 M $GaCl_3$, 20% TBP and 3 M HCl; with 0.385 M $GaCl_3$ the distribution ratio dropped to 0.81, corresponding to 44.8% extraction. $AlCl_3$ was a particularly effective salting out agent and was not itself extracted; this suggested that TBP might be used to recover Ga from Bayer liquor and a process was developed to do this. Lime was first added to the liquor in order to precipitate about half of the alumina; gallium and the remaining aluminum were then coprecipitated by neutralizing with CO_2. This solid fraction was dissolved in HCl; an acid concentration of 3 N was maintained. Ga was extracted with 20% TBP and back-extracted with water. Following precipitation with NH_3, dissolution of the hydroxide and electrolysis, 99% pure Ga was obtained. The overall recovery was greater than 90%.

A gallium recovery of 99.5% from germanium plant solutions has been achieved with an extraction process developed by de Schepper.[91] Iron was present in the plant liquors in much higher concentrations than gallium and has a larger distribution ratio with pure TBP, presenting a significant separation problem. Iron powder was added to reduce most of the Fe(III) to Fe(II), which was not extracted. The best Ga/Fe(III) separation was obtained with 4 M H^+ and a 1:1 organic:aqueous phase volume ratio; the separation factor was 58.

Sulfuric acid has been reported to enhance extraction of gallium from HCl;[92] perchloric and phosphoric acids were less effective and had to be added in greater concentrations to have a similar effect. Nitric acid decreased the extraction.

Indium has not received much study. Some work has been done comparing In with Sn; this is discussed in the Base Metals section.

Thallium

Extraction of Tl(III) from chloride solutions has been investigated by Dodson and coworkers.[93,94] Chloride to thallium ratios in the organic phase were 2.94–3.1, indicating that $TlCl_3$ is extracted. Hexane was used as the diluent. Distribution ratios depended on the free Cl^- concentration; with initial Tl concentrations in the range 2×10^{-6} to 2×10^{-5} M, 0.50 M H^+ and an ionic strength of 0.5 M, there was a maximum K_D near 6×10^{-3} M Cl^-. The K_D value was close to 0.9 at the peak. If the Cl^-:Tl(III) ratio is less than 3:1, not enough ligand is available for $TlCl_3$ to be the principal Tl species. Under these conditions, the distribution ratio depends on the acid concentration, in contrast to its behavior at higher Cl^- concentrations. It was suggested that hydrolysis of $TlCl^{2+}$ accounts for the pH-dependence at low Cl^- concentrations. In later work[94] the effect of the extractant concentration was studied. It was found that the extracted species are $TlCl_3 \cdot 2TBP$ and, at higher organic phase Tl concentrations, $HTlCl_4 \cdot 4TBP$; _n_-octane was the diluent.

Extraction of Tl(III) and Tl(I) from H_2SO_4 was negligible except when halide ions were present.[95] The distribution ratio for extraction of Tl(III) from HCl by 50% TBP in benzene was well above 10 over the range 1–10 M HCl and was close to 100 at 5 M HCl. Values of K_D for HBr solutions were similar. In contrast to the work of Dodson and coworkers,[93] Alian et al.[95] reported that

$TlCl_3 \cdot 3TBP$ is extracted from 3.6 M HCl. Addition of HCl, HBr or KI enhanced Tl(III) extraction from phosphoric acid.

Arsenic

As(V) and As(III) can be readily extracted by TBP.

As(V) has been reported to be completely separated from Se by 100% TBP; the aqueous phase contained 500 gpl H_2SO_4.[96] TBP had an As(V) capacity of 200 gpl. The extracted complex was said to be $TBP \cdot H_2SO_4 \cdot H_3AsO_4 \cdot H_2O$.

As(III) was 100% extracted from a solution containing 4 M HCl and 2 M LiCl; 40% TBP in xylene was used.[97] The extractable species could be either a 1:2 or 1:3 As:TBP adduct; the log K_D vs. log[TBP] plot had a slope of 2.6.

Giganov et al.[98] have reported on the removal of As(V) from a copper electrolyte containing 30.4 gpl Cu, 5.72 gpl As, 26.3 gpl Ni, 0.47 gpl Sb, 1.77 gpl Fe and 170 gpl H_2SO_4. More arsenic was extracted with higher acid concentrations and a greater phase ratio. A multistage separation was necessary.

Antimony

Khorasani and Hakim[99] have found the extraction of Sb(III) from chloride solutions to be quite complicated. From NH_4Cl solutions, the species extracted is hydrated $SbCl_3 \cdot 2TBP$. Hydrated $2LiCl \cdot SbCl_3 \cdot 2TBP$ is extracted from LiCl solutions; in this case the degree of hydration apparently depends on the TBP concentration of the organic phase. Distribution ratios in the $SbCl_3$-NH_4Cl-H_2O-TBP-CCl_4 system increased sharply with the TBP/diluent ratio but were unaffected by the total aqueous Cl^- concentration. With 100% TBP, the K_D value observed was between 1.4 and 1.5. For the $SbCl_3$-LiCl-H_2O-TBP-CCl_4 system, the Sb distribution ratio was a strong function of total chloride, having a value of 20 in 100% TBP with 8.465 M Cl^-. K_D values for the LiCl system were higher at higher TBP concentrations. Under comparable Cl^- and TBP concentrations, Sb was more readily extracted from LiCl than from NH_4Cl.

Tellurium

Tellurium is highly extractable. The capacity of TBP for this element is reportedly 140 gpl when the aqueous phase is HCl or HCl + H_2SO_4;[100] species extracted include $TeCl_4 \cdot 3TBP \cdot H_2O$ or $TeCl_4 \cdot 3TBP \cdot 2H_2O$ from 2-10 N HCl, and $HTeCl_5 \cdot 3TBP \cdot H_2O$ from > 10 N HCl or HCl + H_2SO_4. With 1.0-1.5 M TBP in p-xylene and 5 M HCl, 97% extraction was achieved;[101] the extracted compound was $TeCl_4 \cdot 3TBP$. The effects of various impurities have been studied and it has been reported that Te could be separated from Se, As, Cu, Ag, Bi and Pb in HCl or HCl + H_2SO_4;[102] Te was not separated from Fe or Sb by 80% TBP in kerosene. However, washing and back-extraction removed all impurities except Sb, which was later removed in the refining of molten Te.

Hoh et al.[103] have recently reported on the separation of Te(IV) from Se(IV) in acid chloride media; the organic phase was 30% TBP in kerosene. Processes were developed for several types of mixer-settlers. Solvent extraction separation of Se and Te is an important component of a recently patented process for the recovery of precious metals from anode slime.[104] Recovery rates for Se and Te were 94.4 and 71%, respectively. Gold, silver and copper are also recovered.

Cadmium

Cadmium extraction from a number of media has been reported. Comparative studies on the extraction of Cd and such metals as Cu and Zn are discussed in the Base Metals section.

Extraction of Cd from aqueous HCl by pure TBP is exothermic;[18] the possibility of controlling temperature to obtain high Cd-Co and Cd-Cu separation ratios has been suggested since the extraction of cobalt and copper is endothermic. Species extracted from $MgCl_2$ and $BaCl_2$ are $MgCdCl_4$ and $CdCl_2$, respectively; $Sr(CdCl_3)_2$, $SrCdCl_4$ and $CdCl_2$ were said to be extracted from $SrCl_2$ solutions.[105] These species were solvated with 2 or 3 TBP molecules. Zinc complexes extracted from $MgCl_2$, $BaCl_2$ and $SrCl_2$ had the same compositions as the Cd species.

From solutions containing I^-, Cd was extracted as $3TBP \cdot 2CdI_2 \cdot H_2O$, $2TBP \cdot CdI_2$ and $TBP \cdot H_2O \cdot CdI_2$.[106]

Trace amounts of Cd halides were extracted from H_2SO_4[3] and H_3PO_4;[107] the presence of these acids was said to enhance the separations.

Takezhanov et al.[108] developed a process for separating Cd and Tl from liquor obtained by leaching Pb plant dust. In the presence of I^-, Na_2SO_3 and H_2SO_4, Cd and Tl were extracted by TBP in kerosene; small amounts of As, Sb, Cu and Zn were coextracted. Cd and Tl were separated by back-extraction; controlled amounts of dilute HNO_3-H_2SO_4 stripped Cd.

Mercury

Divalent mercury is strongly or weakly extracted depending on the anion. From 0.1 M TBP in hexane and 10^{-4} M NaI, log K_D was found to be 2; for NaBr, K_D decreased by about an order of magnitude, with another substantial drop in NaCl.[109] The ionic strength was held constant with $NaClO_4$ and experiments performed at various mercuric halide concentrations. Distribution ratios dropped with increased HgX_2 concentrations; the drop was much steeper for HgI_2 than for $HgBr_2$ or $HgCl_2$.

Mercury and selenium can be readily separated by 80% TBP in kerosene.[110] The distribution ratio of Hg was about 11 and was nearly independent of the HCl concentration between 1 and 10 N; the Se coefficient increased with the amount of acid but was only 0.013 at 1 N HCl.

PRECIOUS METALS AND PLATINUM GROUP METALS

Silver

Procedures for the separation of silver from cadmium and from zinc have been developed by El-Kot et al.;[111] tests using HCl, HBr, H_2SO_4 and H_3PO_4 were performed. It was concluded that there are a number of media suitable for extraction of small amounts of silver from cadmium or zinc but that 2 M H_2SO_4 was the best acid for extraction of macroamounts of Ag.

Gold

The gold species extracted from acidic chloride solutions is likely to be $HAuCl_4(TBP)_3$ or $AuCl_3 \cdot 3TBP$.[112] Toluene gave higher distribution ratios than benzene, xylene, MIBK, chloroform or carbon tetrachloride. Lithium and magnesium chloride were effective salting out agents. Recommended concentrations were 3 M HCl and 2 M LiCl or $MgCl_2$. With 3-4 M HCl and 2 M LiCl, extraction was complete when the TBP concentration was 1.83 M or greater. Bismuth, titanium, manganese, antimony, chromate, selenite and tellurite strongly interfered.

The effect of varying the diluent during $HAuCl_4$ extraction has been studied by Bucher et al.;[113] with iso-octane or CCl_4, primarily the 3:1 complex was extracted but with benzene the 2:1 complex was formed.

In a fairly unusual application, TBP has been used for flotation of gold in HCl-thiourea media;[114] Au recoveries as high as 99% were reported.

Ruthenium

Extraction of Ru(VII) from solutions containing NaOH, $NaClO_4$ and NaCl has been reported.[115] Ru(VII) is unstable and could be readily reduced if the NaOH concentration was above 2 M; distribution ratios were given as functions of time and the species actually present in either phase were not characterized.

Casey et al.[116] have presented data on the extraction of Ru(IV) from aqueous HCl. The maximum value of the distribution ratio observed for Ru(IV) was 0.6; Ru was the least extractable of the platinum metals. The maximum K_D value occurred at an HCl concentration of 4 M.

The oxidation state of ruthenium reportedly has a significant influence on its extractability. Ru(IV) was found to be about 200 times less extractable from chloride solution than Ru(II).[117]

Rhodium

Trace amounts of $RhCl_3$ in 2 M NaI and varying acid concentrations were extracted by Stella and DiCasa using TBP in toluene.[118] Distribution ratios depended on the acid used. At the same acid concentration, Rh(III) was more readily extracted from H_2SO_4 than from $HClO_4$. In either acid, the distribution ratio increased with acid concentration; at 6 N K_D was 110.1 for H_2SO_4 and 16.9 for $HClO_4$.

Palladium

Several groups have studied the extraction of palladium from chloride solutions. Belousov and Vol'khin[119] reported that Pd(II) is extracted from HCl as H_2PdCl_4, from LiCl as Li_2PdCl_4 and from $CaCl_2$ as $CaPdCl_4$. Casey et al.[116] found that the maximum K_D value for extraction from HCl solutions was 2.3 at 3 M HCl. At a constant H^+ concentration, LiCl was added in varying amounts to determine the effect of the total chloride concentration on the distribution ratio. A dramatic difference was observed. With solutions with 3 M H^+, K_D had a maximum near 4 M total chloride. A shallow minimum was observed near 2 M Cl^- if the H^+ concentration was 1 M; no maximum was found between 1 and 7 M Cl^-. Above 5 M Cl^-, K_D was greater for the HCl-LiCl system than for HCl.

Rhenium

Rhenium (VII) has been extracted from nitrate solutions by Pruett and McTaggart.[120] The aqueous medium was HNO_3 to which $NaNO_3$ was added for some experiments; the $NaNO_3$ did not affect the distribution ratio. At 0.8 M HNO_3, the distribution ratio was a maximum for TBP concentrations from 10 to 80% and temperatures from 25 to 60°C; Re extraction was increased by lowering the temperature and increasing the TBP concentration. The extraction reaction was reported to be

$$H^+(aq) + ReO_4^-(aq) + 3TBP(org) = HReO_4 \cdot 3TBP(org) \qquad (19)$$

A tetrasolvated species was also suggested at high TBP concentrations. The highest distribution ratio was 10, obtained at 25°C in 0.8 M HNO_3 and 3 M TBP.

The effect of the diluent on the extraction of $HReO_4$ has been extensively studied by Diamond and coworkers.[113,121,122] With a CCl_4 diluent, $HReO_4$ was extracted as both disolvated and trisolvated species; 1,2-dichloroethane, with a high dielectric constant, gave only a trisolvate.[121] A diluent of this type would tend to favor dissociation of the extracted complex. Benzene favored 2:1 solvates because it gave good solvation of the cationic complexes through its π electrons.[113,122]

The compounds extracted from $KReO_4$ and HCl are $[3TBP \cdot H_9O_4]^+ReO_4^-$ and $[2TBP \cdot H_9O_4]^+Cl^-$ when the acid concentration was < 5 M.[123]

Rhenium and molybdenum reportedly can be cleanly separated if the proper acid and acid concentration are chosen;[124] from H_2SO_4, Re is extracted but Mo is not.

Osmium

Osmium is fairly readily extracted from aqueous HCl; it has a maximum distribution ratio of 6 near an equilibrium acid concentration of 3 M.[116]

Iridium

Extraction of iridium is complicated by redox and hydrolysis reactions in the aqueous phase; $IrCl_6^{2-}$ is readily reduced to $IrCl_6^{3-}$, a species which then undergoes hydrolysis reactions. Ir(III) is less than 1% extracted by TBP up to a hydrochloric acid concentration of 8 M;[116] Ir(IV) has a distribution ratio of 2.8 near 3 M HCl. Below 1 M TBP, distribution ratios decreased with diluent in the order _n_-hexane > cyclohexane > carbon tetrachloride. With a constant H^+ concentration of 1 M, addition of $MgCl_2$ resulted in a maximum in K_D at about 4.5 M total chloride; the existence of a maximum might be due to salting out and extraction of HCl.

Platinum

Platinum (II) is reported to be extracted as H_2PtCl_4 from HCl, as Li_2PtCl_4 from LiCl, and as $CaPtCl_4$ from $CaCl_2$.[125] From HCl, the distribution ratio has a maximum value of 18.8 at an equilibrium HCl concentration of 4 M;[116] platinum is more readily extracted from hydrochloric acid than Ru, Pd, Ir or Os.

Platinum Group Metals

Knothe[126,127] compared the extraction of the platinum metals in various oxidation states; for HCl solutions, extractability declined in the order: Ir(IV) ~ Pt(IV) > Pd(IV) > Pd(II) ~ Pt(II) >> Rh(III) ~ Ir(III). The choice of diluent significantly raised or lowered the distribution ratios relative to their values in 100% TBP but a given diluent did not affect each platinum metal in the same way. CCl_4, $CHCl_3$, and _n_-heptane depressed the extraction when combined with TBP in a 1:1 ratio; β,β'-dichlorodiethylether, nitrobenzene and 1,2-dichloroethane enhanced extraction of some species but depressed others. Knothe emphasized the effect of metastable species in the aqueous phase.

Stella and Genova[128] studied the effect of diluent on the separation of chloro complexes of Pt and Pd. Trace concentrations of Pt and Pd could be separated in a single step if 80-90% n-hexane were used as the diluent. CCl_4, $CHCl_3$ and 1,2-dichloroethane were less satisfactory. Pt was preferentially extracted.

RARE EARTHS

Cerium

Processes have been developed for the production of Ce from both primary and secondary sources.

A process for the recovery of Ce from fluoridic ores such as bastnasite has been patented.[129] The ore was first leached with acid to give a $Ce(NO_3)_4$ solution. H_3BO_3 was added in order to complex the fluoride in the form of HBF_4 and thus prevent fluoride from being extracted. Cerium was extracted with TBP and the organic phase washed with 5 N HNO_3. Back-extraction with 6 N HCl containing 25 gpl H_3BO_3 and H_2O_2 was then carried out; hydrogen peroxide reduces Ce(IV) to Ce(III). Cerium was then precipitated from the aqueous phase with oxalic acid and the precipitate calcined to CeO_2. Cerium was obtained in 92% yield and the CeO_2 product was free of fluoride. Without the addition of H_3BO_3, the yield of CeO_2 was only 60% of theoretical and the product contained 2.7% fluoride.

A similar process has been reported for recovery of cerium from monazite sands.[130] TBP was used in the separation of Ce from cerous nitrate solutions resulting from leaching of the sands. These crude cerous nitrate solutions contained 52 gpl Ce_2O_3, 70 gpl NH_4NO_3, 0.9 gpl other lanthanides (as Ln_2O_3) and 0.07 gpl iron group oxides and had a pH of 2.6. Ce was separated from solution as $Na_2SO_4 \cdot Ce_2(SO_4)_3 \cdot 2H_2O$; the sulfate salt still contained lanthanum as an impurity. The sulfate salt was converted to an oxide by treating it with 18% NaOH and drying. The impure oxide (still containing La) was dissolved in concentrated HNO_3; 35% TBP in kerosene was used to extract Ce. The single extraction stage was followed by 2 scrubbing stages and 2 stripping stages. $NaNO_2$ was used to reduce Ce(IV) during stripping. Cerium oxalate was precipitated and ignited to give spectrographically pure CeO_2. The overall yield was not given although it was stated that after scrubbing 86% of the Ce originally present was in the organic phase.

Praseodymium

The effect of various salt mixtures on the extraction of Pr by TBP in CCl_4 has been studied by Genov and Dukov.[131] When the total salt concentration was maintained constant, the distribution ratios were considerably larger for extraction from the mixtures; lanthanum distribution ratios were found to be, for example, 1.70 for 3.2 M $NaClO_4$, 0.05 for 0.4 M $NaNO_3$ and 3.62 for a mixture of 3.2 M $NaClO_4$ and 0.4 M $NaNO_3$. Similar effects were reported for $NaNO_3$-NaSCN and $NaClO_4$-NaSCN mixtures. In the case of the perchlorate-nitrate systems, the increased extraction was attributed to changes in the activity coefficients of the Pr salts in the aqueous phase. For the SCN^--ClO_4^- and SCN^--NO_3^- systems, the enhancement was attributed instead to formation of mixed complexes; the synergic effect was somewhat more pronounced in these cases and it was felt that changes in activity coefficients would not be large enough to explain it. It was reported that for high SCN^- concentrations, the primary extracted complex is $Pr(SCN)_3 \cdot 3TBP$.

Synergic extraction of praseodymium by TBP and Aliquat 336 has been reported by Dukov et al.[132] This extraction system is quite complicated; the number of TBP molecules in the extracted species was found to depend in part on the diluent.

Neodymium

The neodymium species extracted from nitrate solutions is $Nd(NO_3)_3 \cdot 3TBP$; solvated $Nd(ClO_4)_3$ is extracted from perchlorate media.[133]

Synergic extraction by quaternary ammonium salts and TBP has been reported;[134] distribution ratios were about 2 orders of magnitude higher in the synergic system than with TBP alone. The extracted species was $[Nd(NO_3)_4 \cdot TBP]^-[R_3RN]^+$.

Europium

Europium is not strongly extracted by TBP; extraction ratios are reportedly 0.4 from 16 M H_2SO_4, 0.2 from 18 M HF and 0.15 from 11 M HCl.[135] Addition of small amounts of HCl depressed Eu extraction even further; the extraction ratio decreased as more HCl was added. Addition of methanol increased europium extraction, addition of _n_-propanol decreased it and the effect of ethanol was intermediate.

Several systems have been reported in which there is a synergic extraction of europium. These include 1-napthoic acid and 5,7-dichloroxine[135], as well as thenoyltrifluoroacetone.[136]

Extraction of Eu(III) by di-(2-ethylhexyl)phosphoric acid and TBP is antagonistic.[70] From 0.2 M $HClO_4$ + 0.8 M $NaClO_4$, the species believed to be present in the extract included TBP solvates of $Eu(DEHP)_3 \cdot HDEHP$, $Eu(DEHP)_3$ and $Eu(DEHP)_2(ClO_4)$.

Nitrate Media

Extraction of rare earths from nitrate solutions has received considerable attention. Species reported in the organic phase are the trisolvates $La(NO_3)_3 \cdot 3TBP$,[137] $Ce(NO_3)_3 \cdot 3TBP$,[137,138] $Pr(NO_3)_3 \cdot 3TBP$,[138] $Eu(NO_3)_3 \cdot 3TBP$,[139] $Gd(NO_3)_3 \cdot 3TBP$,[138,140] $Yb(NO_3)_3 \cdot 3TBP$[138] and $Lu(NO_3)_3 \cdot 3TBP$.[138]

The variation of extraction ratio with temperature has been investigated by Fidelis for all the elements from La to Lu.[141] For each element, the extraction ratio decreased as the temperature increased from 10 to 40°C and K_D was a linear function of 1/T; at a given temperature, the extraction ratio decreased with atomic number. Changes in entropy were the cause of the increasing stability of the lanthanide TBP species at higher atomic numbers. Pure TBP and 12.5 M HNO_3 were used in these experiments.

Infrared spectroscopy has been used to study organic phase complexes of Ce, Pr, Gd, Yb and Lu.[138] It was found that the rare earth atom and the oxygen atom of the TBP molecule were more weakly bound as the rare earth's atomic number was increased. If this factor were the only one influencing the variation in distribution ratio with Z, the distribution coefficient would be expected to decrease with increasing atomic number. At low acid concentrations, however, the K_D vs. Z curve has a maximum. The explanation for this was sought in the relative stabilities of the various species formed in the aqueous and organic phases.

Kalina et al.[139] have recently reported on the thermodynamics of Eu^{3+} extraction from 2.00 M $LiNO_3$-0.01 M HNO_3 by 0.25 M TBP in dodecane; ΔG and ΔH were 0.08 and -9.22 kcal/mole, respectively, while ΔS was -31.2 cal/mole-K. The equilibrium constant was given as

$$\ln K_{eq} = \frac{4642}{T} - 15.71 \tag{20}$$

for the reaction

$$Eu^{3+}(aq) + 3NO_3^-(aq) + 3TBP(org) = Eu(NO_3)_3 \cdot 3TBP(org) \tag{21}$$

Rare earths have been recovered during processing of phosphates to produce fertilizers.[142] The starting material contained 41.03% CaO, 1.50% MgO, 27% P_2O_7, 1.15% Fe_2O_3, 1.06% Al_2O_3 and 0.08% rare earths; this was decomposed by 55% HNO_3 at 60°. Fluoride was removed by precipitation of Na_2SiF_6; $NaNO_3$ was the reagent added. Rare earths were extracted with TBP and obtained in 97% yield.

Brown and Sherrington[143] have described a commercial process for separation of La, Pr and Nd nitrates using tributylphosphate. The solution initially consisted of Na, La, Pr and Nd chlorides. This was converted to a nitrate solution by adding Na_2CO_3 and boiling to precipitate carbonates. Following centrifugation, the solids were dissolved in concentrated HNO_3 to give a solution with about 460 g/dm^3 rare earth oxides. A 60 stage mixer-settler battery was operated for 100 h in a force-feeding mode then for 145 h in total reflux. Product purities were 99% with respect to La, 98% with respect to Nd and 90% with respect to Pr. At the end of the total reflux period, most of the La was in stages 1-20, most of the Pr in 20-40 and most of the Nd in 40-60.

Highly solvated lanthanide species are involved in extraction from perchlorate solutions; Kolarik[3] has reported that the extracted Eu and Ce species are $Eu(ClO_4)_3 \cdot 5TBP$ and $Ce(ClO_4)_3 \cdot 5TBP$.

Nitrate-Perchlorate Media

A large number of species have been reported to be extracted from mixed nitrate-perchlorate solutions. Extraction of Pr and Yb from solutions with an ionic strength of 3.5 M by 2 M TBP was investigated;[144] the diluent was not stated. In addition to the tri- and hexasolvates $M(A)_3 \cdot 3TBP$ and $M(A)_3 \cdot 6TBP$, the mixed complexes $M(NO_3)_2(ClO_4) \cdot 3TBP$, $M(NO_3)(ClO_4)_2 \cdot 3TBP$, $M(NO_3)(ClO_4)_2 \cdot 6TBP$ and $M(NO_3)_2(ClO_4) \cdot 6TBP$ were believed to be present. M denotes Pr or Yb and A denotes NO_3 or ClO_4.

Cerium was recovered by Douglass and Bauer[145] from a liquor resulting from leaching of bastnasite concentrate in sulfuric acid.

Bench scale procedures were developed that would give 95% Ce extraction or a Ce product 99% pure. However, sulfate ion inhibited the extraction and high Ce recovery and high product purity could not be achieved simultaneously.

Fused Salt Systems

Low-melting hydrated salts have been investigated as replacements for the aqueous solutions used in most solvent extraction systems. $Ca(NO_3)_2 \cdot 4H_2O$ (m.p. 42.7°C) has been used at 45°C in conjunction with an n-paraffin-TBP organic phase;[146] Eu, Y, La, Ce(III) and Ce(IV) were studied along with various actinide elements. The distribution ratio was found to increase with atomic number for the lanthanide elements; it was higher for Ce(IV) than for Ce(III). The distribution ratio increased with the initial TBP concentration and was close to 100 for Eu and Y near 2 M TBP. For yttrium and the lanthanides the distribution ratio was higher in the melt-organic system than in the corresponding aqueous-organic systems. The opposite, however, was true for some other elements, including niobium and molybdenum. The extraction reaction in the melt-organic system was analogous to that occurring in the aqueous-organic system

$$M^{n+}(aq) + nNO_3^-(aq) + mTBP(org) = M(NO_3)_n \cdot mTBP(org) \qquad (22)$$

The solvation number was 3 for both n-paraffin and chloroform diluents. $Mn(NO_3)_2 \cdot 6H_2O$ has a melting point of only 29.9°C and has also been used as the inorganic phase in solvent extraction with TBP.[147] The extraction system was kept at 30°C. With xylene or chloroform as the diluent, the distribution ratio increased with the atomic number; Lu, Tm, Eu, Nd, Ce and La were the lanthanides studied. The Z dependence was much less pronounced with 1-hexanol as the diluent. For 20% TBP in diluent, the distribution ratios were $10^2 - 10^4$ times higher in this fused hydrate system than in 12 M or concentrated HNO_3. For dodecane, xylene, and chloroform, distribution ratios for Tm and Eu were about 10^2 times larger than they were in 1-hexanol. With Ce(III), K_D was of the same order of magnitude for all 4 diluents; Ce(III) was also much less extractable than Tm or Eu.

Synergic Extraction

HDEHP-TBP Systems. Extraction of Tb and Eu from $HClO_4$ by

HDEHP and TBP in a variety of aromatic and aliphatic diluents is reported not to be synergic.[148] The Pr, Gd and Yb species extracted for 0.1-0.3 N HNO_3 by di-(2-ethylhexyl) phosphoric acid and TBP in CCl_4 have been stated to be $M(H_2A)_3 \cdot 2TBP$.[149]

With respect to extraction of 0.02 M La by HDEHP, there is an antagonistic effect in the HDEHP-TBP system.[150] The aqueous phase used in this work contained hydrochloric acid; n-heptane was the diluent.

Alkylammonium Ion-TBP Systems. Extraction of cerium (III) by TBP and alkylammonium ions shows a synergic effect but it is not particularly large.[151] The aqueous phase contained 3 M HNO_3; both trilaurylammonium (TLAN) and trilaurylmethylammonium (TLMAN) nitrates were studied. Toluene and n-hexane were used as diluents. For TLMAN the extracted species is likely to be $(R_3R'N)_2Ce(NO_3)_5$; for TLAN, the distribution data suggested $(R_3NH_4)Ce(NO_3)_7$ but the authors believed this was improbable. Addition of alkylammonium salts to TBP increased the distribution ratios for all cases except TLMAN with n-hexane; for this system there was an antagonistic effect. The enhanced extraction was attributed to formation of mixed complexes. The mixed TMAN-TBP complex was believed to be $2(TLMA) \cdot Ce(NO_3)_5 \cdot TBP$. Even with the synergism, the distribution coefficients are still low. With 0.8 M TBP in either diluent and TLAN or TLMAN concentrations up to 0.3 or 0.1 M, respectively, the highest K_D values reported were in the range of 0.05 to 0.1.

Di-n-Butylbutylphosphonate(DBBP)-TBP Systems. Genov and Georgiev[152] have extracted praseodymium and ytterbium nitrates with di-n-butylbutylphosphonate and TBP using $CC\ell_4$, C_6H_6 and ethylisobutylketone diluents. Under comparable conditions of concentration and diluent, Yb(III) was more readily extracted than Pr(III). The distribution ratios were found to be constant for Pr in CCl_4, C_6H_6 and ethylisobutylketone; DBBP concentrations were varied between 0 and 0.670, 0 and 0.620, or 0 and 0.625 M in the 3 diluents while the TBP concentration was varied from 2 M to 0. For Yb in CCl_4, a constant distribution ratio was also observed while for benzene diluent a slight increase was reported. The distribution ratios for all these systems fell in the range of 0.14 to 0.22. The slight synergic effect was ascribed to formation of mixed complexes of the type $M(NO_3)_3 \cdot 2TBP \cdot DBBP$ and $M(NO_3)_3 \cdot TBP \cdot 2DBBP$.

Dialkylmethylphosphonate (DOMP)-TBP Systems. A fairly strong synergic effect has been reported for dialkylmethyl-phosphonate (DOMP) - systems by Mikhlin et al.[153] The phosphonate extractant used was $CH_3PO(OR)_2$ with R = C_7H_{15}-C_9H_{19}; the average molecular weight was 320. Neodymium and yttrium were the lanthanide elements extracted and m-xylene was the diluent. The aqueous phase contained 2 M NH_4NO_3. Mixed solvates were said to form.

Petroleum Sulphoxide (PSO)-TBP Systems. Mikhlin et al.[153] have also reported on synergic extraction of rare earths by petroleum sulphoxide fraction (PSO)-TBP mixtures. As measured by

$$E = \frac{K_{Dmix}}{K_{D1} + K_{D2}}$$

the synergic effect is greatest when the PSO concentration is 25-35%, when E is between 2.6 and 2.8.

SUMMARY

Virtually every metallic element has received some attention during studies of solvent extraction by TBP, as have a number of semi-metals. Emphasis has naturally been placed on elements with distribution ratios high enough to make commercial applications attractive and on elements likely to yield fundamental insights into the mechanism of extraction with tributyl phosphate. To date, the greatest success of TBP extraction in the non-nuclear area has been in the multistage separation of rare earths. From the point of view of process development, synergic extraction systems have been relatively little explored. Reagent cost is likely to remain a key factor limiting applications of synergic systems but it could prove very fruitful to design synergic reagents which are both cheap and sufficiently selective. Future commercial applications of TBP extraction appear most likely for refractory metals such as Ti, W, Ta and Nb and for rare metals such as Ga, Te and Se. Use of TBP in recovering base metals is also possible.

ACKNOWLEDGMENT

This work was supported in part by the Director of Energy Research, Office of Basic Energy Sciences, Chemical Sciences Program, U.S. Department of Energy, under contract No. W-7405-ENG-82.

REFERENCES

1. Y. Hasegawa, T. Ishii, and T. Sekine: Bull. Chem. Soc. Japan, 1971, vol. 44, pp. 275-77.
2. S. S. Korovin, Yu. M. Glubukov, T. A. Rasskazova, and A. M. Reznik, Izv. Vyssh. Ucheb. Zaved., Khim. Khim. Tekhnol., 1971, vol. 14, pp. 1141-45; Chem. Abstracts, 1971, vol. 75, 155347r.
3. Z. Kolarik: J. Inorg. Nucl. Chem., 1974, vol. 36, pp. 409-14.
4. T. V. Healy, J. Kennedy, and G. M. Waind: J. Inorg. Nucl. Chem., 1959, vol.10, pp. 137-47.
5. G. Brunisholz, W. Hirsbrunner, and J. Aerny: Helv. Chim. Acta, 1972, vol. 55, pp. 2956-61.
6. I. S. El-Yamani, M. Y. Farah, and E. N. Abd El-Messieh: J. Radioanal. Chem., 1978, vol. 45, pp. 365-71.
7. W. J. McDowell: J. Inorg. Nucl. Chem., 1968, vol. 30, pp. 1037-49.
8. A. A. Yadav and S. M. Khopkar: Sep. Sci., 1969, vol. 4, pp. 349-55.
9. J. Aggett and T. E. Clark: J. Inorg. Nucl. Chem., 1970, vol. 32, pp. 2037-43.
10. A. Haggag, W. Sanad, A. Alian, and N. Tadroes: J. Radioanal. Chem., 1977, vol. 35, pp. 253-67.
11. A. A. Palant, V. A. Reznichenko, A. V. Stepanov, and K. N. Davydkin: Izv. Akad. Nauk SSSR, Met. 1980, pp. 83-85; Chem. Abstracts, 1981, vol. 94, 125114n.
12. P. G. Thornhill, E. Wigstol and G. Van Weert: J. Metals, July 1971, vol. 23, pp. 13-18.
13. P. E. Vembe: ISEC 77, CIM Special Volume 21, pp. 761-65.
14. B. Jezowska-Trzebiatowska, A. Bartecki, and S. Kopacz: Zh. Neorg. Khim., 1970, vol. 13, pp. 864-70; Chem. Abstracts, 1968, vol. 68, 108507p.
15. V. I. Levin, M. D. Kozlova, and A. S. Sevast'yanova: Radiokhimiya, 1972, vol. 14, pp. 48-54; Chem. Abstracts, 1972, vol. 76, 159012w.

16. L. Boyadzhiev, and E. Bezenshek: Izv. Otd. Khim. Nauki, Bulg. Akad. Nauk., 1974, vol. 7, pp. 105-11; Chem. Abstr. 1974, vol. 81, 109001z.
17. A. Apelblat, J. Inorg. Nucl. Chem., 1980, vol. 42, pp. 461-63.
18. E. A. Belousov, V. M. Ivanov, N. V. Chugunova, and T. F. Mikheeva: Issled. Khim., Tekhnol. Primen. Radioakt. Veshchestr, 1977, pp. 83-91; Chem. Abstracts, 1980, vol. 93, 55010d.
19. E. A. Belousov, and A. A. Alovyainikov: Zh. Neorg. Khim., 1975, vol. 20, pp. 2686-94; Chem. Abstracts, 1976, vol. 84, 9461t.
20. V. I. Levin, M. Abdukayumov, and M. D. Kozlova: Radiokhimiya, 1972, vol. 14, pp. 58-61; Chem. Abstracts, 1972, vol. 76, 159009a.
21. N. M. Rice and M. R. Smith: Phys. Chem. Process Metall., Inst. Min. Metall., London, 1974, pp. 55-63.
22. M. Bressa, D. Buttinelli, and C. Giavarini: Chim. Ind. (Milan), 1979, vol. 61, pp. 893-97; Chem. Abstracts, 1980, vol. 93, 11341h,
23. G. M. Ritcey, B. H. Lucas, and K. T. Price: Hydromet., 1982, vol. 8, pp. 197-222.
24. Y.-C. Hoh, N.-P. Chou, and W.-K. Wang: Ind. Eng. Chem. Process Des. Dev., 1982, vol. 21, pp. 12-15.
25. Pushparaja and U. Sundersanan: Indian J. Chem., 1980, vol. 19A, pp. 998-1002.
26. V. M. P. Forrest, D. Scargill, and D. R. Spickernell: J. Inorg. Nucl. Chem., 1969, vol. 31, 187-97.
27. A. W. Fletcher, D. S. Flett, J. L. Pegler, D. P. Pepworth, and J. C. Wilson: Adv. Extr. Met., Inst. Min. Met., 1968, pp. 686-711.
28. N. M. Rice and M. R. Smith: Can. Met. Quart., 1973, vol. 12, pp. 341-49.
29. A. Alian and A. El-Kot: J. Radioanal. Chem., 1976, vol. 29, pp. 23-37.
30. T. Sato: J. Appl. Chem. Biotechnol., 1972, vol. 22, pp. 1233-42.
31. J. Aggett, T. E. Clark, and R. A. Richardson: J. Inorg. Nucl. Chem., 1969, vol. 31, pp. 2919-26.
32. H. Woidich, and W. Pfannhauser: Mikrochim. Acta (Vienna), 1973, pp. 279-89.
33. M. Golinski: Nukleonika, 1972, vol. 17, no. 9, pp. 6-16.
34. M. Golinski: in Proc. First Int. Solvent Extr. Conf., pp. 603-615.

35. Yu. K. Tselinskii: Anal. Khim. Ekstr. Protsessy, A. T. Pilipenko, ed., Naukova Dumka, 1970, pp. 39-40; Chem. Abstracts, 1971, vol. 74, 57819f.
36. I. M. Gavrilova, V. M. Klyuchinikov, L. M. Zaitsev, and I. A. Apraksin: Zh. Neorg. Khim., 1973, vol. 18, pp. 1640-42; Chem. Abstracts, 1973, vol. 79, 108698d.
37. A. P. Samodelov: Recl. Trav. Chim. Pays-Bas., 1971, vol. 13, pp. 203-08; Chem. Abstracts, 1971, vol. 75, 101767j.
38. D. F. C. Morris and K. J. Sutton: Bull. Chem. Soc. Japan, 1967, vol. 40, pp. 786-87.
39. L. V. Favorskaya, V. A. Presnetsova, G. N. Vainberger, and L. I. Vorob'ev: Tekhnol. Miner. Syr'ya, M. Kh. Azhikeev. ed., Kaz. Nauchno-Issled. Inst. Miner. Syr'ya, 1972, pp. 173-80; Chem. Abstracts, 1976, vol. 85, 167425j.
40. L. V. Favorskaya, V. A. Presnetsova, and G. N. Vainberger: Tekhnol. Miner. Syr'ya, Kaz. Nauchno-Issled. Inst. Miner. Syr'ya, 1973, pp. 173-80; Chem. Abstracts, 1974, vol. 80, 125395r.
41. L. V. Favorskaya, V. A. Presnetsova, and G. N. Vainberger: 3rd Mater. Konf. Khim. Ekstr., Yu. A. Zolotov, ed., Nauka, 1972, pp. 178-81; Chem. Abstracts, 1973, vol. 78, 48581c.
42. L. V. Favorskaya, V. A. Presnetsova, G. N. Vainberger, A. I. Chikodanov, and L. I. Vorob'ev: Tekhnol. Miner. Syr'ya, Kaz. Nauchno-Issled. Inst. Miner. Syr'ya , 1972, pp. 162-72; Chem. Abstr., 1974, vol. 80, 125387q.
43. A. D. Romanova, and L. V. Favorskaya: Tekhnol. Miner. Syr'ya, Kaz. Nauchno-Issled. Inst. Miner. Syr'ya, 1966, pp. 33-43; Chem. Abstracts, 1968, vol. 68, 15006j.
44. S. A. Semenov, A. M. Reznik, N. I. Gavrilov, and L. D. Yurchenko: Koord. Khim., 1980, vol. 6, pp. 1027-29; Chem. Abstracts, 1980, vol. 93, 102160q.
45. L. V. Favorskaya, V. A. Presnetsova, Yu. M. Putilin, M. K. Baibekov, L. I. Vorob'ev, and A. I. Chikodanov: Zh. Prikl. Khim. (Leningrad), 1970, vol. 43, pp. 1158-60; Chem. Abstracts, 1970, vol. 73, 39255u.
46. E. B. Mikhlin, A. M. Reznik, and I. P. Ryzhova: Zh. Neorg. Khim., 1978, vol. 23, pp. 1319-23; Chem. Abstracts, 1978, vol. 89, 31589r.
47. S. S. Korovin, P. G. Berezhko, and A. M. Reznik: 3rd Mater. Konf. Khim. Ekstr., Ya. A. Zolotov, ed., Nauka, 1972, pp. 172-77; Chem. Abstracts, 1973, vol. 78, 48580b.
48. P. G. Berezhko, D. A. Granat, S. S. Korovin, and A. M. Reznik: Zh. Neorg. Khim., 1973, vol. 18, pp. 2193-97; Chem. Abstracts, 1973, vol. 79, 118836m.

49. A. P. Samodelov: Radiokhimiya, 1969, vol. 11, pp. 447-50; Chem. Abstracts, 1970, vol. 72, 6658d.
50. P. G. Berezhko, S. S. Korovin, A. M. Reznik, and M. I. Starozhitskaya: Izv. Vyssh. Ucheb. Zaved., Khim. Khim. Tekhnol., 1971, vol. 14, pp. 987-89; Chem. Abstracts, 1971, vol. 75, 133593r.
51. K. P. Radhakrishnan, and T. C. Owens: J. Chem. Eng. Data, 1972, vol. 17, pp. 478-82.
52. S. S. Korovin, M. Sh. Zhambulova, V. M. Klyuchnikov, and I. A. Apraksin: Zh. Neorg. Khim., 1972, vol. 17, pp. 207-10; Chem. Abstracts, 1972, vol. 76, 90718m.
53. T. Sekine, T. Hamada, and M. Sakairi: Bull. Chem. Soc. Japan, 1966, vol. 39, pp. 244-46.
54. G. Roland, and B. Gilbert: Anal. Chim. Acta, 1972, vol. 60, pp. 57-64.
55. V. N. Startsev and E. I. Krylov: Tr. Ural. Politekh. Inst., 1966, vol. 148, pp. 147-53, Chem. Abstracts, 1967, vol. 67, 85482q.
56. I. P. Sorokin, and L. V. Denisova: Sb. Tr., Vses. Nauch.-Issled. Proekt. Inst. Titana, 1972, pp. 223-28; Chem. Abstracts, 1974, vol. 80, 17761y.
57. P. H. Tedesco, and V. B. de Rumi: Ind. Quim., 1978, vol. 247, pp. 44-46, Chem. Abstracts, 1978, vol. 89, 218329y.
58. P. H. Tedesco, and V. B. de Rumi: J. Inorg. Nucl. Chem., 1980, vol. 42, pp. 1033-36.
59. Y. Komatsu: J. Inorg. Nucl. Chem., 1980, vol. 42, pp. 265-68.
60. G. Roland, M. Pondant, and G. Duyckaerts: Anal. Chim. Acta, 1976, vol. 85, pp. 331-40.
61. I. V. Vinarov, N. P. Kirichenko, and G. M. Popelyukh: Sov. Prog. Chem., 1975, vol. 41, no. 7, pp. 64-67.
62. P. H. Tedesco and V. B. de Rumi: J. Inorg. Nucl. Chem., 1980, vol. 42, pp. 269-72.
63. T. Sato: J. Inorg. Nucl. Chem., 1979, vol. 41, pp. 1605-07.
64. M. A. Abdukayumov, O. A. Abrarov, V. I. Levin, G. F. Savosteeva and M. Kh. Makhkamova: Radiokhimiya, 1978, vol. 20, pp. 62-67, Chem. Abstracts, 1978, vol. 88, 142386w.
65. J. Aggett and D. J. Udy: J. Inorg. Nucl. Chem., 1970, vol. 32, pp. 2802-05.
66. J. P. Cuer, W. Stuckens and N. Texier: ISEC 1974, vol. 2, pp. 1185-200.
67. V. N. Startsev and E. I. Krylov: Zh. Neorg. Khim. 1967, vol. 12, pp. 526-29; Chem. Abstracts, 1967, vol. 66, 99039s.

68. J.-Y. Sellier, B. Marin, and D. Gourisse: Bull. Soc. Chim. France, 1974, pp. 1877-80.
69. S. S. Korovin and Yu. I. Kol'tsov: Zh. Neorg. Khim., 1969, vol. 14, pp. 1062-64; Chem. Abstracts, 1969, vol. 71, 16376d.
70. Z. Kolarik: J. Inorg. Nucl. Chem., 1972, vol. 34, pp. 2911-20.
71. V. I. Levin, and T. D. Titkova: Radiokhimiya, 1974, vol. 16, pp. 267-68; Chem. Abstracts, 1974, vol. 81, 30319a.
72. B. Jezowska-Trzebiatowska and S. Kopacz: Nukleonika, 1967, vol. 12, pp. 635-42; Chem. Abstracts, 1968, vol. 69, 90372z.
73. A. A. Palant and V. A. Reznichenko: Zh. Prikl. Khim., 1976, vol. 49, pp. 883-85; Chem. Abstracts, 1976, vol. 85, 23898t.
74. I. I. Baram: Zh. Prikl. Khim. (Leningrad), 1969, vol. 42, pp. 1411-13, Chem. Abstracts, 1969, vol. 71, 74763p.
75. V. I. Lakshmanan, and B. C. Haldar: J. Indian Chem. Soc., 1970, vol. 47, pp. 72-78.
76. T. S. Urbanski, M. Chojecki, and R. Kaczynska: J. Radioanal. Chem., 1976, vol. 30, pp. 369-76.
77. J. J. Cruywagen, and H. A. C. McKay: J. Inorg. Nucl. Chem, 1970, vol. 32, pp. 255-65.
78. A. N. Zelikman, G. M. Vol'dman, and V. S. Kagerman'yan: Zh. Neorg. Khim., 1972, vol. 17, pp. 783-89; Chem. Abstracts, 1972, vol. 76, 159061m.
79. V. S. Shmidt, E. A. Mezhov, Z. N. Tsvetkova, and V. N. Shesterikov: Radiokhimiya, 1973, vol. 15, pp. 126-27; Chem. Abstracts, 1973, vol. 78, 115814w.
80. Y. W. Sanad, A. Haggag, and N. Tadros: J. Radioanal. Chem., 1978, vol. 42, pp. 59-73.
81. M. A. Zakharov, S. S. Korovin, V. M. Klyuchnikov, and I. A. Apraksin: Zh. Neorg. Khim., 1973, vol. 18, pp. 1916-20; Chem. Abstracts, 1973, vol. 79, 129744f.
82. L. V. Favorskaya, G. A. Nikiforova, and E. N. Pankova: Tekhnol. Miner. Syr'ya. M. Kh. Azhikeev, ed., Kaz. Nauchno-Issled. Inst. Miner. Syr'ya, 1972, pp. 188-95; Chem. Abstracts, 1976, vol. 85, 180612n.
83. N. S. K. Prasad, T. N. Ranganathan, and M. S. Sastri: Indian J. Technol., 1966, vol. 4, pp. 268-71.
84. G. M. Vol'dman, A. N. Zelikman, G. N. Ziberov, and V. S. Kagerman'yan: Tsvetn. Met., 1978, pp. 46-49; Chem. Abstracts, 1978, vol. 89, 28300j.
85. A. N. Zelikman, G. M. Voldman, V. K. Rumyantsev, G. N. Ziberov, and V. S. Kagerman'yan: U.S. Patent, 3,969,478, 1976.

86. Yu. N. Silaev, S. P. Kokurkina, A. M. Reznik, and Yu. M. Glubukov: Izv. Vyssh. Uchebn. Zaved., Khim. Khim. Tekhnol., 1980, vol. 23, pp. 534-36; Chem. Abstracts, 1980, vol. 93, 81369x.
87. A. M. Reznik, L. D. Yurchenko, L. A. Zekel, M. P. Nosovskii, and M. Ya. Shpirt: Tr. Yubileinoi Konf., A. N. Bashkirov, ed., Mosk. Inst. Tonkoi Khim. Tekhnol. Moscow, 1972, pp. 315-16; Chem. Abstracts, 1974, vol. 81, 80888f.
88. I. P. Alimarin, N. I. Ershova, T. A. and Bol'shova: Vestn. Mosk. Univ., Ser. 2: Khim., 1967, vol. 22, pp. 51-54; Chem. Abstracts, 1967, vol. 67, 121951c.
89. D. F. C. Morris, B. D. Andrews and E. L. Short: J. Inorg. Nucl. Chem., 1966, vol. 28, pp. 2436-37.
90. T. R. Bhat and S. Sundararajan: J. Less-Common Metals, 1967, vol. 12, pp. 231-38.
91. A. de Schepper: Hydromet., 1979, vol. 4, pp. 285-90.
92. V. Pfeifer: J. Radioanal. Chem., 1970, vol. 6, pp. 47-55.
93. R. M. Walters and R. A. Dodson: Solvent Extraction Chemistry, D. Dyrssen, J.-O. Liljenzin and J. Rydberg, eds., North-Holland, 1967, pp. 71-78.
94. H. M. Widmer and R. W. Dodson: J. Phys. Chem., 1970, vol. 74, pp. 4289-91.
95. A. Alian, A. El-Kot, A. Haggag, W. Sanad, and N. Tadros: J. Radioanal. Chem., 1975, vol. 26, pp. 39-49.
96. A. M. Tserekova, G. P. Giganov, G. K. Kolomina, and V. I. Fokina: Sb. Nauch. Tr., Vses. Nauch.-Issled. Gornomet. Inst. Tsvet Met., 1970, vol. 19, pp. 119-24; Chem. Abstracts, 1971, vol. 74, 33756g.
97. S. Kalyanaraman, and S. M. Khopkar: Talanta, 1977, vol. 24, pp. 63-65.
98. G. P. Giganov, V. F. Travkin, and N. P. Anikina: Tsvetn. Met., 1978, pp. 27-29; Chem. Abstracts, 1978, vol. 89, 166662e.
99. S. S. M. A. Khorasani, and Q. A. Hakim: Pak. J. Sci. Ind. Res., 1966, vol. 9, pp. 220-23.
100. G. P. Giganov, I. A. Sputnova, and F. I. Vershinina: Zh. Neorg. Khim., 1967, vol. 12, pp. 755-61; Chem. Abstracts, 1967, vol. 67, 26348c.
101. L. I. Bakunina and V. I. Murashova: 7th Ural. Konf. Spektrosk., 1971, vol. 2, pp. 98-100; Chem. Abstracts, 1973, vol. 78, 37480t.
102. T. Kh. Tserekov, G. P. Giganov, and G. I. Pashkov: Tsvet. Metal., 1972, pp. 60-63; Chem. Abstracts, 1973, vol. 78, 19131a.

103. Y.-C. Hoh, C.-C. Chang, W.-L. Cheng, and I-S Shaw: Hydromet., 1983, vol. 9, pp. 381-92.

104. W. K. Wang, Y.-C. Hoh, W.-S. Chuang and I-S. Shaw: U.S. Patent 4,293,332, 1981.

105. L. Yu. Zakharova and E. A. Belousov: Zh. Fiz. Khim., 1975, vol. 49, p. 3019; Chem. Abstracts, 1976, vol. 84, 65809z.

106. Yu. I. Kol'tsov, N. M. Kuz'min and G. I. Zhuravlev: Zh. Neorg. Khim., 1970, vol. 15, pp. 179-81; Chem. Abstracts, 1970, vol. 72, 71215w.

107. A. Alian, A. Badran, M. S. El-Bassiouny, and A. El-Kot: J. Radioanal. Chem., 1975, vol. 27, pp. 411-20.

108. S. T. Takezhanov, L. S. Getskin, G. P. Giganov, A. S. Kulenov, and G. L. Pashmov: Sb. Nauch. Tr. Vses. Nauch. Issled. Gorno-Met. Inst. Tsvet. Metal., 1968, vol. 12, pp. 5-10; Chem. Abstracts, 1968, vol. 69, 110444e.

109. T. Sekine and T. Ishii: Bull. Chem. Soc. Japan, 1970, vol. 43, pp. 2422-29.

110. A. M. Tserekova and G. P. Giganov: Sb. Nauch. Tr. Vses. Nauch. Issled. Gorno-Met. Inst. Tsvet. Metal 1968, vol. 12, pp. 32-35; Chem. Abstracts, 1969, vol. 70, 23547t.

111. A. El-Kot, A. Ghoneim, and A. Alian: J. Radioanal. Chem., 1980, vol. 59, pp. 351-59.

112. A. A. Yadav and S. M. Khopkar: Sep. Sci., 1970, vol. 5, pp. 637-43.

113. J. J. Bucher, R. Zuehl, and R. M. Diamond: J. Inorg. Nucl. Chem., 1975, vol. 37, pp. 211-20.

114. A. S. Bobrova, E. M. Esmont, and A. S. Chernyak: Nauch. Tr., Irkutsk. Nauch.-Issled. Inst. Redk. Tsvet. Metal., 1972, pp. 119-21; Chem. Abstracts, 1973, vol. 79, 56310z.

115. G. M. Dakar, B. Z. Iofa and An. N. Nesmeyanov: Sov. Radiochem., 1975, vol. 17, pp. 370-73.

116. A. T. Casey, E. Davies, T. L. Meek, and E. S. Wagner: Solvent Extraction Chemistry, D. Dyrssen, J.-O. Liljenzin and J. Rydberg, eds., North-Holland, 1967, pp. 327-34.

117. E. Hallaba, R. A. I. Azzam, and M. Anissa: Z. anorg. allgemeine Chem., 1963, vol. 324, pp. 233-40.

118. R. Stella and M. DiCasa: J. Radioanal. Chem., 1973, vol. 16, pp. 183-90.

119. E. A. Belousov and N. N. Vol'khin: 3rd Zhidk. Ekstv., Tr. Vses. Nauch.-Tekh. Soveshch., P. G. Romankova, ed., Khimiya, 1969, pp. 402-403; Chem. Abstracts, 1970, vol. 73, 90241p.

120. D. J. Pruett, and D. R. McTaggart: J. Inorg. Nucl. Chem., 1981, vol. 43, pp. 2109-12.

121. J. J. Bucher and R. M. Diamond: J. Phys. Chem., 1969, vol. 73, pp. 675-83.
122. J. J. Bucher and R. M. Diamond: J. Phys. Chem., 1969, vol. 73, pp. 1494-500.
123. V. I. Bibikova, K. V. Marunova, and K. D. Gavrilova: Nauch. Tr., Nauch.-Issled. Proekt. Inst. Redkometal. Prom., 1972, vol. 38, pp. 19-23; Chem. Abstracts, 1974, vol. 81, 96859h.
124. N. Iordanov and St. Mareva: C. R. Acad. Bulg. Sci., 1966, vol. 19, pp. 913-16; Chem. Abstracts, 1967, vol. 66, 32382e.
125. E. A. Belousov and N. N. Vol'khin: Izv. Vyssh. Ucheb. Zaved., Tsvet. Met., 1970, vol. 13, pp. 66-71; Chem. Abstracts, 1970, vol. 73, 102551e.
126. M. Knothe: Z. anorg. allgemeine Chem., 1980, vol. 470, pp. 216-26.
127. M. Knothe: J. Radioanal. Chem., 1981, vol. 63, pp. 207-16.
128. R. Stella and N. Genova: Radiochem. Radioanal. Letters, 1974, vol. 16, pp. 273-82.
129. L. Hafner: German Patent, 2,633,115, 1977, Chem. Abstracts, 1977, vol. 86, 124825t.
130. F. A. Saleh: Z. anorg. allgemeine Chem., 1966, vol. 347, pp. 205-214.
131. L. Genov and I. Dukov: Acta Chim. Acad. Scient. Hung., 1976, vol. 89, pp. 297-305.
132. I. Dukov, G. Kassabov, and L. Genov: Monatsh. Chem., 1979, vol. 110, pp. 335-42.
133. L. G. Tebelev and R. F. Melkaya: Zh. Neorg. Khim., 1973, vol. 18, pp. 2814-18; Chem. Abstracts, 1974, vol. 80, 20071d.
134. E. B. Mikhlin, A. M. Rozen, T. M. Norina, V. N. Nikonov, T. A. Afonina, and A. V. Tumanov: Radiokhimiya, 1977, vol. 19, pp. 294-301; Chem. Abstracts, 1977, vol. 87, 91491w.
135. S. A. Marei, A. Badran, A. Haggag, and E. Hanna: J. Radioanal. Chem., 1979, vol. 53, pp. 59-70.
136. S. A. Pai, J. N. Mathur, P. K. Khopkar and M. S. Subramanian: J. Inorg. Nucl. Chem., 1977, vol. 39, pp. 1209-11.
137. A. Apelblat, and A. Hornik: Solvent Extraction Chemistry, D. Dyrssen, J.-O. Liljenzin and J. Rydberg, eds., North-Holland, 1967, pp. 296-304.
138. L. Genov and M. Sachariewa: Monatsh. Chem., 1968, vol. 99, pp. 1976-81.
139. D. G. Kalina, G. W. Mason, and E. P. Horwitz: J. Inorg. Nucl. Chem., 1981, vol. 43, pp. 159-63.
140. G. A. Yagodin, V. V. Sergievskii, and L. V. Evdokimova: Izv. Vyssh. Uchebn. Zaved., Khim. Khim. Tekhnol., 1980, vol. 23, pp. 1388-92; Chem. Abstracts, 1981, vol. 94, 53724m.

141. I. Fidelis: J. Inorg. Nucl. Chem., 1970, vol. 32, pp. 997-1003.
142. M. N. Nabiev, T. V. Zhilyaeva, E. S. Gureev, Sh. S. Namazov, I. I. Orestova, and A. A. Kist: VINITI 3600-77, 1977; Chem. Abstracts, 1979, vol. 90, 123920r.
143. C. G. Brown and L. G. Sherrington: J. Chem. Tech. Biotechnol., 1979, vol. 29, pp. 193-209
144. L. Genov and M. Sachariewa: Monatsh. Chem., 1973, vol. 104, pp. 470-76.
145. D. A. Douglass and D. J. Bauer: Report of Investigations 5513, U.S. Bureau of Mines, Washington, D.C., 1959.
146. Y. Aratono and E. Akatsu: J. Inorg. Nucl. Chem., 1974, vol. 36, pp. 1141-46.
147. E. Akatsu and Y. Aratono: Anal. Chim. Acta, 1972, vol. 62, pp. 325-35.
148. A. T. Kandil and K. Farah: J. Inorg. Nucl. Chem., 1980, vol. 42, pp. 277-80.
149. L. Genov and G. Georgiev: God. Vissh. Khim.-Tekhnol. Inst. Sofia, 1978, vol. 23, pp. 211-18, Chem. Abstracts, 1979, vol. 90, 77181a.
150. Y. Hirashima, M. Mugita, and J. Shiokawa: J. Inorg. Nucl. Chem., 1976, vol. 38, pp. 1199-202.
151. Z. Kolarik, R. G. Puzic and Z. B. Maksimovic: J. Inorg. Nucl. Chem., 1969, vol. 31, pp. 2485-98.
152. L. Genov and G. Georgiev: Monatsh. Chem., 1969, vol. 100, pp. 1240-45.
153. E. B. Mikhlin, A. M. Rozen, T. M. Norina, V. N. Nikonov and T. A. Afonina: Russ. J. Inorg. Chem., 1976, vol. 21, pp. 1018-21.

SOLVENT EXTRACTION OF METALS - METAL TRANSFER RATES AND CONTACTOR DESIGN

Thomas W. Chapman

Chemical Engineering Department
University of Wisconsin
Madison, Wisconsin, U.S.A.

Solvent extraction of a metal usually occurs by means of a reversible, heterogeneous reaction. The rate of metal transfer may be determined by either intraphase mass-transfer rates or reaction kinetics. The nature of the rate-controlling step determines equipment design criteria, especially when several metals can be extracted and selectivity is a concern.

Mass transfer rate models are developed that incorporate the essential features of specific system chemistry. A simple experimental tool, a growing-drop extraction cell, is described. Such a cell can be used to measure slow reaction rates or to confirm a mass transfer model for a system with fast kinetics. Examples of systems studied include copper extraction from acid sulfate with Kelex and LIX 64N, copper and zinc chloride extrasction by TIOA, and copper and nickel extraction from ammonium sulfate with LIX 64N.

Quantitative models of equilibria and rates permit simulation of various types of contactors. The effects of contactor parameters on extraction efficiency and selectivity are demonstrated.

There are two general types of equipment that are used for the continuous solvent extraction of metals, staged contactors and differential contactors. The distinction between these two types is based as much on the way they are conceptualized in process design as in the details of their physical configuration. For example, an in-line static mixer with cocurrent flow of the immiscible phases could be viewed either as a stage or as a tubular reactor, depending on the level of analytical characterization adopted by the designer. And a short countercurrent column that exhibits extensive axial mixing might be viewed as a single stage under some circumstances. In both types of contactor local interphase mass transfer proceeds in the direction that brings the bulk phases toward equilibrium. The difference between our conceptions of the two types is that a contacting stage is described in terms of the streams that would leave the contactor if the exiting phases reached equilibrium whereas the description of a differential contactor includes a continuous spatial variation of bulk-phase compositions and inherently involves metal-transfer rate parameters as well as equilibrium considerations.

Most of the large solvent extraction installations in the metals industry, namely those extracting copper from sulfuric acid leach liquors, use countercurrent cascades of large mixer-settler units to accomplish both extraction and stripping. Because of the extraordinary selectivity and strongly favorable equilibrium properties of the copper-chelating extraction agents (LIX, etc.) it has been possible to describe these units not only in terms of equilibrium stages but even as pseudo-binary extraction processes in which interstage changes in pH or free reagent level are neglected or treated as a minor factor. Although interfacial reaction rates are known to be slow, the influence of rate factors could be handled in a fairly simple, empirical manner by requiring long residence times (e.g. 2 min.) in each mixer in order to achieve high stage efficiencies. Rates of iron extraction are apparently so slow that detailed reactor design has not been required in these systems to insure required selectivity. Thus, the remarkable properties of the copper extraction reagents have made the engineering of what is inherently a very complex system almost trivial. In considering the prospects for new hydrometallurgical extraction processes, one must expect such elegant simplicity to be the exception rather than the rule.

In most cases of hydrometallurgical process development one encounters a leach liquor containing several metals, two or more of which should be separated and recovered economically. Except for the acid copper system there are not yet available sufficiently selective extractants that will allow direct and complete extraction of the individual metals one by one. In this case the challenge to the engineer is much greater, and he must explore a number of process possibilities to identify feasible routes and optimal designs.

The efficient analysis of complex processes requires the use of quantitative models that combine physical laws and experimental data. The purpose of this paper is to summarize the development of some models that are useful in the description of metal extraction systems, particularly the solvent extraction units. The simplest realistic model of solvent extraction processes is the equilibrium stage model. This model requires proper description of the two-phase equilibria but avoids consideration of rate phenomena. It is adequate for describing equipment that is staged, such as mixer-settlers, and has sufficient mixing and residence time for the exiting streams to reach equilibrium. More detailed analysis of individual stage efficiencies and the description of differential contactors, such as countercurrent columns, requires treatment of the local interphase mass transfer rate and flow configurations as well as the equilibrium properties. This paper reviews both equilibrium and rate models for several specific systems and shows how such models may be used to size equipment as well as to compare the performance of different types of contactors with respect to both efficiency and selectivity.

EQUILIBRIUM MODELS

A number of workers have attempted to correlate solvent extraction equilibria in the form of a metal equilibrium curve or a distribution coefficient as an empirical function of system composition variables such as pH or extractant level. The difficulty with this approach is that the independent variables of such a correlation are not really independent variables within a process. Solution pH will change from point to point in a process, just as metal concentrations do, according to mass balances, charge balances, and solution chemistry. Our approach has been to describe the species distribution by using mass-action-law equilibrium constants to characterize all the reversible equilibria, both homogeneous and heterogeneous, that determine the phase equilibria. The symmetry of this approach does not require one to view any particular species concentrations as independent variables. A second advantage is that parameters determined in single metal extraction systems can be used to estimate the simultaneous distribution equilibria in a multiple-metal system.

The simplest case one might consider is the extraction of metal A as a divalent cation by an acidic extractant HR according to the reaction

$$A^{+2} + 2\ HR \rightleftarrows AR_2 + 2\ H^+ \qquad (1)$$

If no complexation, association, or buffering reactions occur simultaneously, the equilibrium distribution of metal A should be fairly well described by the mass-action equilibrium condition

$$(AR_2)\ (H^+)^2/(A^{+2})(HR)^2 = K_A \qquad (2)$$

where the equilibrium constant K_A should be a function of temperature but not dependent on composition to any great extent.

Unfortunately, few real extraction systems follow Eq. 2 directly in terms of analytical concentrations. The concentrations appearing there are actual species concentrations that may differ from apparent or analytical concentrations because of reversible homogeneous reactions that allow a component to exist as several different chemical species. The equilibria of the competing reactions must be characterized so that the actual concentrations appearing in Eq. 2 can be calculated. Activity coefficient variations will cause the equilibrium constant to vary with composition, but this variation should be minor if all important chemical interactions are identified and accounted for my means of stability constants.

For the case of copper extraction from sulfuric acid solutions, where no significant complexation alters ionic concentrations, Eq. 2 does represent the extraction equilibrium fairly well. Hoh and Bautista[1] obtained values for K_A on the order of 0.05 for LIX 64N, and Bauer and Chapman[2] found K_A to be 88.6 for Kelex 100 at 25°C. In the latter case Eq. 2 was not successful in fitting the copper distribution under stripping conditions until the extraction of sulfuric acid by Kelex was accounted for by a second mass-action equilibrium expression. At low pH the acid reacts with Kelex and crowds the copper out of the organic phase. Figure 1 compares the predictions of Eq. 2 alone and in conjunction with the acid extraction equilibrium expression with experimental stripping data. The two-reaction model is quite successful in predicting the system behavior.

Other metal extraction systems are not so easy to model. One might expect extraction of copper or nickel from acidic sulfate solutions by di-(2-ethylhexyl) phosphoric acid (DEHPA) to be directly analogous to the cases mentioned above and thus described by Eq. 2. Troyer[3] found, however, that these systems are more complex in that the reagent is dimerized and probably associated to high degrees of polymerization, and the extracted complex is $AR_2(HR)_2$. Furthermore, in the pH range of interest a number of simultaneous equilibria must be considered in order to compute ionic concentrations accurately:

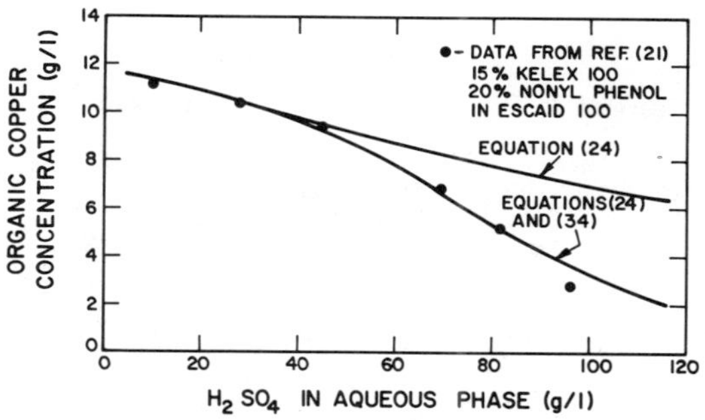

Fig. 1. Predicted and experimental organic copper levels at equilibrium in stripping of Kelex 100. Curve labeled Eq. 24 is based on one equilibrium constant; the other includes sulfuric acid extraction equilibrium.[2]

metal hydrolysis forming a hydroxy complex, bisulfate dissociation to sulfate, metal sulfate ion-pair formation, and the extraction of any cations such as sodium that are added as a base to drive the extraction. In spite of these complications, Troyer was able not only to fit the individual copper and nickel extraction data but also to predict accurately with the same set of parameters the simultaneous distribution of the two metals. Figure 2 shows the experimental and predicted organic copper levels obtained with DEHPA when specific concentrations of NaOH were added to solutions of copper and nickel sulfate. Figure 3 compares the predicted nickel extract levels with the observed values for the same experiments.

Another example where numerous simultaneous equilibria must be considered is the extraction of chloride complexes by tri-isooctyl amine (TIOA). Thiel[4] studied extraction of cupric and zinc chlorides from acidic NaCl solutions by TIOA in xylene and fit the distribution equilibria by considering not only the stepwise formation of metal chloride complexes and the extraction reaction but also the extraction of HCl to form an amine salt and the subsequent dimerization of the organic salt. Although the model was successful in fitting the single metal extraction data and in predicting the extraction of zinc in the presence of copper, the single-metal equilibrium parameters were not adequate to predict the copper equilibrium in the presence of zinc. One suggested cause for this apparent failure of the model is the possibility of spontaneous copper reduction to cuprous in the organic phase.

Finally, another example in which complex-formation equilibria must be modelled is the extraction of metals from ammonia solutions.

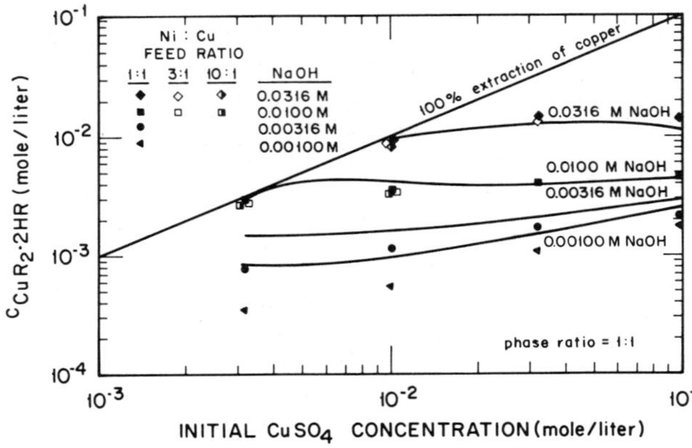

Fig. 2. Predicted and observed organic copper concentrations in equilibrium with 0.10 M DEHPA in xylene for different amounts of base added and several copper to nickel ratios. Curves are predicted by the model for the indicated initial NaOH concentration.

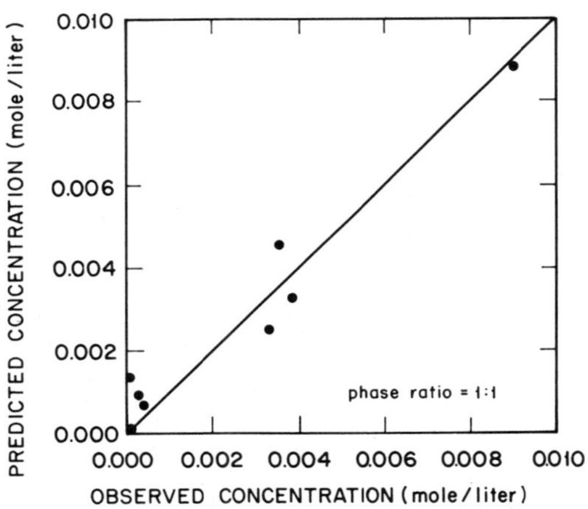

Fig. 3. Comparison of the predicted and experimental organic nickel concentrations for simultaneous extraction of nickel and copper. Data from experiments indicated in Fig. 2.

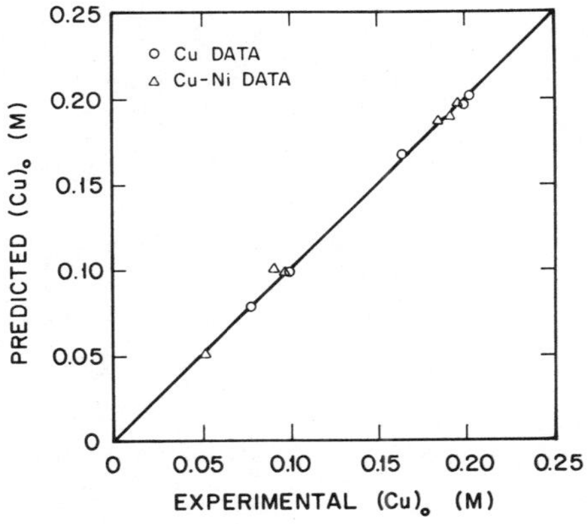

Fig. 4. Comparison of equilibrium model predictions with data for organic copper concentration in extraction of nickel and copper from ammonia solution by LIX64N.

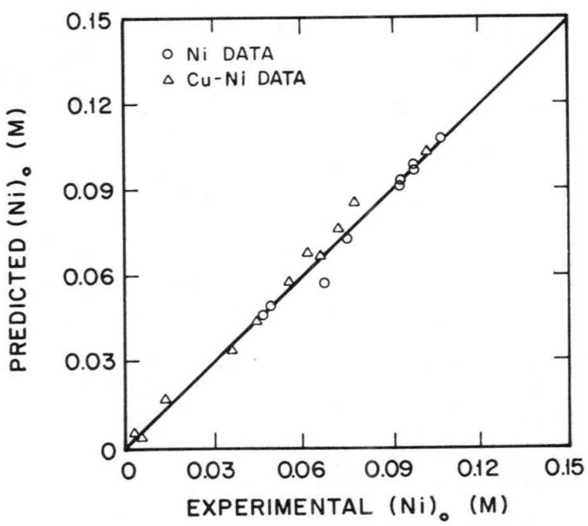

Fig. 5. Comparison of equilibrium model predictions with data for organic nickel concentration in extraction by LIX 64N from Cu-Ni-ammonium sulfate solutions.

DeRuiter[5] has studied the extraction of copper and nickel from ammonia-ammonium sulfate solutions by LIX 64N. His model included equilibrium constants for the following reversible reactions: coordination of copper by four and five ammonia ligands, coordination of nickel by five and six ammonia ligands, the extraction of each of the cations releasing ammonia, hydrolysis of ammonia to ammonium, and extraction of ammonia both by physical solubility and by chemical association with LIX. The effort was quite successful as shown in Fig. 4 and 5, which are plots of predicted versus measured organic metal concentrations for copper and nickel, respectively, both individually and in mixed solutions. Algebraic combination of the copper-ammonia stability constants and the extraction equilibrium constant obtained in this system allows calculation of the K_A that is defined in Eq. 2. It is interesting and gratifying to find that the value obtained from DeRuiter's model is 0.02, a value that agrees quite well with that obtained for LIX 64N in acidic systems.[1]

EQUILIBRIUM STAGE CALCULATIONS

After the chemistry of an extraction system is understood and an equilibrium model has been developed that accounts for all significant species and their reactions, one is able to undertake the first level of process design--equilibrium stage calculations. Such calculations provide a first-approximation estimate of the effects of various operating variables on the expected performance of a separation process; they may in fact simulate fairly well the actual performance of a staged contactor such as a cascade of mixer-settlers if reaction kinetics and mass-transfer rates are fast enough to provide high stage efficiencies.

Several years ago we published[6] a formulation of the equilibrium stage calculation that combined an equilibrium model of the type suggested above with appropriate mass balances to describe a countercurrent cascade of equilibrium stages. The mathematical problem that results consists of a system of simultaneous, nonlinear algebraic equations which may be solved for the species concentrations leaving each of the stages. An iterative linearization/matrix inversion algorithm was suggested for solving this problem. Solution for any particular system allows one to investigate the effects of number of stages, feed stream concentrations, equilibrium constants, and phase ratio on extraction efficiency and on selectivity in multiple metal systems. Brana[7] has considered the more general case where interstage flow may not be purely countercurrent.

As an example of the type of information that can be generated from the equilibrium stage model, Table 1 shows the calculated effect of phase ratio on the simultaneous extraction of two metals, A and B, each of which follow Eq. 2 with K_A=100 and K_B=10. The calculation has been done for three countercurrent equilibrium stages. The results indicate that in such a system there should be an optimal phase ratio, probably somewhere around 1.0 for the feed concentrations considered here, which would balance the amount of extraction of A against the degree of contamination with B. Such preliminary estimates would be useful in guiding an experimental development program.

A more interesting example is the case of zinc and copper chloride extraction by TIOA. Thiel[4] applied the equilibrium stage algorithm to his two-metal equilibrium model. Because of the numerous species that must be considered in this system, 16 equations must be solved for each stage. A typical result of the model calculation is shown in Table 2 which gives total copper and zinc concentrations in each stream of a four-stage countercurrent cascade. Because zinc is extracted so much more strongly than copper(II) by TIOA, a very good separation is predicted. An interesting effect, however, is that the copper level in both phases goes through a maximum between the third and fourth stages. If a copper purge were required, this might be the place to take it.

Thiel used his model to calculate a zinc/copper separation factor and explored the predicted effects of chloride level and feed pH as well as phase ratio on the separation factor. He also checked the sensitivity of the process calculation to the values of the equilibrium model parameters and found that statistical uncertainties in his equilibrium constants were not critical.

Besides allowing preliminary process calculations an equilibrium stage model provides a benchmark against which to compare pilot-plant data. By comparing data with the equilibrium stage simulation one could infer practical rate information. For example, if the process were being conducted in mixer-settlers, some estimates of stage efficiency could be made and correlated with residence time, internal phase ratio, power input, etc. to obtain guidelines for scale-up. If the extraction were being done in a packed column, the height equivalent to a theoretical stage could be estimated and similarly used as a rough guide in scale-up.

If only one metal is being extracted, identification of such efficiency or rate factors on an ad hoc basis would be reasonable.

Table 1. Effect of phase ratio (R=O/A) on the relative extraction of two metals. Three countercurrent equilibrium stages. $K_A = 100$, $K_B = 10$. Feed conditions: $C_{A^{+2}} = C_{B^{+2}} = 0.03M$, $C_{HR} = 0.06M$, $C_{H^+} = 10^{-4}M$, $C_{AR_2} = C_{BR_2} = 0$.

Phase ratio, R	A extracted (%)	B extracted (%)
2.0	99.97	81.90
1.7	99.89	62.55
1.5	99.72	46.45
1.3	99.16	29.19
1.2	98.22	20.83
1.1	95.38	14.14
1.0	89.14	10.63
.9	81.03	8.84
.5	45.41	4.57
.1	9.09	0.91

Table 2. Calculated separation of copper and zinc by TIOA in four countercurrent equilibrium stages.
Aqueous feed: 0.80M copper, 0.20M zinc, pH = 1.0
Organic feed: No metals present, 0.20M TIOA•HCl
Phase ratio = 2.0 (O/A)

Aqueous Conc. (M)			Organic Conc.(M)		
Stage No.	Copper	Zinc	Stage No.	Copper	Zinc
Feed	0.8000	0.2000			
1	0.8000	0.1999	Product	0.0010	0.0964
2	0.8014	0.1974	2	0.0011	0.0963
3	0.8456	0.1326	3	0.0017	0.0951
Product	0.7979	0.0073	4	0.0239	0.0627
			Feed	0.0000	0.0000

In fact, the design criterion for Cu-LIX systems, a specified mixer residence time, is presumably generated in such a manner, numerous studies of reaction kinetics notwithstanding. On the other hand, if several metals are extracted simultaneously, the identification of a stage efficiency may not be straightforward. Experiments by Ritcey[8] and mass-transfer calculations[6,9] show that two metals can exhibit quite different efficiencies within the same contactor and that relative efficiencies can change dramatically with contactor type. In situations where extraction selectivity is a concern and where new types of contactors are being considered, detailed analysis of rate effects is needed. Metal extraction rate phenomena can become very complex so that a theoretical model that provides a rational framework for correlating data and extrapolating available information to different contactor designs and different operating conditions should be very useful.

EXTRACTION RATE MODELS

Because metal extraction reactions occur heterogeneously their rates will be proportional to liquid-liquid interfacial area. These rates may be controlled either by heterogeneous reaction kinetics, by the rates of transport of reactants to the surface or of products away, or by a combination of these processes. Furthermore, the rate of a particular extraction may be influenced by homogeneous reactions or even competing heterogeneous reactions that can alter interfacial species concentrations. With so many factors entering into the rate of the process, it is important to identify which are controlling and to develop a rate model that can be used in contactor design calculations.

To determine whether a particular extraction reaction is fast or mass-transfer limited, one must have available a mass transfer model that characterizes the effect of the system chemistry on transport rates and on shifts in interfacial species concentrations from local bulk values. Such a model can then be applied to a well defined experimental system in which interfacial area and physical mass transfer coefficients are known. If extraction rates are measured in such a system, the transport model can be used to determine whether interfacial concentrations maintain equilibrium; if so, the extraction reaction is considered to be fast. If interfacial equilibrium is not maintained during extraction, calculated interfacial concentrations may be used in conjuction with observed fluxes to determine a heterogeneous reaction rate model.

Although we have used a penetration model to estimate the effect of ionic migration on solvent extraction[10], the only mass transfer model that is really practical for equipment design is the stagnant film model. In using this model, one assumes that the effective stagnant film thickness in a given situation can be estimated from standard mass-transfer-coefficient correlations. Then the specific system chemistry is accounted for by solving a two-film diffusion problem that includes all relevant species and reaction terms. The net result is a relationship between actual metal flux and the maximum flux that would be expected for simple physical transport of a metal species across one of the boundary layers. The ratio of these fluxes will depend on numerous factors including bulk concentrations, the parameters of the equilibrium model, and parameters from the heterogeneous kinetics model if a fast interfacial reaction is not assumed. If more than one extraction reaction can occur, the relative rates of extraction will be computed, which is directly useful in estimating selectivity effects.

The formulation of such mass transfer models was outlined for the metal chloride-TIOA system in a recent paper.[11] The approach is to write steady-state continuity equations for each species in the films on either side of the interface. Convection is neglected in the stagnant-film model, but most of these continuity equations contain homogeneous reaction rate terms that account for local production or disappearance of solute species. The continuity equations can be combined to cancel the homogeneous reaction terms and, if the homogeneous reactions are fast, one need solve only diffusion equations for combinations of species along with the local equilibrium equations to get species profiles.

For the cupric chloride-TIOA system the result of the calculation may be expressed as

$$\overline{N}_{Cu} = \overline{N}_{Cu}\ (X_1,\ X_2,\ X_3;\ K_1^*,\ K_2^*,\ \ldots\ K_6^*) \qquad (3)$$

where $\overline{N}_{Cu} = \delta\overline{N}_{Cu}/D\ c_\infty$, N_{Cu} is the actual interfacial flux of copper, D is its diffusivity, δ is the effective aqueous film thickness, and c_∞ is the analytical concentration of copper in the bulk aqueous phase. The dimensionless flux $\overline{N}_{Cu}$ may be viewed as a fraction of the maximum possible copper flux that could be obtained if all forms of copper in the bulk phase were free to diffuse to the interface and the interfacial concentration were held at zero. The variables affecting $\overline{N}_{Cu}$ are bulk concentration ratios, X_i, and dimensionless ratios of six equilibrium constants to the bulk copper concentration, K_i^*.

Typical values of $\overline{N}_{Cu}$ are shown in Fig. 6. Curve a shows the predicted variation of copper flux with X_3 which represents the degree of copper loading in the organic phase. Note that the dimensionless flux is much less than one in spite of the fact that the interfacial reaciton is assumed to be fast. This calculation shows the flux dropping to zero when the bulk organic phase reaches equilibrium loading; the flux goes negative at higher loadings which indicates a finite rate of stripping.

Curves b and c show the result of a more complete calculation of copper flux in the presence of zinc extraction. The zinc retards the copper extraction rate by competing for reagent. In curve b the amine is partly loaded with zinc, and some zinc is being stripped. In curve c zinc is being extracted; it retards the copper extraction by consuming amine near the interface. This effect is significant even though the level of zinc is only 1/10 that of copper. In these calculations all diffusivities have been assumed equal as have the two film thicknesses to reduce the number of parameters in the model.

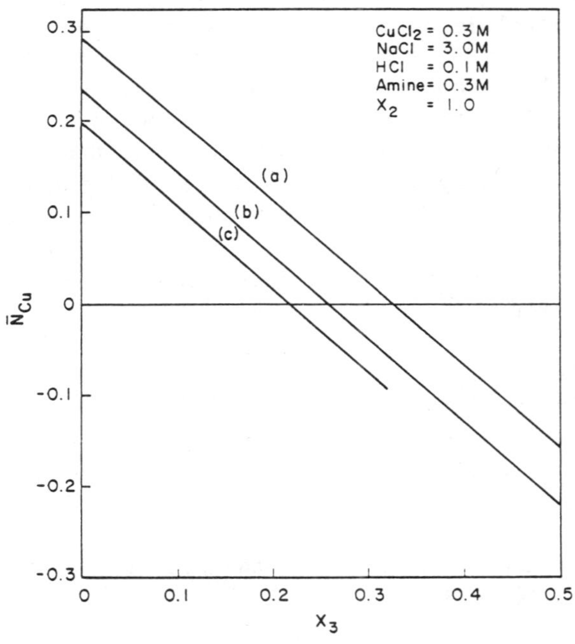

Fig. 6. Effect of the presence of zinc upon the flux of copper in extraction by TIOA. Curve a: no zinc. Curve b: 0.05 M Zn in the organic phase. Curve c: 0.03M $ZnCl_2$ in the aqueous phase.

Although such mass transfer models can be developed on the basis of an adequate description of the solution chemistry, one must conduct rate experiments to determine whether the interfacial boundary condition on the species concentration distribution should be equilibrium or a kinetics model for the interfacial reaction. The following section discusses such measurements.

EXTRACTION RATE MEASUREMENTS

An experiment for measuring extraction rates should offer the following attributes: known interfacial area, well characterized mass transfer properties (i.e. a known film thickness), and a uniformly accessible surface which is frequently renewed so that surfactants can not build up and block the reaction. We have found that the growing drop mass transfer cell developed by Bauer[12] meets these conditions and provided a convenient method for studying extraction rates in a variety of systems.

The growing-drop apparatus was developed in analogy with the dropping mercury electrode of polarography. A diagram is shown in Fig. 7. A known volume of organic phase is placed in the upper chamber, and a steady flow of aqueous phase is introduced from the mouth of a capillary tube. An aqueous-organic interface of minimal area is maintained in the neck of the cell between the upper and lower compartments. The aqueous phase flowing from the capillary tip forms a growing drop that does not dislodge for a second or two. While it is growing, the flow in each phase is essentially radial. This flow establishes a simple convective pattern and allows theoretical calculation of the mass transfer coefficients. After the drop reaches a certain size, it detaches from the capillary and falls to the receiver neck. The amount of mass transferred during the drop fall is negligible, and that at the neck interface is small but can be estimated with an unsteady-diffusion model. Thus the theoretical mass transfer properties of the cell can be calculated and have been confirmed by calibration with a physical solute transfer system.[12]

From the transport analysis of the drop the average film thickness on the drop is estimated to be

$$\delta = 0.458\ (D\pi\ t_d)^{1/2} \tag{4}$$

where t_d is the drop lifetime. The average film thickness in the receiver neck is of the same order of magnitude (0.05 mm), but the

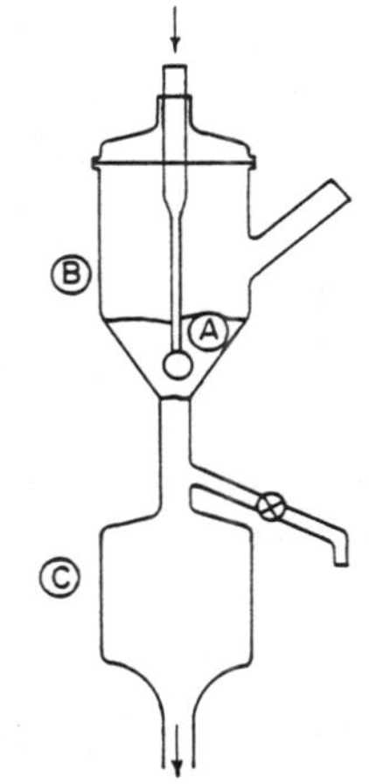

Fig. 7. Diagram of the growing drop mass transfer cell.

relatively small area of the neck limits the relative amount of mass transferred in the neck to about 15% of that on the drop.

A macrocopic mass balance on metal concentration in the organic phase, which is measureable by sampling and analysis, relates the rate of metal accumulation to the average rate of interfacial transport. The measured rate, combined with the theoretical values of the film thicknesses, can be compared with the predictions of the two-film model, which includes all of the details of the species equilibria, to determine whether the interfacial reaction is fast. If the observed rate of extraction is slower than that predicted by the mass transfer model solved with equilibrium at the interface, one can conclude that the reaction is at least partly controlled by reaction kinetics. In this case the mass transfer model can be used to calculate from the measured flux the interfacial species concentrations. These in turn can be used to develop an empirical heterogeneous kinetics model that can be used as a boundary condition on the mass transfer model in equipment-design calculations where the mass transfer coefficients may be different from those in the growing-drop cell.

Bauer[12] used the growing-drop cell to measure the kinetics of copper extraction from acid solution by Kelex, and the validity of his results was confirmed by Tunison [13,14] who measured copper extraction efficiency in a Kenics mixer. In this case the mass transfer resistance was negligible because the heterogeneous reaction rate was so slow.

At the other extreme Tse[11,15] investigated two systems in which the extraction is mass-transfer-limited: extraction of cupric and zinc chlorides by TIOA. One of Tse's copper extraction experiments with copper is represented by Fig. 8 which shows the amount of copper extraction in the growing-drop cell as a function of time. The points in Fig. 8 are experimental points and the curve is a theoretical expression based on the equilibrium and mass transfer models indicated above. The agreement between data and theory leaves no doubt regarding the validity of the mass-transfer model nor the fact that this reaction is indeed fast.

Tse also measured the rates of simultaneous extraction of copper and zinc in the growing-drop cell. The result is shown in Fig. 9. It is seen that both the computed and experimental amounts of copper

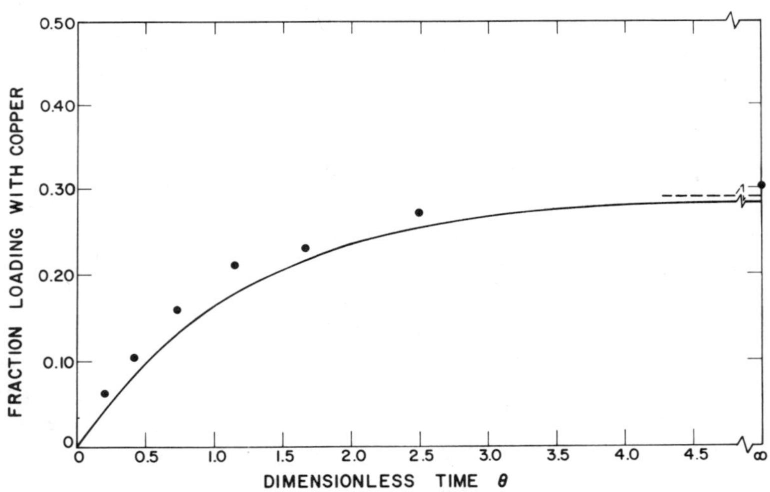

Fig. 8. Comparison between experimental and theoretical extraction of copper chloride by TIOA in the growing drop cell. (aqueous phase: 0.1M $CuCl_2$, 2M NaCl, 0.1M HCl; initial organic phase: 0.2M TIOA-HCl)

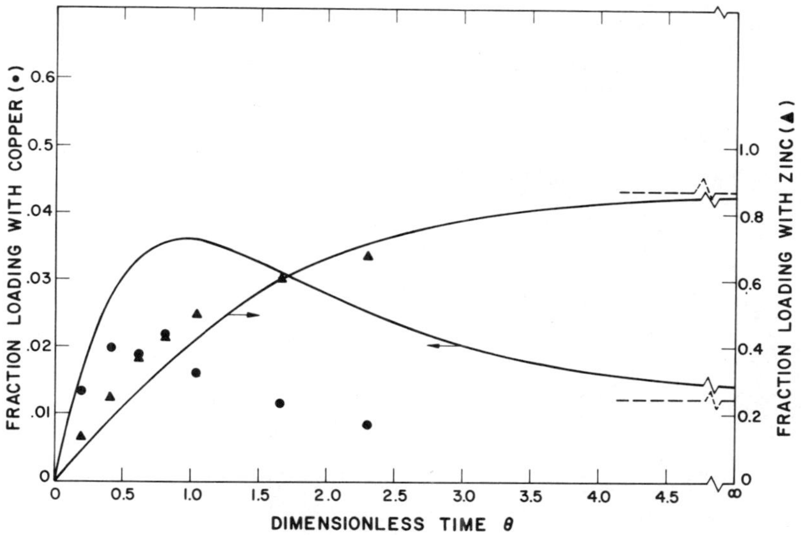

Fig. 9. Comparison between experimental and theoretical extraction of copper and zinc chlorides by TIOA in a growing drop cell. (aqueous phase: 0.099M $CuCl_2$, 0.025M $ZnCl_2$, 2M NaCl, 0.05M HCl; initial organic phase: 0.1M TIOA.HCl)

pass through a maximum at short times. Although agreement for the zinc data was quantitative, the predicted amount of copper extraction was overestimated throughout the duration of the experiment. We believe this discrepancy to be not a defect in the mass transfer model but another manifestation of some chemical interaction that modifies the copper equilibrium in the presence of zinc.

Vargas[16] obtained some preliminary results on the rate of copper extraction by LIX64N from acid and from ammonia solutions. It was found that the copper extraction rate increases with pH, becoming mass-transfer limited in the alkaline ammonia solutions.[11] More recently DeRuiter[5] has not only developed equilibrium models for copper and nickel extraction from ammonium sulfate solutions but also derived mass transfer models for these systems and checked them with the growing-drop cell. A schematic of his two-metal mass transfer model is shown in Fig. 10 which indicates the species considered. Results of two of DeRuiter's copper extraction runs are shown in Fig. 11. The experimental data agree quite well with the computed curves which indicates that the copper extraction here is mass-transfer limited.

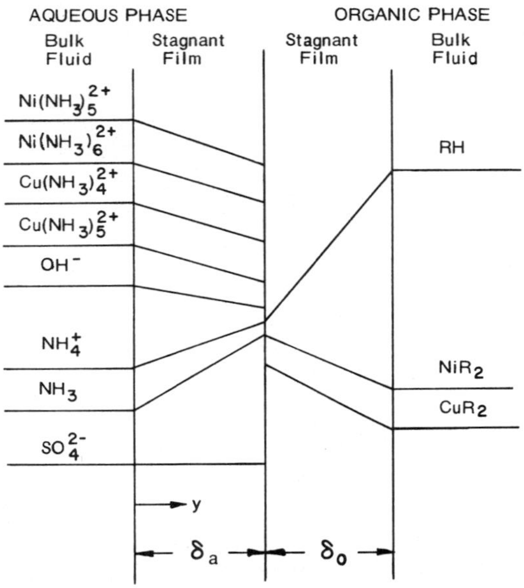

Fig. 10. Schematic drawing of species profiles near the interface during extraction of copper and nickel from ammonia solution.

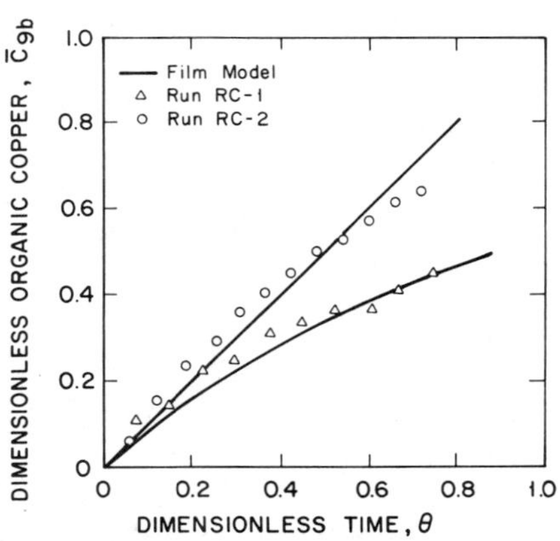

Fig. 11. Comparison of copper extraction data in the growing drop cell with mass-transfer model predictions for the ammonia-LIX 64N system.

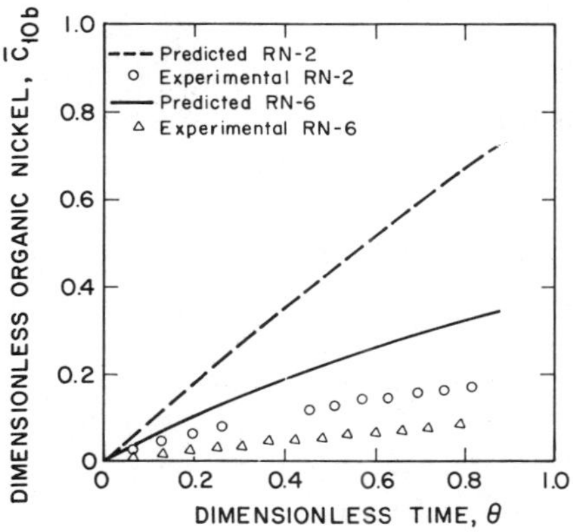

Fig. 12. Comparison of nickel extraction data in the growing drop cell with mass-transfer model predictions for the ammonia-LIX 64N system.

Comparisons of nickel extraction data with mass-transfer model predictions are shown in Fig. 12. It is seen that the nickel extraction rate is considerably lower than expected for a fast reaction. DeRuiter was able to calculate interfacial species concentrations during the extraction and to determine an empirical nickel-LIX64N kinetics model. In this case the rate is not so slow that mass transfer effects are completely negligible.

Finally, DeRuiter measured the rates of simultaneous extraction of copper and nickel from ammonia solution by LIX64N in the growing-drop cell. He found that the rate of copper extraction is fast and fairly well estimated by the mass-transfer model. On the other hand nickel extraction is very slow, slower even than predicted by the single-metal kinetics model. This is an interesting effect that bears further scientific investigation, but the present observation already offers some important process design implications.

An experimental tool such as the growing-drop cell, combined with a chemically consistent mass-transfer model, can provide information about the local rate behavior of any given solvent extraction system. The last step in a rational approach to process development is to incorporate the rate information in macroscopic models of various contactor designs to elucidate the relationships between contactor configuration, size, and operating conditions and its resulting preformance in terms of extraction efficiency and selectivity. The next section considers the problem of contactor simulation.

CONTACTOR DESIGN CALCULATIONS

The essence of a macroscopic two-phase contactor model is quite simple in that the change in the concentration of any component in either phase from inlet to outlet is directly related to its average transport rate across the phase boundary. The difficulty in developing a quantitative, predictive model of any particular unit usually arises in trying to estimate that average transport rate. As we have noted above, the total transport rate of a metal will be proportional to the total interfacial area in the equipment, which is a quantity that depends on many physical factors. Furthermore, the local metal flux is also a complicated function of many variables, the most important of which are the local bulk concentrations and the local film mass transfer coefficients. To make this problem tractable, one usually assumes that the mass transfer coefficients (and the interfacial area per unit volume) are constant over the volume of the contactor. This assumption reduces the problem to one of characterizing macroscopic flow patterns of the phases, and sometimes their large-scale mixing behavior, in order to trace the variation of bulk concentrations with respect to position in the contactor.

With mixer-settlers one can simplify the modelling problem and by-pass the rate question completely by assuming that the mixing and residence time in each mixer are sufficient for the exiting streams to reach equilbrium. This limiting case, the equilibrium-stage model described above, is useful because it is often approached in practice by systems with fast kinetics. Equipment that conforms to the equilibrium stage case can be designed and simulated quite readily in terms of mass balances and equilibrium parameters and without reference to rate considerations. The disadvantages of such systems are that they may require excessively large volumes and solvent inventories to approach the equilibrium case and that unexpected or undesirable selectivity variations may be encountered in multiple metal systems that do not reach equilibrium.

To analyze efficiency and selectivity effects in real mixer-settler units that do not attain stagewise equilibrium, one must incorporate rate phenomena into the stage model. The next level of modelling of a stage past the equilibrium assumption comes from assuming that each of the bulk phases within the mixer is sufficiently well mixed that the bulk concentrations are uniform and everywhere equal to their exit values. This is the well known CSTR model of chemical reactor design, and it implies that the interfacial metal flux is uniform over the interfacial area in the mixer and corresponds to the local value determined at the exit bulk concentration levels.

Mathematically the simulation of this type of staged process is still an algebraic problem as is the equilibrium stage calculation. A mass balance containing the total rate of extraction in each stage replaces the equilibrium expression as a constraint in the model formulation. We have discussed the treatment of this problem in earlier papers.[6,17] For a simple, mass-transfer-limited extraction that follows Eq. 1, the stage efficiency depends on both the equilibrium model parameters as well as the quantity $k_1a\tau$, where k_1 is the aqueous film mass transfer, a is the interfacial area per unit volume, and τ is the residence time (volume/flow rate) of the aqueous phase in the mixer. The product k_1a is a volumetric mass transfer coefficient and is a quantity that is frequently measured and correlated as a function of flow parameters in various mass transfer devices. If adequate equilibrium and two-film mass transfer or kinetics models were available, a knowledge of the k_1a factor would make it possible to relate extraction performance to mixer size and throughput for a cascade of non-equilibrium stages. Thus, the k_1a factor is the quantity that one should compute from pilot plant data as a design tool for scale-up. For the case of a very slow heterogeneous reaction the k_1 in this efficiency factor would be replaced by a reaction rate constant so that the critical equipment parameter would be just a.

The next refinement in modelling of mixer-settlers must test the validity of the CSTR approximation. One question is whether significant extraction occurs in the settlers such that their exiting stream concentrations differ from those leaving the mixers. Another question is whether the mixers themselves are uniform or whether there may be flow segregation and short-circuiting. To answer these difficult questions in a thorough manner would require a detail mapping of flow patterns through the equipment and solution of differential transport equations to determine concentration profiles as functions of position. At this level the mixer-settler is no longer being viewed as a staged, or lumped-parameter, process. It can no longer be modelled by just algebraic equations; it must be viewed as a differential, or distributed-parameter, contactor.

Unfortunately, most types of contactors that are alternatives to mixer-settlers must be treated as differential contactors, even in their most primitive design models, unless they contain some physical unit that obviously is well mixed. Models of differential contactors, such as packed columns, can be simplified by assuming that the macroscopic flows and bulk composition variations in the equipment are one-dimensional and that no bulk mixing occurs in the direction of flow. This is the plug-flow approximation of chemical reactor design. It is the opposite extreme from the CSTR model with respect to back-mixing efficiency. Fortunately for the sake of design calculations most real contactors (and reactors) conform rather well in their geometry and flow charcteristics to one or the other of these idealized models.

A good example of a contactor that is properly modelled as a differential plug-flow device is the Kenics static mixer studied by Tunison[14] and by Tse.[15] The static mixer is simply a pipe containing fixed helical mixing vanes through which the two phases are pumped. Drops are broken up and the phases are well mixed by the vanes, but axial dispersion is not great. Macroscopicly concentrations may be viewed to vary only in the axial direction, and axial mixing may be neglected. A one-dimensional differential mass balance may be written to relate the rate of change of concentration to the local rate of mass ransfer, and integration yields the axial concentration profiles.

Tunison measured the degree of extraction of copper from acid solution by Kelex 100 in a Kenics mixer. He also estimated interfacial area per unit volume of contactor as a function of flow conditions from photographic measurements of drop sizes. Copper extraction efficiencies were low because of the slow reaction kinetics, but Tunison showed that Bauer's rate model from the growing drop cell correctly predicted the rate dependence on composition. With that functionality known, Tunison was able to fit his extraction efficiency data with a single equipment parameter, which was essentially the interfacial area in the contactor.

Tse used the same apparatus to extract continuously cupric and zinc chlorides with TIOA. In these systems the rates are mass-transfer limited so that high extraction efficiencies were observed. As with the stirred tank model the differential contact model predicts conversion to be a function of k_1a. By comparison of his extraction data with the model Tse was able to determine the equipment parameter k_1a for the Kenics mixer as a function of the flow rate through the mixer. The k_1a correlations from the copper and from the zinc measurements were essentially the same, which indicates that the values should be applicable to other extraction systems. Again, the k_1a parameter is the principal equipment parameter which is needed for design and scale-up if the equilibrium and rate models are already known.

Besides guiding the characterization of a given piece of equipment for use with a particular extraction system, the idealized

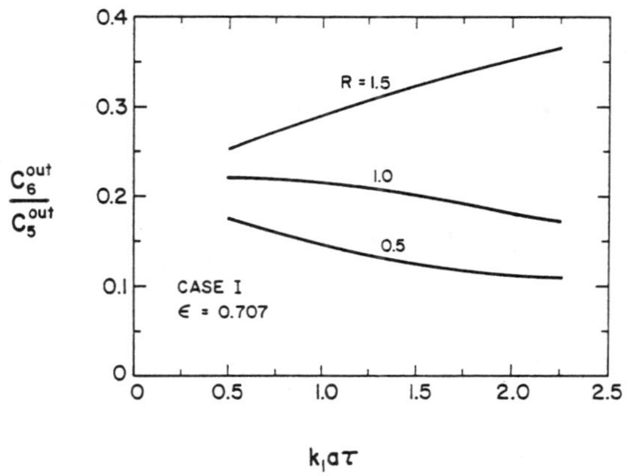

Fig. 13. Effect of phase ratio R on selectivity of extraction. C_6/C_5 is the ratio of metal B to metal A in the exit organic stream from a countercurrent column with equal concentrations of A and B in the aqueous feed. $k_1a\tau$ is number of transfer units in the column.

macroscopic contactor models that incorporate the essential features of solvent extraction chemistry as suggested here can be used to make preliminary comparisons of alternative contactor types. Altway et al.[17] have presented sample calculations for two-metal extraction and stripping systems that compare several types of contactor. The system chemistry was taken to be represented by Eqs. 1 and 2 for metals A and B with $K_A = 100$ and $K_B = 10$. The reactions were taken to be mass-transfer limited.

Figure 13 presents an interesting result for extraction in a countercurrent column that is fed with equal aqueous concentrations of metals A and B. The figure shows the ratio of the metal concentrations in the loaded organic (BR_2/AR_2) as a function of the predominant equipment parameter $k_1a\tau$ and the phase ratio. As the column gets larger or more efficient (large $k_1a\tau$), the selectivity for extraction of A over B gets worse unless the organic/aqueous phase ratio is limited. (In this case R = 1.0 corresponds to a stoichiometric amount of reagent sufficient to extract metal A only.)

Calculated degrees of extraction for this system in three different types of contactor are shown in Table 3. A counter current column and a cascade of three stirred tanks, with comparable values of $k_1a\tau$ for the two units, are compared with the predicted performance of three equilibrium stages. The entries in the Table are the

Table. 3. Calculated conversions in the simultaneous extraction of two metals in three types of contactors as functions of phase ratio (R = O/A). Entries are fractional extraction of A/fractional extraction of B.

Phase ratio	Countercurrent equilibrium stages (3 stages)	Countercurrent stirred tanks (3 stages) $k_1 a\tau = 1.5$	Countercurrent column $k_1 a\tau = 1.5$
0.5	0.454/0.046	0.364/0.058	0.471/0.049
1.0	0.891/0.106	0.492/0.110	0.759/0.160
1.5	0.997/0.465	0.534/0.133	0.908/0.359
2.0	0.999/0.819	0.553/0.147	0.963/0.578

exit values of AR_2 and BR_2 conversion at different phase ratios. It is seen that the countercurrent column is generally more efficient than the cascade of stirred tanks of comparable size. Also, the selectivities in the three types vary somewhat so that experimental data on selectivity in one contactor may not be valid for another type of equipment. These calculations may explain some of the experimental data of Ritcey and Lucas[8] who observed different selectivity in various contactors for extraction of the same multiple-metal system.

SUMMARY

The complicated chemistry of metal extraction systems introduces a number of factors that do not appear in conventional solvent extraction and make process development and optimization difficult tasks. This paper suggests a systematic approach to the analysis of metal extraction systems that is intended to make efficient use of experimental effort and to provide accurate and useful process models that will lead to efficient operations.

The suggested approach is as follows:

1. For a particular extraction system identify all of the important species and reactions.
2. Experimentally determine the parameters needed to describe the reaction and phase equilibria in the system in the region of interest.

3. Develop a two-film mass transfer model that is consistent with the chemistry and the reaction equilibria and accounts for all the significant species in the system.
4. Measure extraction rates in the growing-drop cell or some other well characterized apparatus to check the mass-transfer model.
5. If measured rates fall below those predicted by the mass-transfer model, develop a heterogeneous kinetics correlation in terms of interfacial species concentrations.
6. Incorporate the specific equilibrium and rate models of the system in various idealized contactor models to identify possible idiosyncracies of the system and to determine which type of contacting equipment should best satisfy process objectives.
7. Conduct pilot-scale tests of the system in a chosen contacting device to test the contactor model and to determine specific equipment parameters for scale-up.

If this approach is successful and a development engineer can obtain meaningful models and parameter values at each step in the investigation, then he should have an excellent understanding of the system characteristics and be prepared to design an efficient and effective solvent extraction process.

REFERENCES

1. Y. C. Hoh and R.G. Bautista: Met. Trans. 1978, vol. 9B, p. 69.
2. G. L. Bauer and T. W. Chapman: Met. Trans. 1976, vol. 7B, p. 519.
3. S. D. Troyer: M.S. Thesis, University of Wisconsin, Madison, WI, 1975.
4. M. W. Thiel: M.S. Thesis, University of Wisconsin, Madison, WI, 1974.
5. R. A. DeRuiter: M.S. Thesis, University of Wisconsin, Madison, WI, 1981.
6. T. W. Chapman: AIChE Symp. Ser., 74, 1978, No. 173, p. 120.
7. F. J. Brana-Mulero: Ph.D. Thesis, University of Wisconsin, Madison, WI, 1980.
8. G. M. Ritcey and B. H. Lucas: I. Chem. E. Ser., No. 42, 1975, p. 12.1.
9. A. Altway: M.S. Thesis, University of Wisconsin, Madison, WI, 1979.
10. M. E. Tunison and T. W. Chapman: IEC Fundamentals, 1976, vol. 15, p. 196.

11. S. Tse, E. S. Vargas and T. W. Chapman: Hydrometallurgical Recovery of Metals from Ores, Concentrates, and Secondary Sources, M. C. Kuhn, ed., p. 209, SME-AIME, 1981.
12. G. L. Bauer: Ph.D. Thesis, University of Wisconsin, Madison, WI, 1974.
13. M. E. Tunison: Ph.D. Thesis, University of Wisconsin, Madison, WI, 1976.
14. M. E. Tunison and T. W. Chapman: AIChE Symp. Ser., vol. 74, 1978, p. 112.
15. S. W. S. Tse: M.S. Thesis, University of Wisconsin, Madison, WI, 1978.
16. E. S. Vargas: M.S. Thesis, University of Wisconsin, Madison, WI, 1979.

EQUIPMENT SELECTION AND DESIGN FOR METALS SOLVENT EXTRACTION

Carl Hanson

Schools of Chemical Engineering
University of Bradford
Bradford, West Yorkshire BD7 1DP United Kingdom

INTRODUCTION

The last decade has seen the application of solvent extraction in the minerals processing field develop rapidly and on a large scale. In the main, previously developed classes of contactor have been employed, although selection between them has been influenced by characteristics of hydrometallurgical applications not necessarily met in other industrial processes using solvent extraction. These include:

(a) The very scale of the hydrometallurgical applications, which can involve volumetric throughputs much in excess of other industrial uses of the technique.

(b) The fact that the processes involve complex formation between the metal and the extractant, thus possibly introducing a significant chemical kinetic resistance into the mass transfer process.

(c) The fact that maintenance of one phase as the continuous (i.e. aqueous in organic or organic in aqueous dispersion) may have significant practical advantage in terms of phase separation rate, losses of reagent by entrainment, contamination of product by entrainment, etc.

(d) The probability of fluctuations in feed quality, both in terms of concentrations of the various metals and of solids in suspension, due to variations in the ore being processed and the performance of up-stream units such as clarifiers.

To the above must be added the requirement of reliability, both in terms of confidence that design information is adequate for a particular contactor to be specified such that it will meet production demands, and that it will be reliable in operation, maintaining production bearing in mind that many hydrometallurgical installations are in relatively inaccessible locations.

It must be recognized that solvent extraction is but one operation in a much larger total process and that the capital cost of the solvent extraction contactors may be only a modest proportion of the total. Confidence in the reliability of design and performance may therefore justifiably outweigh minimization of capital cost. While such considerations were certainly dominant in the selection of contactors for earlier plants, the capital cost of the solvent extraction unit, with its associated materials inventory, is not inconsiderable and has given incentive for the development of improved and more compact designs.

Finally, reference must be made to safety. Virtually all solvent extraction processes involve flammable liquids and it is important for the fire hazard to be recognized at the stage of contactor selection and design. In addition, most solvent systems are toxic to at least some extent and consideration has to be given to the ecological consequences of solvent loss to the environment by means such as entrainment in raffinate streams.

SOLVENT EXTRACTION EQUIPMENT

Classification

Virtually all applications demand the equivalent of multi-stage counter-current contacting of the two phases. Commercial units can be divided into two broad categories: stagewise and differential.

Stagewise contactors are divided into a number of separate stages. In each of these the phases are intimately mixed to promote mass transfer before being separated and passed to the adjacent stages so as to give counter-current flow between the stages. The residence-time allowed is usually sufficient for the phases to come more-or-less to equilibrium and so the physical stages in the equipment approximate to the steps in the classic McCabe-Thiele type graphical construction for determining the number of stages required to effect a particular separation. Equipment of this type is typified by various forms of mixer-settler.

In differential contactors there is mass transfer throughout the unit and phase disengagement only at one end. The two phases are never brought to equilibrium and a driving force for mass transfer exists at all points except, possibly, in the disengagement zone. Such contactors are typified by various forms of liquid extraction column. They do not have any stages which can be physically identified with the steps in a McCabe-Thiele type construction. Whilst their heights are sometimes identified in terms of the number of theoretical stages required for the separation coupled with the height equivalent to a theoretical stage, the transfer unit approach is more appropriate for a full analysis.

Inevitably, there are a few designs of contactor which have a character intermediate between these two main categories.

The principal types of liquid-liquid contactor have been described in detail elsewhere[1] and an overall review is outside the scope of this present paper. The following are simply brief description of some of the more important units which have been employed in hydrometallurgical applications.

Stagewise Contactors

Simple mixer-settlers have been developed which require no inter-stage pumping, the hydrostatic driving force for flow coming from density differences caused by interdispersion of the phases. If the throughput is not too large, a multi-compartment unit can be built in a single rectangular vessel by use of suitable partitions. This class of unit is typified by the so-called Windscale design of mixer-settler, originally developed to meet the needs of the nuclear industry. They are simple to construct and reliable in operation. However, since the hydrostatic driving force is proportional to the density difference between the phases and the depth of the liquor, they are bulky in relation to their throughput, having long residence-times and consequently high inventories. They are suitable only for low throughput applications in which residence-time is unimportant and mechanical simplicity coupled with minimum capital cost are desired.

To overcome the limitations of the simple gravity mixer-settler, several designs of pump-mix mixer-settler have been developed. These incorporate some form of pumping impeller to provide the driving force for flow. This impeller may also cause interdispersion of the two phases or there may be a second impeller for this purpose and/or for maintaining the dispersion. The volumetric throughput then becomes independent of the depth of liquor, allowing the use of shallow settlers to minimize solvent inventory, the throughput of a settler depending primarily on the area of contact between the two phases.

Three well-known forms of pump-mix mixer-settler which have been adopted for hydrometallurgical applications are illustrated in Fig. 1 and show some of the main variations found in practice. The General-Mills design was developed as a service to users by the reagent manufacturer, now Henkel, Inc. It involves a cylindrical mixing compartment fitted with longitudinal baffles. The feed streams are introduced into the base and are both pumped and interdispersed by a top-shrouded turbine-type impeller situated immediately above the entrance. The dispersion leaves the top of the compartment and flows via a suitable expansion section into a rectangular settler. A picket fence is located near the entrance to aid distribution of the dispersion across the width of the settler.

The Davy Powergas design was developed by the company of that name (now Davy McKee). It uses a rectangular-sectioned mixer compartment into which the two phases are introduced via a draught tube. They are drawn up and interdispersed by a specially designed shrouded impeller having swept vanes which, but use of the draught tube, is located more-or-less centrally in the compartment. A horizontal vortex breaker is fitted just below the surface of the liquor. The dispersion leaves from the top and is led via a suitable channel into the settler, where it is introduced onto the active interface. The settler is of the shallow horizontal type, usually with a picket fence to aid distribution.

The Israeli Mining Industries (I.M.I.) design incorporates a number of interesting features. The two feeds are introduced separately into a cylindrical mixing compartment where they are interdispersed by a simple impeller. Dispersion is drawn out of the mixing compartment up a conical draught tube by means of an axial-flow pumping impeller located in the bottom of the tube and mounted on the same shaft as the other impeller. It then flows via a transfer channel into the centre of a cylindrical settler. This incorporates an anti-turbulence baffle. The two settled phases are withdrawn peripherally.

Differential Contactors

Most differential contactors are based on some form of vertical column. The diameter is determined by the throughput to be handled

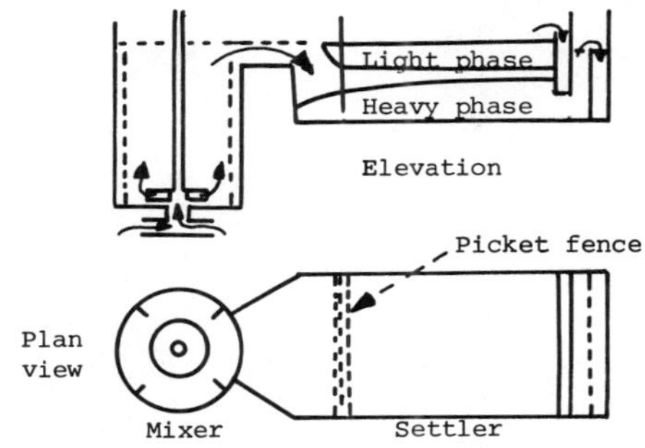

Principal features of General Mills mixer-settler

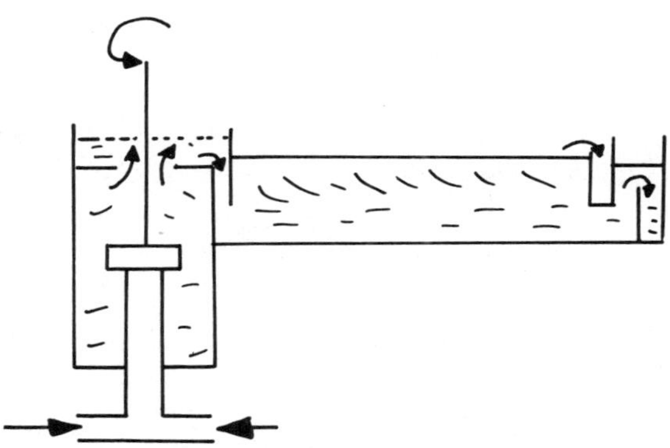

Original Davy Powergas (now Davy McKee) mixer-settler

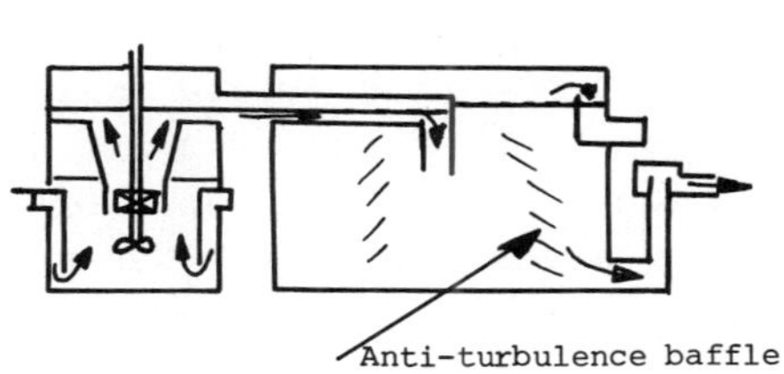

Essential features of I.M.I. mixer-settler

Fig. 1. Typical Pump-mix Mixer-Settlers

and the requirement to operate with velocities below the flooding point, i.e. velocities must be sufficiently low for the phases to pass counter-current under the action of gravity on the density difference. The height depends on the number of transfer units required for a particular separation and the rate of solute transfer between the phases. The latter depends on the mean drop size and the mass transfer coefficient, as well as on the driving force. The mass transfer coefficient is a function of the degree of turbulence in the two phases.

The simplest forms of differential contactor are various types of unagitated column, e.g. spray columns, packed columns and simple perforated plate columns. The first two of these have relatively low efficiencies, i.e. high values of height of transfer unit (HTU), since there is little to aid mass transfer. Packed columns of this type were used in some early nuclear installations because of their extreme simplicity and the absence of mechanical moving parts. However, the heights required were considerable.

The simple perforated tray column is a possible exception to these comments. It has been applied successfully on a very large scale in the petrochemicals industry. The coalescence and redispersion of the dispersed phase which takes place at each tray promotes mass transfer. However, it does not appear to have been utilized for metals.

The demands of the petroleum industry for high-capacity columns of reasonable efficiency led to the development in the 1950's of a number of mechanically agitated column contactors such as the Scheibel Column, the Rotating Disc Contactor (RDC) and the Oldshue-Rushton (Mixco) Column, all of which have some form of impeller mounted on a central rotating shaft to improve phase interdispersion and promote mass transfer by increasing turbulence. While all have been tested on a small scale with metals systems, only the Oldshue-Rushton appears to have been adopted for any large-scale production facility and no documentation has appeared on its performance in this application. More recently introduced rotary agitated columns include the Asymmetric Rotating Disc Contactor (ARD) and the Kühni Column. The latter, in particular, has been adopted for a variety of applications although none is known to have been more than peripheral to hydrometallurgy. This also applies to the RTL Raining Bucket Contactor.

The alternative means of introducing mechanical energy is by way of pulsation. Pulsed packed columns have been built on a large scale (up to some 4m diameter) and have found considerable favor in recent years, although not for metals extraction. Pulsed perforated plate columns, however, have long been a favorite form of contactor in the nuclear industry. Their particular attraction for this application is the fact that the actual pulsating device can be located remote from the column. Such columns have a number of modes of operation depending on pulse frequency and amplitude and specific throughput. Considerable experience has been gained with pulsed perforated plate columns and much has been documented in the literature.

The fairest means of comparing the relative performance of different column contactors has been a matter for debate. Minimum HTU is not adequate in itself since capacity is also a matter for concern in industrial installations. Ponikaroo[2] proposed an efficiency function $\eta = T/H$, where T is specific throughput (i.e. throughput per unit cross sectional area) and H is HTU. Based on the data then available in the literature, he concluded that pulsed perforated plate columns had the highest value of η. Such units have the added advantage of being able to handle considerable solids in suspension and a pulsed perforated plate column has been used on a pilot scale for solvent-in-pulp operations.[3]

In the above cases, the pulsation is applied to the continuous liquid phase. The alternative is to pulsate perforated plates through the liquids and two forms of pulsating plate columns have been developed. While in use industrially, they do not appear to have been favored, as yet, for metals extraction.

Finally, reference should be made to centrifugal contactors. These have the attractions of compactness and low residence times, plus an ability to handle systems having poor phase separation characteristics. These are desirable features for systems of low stability and/or high value. They are most frequently found in the pharmaceutical industry but have been applied in hydrometallurgy. Thus the Podbielniak contactor is in use in some metals extraction plants, while the Robatel Centrifugal Contactor was developed very much to meet the needs of nuclear fuel reprocessing. Some designs of centrifugal contactor are truly differential in character, e.g. the Podbielniak, whereas others are stagewise, e.g. the Robatel. Centrifugal contactors are obviously precision machines and some potential users, often in somewhat inaccessible locations, have been deterred by thoughts of maintenance requirements and possible difficulties with spares. It is not known whether such fears are justified. The other potential uncertainty is the ability of such units to handle the small amount of suspended solids found in the feed to the solvent extraction section of most hydrometallurgical installations.

Contactor Selection

Any of the contactors described can be used successfully for simple processes on a modest scale. In this context, simple processes require only a small number of theoretical stages, operate with a phase ratio near 1:1, involve phases having favorable properties and, in particular, rapid extraction and phase separation characteristics. Outside this range, the process begins to impose constraints. In an earlier publication,[4] a logic tree was proposed based on phase separation characteristics, number of theoretical stages required to achieve the separation, throughput, and constraints of available headroom or area. This is reproduced as Figure 2. Whilst simplistic, it illustrates a method of approach. Further technical parameters which might be introduced are (a) the kinetics of extraction, (b) the variation expected in phase flow ratio due to solute transfer, and (c) the solids content of the feed. Some important metals extraction processes exhibit relatively slow kinetics, e.g. copper extraction using the first generation of commerical extractants. This gives strong incentive for the use of mixer-settlers since the mixer residence-time can be adjusted to suit the system. Residence-times in columns depend on hold-up and are short, so giving high values of HTU for systems with slow kinetics even though driving forces should be larger than with simple mixer-settlers. Substantial variations in phase flow ratio can be difficult to accommodate in columns, one section operating near to flooding while the other sections have low hold-up and consequently poor efficiency. This can be accommodated to some extent in certain designs by variation in the free cross-sectional area in partitions betweeen compartments. In mixer-settlers, however, it presents little problem.

In addition to the technical parameters, adequate weight must be given to economic and strategic aspects. As explained previously, the capital cost of the solvent extraction contactor is unlikely to dominate the overall economics of the project. Features such as a greater degree of design or operational reliability, greater flexibility or a shorter period required for delivery and erection of the plant may more than justify choice of a contactor of somewhat higher cost than the minimum. Again, certain designs are proprietary and

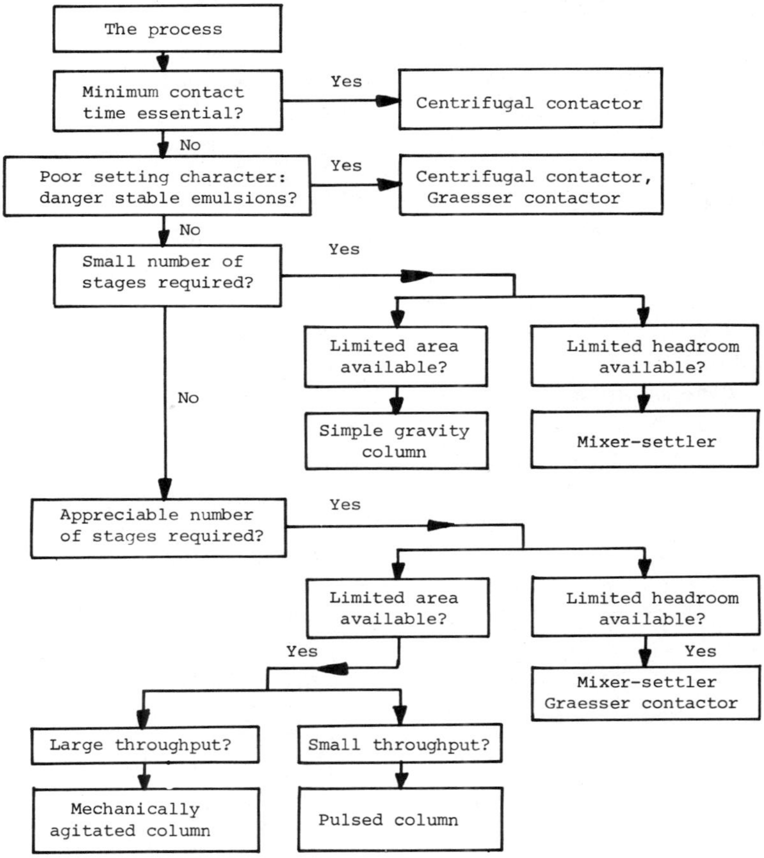

Figure 2. Selection Tree (with permission, Ref. 4)

only available from a contractor whereas others are not associated with any particular contractor. The decision on whether-or-not to design the unit in-house may therefore restrict the choice. Finally, consideration must be given to safety and loss prevention. Most commercially important solvents are flammable and so present a fire hazard. A large air-solvent interface such as features in classic gravity settlers must be considered a major disadvantage, particularly for low flash point systems. Quite apart from the fire hazard, it can lead to significant losses of diluent by evaporation.

Most plants so far built for metals solvent extraction have employed one form or another of mixer-settler. In many cases this has been inevitable in terms of the systems exhibiting slow kinetics or the throughputs required being considerably in excess of those so far handled in columns. Given some pilot-plant data, a mixer-settler can be designed for almost any application with reasonable confidence that the processing objectives will be met. As a class of contactor they are tolerant of over-design and this can be used to overcome design uncertainties, albeit at the expense of greater capital cost. The same cannot be said for columns. Scale-up is less well explored, certainly in terms of experience documented in the open literature. Performance is very dependent upon hydrodynamics and the extent of departure from the normal assumption of counter-current plug flow of each phase varies with scale. The departure is known as axial mixing and comprises a range of phenomena which all contribute to a spread of residence-times. Methods exist whereby allowance can be made at the design stage for the effects of axial mixing but their application demands some knowledge of its extent at the scale in question. Thus design confidence, possibly coupled with some conservatism, has kept the mixer-settler dominant in hydrometallurgy. Yet columns have been widely used on a large scale in other industries and it is important that the success of mixer-settlers in hydrometallurgy does not blind users to their limitations nor to the existence of alternatives.

The characterization of liquid-liquid systems required for contactor selection has been discussed by Barnea[5], with particular reference to hydrometallurgical applications.

MIXER-SETTLERS

Design and Scale-Up

The residence-time required in the mixer to give the required approach to equilibrium is usually established in a pilot plant. That same residence time is then adopted for the full-scale unit and this, in conjunction with the throughput required, effectively determines the volume of the mixer compartment. This approach presupposes that the same hydrodynamic (and therefore mass transfer) conditions will be maintained on the two scales. It is not always recognized that this is virtually impossible to achieve in practice. Two scale-up criteria have been adopted: (a) to maintain equal impeller tip speed at a fixed ratio of impeller/tank diameter, (b) to maintain equal power input per unit volume, also at a fixed ratio of impeller/tank diameter. Use of the former leads to a progressive reduction in power input per unit volume with increase in scale, whereas use of the latter gives a progressive increase in tip speed. Work using the extraction of copper into LIX 64N and covering a very wide range of scale[6] showed use of equal tip speed to lead to a progressive reduction in stage efficiency. This did not occur with power input per unit volume maintained constant, although it is not known whether the increase in tip speed produced an increase in entrainment losses. It does seem that equal power input per unit volume is the safest criterion to adopt, at least in terms of maintaining stage efficiencies. Equal tip speed, however, seems a reasonable parameter for changing impeller diameter in a tank of fixed size.

The area of settler required for a particular application has to be determined from a pilot plant investigation designed to produce the characteristic curve, i.e. plot of dispersion band depth as a function of specific throughput (throughput of dispersed phase per unit area). Such curves are functions of temperature and vary according to which of the phases is dispersed. While translation of such data to a larger scale involves the assumption that the different scale mixers will produce dispersions having the same settling characteristics, nevertheless use of this approach has been successful in practice giving, if anything, a degree of overdesign. The area determined in this way then has to be converted into a length/breadth ratio, assuming the conventional rectangular format is to be adopted. While there is a temptation to retain the same format as the pilot plant or one considered conventional (say 4/1), it must be recognized that linear velocities will increase with scale if depths are kept constant and the same length/breadth ratio adopted. This can ultimately lead to excessive entrainment of the dispersed phase in the exit continuous phase,[6] as illustrated for the Cu/LIX 64N system in Fig. 3.

Operational Problems

Apart from inadequate performance in terms of stage efficiency, the most common operational problems with mixer-settlers are poor phase stability (i.e. the phases are prone to invert) and high entrainment levels. The two phenomena are linked in some respects in that conditions leading to poor phase stability are associated with high levels of entrainment.

Entrainment itself can arise from the design and operation of both the mixer and the settler. In the case of the mixer, too high an agitator speed is one obvious cause, it then taking the form of fine droplets of dispersed phase. However, inadequate agitation can lead to stratification of the dispersed phase in a mixer with high localized hold-ups in some regions. This can cause phase inversions

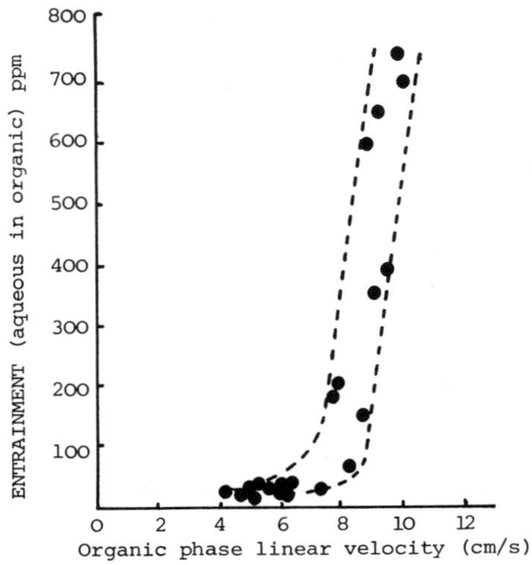

Fig. 3. Entrainment and linear velocity (from Ref. 6)

(localized or general) which will be accompanied by high levels of entrainment, this time of droplets of the notionally continuous phase in the settled dispersed phase leaving the unit. It is important that an impeller speed is used which is sufficient to avoid this situation. From the point of view of entrainment, too low an agitation rate can be as dangerous as excessively high values. The positioning of the phase inlets in the mixer compartment affects the minimum power input required to avoid this phenomenon.[7] Yet another possible cause of entrainment is ingress of air to the dispersion in the mixer, e.g. from a vortex.[8] This is broken into fine bubbles which adhere to the drops giving them additional buoyancy.

The mode of introduction of the dispersion into the settler can itself produce entrainment. It is important that the dispersion is not forced to pass through a layer of the coalesced dispersed phase, otherwise entrainment of continuous phase in the coalesced disperssed phase may result.

Turning finally to the settler itself, reference has already been made under scale-up to excessive linear velocities of the settled phases in the settler as a possible cause of entrainment. This will be of the original dispersed phase in the continuous. Efficient operation of a settler demands an absence of turbulence which will naturally inhibit the sedimentation process. Entrance effects are important in this respect and picket fences or other forms of simple anti-turbulence baffle placed near the entrance, and possibly also at intermediate points, can considerably enhance performance.

The presence of solids in suspension can influence both settler performance and entrainment levels. A recent quantitative study[9] suggests it reduces dispersion band depths but increases entrainment. The latter is additional to possible solvent losses with crud, this being a generic term for solids accumulating within solvent extraction plants, often in the interfacial region. The sources and behavior of crud have been the subject of a recent review.[10]

Some Recent Developments

The main disadvantage with mixer-settlers is their bulk, with correspondingly high hold-ups of the expensive extractants used in solvents for hydrometallurgical processes. Much recent effort has been aimed at the development of more compact designs.

In terms of the mixer, analogy has been drawn with reactor design principles to argue for either a multi-compartment approach or a unit giving co-current flow analogous to a tubular reactor. The Holmes and Narver Inc. low-profile mixer-settler, adopted in at least one recent plant,[11] aimed primarily at a reduction in civil engineering costs but does include, inter alia, a form of multi-compartment mixer. The principle was explored in detail, including full-scale tests, for a new plant proposed in Zambia.[6] To gain the full potential benefit, it is essential to minimize backmixing between the compartments as this reduces the driving force and defeats the objective. While a multi-compartment design is inevitably more complex, its adoption for very large scale units such as some of those used for hydrometallurgical applications may allow use of standard impeller drive mechanisms rather than requiring special units, thus effecting some economy.

Attempts at developing a mixer with characteristics analogous to a tubular reactor have centered on the use of static mixer elements.[12,13] Perhaps the best known of these is the Kenics mixer, although units with the same objectives are manufactured by Sulzer and a number of other companies. Mass transfer performance appears good and it is claimed that such units produce dispersions having very uniform drop sizes. Apart from a possible reduction in overall volume, static mixers do give a totally enclosed design. To realize their full potential in this respect they need to be coupled with sealed settlers, something which it is not easy to combine with control of interface level and hydraulic independence of the stages.

However, it is understood that at least one commercial plant is now in operation with a multi-stage mixer-settler comprising static mixers coupled to cylindrical and totally-enclosed settlers.

The large settlers which characterized the early plants were an obvious target for improvement. One approach has been exploitation of preferential wetting by introduction of suitable surfaces. A special knitted mesh was developed for this purpose involving alternate strands woven from stainless steel and a suitable plastic (PTFE).[14] This was extremely successful at a pilot plant scale but does not appear to have been adopted for commercial use. Two possible difficulties can be envisaged: slow blockage by solids in suspension in the liquids and possible changes in wetting characteristics with use over long periods.

An alternative approach, and one showing great promise, is the use of a low frequency pulsed electric field.[15] This is only applicable to water-in-oil dispersions and, even then, only if the conductivity of the oil phase is low. However, given these conditions, it is effective, simple and consumes virtually no energy.

A very novel approach has been adopted in the new Davy McKee CMS contactor.[16] While stagewise in operation, this virtually eliminates the individual settler compartments. The principles are shown in Fig. 4. The two phases are mixed and interdispersed in a zone between two sets of baffles or grids. By adjustment of the interface level in this zone prior to start-up (by hydraulic balance),

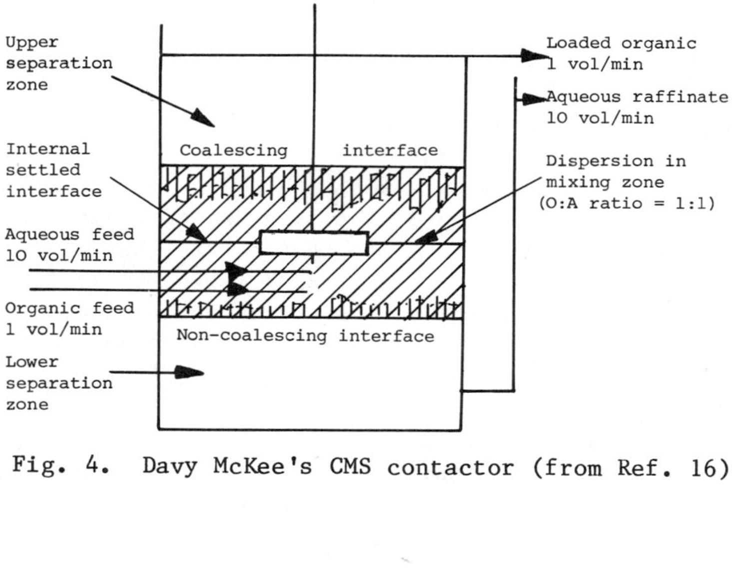

Fig. 4. Davy McKee's CMS contactor (from Ref. 16)

it is possible to vary the relative hold-up of dispersed phase in the dispersion in the mixing zone independently of the overall flow ratio. This provides an additional control parameter and should increase versatility. The phases need to disengage outside the mixing zone. In appropriate cases, this can be accelerated by use of an electrostatic field as described above.

Finally, a reference should be made to the Krebs mixer-settler, introduced commercially in recent years. While based on a single mixer and a separate settler, this has a number of interesting features and the design has been adopted for a number of recent large commercial plants. The pump-mix unit incorporates a conical draught tube to allow some classification to take place. Further settling occurs in a long channel which carries the dispersion to the opposite end of the settler, from which it flows back in the direction of the mixer. Because of this, in a multi-compartment plant all the mixers are at the same end, a feature of some attraction to the operators.

Design and testing methods for mixer-settlers have been detailed recently by Barnea.[17]

OTHER CONTACTORS

Column contactors have been little used in hydrometallurgy as yet except for pulsed perforated-plate columns in the nuclear industry. The other exception is the use of mechanically-agitated columns such as the Kühni column for the purification of phosphoric acid. Information on the selection, testing and scale-up of column contactors is available elsewhere.[18]

CONCLUSIONS

Up to the present time, virtually all hydrometallurgical operations based on solvent extraction have employed mixer-settler type contactors. Developments have taken place which are leading to a reduction in the size of such units and are thus minimizing the disadvantage associatd with having to provide phase separation at each stage. However, sight must not be lost of the fact that developments have also been taking place with commerical extractants. Some of the features which originally made the use of mixer-settlers inevitable are no longer present and it is possible that the next decade may see greater potential for the use of differential contactors.

REFERENCES

1. P. J. Bailes, C. Hanson and M. A. Hughes: Chemical Engineering, 83 (2), 86 (1976).
2. I. I. Ponikaroo et al.: Int. Chem. Eng., 2, 546 (1962).
3. G. M. Ritcey and A. W. Ashbrook: Chapter 5 in Solvent Extraction: Principles and Applications to Process Metallurgy, Part II (Elsevier, Amsterdam, 1979).
4. C. Hanson: Chemical Engineering, 75 (18), 98 (1968).
5. E. Barnea: Hydrometallurgy, 5, 15 (1979).
6. J. R. Orjans, C. W. Notebaart, J. C. Godfrey, C. Hanson, and M. J. Slater: Proceedings International Solvent Extraction Conference ISEC '77, p. 340 (C.I.M., Montreal, 1979).
7. J. C. Godfrey and V. Grilc: Proceedings 3rd European Conference on Mixing, paper C3, York, April 1979 (BHRA, 1980).
8. J. C. Godfrey, C. Hanson, M. J. Slater, and S. Tharmalingam: Fundamental Aspects of Hydrometallurgical Processes, p. 127, A.I.Ch.E. Symposium Series, Vol. 74 (1978).
9. J. Gledhill: M.Phil. Thesis, University of Bradford, United Kingdom (1980).
10. G. M. Ritcey: Hydrometallurgy, 5, 97 (1980).
11. A. D. Kennedy and C. L. Pfalzgraff: Proceedings International Solvent Extraction Conference ISEC '77, p. 333 (C.I.M. Montreal, 1979).
12. M. E. Tunison and T. W. Chapman: Fundamental Aspects of Hydrometallurgical Processes, p. 112, A.I.Ch.E. Symposium Series, Vol. 74 (1978).
13. J. C. Merchuk, R. Shal, and D. Wolf: Ind. Eng. Chem. Proc. Des. Dev., 19, 91 (1980).
14. I. D. Jackson, J. B. Scuffham, G.C.I. Warwick and G. A. Davies: Proceedings International Solvent Extraction Conference ISEC '74, p. 567 (S.C.I., London, 1974).
15. P. J. Bailes and S.K.L. Larkai: Trans. I. Chem.E., 59, 229 (1981) and 60, 115 (1982).
16. J. B. Scuffham: The Chemical Engineer (370), 318 (1981).
17. E. Barnea: Hydrometallurgy, 5, 127 (1980).
18. H.R.C. Pratt and C. Hanson: Chapter 16 in Handbook of Solvent Extraction, Ed. T. C. Lo, M.H.I. Baird and C. Hanson (Wiley, New York, 1983).

COMPOSITION CONTROL OF EXTRACTORS

Lawrence E. Burkhart

Ames Laboratory and Department of Chemical Engineering
Iowa State University
Ames, Iowa 50011

INTRODUCTION

Solvent extraction developed--with a few notable exceptions, particularly in the petroleum field--as a commercially important process operation only after intensive work on extraction processes for nuclear fuel reprocessing during World War II. Since then it has found increasing application in the separation of metals, aromatics, and more recently, the treatment of waste water. Whereas the early applications depended on simple differences in the solubility of a material distributed between two immiscible solvents, newer processes largely involve chemical reactions, temperature and pH sensitivity, and valence control to effect changes in distribution in multi-solute systems. Many of these new processes are drawn from our growing knowledge of more sophisticated extractants which form complexes with the solute to be extracted and are highly selective. It is this latter property that makes them especially useful in hydrometallurgical applications where one or two metals must be recovered from an impure, multicomponent system. Also, as these new developments in solvent extraction chemistry have appeared, the range of applicable metal extraction processes has broadened--from valuable metals in small volumes of solution to large flows of less valuable metals present in low concentrations.

Added to these complexities is the relatively long time constant of some extractors. Accurate control can be economically feasible and, in fact, sometimes absolutely necessary. Because of the time constants involved, feedforward control may be an attractive strategy and in some cases even simple adaptive control has been useful. Accurate and/or reliable sensors, sometimes hard to obtain, are important in these applications. In lieu of better sensors, filtering or estimation techniques may provide better inputs to control algorithms.

TYPES OF EXTRACTION EQUIPMENT

Solvent extraction equipment can be divided into two categories --stagewise contactors and continuous contactors. Most stagewise contactors are of the mixer-settler type. They consist of separate units connected in a cascade arrangement. The feed enters at one end of the cascade and flows counter current to an extracting solvent which enters at the opposite end. Each unit of the cascade has a mixing section and a settling section. The two phases are vigorously mixed and allowed to flow into the settling chamber where coalescence and phase disengagement takes place. The separated liquids flow out of the settler, one phase to each of two adjacent stages so that countercurrent flow is achieved. Flow within the settling chamber may be cocurrent or countercurrent and some designs have provisions for recycling part of either or both phases back to the mixer to provide a desirable phase ratio in the mixer. When properly designed, the performance of each unit can approach closely that of a theoretical stage. Thus cascade design from theoretical calculations is relatively straight forward.

The second category is differential or continuous contactors. They are column-type units where the heavy phase enters at the top of the column, the light phase enters at the bottom and, as they flow countercurrent to each other, the two liquids are mixed intimately to obtain good contact. This category can be sub-divided into columns that have external agitation and those that do not. The latter type may be packed with a solid packing material or may contain sieve plates with downcomers placed at intervals along the length of the column. Those with external agitation have stirrers, paddles, or discs at periodic intervals - or the liquid contents may be pulsed up and down through packing material or through sieve plates. Although they can be used for large flows, such applications require large column diameters and, if external agitation is supplied, the required power input can become large. Large diameter columns are also subject to channeling, which results in poor mixing --thus complicating a scale-up procedure that is already less reliable than that for mixer-settlers.

A final, admittedly small category, is the centrifugal extractor. In conceptual design it may resemble either a mixer-settler (with mixing and settling zones) or a perforated plate column. It is the most highly proprietary type of unit and little published performance data exist. They have, however, been found to be highly successful in particular applications. A centrifugal extractor has a very low hold-up, a small solvent inventory, a very short residence time and a fast response to process changes. It's importance in hydrometallurgy is debatable. It could, in principle, be used for separations based on kinetic rate differentials for the extraction reactions of different species.

FACTORS AFFECTING EXTRACTOR OPERATION

Simple single-solute extraction processes are governed by the equilibrium solubility distribution between two immiscible phases. This establishes the maximum separation that can be achieved and is also a factor in determining the rate of mass transfer in an extractor according to the familiar relation,

$$N = ka\ (y^* - y) \tag{1}$$

where N is the rate of mass transfer in moles/unit time, a is the interfacial area and $(y^* - y)$ the concentration driving force. The value of y^* is set by the equilibrium distribution. For simple systems this equilibrium information can be expressed as a single curve relating the concentration of solute in one phase to that in other phases over the concentration range of interest. For multicomponent systems with interacting solutes the equilibrium distribution of a given solute is also a function of the concentrations of the other components and may also depend upon the solvent loading. It may also be strongly temperature dependent. It is much more difficult to grasp fully the condition of equilibrium in these complex systems when one is trying to get a "feel" for the dynamics of the process.

For a multistage extractor the overall separation obtained may be expressed by an overall material balance as,

$$\frac{X_f - X_r}{Y_c - Y_s} = \frac{S}{F} \tag{2}$$

where S is the solvent flow rate, F is the feed flow rate, $(X_f - X_r)$ is the difference in concentration of the feed stream in and out of the extractor, and $(Y_c - Y_s)$ is the difference in the solvent concentration as it passes through the extractor. The extract concentration, Y_c, and the raffinate concentration, X_r, are dependent upon the flow rates and the feed and solvent concentrations. A change in any of the independent terms in Eqs. 1 and 2 will affect the composition of the product. To a first approximation, Eq. 1 may be thought of as representing internal effects and Eq. 2 as representing external effects.

It is instructive to examine each of the terms in these equations in light of their potential influence on column control. In elementary treatments of extraction dynamics the distribution coefficient is considered unimportant unless there are significant temperature changes with time. But in those cases where the extractants react chemically with the solutes they remove, the equilibrium distribution can depend upon oxidation state, pH, or other elements of the chemical environment in each phase. Further, for multicomponent systems at low solute concentrations the distribution of each species may be independent of one another, but at higher concentrations solute interaction takes place and the equilibrium relations become more complex. A typical example is the extraction of copper with the hydroxy oxime LIX reagents developed by General Mills. Copper ions in the aqueous phase are extracted by the LIX reagent in an aliphatic diluent according to the reaction,

$$Cu^{2+} + \underline{2RH} \quad \underline{CuR_2} + 2H^+$$

where the organic phase is underlined and the reagent is designated as R. The process is based on the ability of the LIX to exchange selectively cupric and hydrogen ions. At low acidities the copper is extracted into the organic phase, but as this extraction takes place hydrogen ions are added to the aqueous phase, causing the pH to decrease and shifting the reaction to the left. Thus, the equilibrium is a function of the pH. Extracted copper can be stripped and solvent regenerated in a strongly acidic solution. If one wished to extract concentrated copper solutions it would be necessary to neutralize some of the acid formed during the extraction reaction.

In the nuclear industry, the separation of plutonium from uranium by tributyl phosphate (TBP) is a function of the oxidation state of the uranium and the plutonium ions. These two metals in their high oxidation state (U^{+6} and Pu^{+4}) in an aqueous feed are coextracted from a spent fuel solution by TBP but if the Pu^{+4} is then selectively reduced to Pu^{+3} by an appropriate reducing agent (e.g. ferrous sulfamate or hydroxylamine) the plutonium can be back-extracted with an aqueous strip solution to effect a separation between the two elements since Pu^{+3} is insoluble in the TBP.

The mass transfer coefficient and the interfacial area terms in Eq. (1) are affected by any changes in process conditions that change the degree of mixing or the relative amounts of the two phases in the extractor. An obvious source of this type of disturbance is any change in the energy input to externally agitated extractors such as pulse columns or rotating disc (RDC) contactors. The change can be accidental or it may be intentionally introduced as part of the overall control strategy. Less obvious changes that can occur during process operation are the accumulation of impurities, solvent degradation products, or finely divided solids that concentrate at interfaces and reduce the rate of coalescence. This affects the dispersed phase inventory in the column and so changes the interfacial area. It may also lead to flooding.

Finally, the more commonly considered disturbances in extraction systems are contained in Eq. 2, which shows that the product concentrations will change with variations in inlet concentrations and flow rates. These external effects give rise to the basic control systems that are applied to extractors. They will be discussed in the next section. However, some of the internal effects cited earlier may be the source of the most difficult control problems.

BASIC CONTROL PHILOSOPHY

Extraction processes have not been considered as prime targets for control engineers. As a rule, they are fairly stable. Many of the most widely used extractors in the past have had large inventories relative to process flows, and thus slow response times. The tendency has been to use local control of inlet flows, alarms to indicate abnormal conditions, off-line analyses, and operator experience in correcting for process deviations.

In controlling product composition one needs to consider hydrodynamic stablity, changes in mass transfer rates, and chemical stability. The classic control strategy for an extraction column, shown in Fig. 1, consists of two loops - one for flow control and one for interface control. The inlet solvent flow is determined by ratio control with the flow rate of the entering feed stream, and the interface level is controlled by varying the hydrostatic pressure

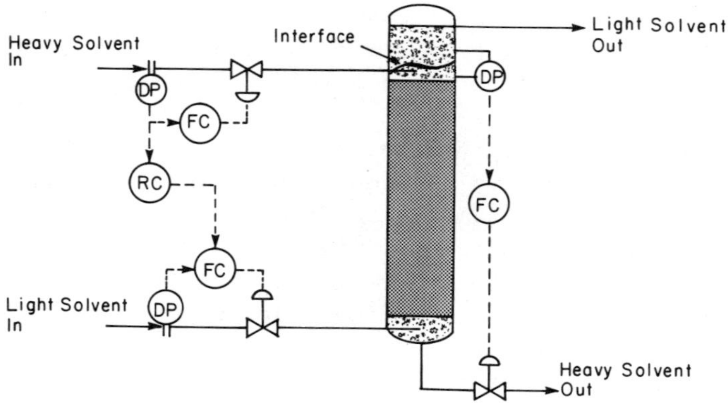

Fig. 1. Conventional control loop for an extractor.[3]

in the column via a control valve in the extract line. If, in addition, there is an on-line product composition measurement (not shown in Fig. 1) of the product stream then this signal is used in a cascade fashion to vary the set-point of the ratio controller. The reason for this arrangement can be seen from the overall column material balance, Eq. 2. Maintaining a constant flow ratio maintains, in principle, a constant product composition if nothing else changes and if the system is insensitive to total flow rate changes over a reasonable range of values. If the feed concentration changes, the flow ratio can be modified by the signal from the product composition analyzer.

For mixer-settlers the basic arrangement is similar except that the interface control is different. For mixer-settlers in which there is a hydrodynamic equilibrium between the stages, the interfaces in each settler are set by overflow weir adjustments; for extractors which are designed to be hydraulically independent of each other, the interface position is determined by pumping rates between the individual stages.

The Falconbridge Matte Leach Process, described by Vembe[1], for the removal of iron, copper, and cobalt impurities from a nickel feed solution is a good example of this "standard" control strategy. It is shown schematically in Fig. 2. The nickel rich feed enters a two stage mixer-settler battery at the upper left where Fe^{+3} is extracted with TBP. The feed, with the iron removed, then enters a three stage extractor where Cu^{+2} and Co^{+2} are extracted with tri-iso-oxtyl-amine (TIOA). The purified nickel stream is then sent

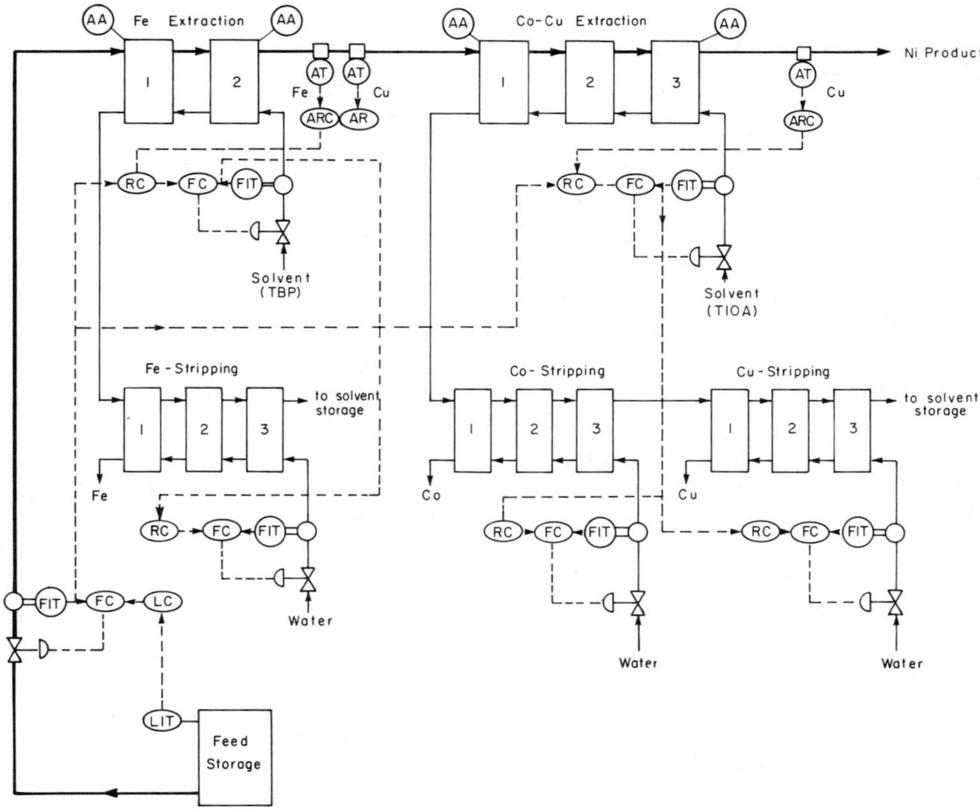

Fig. 2. The Falconbridge Matte-Leach plant solvent extraction circuit[1]; a good example of "standard" control strategy in a solvent extraction process.

to a cooling crystallizer where $NiCl_2 \cdot 4H_2O$ crystals are produced. The initial feed rate to the extractors is regulated by a level controller on the feed tank. The solvent flows to the two extractors and are then ratioed to this feed flow rate. Signals from on-line product composition analyzers (Fe in the first battery and Cu in the second) are used in cascade to adjust the set-points of each flow ratio controller to compensate for concentration fluctuations.

In the two strippers the water streams are ratioed to their respective organic feeds. The entire control hierarchy is thus based on the initial feed as the only "wild" flow rate in the system. The signal from the Fe analyzer adjusts the solvent-to-feed ratio in the first extractor to maintain a high Fe loading in order to minimize HCl losses in the stripper. In practice, Vembe reports that the two copper analyzers serve largely as fault detectors --the first to detect a malfunction in the operation preceeding the extractors and the second to monitor the purity of the nickel product stream from the extraction system.

The last stage before the product leaves the system is operated aqueous continous to reduce the loss of nickel chloride in entrained aqueous droplets in the organic stream fed to the stripper. Also, the first stage is operated organic continuous to minimize the loss of organic phase as entrained droplets in the aqueous stream. These latter points illustrate important points in the design of the instrumentation and control system. Although not as interesting from an academic or research point of view, the need for adequate fault detection is very important economically since the propagation of a disturbance resulting from a major fault or failure in the process can be expensive and time-consuming in terms of product loss, plant shut-down time, and recycling costs. One other point should be mentioned. Phase ratio detectors are also used to be sure that the correct phase is continuous in the mixers. Besides insuring that droplet entrainment losses are minimized, as in this case, coalescence in the settlers in many extraction systems is very sensitive to which phase is continuous in the mixers--thus phase inversion may cause flooding.

CONTROL OF A PROCESS WITH COMPLEX EQUILIBRIA

When the extraction process under consideration has interacting solutes it is often very difficult to develop a reliable control strategy without understanding the dynamics of the system. A good example for illustration is the work of McCutcheon, et al.[2] on the control of a pulsed extraction column for separating plutonium and uranium in a plutonium scrap recovery plant. The process is shown in Fig. 3. It differs from more established flowsheets for plutonium-uranium separations in that the concept of plutonium recycle is used. To maintain a high plutonium concentration in the column, about 65% of the plutonium product is recycled. This provides a high plutonium product concentration and reduces the heat load on the product evaporators which follow the solvent extraction unit.

The basic three-unit process is carried out in three pulsed liquid-liquid extraction columns. Tributyl phosphate (TBP) in a of a disturbance resulting from a major fault or failure in the process can be expensive and time-consuming in terms of product loss, plant shut-down time, and recycling costs. One other point should be mentioned. Phase ratio detectors are also used to be sure that the correct phase is continuous in the mixers. Besides insuring that droplet entrainment losses are minimized, as in this case, coalescence in the settlers in many extraction systems is very sensitive to which phase is continuous in the mixers--thus phase inversion may cause flooding.

CONTROL OF A PROCESS WITH COMPLEX EQUILIBRIA

When the extraction process under consideration has interacting solutes it is often very difficult to develop a reliable control strategy without understanding the dynamics of the system. A good example for illustration is the work of McCutcheon, et al.[2] on the control of a pulsed extraction column for separating plutonium and uranium in a plutonium scrap recovery plant. The process is shown in Fig. 3. It differs from more established flowsheets for plutonium-uranium separations in that the concept of plutonium recycle is used. To maintain a high plutonium concentration in the column, about 65% of the plutonium product is recycled. This provides a high plutonium product concentration and reduces the heat load on the product evaporators which follow the solvent extraction unit.

The basic three-unit process is carried out in three pulsed liquid-liquid extraction columns. Tributyl phosphate (TBP) in a hydrocarbon diluent is used as the solvent. Uranium and plutonium in their high oxidation states (U^{+6},Pu^{+4}) are first co-extracted. This product from the first column then becomes the feed to the second column, a partitioning column. Separation is achieved by selectively reducing the plutonium with hydroxylamine. Although the plant was operational, success in maintaining operational stability for extended periods of time had been marginal. At intervals, significant amounts of plutonium would break through into the organic waste stream (CCX), pass into the third column and plug it. The unit would then have to be cleaned and the process restarted.

The problem was attacked by process simulation. Five key reactions, listed in Fig. 4, were identified by careful analysis of the process. These reactions were then used in a mathematical simulation of the partitioning column. The Pu^{+4}-hydroxylamine

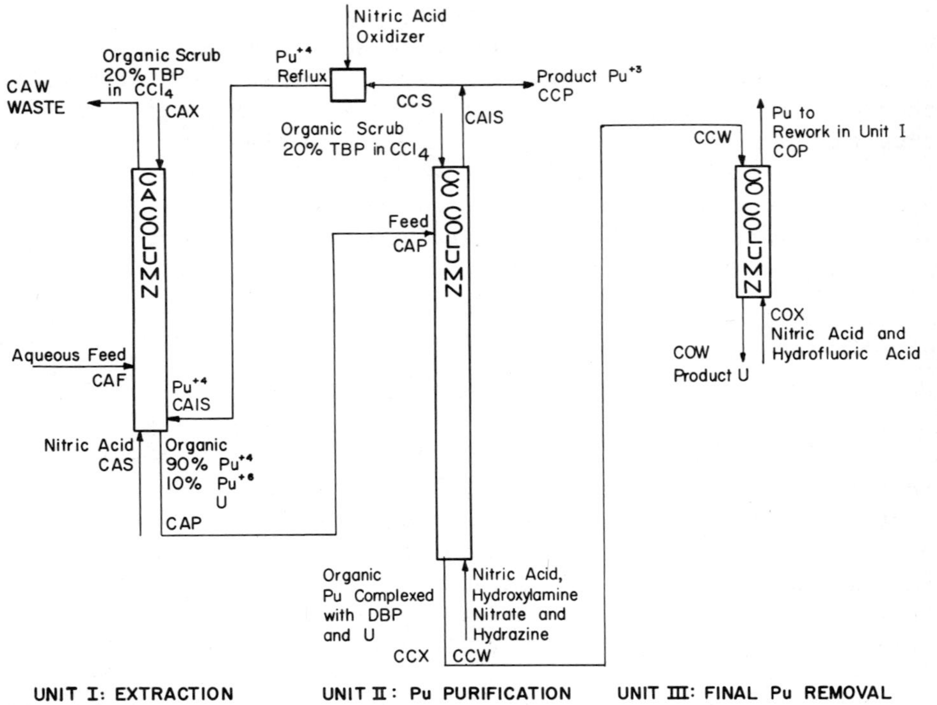

Fig. 3. Plutonium scrap recovery flowsheet[2]; an example of a multicomponent solvent extraction process with interacting solutes.

hydrocarbon diluent is used as the solvent. Uranium and plutonium in their high oxidation states (U^{+6},Pu^{+4}) are first co-extracted. This product from the first column then becomes the feed to the second column, a partitioning column. Separation is achieved by selectively reducing the plutonium with hydroxylamine. Although the plant was operational, success in maintaining operational stability for extended periods of time had been marginal. At intervals, significant amounts of plutonium would break through into the organic waste stream (CCX), pass into the third column and plug it. The unit would then have to be cleaned and the process restarted.

The problem was attacked by process simulation. Five key reactions, listed in Fig. 4, were identified by careful analysis of the process. These reactions were then used in a mathematical simulation of the partitioning column. The Pu^{+4}-hydroxylamine

Pu(IV) - hydroxylamine

$$4Pu(IV) + 2NH_3OH^+ \rightarrow 4Pu(III) + 6H^+ + N_2O + H_2O$$

$$2Pu(IV) + 2NH_3OH^+ \rightarrow 2Pu(III) + 4H^+ + N_2 + 2H_2O$$

Pu(III) - nitric acid

$$2Pu(III) + 3H^+ + NO_3^- \xrightarrow{HNO_2} 2Pu(IV) + H_2O + HNO_2$$

Pu(IV) - hydrazine

$$Pu(IV) + \frac{1}{4} N_2H_5^+ \rightarrow Pu(III) + 1\frac{1}{4} H^+ + \text{gaseous nitrogen products}$$

Nitrous acid - hydroxylamine

$$NH_3OH^+ + HNO_2 \rightarrow N_2O + 2H_2O + H^+$$

Fig. 4. Key reactions in the partitioning column.

reduction reaction is responsible for the partitioning. This reaction is opposed by the autocatalytic oxidation of Pu^{+3} back to Pu^{+4}, in which nitrous acid is both a catalyst and a product. Hydrazine is added to the column to destroy the nitrous acid and thus suppress the autocatalytic re-oxidation of plutonium. Hydrazine also reduces some of the plutonium, but the Pu^{+4}-hydrazine reaction is more important for its removal of hydrazine from the column than for its action as a plutonium reductant. Finally, some of the nitrous acid is also destroyed by reaction with hydroxylamine. Little rate data were available on some of the reactions and best estimates had to be made.

The model predicted known steady-state output values, the influence of process variable changes on the plutonium profile, the apparent suddenness of partitioning failure, and explained the logic of empirically developed manual control procedures which had been used by the plant operators. The high concentration of plutonium in the partitioning column and the autocatalytic nature of the re-oxidation reaction provided the answer. Once triggered, the autocatalysis spreads rapidly throughout the column. The plutonium profiles, shown in Fig. 5 and 6, drop farther down in the column as the aqueous solvent rate is decreased (a procedure which had been used in the plant to increase the Pu concentration in the product stream) and ultimately reach the bottom of the column. The plutonium level in this stream is then too high to be handled by the third column, and the third column plugs.

The levels of nitrous acid which could have been reasonably assumed to originate within the column were, however, insufficient to account for this behavior. Examination of the flowsheet revealed two potential external sources. In the feed stream to the co-decontamination column, nitrous acid is produced by the dissolution

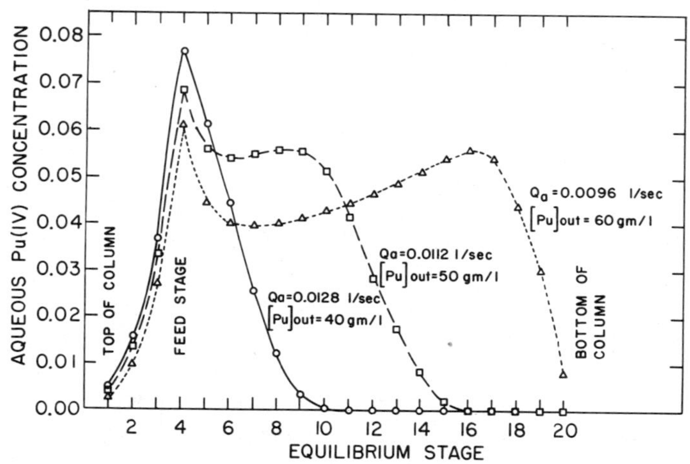

Fig. 5. Broadening of the aqueous Pu^{+4} profile.

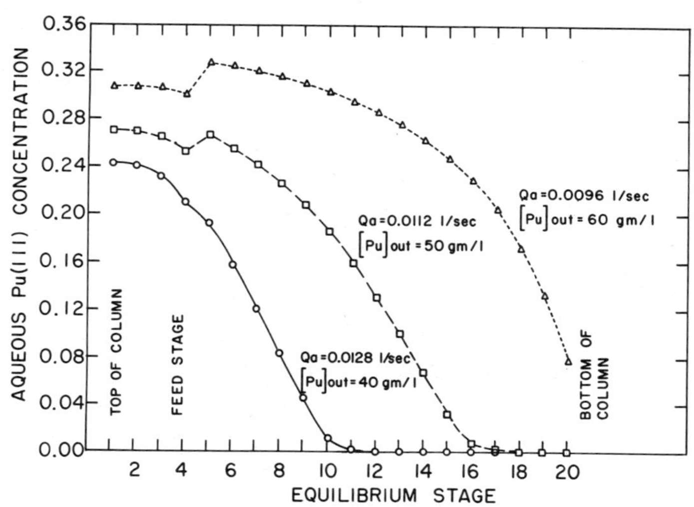

Fig. 6. Broadening of the aqueous Pu^{+3} Pu^{+3}

of the feed material. In the recycle stream between the first and second column, nitrous acid is generated when Pu^{+3} is oxidized back to Pu^{+4}. The nitrous acid from both of these sources enters the aqueous stream in the co-decontamination column and, because it forms a strong complex with TBP, is picked up quickly by the organic phase which becomes the feed to the partitioning column. Subsequent plant measurements confirmed these two external nitrous acid sources and also confirmed high levels of nitrous acid in the feed to the partitioning column.

A control scheme was then proposed which was designed to hold the tip of the plutonium profile well up from the bottom of the column so that manipulation of the aqueous solvent could still be used to compensate for process disturbances without risk of contaminating the uranium product stream at the bottom of the column.

In this case, process simulation was essential to understanding the nature of the problem and in developing a solution. Besides the complex solute interactions, internal data from the plant were difficult to obtain because of the high radioactivity levels and the remote location of the equipment. Properly used, process simulation is a powerful tool in understanding how to control a system properly.

OTHER CONTROL METHODS

Additional control beyond the ratioing of the two input streams and feedback control on product composition may be needed because of some specific characteristic of the physical system.

A few examples will show the type of difficulty that might be encountered. The problem of interfacially active impurities collecting at the interface plagues columns just as it does mixer-settlers. One case involves the recovery of a valuable catalyst by simultaneous chemical reaction and solvent extraction in a large rotating disc contactor operating very near its flooding point in order to obtain maximum efficiency. Finely divided solids formed during the recovery reaction in the column collected at the interfaced and retarded coalescence. The problem was aggravated by traces of a surface active impurity that entered in the feed stream. As the organic holdup increased toward flooding a signal from the pressure sensor at the bottom of the column slowed the speed of the rotating discs. This action, however, increased the catalyst losses in the raffinate and required a decrease in the feed flow rate to maintain catalyst recovery. In this particular example the speed of rotation was quite slow and a change of 1 or 2 rpm was sufficient for control purposes.

Changing the degree of agitation is not always a good control measure. In pulsed columns a variation in the pulse frequency gives a rapid output response but attempts to use this technique for control have not been too successful. Pulsed columns are operated near their flooding points and the shift in holdup distribution following a sudden change in pulse frequency can create a large accumulation of dispersed phase in one part of the column, cause phase reversal, and produce "cyclic local flooding" which persists until the slug of dispersed phase has worked its way to the end of the column.

REFERENCES

1. P. E. Vembe: Proceeding, 1977 International Solvent Extraction Conference, Toronto, Canada.
2. E. B. McCutcheon, L. E. Burkhart, and R. E. Felt: Milwaukee Meeting, Instrument Society of America, 1975.
3. F. G. Shinsky: Process-Control Systems. McGraw-Hill Book Company, NY, 1979.

USE OF STATISTICAL METHODS OF EXPERIMENTAL DESIGN IN OPTIMIZING HYDROMETALLURGICAL PROCESSES

A. R. Burkin

Department of Metallurgy and Materials Science
Imperial College
London SW7 2BP U. K.

INTRODUCTION

Many hydrometallurgical processes take place under conditions where use of the theory of the thermodynamics of aqueous solutions is difficult or practically impossible. This can make it necessary to try to optimize the working of the process by trial and error during its development. While working on attempts to use a computerized Databank system[1] to increase the possibility of using thermodynamic information as a guide to process development, it was decided to investigate how the use of statistical methods of experimental design could be employed for this purpose also. Dr. A. J. Monhemius had developed an interest in that subject while working in Brazil and guided Dr. M. J. Pinto Dias[2] in the early stages of her work on its application to the Jarosite process. Dr. Styliani Agatzini[3] took the theoretical treatment further during her work on the Goethite process and subsequently the method has been used by many workers in the Hydrometallurgy Group. All are greatly indebted to Dr. A. C. Atkinson of the Department of Mathematics, Imperial College for the special course of lectures which he gave for them and for his assistance and advice over a long period. The role of the present author has been to watch skeptically the attempts of a young, very gifted group of workers to apply rigorously mathematical methods to the design of hydrometallurgical processes. It seems now that they have been successful and any questions on statistical matters should be directed to them.

The jarosite and goethite processes for removing iron from process liquors were selected for the first two studies because the optimum conditions for using them in a commercial plant are known and because they work in concentrated solutions containing several salts. The starting point of each investigation was to set out information which could be expected to come from a chemical laboratory in which experiments had identified a process of potential interest for iron removal. The variables which might be expected to be important were selected and by use of a fractional factorial design a few screening experiments were carried out to identify those variables which were significant.

Factorial designs at two levels, unreplicated in some cases replicated in others, were then arranged and the experiments carried out. Empirical equations were set up for the two processes and were used to draw three-dimensional and contour plots of the responses studied. In each process the conditions which are used in practice were among those predicted to be about optimum.

Among the advantages of this method of experimentation is the fact that several variables are changed together so that interactions betweeen variables are identified, and these can shed light on the chemical behavior of the system. In addition, the number of experiments required to optimize conditions is greatly reduced compared with conventional methods.

THE JAROSITE PROCESS

The following variables (with their code letters) were investigated at two levels (low or −, and high, +). Fe^{3+} concentration, A; H_2SO_4 concentration, B; Precipitating reagent concentration, C; Nature of precipitating reagent, D (low level Na^+ only, high level NH_4^+ only); Excess sulphate added as zinc sulphate, E; Time of reaction, F; Dummy factor, G.

The 8 screening experiments required to investigate which of the 7 variables (or factors) were significant were carried out at 100°C according to a Plackett-Burman design, Table 1.

It had already been accepted that a high temperature was required to precipitate the iron as jarosite. The other factors which were found to be significant as a result of the experiments were I) concentration of sulphuric acid; II) concentration of Fe^{3+}; III) excess sulphate concentration and IV) the nature of the precipitating reagent, $(NH_4)_2SO_4$ being better than Na_2SO_4, other conditions being the same.

Table 1. Screening Experiment

Experiment Number	A	B	C	D	E	F	G
1	+	+	+	-	+	-	-
2	+	+	-	+	-	-	+
3	+	-	+	-	-	+	+
4	-	+	-	-	+	+	+
5	+	-	-	+	+	+	-
6	-	-	+	+	+	-	+
7	-	+	+	+	-	+	-
8	-	-	-	-	-	-	-

Variables Studied: A–G

Factorial Design No. 1

Since equilibrium conditions of precipitation were to be studied it was necessary to use a reaction time of adequate length and for practical reasons the time chosen was 37 hours. The temperature was 100°C and the variables to be studied were those shown in Table 2, which gives also their levels, chosen to be close to those in a real leach solution. The low and high levels of precipitating agent were the stoichiometric amounts theoretically required to precipitate the low and high levels respectively of ferric iron.

A duplicated full factorial design was planned in order to have an independent estimate of the experimental error. To cancel out the effect of any uncontrolled variables the experiments were randomized using a table of random numbers. The experiments were numbered in standard order for the duplicated full factorial as follows: (1), (1), a,a,b,b,ab,ab, c,c, ... abcd, abcd. The randomized order was then followed by splitting the duplicated full factorial into two blocks, each corresponding to a full factorial.

Table 2. Factorial Design No. 1. Variables and their Levels

Factor	Variable	Low Level	High Level	Units
A	Fe^{3+} concentration	2	10	g/L
B	H_2SO_4 "	0	3.9	g/L
C	$ZnSO_4$ "	0	0.5	moles/L
D	$(NH_4)_2SO_4$ "	0.8	4.0	g/L

It can be questioned whether this splitting into two complete factorial "units" of experiments which would be carried out in two consecutive weeks eliminates block effects which might result from uncontrolled variables affecting the system during one week and not the other. However precipitation reactions of this kind are generally not kinetically reproducible unless seeded so that the major source of error in the experiments was likely to be the system itself rather than external uncontrolled variables.

The initial compositions of the solutions used are given in Table 3, together with the code of each treatment combination. A small type letter indicates the presence of the capital letter variable at high level; the absence of the letter indicating the variable to be at low level. A treatment combination coded with (1) has all variables at low level.

After hydrolysis the liquid and solid phases were separated by filtration, the pH of each solution was measured and it was analysed for iron. An X-ray powder photograph was taken for each sample of solid.

Table 3. Factorial Design No. 1. Initial Compositions of the Solutions: A, Fe^{3+}; B, H_2SO_4; C, $ZnSO_4$; D, $(NH_4)_2SO_4$

Treatment Code	Fe^{3+} (i)	g/L duplicate	H_2SO_4 g/L	$ZnSO_4$ mole/L	$(NH_4)_2SO_4$ g/L
(1)	1.96	1.93	0	0	0.8
a	9.80	9.90	0	0	0.8
b	1.84	1.88	3.9	0	0.8
ab	10.00	8.86	3.9	0	0.8
c	1.98	1.86	0	0.5	0.8
ac	9.70	9.00	0	0.5	0.8
bc	1.94	1.84	3.9	0.5	0.8
abc	9.80	9.36	3.9	0.5	0.8
d	1.94	1.93	0	0	4.0
ad	9.50	9.66	0	0	4.0
bd	1.90	1.93	3.9	0	4.0
abd	9.50	9.76	3.9	0	4.0
cd	1.98	1.82	0	0.5	4.0
acd	10.00	9.20	0	0.5	4.0
bcd	1.90	1.91	3.9	0.5	4.0
abcd	9.50	9.76	3.9	0.5	4.0

The results are not given here but they showed a very large discrepancy between the amounts of iron precipitated in the duplicate experiments of some treatment combinations, particularly b. These treatment combinations were repeated but still reproducibility was poor. This was found to be due to the fact that in some experiments a phase other than jarosite was precipitated also.

The solid hydrolysis products gave X-ray diffraction lines corresponding to those of ammonium jarosite, $NH_4Fe_3(SO_4)_2(OH)_6$ and a solid solution of ammonium and hydroxonium jarosite, in which some NH_4^+ is replaced by H_3O^+. This formation of a solid solution is well known and has been studied in detail, and is indicated by a shift in some of the d-spacings. Some solids showed also lines indicating the presence of glockerite, $Fe_4(OH)_{10}SO_4$ and a few showed diffuse bands characteristic of poorly crystalline α-goethite.

All of the differences in the type of precipitate obtained from each duplicated hydrolysis experiment are reflected in the variance of each of the treatment combinations. The variance of each of them is given in Table 4 and this allows the homoscedasticity test to be carried out, in order to find out if the results obtained constitute a sample from a normally distributed population of results. This test is of fundamental importance as the statistical analysis of the results will be based on assumptions which are valid for statistically normally distributed populations. Although some of the variations are very large they still belong to a normally distributed population of results. The highest value of S_I^2 found was 139.445 and the sum of all S_I^2 was 506.025. The ratio 139.445/506.025 = 0.2756 which is less than 0.4546, the value given as relevant for 16 independent estimates of a result, each calculated from a sample of 2 with $\alpha = 0.05$. Thus the variances were homogenous.

To study statistically the effect of each factor on the precipitation of Fe^{3+} as well as the significance of any possible interactions it was necessary to define exactly the dependent variable, percentage of Fe^{3+} precipitated. The statistical analysis showed that if the dependent variables was taken as the total amount of Fe^{3+} precipitated, irrespective of the type of solid phase obtained, then the sign and significance of the effects of some factors and interactions were opposite to those obtained if the dependent variable was defined as being euqal to the percentage of Fe^{3+} precipitated as a crystalline phase. The term crystalline phase covers precipitation of Fe^{3+} as either jarosite solid solution or as glockerite.

Table 4. Factorial Design No. 1. Homoscedasticity Test, α = (0.05)

Treatment Code	Results (i)	Results (ii)	Average Fe Precipitation %	Variance of each treatment $S_i^2 = (y^1-y^2)^2/2$
(1)	91.3	90.7	91.10	0.180
a	52.7	50.0	51.35	3.645
b	14.1	1.6	7.85	78.125
ab	35.8	19.9	27.85	126.405
c	97.5	96.2	96.85	0.845
ac	70.6	62.8	66.70	30.420
bc	81.4	64.7	73.05	139.445
abc	53.4	41.2	47.30	74.420
d	93.8	95.9	94.85	2.205
ad	70.5	67.4	68.95	4.805
bd	62.6	67.9	65.25	34.045
abd	50.6	52.4	51.50	1.620
cd	97.0	96.7	96.85	0.045
acd	86.4	90.2	88.30	7.220
bcd	90.0	83.8	86.90	19.220
abcd	69.2	71.8	70.50	3.380
			TOTAL	506.025

Further statistical analysis of the results indicated that significant interactions occurred between the variables studied so that a model based on main effects only was not suitable for this system. A full-scale second order model was required. Several routes were possible to obtain this by use of different second order designs. One of them would involve expansion of the first order design by addition of extra data points forming a rotatable second order design. However, this would expand symmetrically from the center point of the original two level factorial experimental design and would certainly give a mixture of phases for some treatment combinations, which was not acceptable.

Another solution would be to carry out a three or four level factorial experimental design on the four factors. This would lead to the need to carry out many more experiments than before and it was decided to carry out another fully duplicated two-level factorial experimental design choosing the levels so as to interlink with the levels of the previous factorial design, and locating them in a region of more favorable conditions for the precipitation of jarosite. These seemed to be at the higher levels used in the previous factorial design.

Factorial Design No. 2

The levels of the variables used are shown in Table 5. The hydrolysis reactions were again carried out at 100°C.

Only jarosite solid solution precipitated from the solutions used in factorial design No. 2 so that there was no need to distinguish between total precipitation of iron and precipitation of it as jarosite in order to carry out a statistical study. To study the main effects and interactions of the factors on the amount of jarosite precipitated, a Yates analysis and an analysis of variance were carried out on the total percentage of Fe^{3+} precipitated. All main effects were found to be statistically significant.

The effects of factors A and B were negative; that is the amount of jarosite precipitated decreased when the concentrations of Fe^{3+} and of free H_2SO_4 were increased. This is the opposite of what was found in Factorial Design No. 1 and shows how the change in levels can affect the sign of the effects. The effects of variables C and D remained positive. In Design No. 2 only interactions AB and AD were found to be significant and they were positive but of much smaller magnitude than the main effects. In Design No. 1 almost all interactions were significant and a number of them were of the order of magnitude of the main effects. The existence of only a small number of significant interactions in Design No. 2 means that the reponse surface (% of Fe^{3+} precipitated) is only slightly curved as opposed to the region covered in Design No. 1. It is of interst that interaction BC is not statistically significant. This indicates that the equilibrium HSO_4^-/SO_4 is buffered at these relatively high concentrations of H^+ and total SO_4^{2-}.

Table 5. Factorial Design No. 2. Variables and their Levels

Factor	Variable	Low Level	High Level	Units
A	Fe^{3+} concentration	6	14	g/L
B	H_2SO_4 "	1.92	5.76	g/L
C	$ZnSO_4$ "	0.25	0.75	moles/L
D	$(NH_4)_2SO_4$ "	2.40	5.60	g/L

It was shown using the homoscedasticity test that all the experimental points had the same population variance and the statistically significant parameters in the mathematical model (regression equation) were related to the effects found from Yates analysis. The equation found was

$$Y = 63.36 - 11.58X_1 - 9.43X_2 + 2.31X_3 + 5.88X_4 + 2.43X_1X_2 + 1.05\,X_1X_4 \pm \delta \quad (1)$$

where Y is the predicted value of the percentage of Fe^{3+} precipitated as jarosite. X_1 X_2 X_3 and X_4 are "coded" variables related to the so-called "natural" variables A B C and D expressed in the concentration units shown in Table 5, by the following equations:

$$X_1 = \frac{[Fe^{3+}] - 10.07}{3.95}; \quad (2a) \qquad X_2 = \frac{[H_2SO_4] - 3.84}{1.92} \quad (2b)$$

$$X_3 = \frac{[ZnSO_4] - 0.50}{0.25}; \quad (2c) \qquad X_4 = \frac{[(NH_4)_2SO_4] - 4.00}{1.60} \quad (2d)$$

The presence of $\pm\,\delta$ in the equation gives the 95% confidence limits of the predicted value of Y.

The experimental values of Fe^{3+} precipitation obtained in Factorial design No. 2 are compared with the values predicted from equation in Table 6. Agreement is good.

Factorial Design No. 3

To study the effect of temperature on the precipitation of jarosite when the experimental conditions were kept the same as those which led to the precipitation of jarosite only, it was decided to carry out another duplicated two-level factorial design, keeping the number of factors equal to 4 so as to avoid a large increase in the number of experiments. Since temperature must be one factor, it was decided to fix the concentration of $(NH_4)_2SO_4$ at the stoichiometric amount required to precipitate Fe^{3+} as ammonium jarosite in each treatment combination. By doing this it would be necessary to carry out only half of the total number of experiments, since the other half corresponded to experimental conditions studied in factorial design No. 2. In order to try to avoid the precipitation of phases other than jarosite (hematite for example) a high level temperature of 130°C was chosen.

Table 6. Factorial Design No. 2. Comparison Between Experimental and Predicted Values

Treatment Code	Results (i)	Results (ii)	Average of Results	Predicted Value
(1)	81.65	81.15	81.40	79.79
a	48.75	48.43	48.59	49.23
b	58.82	52.28	55.55	56.16
ab	33.95	33.55	33.75	35.28
c	84.67	83.17	83.92	84.38
ac	60.29	51.40	55.85	54.42
bc	56.84	62.27	59.56	60.90
abc	39.61	42.81	41.21	39.91
d	88.51	88.05	88.28	89.11
ad	64.16	61.36	62.76	63.42
bd	68.64	67.37	68.01	65.76
abd	50.64	49.54	50.09	49.50
cd	93.27	93.97	93.62	93.98
acd	68.64	67.19	67.92	68.29
bcd	68.88	69.52	69.20	70.50
abcd	52.89	55.20	54.05	54.24

The statistical analysis of the results of this factorial design are not given here. The effects of temperature on the precipitation of iron is very great and the kinds of interactions between the variables can be interpreted in terms of the interactions between the ionic species present in the solutions, and thus on the thermodynamics of the system. In particular it permits selection, by comparison with thermodynamic data for species, of the species to which a particular interaction might correspond. However, it does not rule out the possibility that other species, although not changing their concentrations very much between 100° and 130°C, might also exist in amounts comparable to these of the species to which the interaction was attributed.

CONCLUSION

The methods of designing groups of experiments which are discussed above make it possible to use several other methods of statistical analysis which are more than can be dealt with in the space of this paper. They can also be used to extend the scope of the overall investigation. For example, Dr. Agatzini studied the precipitation of iron as goethite from a solution containing nickel, cobalt and ferrous iron such as might be obtained by sulphuric acid leaching of a reduced nickel laterite ore. The first stage of the work was similar in nature to that carried out on the jarosite process, although the fact that no other phase containing iron was precipitated made it possible to expand a first order into a rotatable second order design. In the second stage of the work the nature of the response studied was extended to study the amounts of nickel and of cobalt coprecipitated with the goethite.

REFERENCES

1. A. R. Burkin: The Industrial Use of Thermochemical Data, T. I. Barry, ed., The Chemical Society, London (1980).
2. M. J. Pinto Dias: A statistical approach to the precipitation of iron as jarosite, Ph.D. thesis, University of London, 1979.
3. S. Agatzini: A statistical approach to the precipitation of iron as goethite, Ph.D. thesis, University of London, 1981.

APPLICATIONS OF ION EXCHANGE IN HYDROMETALLURGY

Michael Streat

Department of Chemical Engineering
and Chemical Technology
Imperial College
London SW7

INTRODUCTION

A list of metals that have been recovered and purified commercially by ion exhcange is given in Table 1. In some cases, the scale of operation is relatively small, e.g. the rare earth elements, the transuranic elements, the platinum group metals, though the value of metal recovered is usually extremely high. Ion exchange is particularly suitable for high cost, low throughput purification processes. On the other hand, recovery of trace amounts of metals from waste streams accounts for many applications, e.g. chromium from spent metal plating solutions and copper and zinc from the wastes arising in the rayon and synthetic fibre industry. However, the largest single application in hydrometallurgy is the recovery and concentration of uranium from naturally-occurring ore-bodies. At the present time, about 30,000 tons/annum of uranium ore concentrate are processed as yellow cake (ammonium diuranate) in the Western World. Principal producers are the United States of America, Canada, South Africa and Australia. Traditionally, fixed bed ion exchange plants were installed to meet the early demand of the nuclear industry, though more recently, liquid extraction using liquid ion exchange reagents has found wide acceptance in many locations. Uranium plants are typically capable of processing 100-1000 m^3/hr of pregnant solution and it is the uranium industry that has been responsible for the many innovations that have occurred in the development of ion exchange technology in recent

Table 1. Metals Recovered and Purified Commercially by Ion Exchange

1. Uranium	8. Chromium
2. Thorium	9. Copper
3. Rare earths	10. Zinc
4. Transuranic elements	11. Rhenium
5. Gold	12. Nickel
6. Silver	13. Cobalt
7. Platinum group metals	

years. These developments are now beginning to pervade into new areas of application for ion exchange in hydrometallurgy and these will be mentioned in this paper.

ION EXCHANGE RESINS

Early ion exchange materials were naturally-occurring clays and minerals with inorganic structure, usually of the alumino-silicate type. However, the great advance in ion exchange development took place during the 1940's when phenol-formaldehyde condensation polymer-based materials were first synthesized. Synthetic organic materials are now almost universally used for all ion exchange applications. The early resinous structures were superseded by the development of polyelectrolyte gel materials based on the copolymerization of styrene and divinyl benzene. This provided a 3-dimensional polymer backbone structure onto which ionogenic functional groups could be attached easily by organic synthesis. The properties of the backbone or matrix structure are extremely important and there have been several further advances in recent years. Many ion exchange materials are now synthesized on a polyacrylic backbone structure and whereas the polyelectrolyte gel materials have no clearly defined physical structure when viewed under the electron microscope, new "macroporous" or "macroreticular" structures have been developed with clearly defined pores, analogous to conventional adsorbent materials such as silica gel, alumina, molecular sieve, etc. The method of synthesis is flexible and permits the production of ion exchangers as planar sheets (membranes) and as truly spherical beads or in exceptional cases as irregular particulate material. Since the use of ion exchange beads is most highly developed in hydrometallurgy at present, this paper

Table 2. Types of Ion Exchange Resin

Type	Functional Group	Capacity m.equ/g	pH Range
Strong acid	$R - SO_3H$	5.0	0 -14
Weak acid	R - COOH	10.0	4 - 14
Strong base	$R - CH_2N^+(CH_3)_3$	4.0	0 - 14
Weak base	$R - CH_2N(CH_3)_2$	~5.0	0 - 5

is confined to a discussion of modern trends and developments in the use of particulate material.

Ion exchange materials in common use fall into four categories as shown in Table 2. Strong acid resins are produced by introducing sulphonic acid groups into the 3-dimensional hydrocarbon backbone structure. Weak acid resins are normally based on a carboxylic acid functional grouping and this is best achieved by synthesis of a polyacrylic polymer. Strong base resins are formed by Friedal-Craft reaction to introduce a quaternary ammonium grouping onto the hydrocarbon backbone. Similarly, weak base resins are formed by the inclusion of a tertiary ammonium group. Strong acid and strong base resins can exchange ions over the entire pH range and are therefore capable of splitting salts, e.g.

Cation exchange (strong acid) $RH^+ + NaCl = \overline{RNa^+} + HCl$ (1)

Anion exchange (strong base) $ROH^- + NaCl = \overline{RCl^-} + NaOH$ (2)

The overbars denote the ion exchange resin phase. Weak acid resins respond on dissociation of the carboxylic acid grouping and usually exchange in alkaline solution, e.g.

$$\overline{R\ COOH} + NaHCO_3 = \overline{R\ COONa} + H_2O + CO_2 \qquad (3)$$

Weak base resins require protonation of the functional group, in acid media and are therefore capable of exchanging anions in acid solution.

$$\overline{R - CH_2N(CH_3)_2} + HCl = \overline{R - CH_2N(CH_3)_2 \text{---} H^+ \; Cl^-} \qquad (4)$$

free base form hydrochloride form

$$2\,\overline{R - CH_2N(CH_3)_2 \text{---} H^+Cl^-} + UO_2(SO_4)_2^{--} = \qquad (5)$$

$$\overline{[R - CH_2N(CH_3)_2 \text{--} H^+]_2 \; UO_2(SO_4)_2^{--}} + 2Cl^-$$

One advantage of weakly acidic or basic resins is the relative ease of regeneration. It is important to observe that ion exchange resins must be regenerated with a fresh chemical reagent, usually an acid or base, and at best 1 equivalent of regenerant is consumed for every 1 equivalent of metal sorbed (this assumes 100% regenerant efficiency). Normally excess regenerant is required and it transpires that regenerant consumption represents the most significant direct operating cost of ion exchange processes. However, weak acid or base resins can be regenerated with almost stoichiometric (100%) efficiency.

SELECTIVITY

Ion exchange resins, i.e. anion and cation, show substantial selectivity or preference for one ionic species over another. Selectivity depends largely on the resin structure and specifically on the ionic radius and ionic charge density of the species in question. Ion exchange resin beads (polyelectrolyte gels) can swell and contract in aqueous solution. Note that linear polystyrene sulphonic acid is totally water soluble, whereas the cross-linked resin structure cannot dissolve but will tend to imbibe water to surround the fixed ionogenic hydrophilic sites. In fact, enormous osmotic swelling pressures can develop within the resin structure when beads (say 5 mequ/g capapcity, i.e. about 2 M) are immersed in a dilute aqueous solution, (say 0.001 M). A delicate balance of osmotic forces governs the selectivity sequence of both cation and anion resins. As a general rule, a small hydrated ion is preferred to a large hydrated ion in resins of normal cross-linking (8% divinylbenzene). Also an ion of higher charge density will tend to displace an ion of smaller charge density so that divalent ions such as Mg^{++}, Ca^{++}, etc. are preferred to Na^{+}, K^{+}, etc. A selectivity coefficient can be defined by applying the laws of mass action to the ion exchange process.

$$z_B \overline{R\,A^{z_A+}} + z_A\, B^{z_B+} = Z_a\, \overline{R\,B^{z_B+}} + z_B\, A^{z_A+} \tag{6}$$

$$K_A^B \left(\frac{C}{\overline{C}}\right)^{z_A - z_B} = \frac{(y_B)^{z_A}(x_A)^{z_B}}{(x_B)^{z_A}(y_A)^{z_B}} \tag{7}$$

K_A^B is not a thermodynamic coefficient, rather a rational selectivity coefficient which can be used to quantify the preference of one ion over another. The selectivity scale for some cations and anions is given in Table 3 and Table 4 for the purposes of illustration.

Table 3. Selectivity of Cations in the Hydrogen Cycle of a Sulponic Acid Cation Exchange Resin.

Cation	Selectivity Coefficient with Respect to Hydrogen Ion
Li^{+}	0.8
Na^{+}	2.0
K^{+}	3.0
NH_4^{+}	3.0
Mg^{++}	26.0
Ca^{++}	42.0

Table 4. Selectivity of Anions in the Hydroxide Cycle of a Quaternary Anion Exchange Resin

Anion	Selectivity Coefficient with Respect to Hydrogen Ion
HCO_3^{-}	4.0
Cl^{-}	11.0
NO_3^{-}	40.0
SO_4^{--}	50.0

Ion exchange resins are not specific and will tend to sorb all the species from aqueous solution, though each of the species will be adsorbed to a different degree. It follows, therefore, that a conventional strong acid resin is not metal specific for the purposes of hydrometallurgical recovery. For example, a sulphuric acid solution containing traces of copper, cobalt, zinc, ferric and ferrous iron, magnesium, calcium is unlikely to be amenable to copper recovery by a strong acid cation resin. Selectivity of copper over the other species would be small and quite unattractive.

Recently, however, interest has increased in metal selective resins, particularly for the type of solution mentioned above.[1] Chelating functional groups have long been introduced into ion exchange materials[2] and a large number of possibilities exist. Resin structures containing imino diacetic acid type groups have been studied extensively for the recovery of the transition group metals from aqueous sulphuric acid solutions. Typical resins are Dowex A-1, Lewatit TP 207, Amerlite XE 318, and Zerolit S1006. The structure is given below:

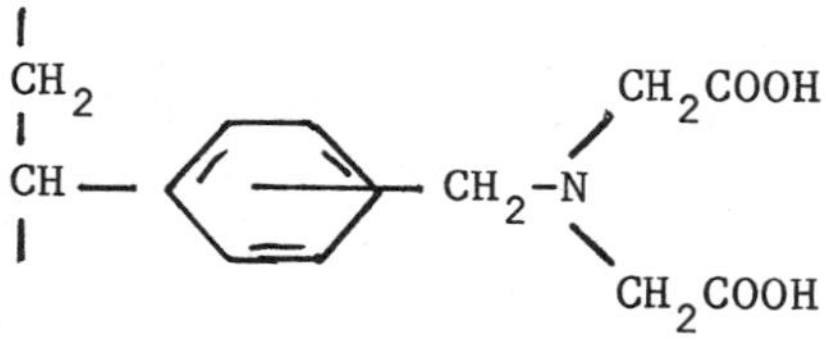

In aqueous sulphuric acid solutions with a pH value about 1.5, the selectivity of copper over ferrous and ferric iron is good[3] and the possibility of a separation process exists. The copper chelate structure is possibly:

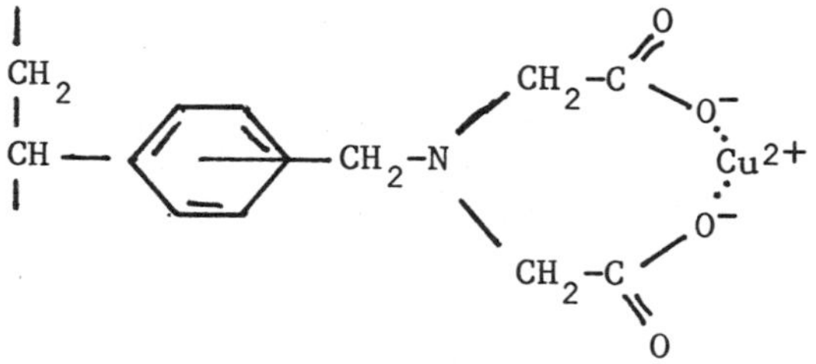

At higher pH values, the selectivity for copper falls off and at pH > 3 iron is sorbed and at pH > 6, Ca^{++}, Mg^{++}, Al^{+++} are also sorbed. However, the development of chelating resins is a potential growth area in ion exchange resin development and there are indications of several new products emerging for particular applications (copper and zinc in particular).

Another promising development is the impregnation of hydrocarbon beads with well-established reagents capable of selective metal extraction.[4] Warshawsky has impregnated macroporous hydrocarbon beads with reagents such as Di-2EHPA, Alamine 336, LIX reagents. The products appear to have all the properties of the liquid reagent and there is the indication that solvent loss, due to miscibility of the reagent with the aqueous phase, is reduced. Large-scale evaluation of these materials has not yet taken place so that a commercial assessment is premature. However, this type of research is encouraging and could lead to a family of new reagents with wide application in hydrometallurgy. Separately Bayer have developed "Levextrel" materials by an independent route.[5] Here, reagents such as tri-n-butyl phosphate (TBP) are embodied within the pore structure of a macroporous hydrocarbon bead. Some success has been achieved for these materials in the recovery of trace uranium from nuclear waste streams. Other reagents have been tried with limited success and it is too early to predict whether these materials will make any immediate impact on the scope for hydrometallurgical processes. In fact, anion exchange processes appear to be the most attractive for metal separation and recovery. Uranium recovery is an anion exchange process (and will be discussed later), and it is already apparent that anion separation processes can be achieved from chloride solutions.[6] Other examples are chromium which is usually recovered as the chromate anion from metal plating solutions.

ION EXCHANGE KINETICS

Liquid extraction and solid ion exchange have much in common. They are both interfacial mass transfer processes and whereas it is necessary to create droplets to provide interfacial area in the former case, ion exchange resin beads can be considered as "solid" drops of reactive material. In both cases, the distribution coefficient for metals as anions or cations is often large. However, the rate of exchange or mass transfer is profoundly different. Liquid extractants are rapid to attain equilibrium due to convective circulation within the droplets causing a continual levelling of the concentration gradient. In solid ion exchange resins, the rate of exchange is often slow and is governed by diffusion from the bead edge to the center of the particle. The rate of exchange is proportional to the concentration driving force within the bead and on the diffusion coefficient of the metals or complexes within the resin phase. Under these conditions the rate of exchange is independent of solution flow-rate. Ion exchange particles in commercial use are usually about 0.5-1 mm in diameter and the rate of exchange is inversely proportional to the square of the bead diameter. In very dilute solutions (less than about 0.05 M) liquid film diffusion is significant and the rate of exchange is also slow and is then dependent on solution flow-rate and the driving force between bulk solution and the interface of the particle. The fact that ion exchange is a relatively slow kinetic process means that it is customary to use a cascade of fixed bed ion exchange columns in series and to operate at flow-rates which ensure that the rate of exchange is compatible with the mass transfer of ions at any cross-section of the reacting portion of the bed. In fixed bed columns, a flow-rate of 20 gal/min ft^2 (0.02 $m^3/s\ m^2$) is commonly adopted as a yardstick, since this is an acceptable compromise between break-through and excessive pressure drop due to flow.

DESIGN OF ION EXCHANGE EQUIPMENT

Ion exchange is traditionally a fixed bed process. The disadvantage of a fixed bed process is that it is cyclic in operation and that at any one instant only a relatively small part of the hold-up of resin is doing useful work. Fixed columns are normally contained in pressure vessels and provided with ancillary valves and pipework to control the flow. Columns are rarely used singly and are normally cascaded in order to give continuous operation as far as the feed solution is concerned. Typical examples of this are the processing of a pregnant uranium leach liquor. Here, three columns in series are used and with backwashing and split elution there are eight separate stages in a complete cycle. In many installations, four columns are used, one serving as a stand-by unit. In a large uranium mill, there could be several such lines of ion exchange columns in parallel and upwards of 20 columns in service is not uncommon in a typical plant (e.g. in South Africa). A large plant will require many automatic valves and timers and therefore the control system is complex. In essence, a continuous process is maintained by keeping the ion exchange resin stationary and moving the pregnant feed point from column to column. This, of course, has the advantage that the resin is not subject to any arduous conditions and it is therefore possible for the resin to operate satisfactorily for periods of up to 20 years. Though the capital cost of the installations is large, the operating costs are usually low. Replacement of resin is negligible and the main direct operating cost is the regenerant chemical. In the uranium industry this is nitrate ion (HNO_3/NH_4NO_3), though nowadays both hot or cold sulphuric acid is preferred.

Whereas the fixed bed column is essentially simple, the disadvantage of a fixed bed plant lies in the complexity of the ancillary equipment. Each column in a cascade will require several automatic control valves and associated equipment involving process timers. The hold-up of resin is large and this can be a problem if poisoning is a threat, e.g. in the South African uranium industry. The cyclic variation of product concentration in many applications can be a disadvantage and often leads to extensive tankage and storage of bulked solutions. By analogy with other chemical engineering separation processes, ion exchange in fixed beds is a multiple batch process. There has always been interest in the development of continuous ion exchange processes and early attempts to improve the process technology date back to the nineteenth century.[7] The patent literature abounds with novel and ingenious techniques for the operation of continuous ion exchange equipment. However, advances have occurred only in recent years and these are associated with the expansion of the uranium processing industry. Examples of the development of ion exchange equipment are given in Table 5, and it may be seen that there are three major categories: (a) moving fixed beds; (b) jigged and agitated beds; and (c) fluidized beds.

The effectiveness of the fixed bed column has been proved in water treatment, e.g. demineralization, dealkalization, softening, etc., and it is thus not surprising that early attempts to improve

Table. 5. Examples of ion exchange equipment.

Type of Equipment	Example
Fixed Bed	
Cascade of Fixed Beds	
Continuous Ion Exchange	
Moving Fixed Bed	Porter-Arden
Jigged Bed)	Weiss et al.
Agitated Bed)	Resin-in-pulp
Fluidized Bed	Swinton/Weiss
	Cloete-Streat
	USBM
	Himsley
	NIM

the process technology centered on the concept of a countercurrent moving fixed bed process. The intention was to retain the high efficiency of the ion exchange process but simplify the number of columns, valving, pipework, controls, etc.

The idea of moving fixed beds was also adopted in the uranium industry and an early example of continuous ion exchange was the Porter-Arden process operated in the Blind River region of Ontario, Canada, in the late 1950's. Here, exhausted resin was transported to an empty conventional column by hydraulic conveying. The resin was then backwashed and transported to a bank of elution columns, fresh resin having previously been transported to the vacant extraction column. The movement of resin around the fixed bed installation was essentially counter-current though the process was not fully continuous by modern standards. Installations of this type were the forerunners of much of the continuous ion exchange development in the uranium industry.

At about the same time, considerable work was in progress on the handling and treatment of uranium leach pulps. The possibility of recovering uranium from unfiltered pulps offered great savings in cost and encouraged some outstanding development work. Clearly, fixed bed equipment was unsuitable for unclarified pulps and thus a radical change in ideas was required. Separate development took place in the U.S.A. and Australia culminating in the concept of jigged or agitated beds of resin. Here, the ion exchange particles were separated momentarily by pulsation, vibration or agitation. Notable developments were the resin-in-pulp contactors. Typical of these were the Anaconda Blue Water Plant in Grants, New Mexico, and the Atlas Minerals Moab Mill in Utah, U.S.A. The equipment is crude by modern standards and usually involved a cascade of baskets containing ion exchange resin retained by stainless steel mesh.[8] The porosity of the mesh retained the resin but allowed ingress of a flowing leached uranium ore pulp. The baskets were agitated by reciprocating action to avoid clogging of the bed. The process operates by alternating pregnant feed, wash and eluant solutions. Though there is no counter-current movement of the particles, this process demonstrated that an "expanded" bed contactor could recover and separate uranium from bulk impurities.

A similar concept was developed separately in Australia by Swinton, Weiss and co-workers.[9] They devised a moving agitated or jigged bed for the treatment of high solids concentration uranium ore pulps. Continuous pulsation of the bed (about 200 cycles/min) caused momentary separation of the particles and this allowed the slurry to pass through the bed in upflow. A screen prevented carry over of the ion exchange particles. In many respects, this contactor was revolutionary since it attempted to solve the problem of slurry handling and countercurrent flow of resin simultaneously. In the event, the project was dogged with problems associated with the top screen. This screen would blind with wood pulp and other gangue and thus the pressure drop became excessive. The idea did not proceed to full-scale development.

The concept of a multistage plate-type column is well known in chemical engineering and forms the basis of many separation processes in vapor-liquid, gas-liquid and liquid-liquid operations. The development of a vertical multistage fluidized bed column has always been appealing in ion exchange. The earliest attempts used a vertical column containing perforated distributor plates and downcomers. Ion exchange resin was supported on the distributor plate in the no-flow condition and during upward flow the particles passed over a weir and into the downcomer. Resin moved in down-flow from stage-to-stage and could be removed from the base of the column. The concept was feasible in principle, and pioneering work was performed by Swinton and Weiss[10] in 1953. However, the operation of this column and others like it was hydraulically unstable. Control of resin holdup and by-passing of the pregnant feed were problems. Start-up was not easy since the pressure drop in the downcomers was invariably less than through the resin bed and thus by-passing was readily initiated. This early work with fluidized beds faded in the 1950's and was ultimately revived in the early 1960's by Cloete and Streat at Imperial Colege.[11] In essence, the Cloete-Streat ion exchange contactor is a multistage fluidized bed containing perforated distributor plates and <u>no</u> downcomers. The hole size in the plates is greater than the maximum resin particle size. Countercurrent flow of resin is achieved by controlled cycling of the liquid feed.

In this way, a precisely determined fraction of the resin holdup per stage can be transferred. In practice, it is desirable to transfer the entire resin holdup per cycle. This column has been developed for application in the uranium industry and the greatest advances have occurred in South Africa. Work at the National Institute for Metallurgy in Johannesburg[12,13] proved the effectiveness of the Cloete-Streat column for uranium recovery from acid leach solutions. Pilot-plant experiments confirmed the scale-up of the equipment and showed that it was possible to produce uranium concentrate of similar quality to that produced by liquid-liquid extraction.[14] Full-scale plants have now been built in South

Africa. NIMCIX plants have been built and commissioned at Blyvooruitzicht gold mine and Stilfontein gold mine and it is intended to build further plants at Harmony gold mine and East Rand gold mine. These plants are all based on the Cloete-Streat concept and a further full-scale plant has been built by Davy McKee for Johannesburg Consolidated Investments at Randfontein. In the U.S.A., Bateman Uranium Corporation are building a NIMCIX plant at United Nuclear-Homestake Partners mill in New Mexico for treatment of unclarified liquors from a tailings pond. Extraction columns up to 5 m diameter containing eight stages of total height about 20 m are in operation. Separately, a full-scale plant was built to treat spent hydrochloric acid using three separate Cloete-Streat columns. A simple iron-zinc separation was achieved by anion exchange prior to acid recovery in a Lurgi pyrohydrolysis reactor.[15]

Independent work in the U.S.A. led to the development of the USBM contactor.[16] This is a multistage countercurrent fluidized bed reactor and operates by controlled cycling. In many respects it is similar to the Cloete-Streat contactor and differs only in plate design, mode of solids transfer, and resin withdrawal. This column has all the advantages of a multistage fluidized bed and has found application for uranium recovery at various locations in the U.S.A.

More recently, a slightly modified form of the multistage fluidized bed has been developed by Himsley.[17] The Himsley column consists of vertical stages containing resin and a single fluid inlet. Pregnant feed in upflow passes from stage to stage continuously and during operation the feed flow is not stopped. Resin is transferred sequentially from stage to stage by using an auxiliary pump to transfer particles down the column. An empty stage at the base of the column allows the process to start. Fresh resin is transferred hydraulically to the top stage. The process requires a series of control valves and timers to sequence the operations. Also, a screen in each stage avoids the carry-over of resin particles during transfer. This column is now in operation for uranium recovery at Agnew Lake, Ontario, Canada, and at Vaal Reefs Gold mine in South Africa.

Though most fluidized bed development is in the form of vertical columns it is also possible to operate the system in a series of horizontal tanks. Possibly the largest CIX plant in the world operates at Rössing Uranium Mine in Namibia. This plant, designed by Porter, comprises an array of horizontal interconnected tanks, each acting as an effective fluidized bed. The resin beads are progesssively moved from tank to tank by airlifts in countercurrent to the flow of pregnant solution. The total pregnant flow to the plant is about 3500 m^3/hr and each contacting tank is 6 m square by 3.5 m deep. The initial plant had 4 lines of 5 contacting tanks in each. Careful design of resin traps has been included in the plant; these comprised a sophisticated rotating trommel and this has now permitted a sixth contactor to be installed in each line. Complete details of this plant have not yet been published.[18,19]

FUTURE DEVELOPMENTS

The impact of continuous ion exchange in the uranium industry during the 1970's has been considerable. There had been an increasing trend towards liquid-liquid extraction (often termed liquid ion exchange) in North America and South Africa, and many new plants used custom-designed liquid extraction equipment. Liquid extraction often produces a rather better quality concentrate at relatively low operating costs. There are however problems, since liquid extraction is not suitable for low uranium concentration feed solutions, e.g. <100-200 ppm U_3O_8. Solvent loss is significant, leading to operating problems and also to effluent problems. Dissolved reagent can attack rubber-lined vessels and lead to severe corrosion. Solid ion exchange overcomes most of these difficulties and is particularly suitable for applicaiton at low pregnant concentrations. This will become more important in the future as low-grade uranium deposits are recovered, e.g. Rössing ore contains about 0.035 weight percent U_3O_8. The Rössing flowsheet incorporates both ion exchange and solvent extraction and employs the consierable advantages of both techniques. CIX is used to upgrade the low concentration feed (0.15 g U_3O_8 per litre) and produce an eluate feed of constant composition containing 3-4 g/l U_3O_8. This eluate feeds the solvent extraction plant which performs the effective separation of uranium from the trace impurities. The final strip solution contains about 10 g/l U_3O_8 and after precipitation and calcination produces a product containing approximately 97% U_3O_8 by weight.

The combination of ion exchange and liquid extraction is not new in the uranium industry and many earlier process plants have used this technique. The Eluex plants in North America, e.g. Gas Hills, Wyoming used a resin-in-pulp plant followed by solvent extraction. However, the ion exchange process was crude, usually of the agitated basket type. The Porter system at Rössing is an interesting advance on this technology. There is likely to be rapid development of fluidized bed columns for resin-in-pulp applications in the expanding uranium production industry. Some evidence already exists in small-scale work that the Cloete-Streat column can handle slurry feeds containing up to 15% by weight solids.[20] Larger-scale development work will no doubt prove that resin-in-pulp operations can be performed in fluidized bed columns. The great advantage of the fluidized bed is the low capital cost, low operating cost, small space requirement, simple instrumentation and control compatability with conventional solvent extraction equipment. It is likely that CIX will make in-roads into other branches of hydrometallurgy, notably the recovery of copper, nickel, zinc, manganese, chromium and it is possible that the precious metal industry, particularly platinum, will attempt to develop CIX in conjunction with solvent extraction.

Future advances in other areas of application are difficult to predict. The water industry treats a low cost commodity and is therefore less likely to be attracted by innovations which have only a modest effect on overall economics. Despite the fact that there have been many attempts to introduce CIX into water treatment, the extent of new process technology is not widespread. In other industries, involving high cost commodities such as pharmaceuticals and biological products, the scale of operation is often small and thus the advantage of CIX is lost. The uranium industry is likely to provide further incentives for development in the next decade, since the current predictions are that uranium production will be boosted from 30,000 te/annum now to about 80,000 te/annum by 1990.

REFERENCES

1. R. R. Grinstead, W. A. Nasutavicus, R. M. Wheaton, and K. C. Jones: Extractive Metallurgy of Copper, A.I.M.E., NY, 1976.
2. F. Helfferich: Ion Exchange, McGraw Hill, NY, 1962.
3. D. Naden and J. P. Evans: J. Appl. Chem. Biotechnol., 1974, vol. 24, pp. 687.
4. A. Warshawsky and Patchornik: Proc. International Conference on Ion Exchange, Cambridge, Society of Chemical Industry, London, paper 38, 1976.
5. R. Kroebel and A. Meyer: Hydrometallurgy Symposium, Manchester, Instn. Chem. Engr. Symposium Series No. 42, paper 24, 1975.
6. A. K. Haines, T. H. Tunely, W.A.M. Te Riele, F.L.D. Cloete, and T. D. Sampson, J. South African Min. Metall., 1973, vol. 74, p. 149.
7. F. O. Matthiessen: U.S. Patent 329,185, 1885.

8. R. F. Hollis and C. K. McArthur: Proceedings First U.N. Conference on Peaceful Uses of Atomic Energy, Geneva, vol. 8, p. 54, 1955.
9. T. V. Arden, J. B. Davis, G. L. Herwig, R. M. Stewart, E. A. Swinton, and D. E. Weiss: Proceedings of Second U.N. Conference on Peaceful Uses of Atomic Energy, Geneva, vol. 3, p. 396, 1958.
10. E. A. Swinton and D. E. Weiss: Australian J. Appl. Sci., 1953, vol. 4, p. 316.
11. F.L.D. Cloete and M. Streat: British Patent 1,070,251, 1962.
12. A. K. Haines: J. South African Inst. Min. and Metal., 1978, vol. 78, p. 303.
13. W. M. Craig, A. K. Haines, D. I. Nicol, C. M. Olen, M.E.E. Douglas, and G. D. Louw: J. South African Inst. Min. and Metal, 1978, vol. 78, p. 316.
14. A. K. Haines, W. M. Craig, A. Faure, A. R. Hendriksz, W. J. Willis, and D. I. Nicol: Eleventh International Mineral Processing Congress, Cagliari, paper 32, 1975.
15. A. K. Haines, T. H. Tunley, W.A.M. te Riele, F.L.D. Cloete, and T. D. Sampson: J. South African Inst. Min. and Metal., 1973, vol. 74, no. 4, p. 149.
16. D. R. George, J. R. Ross, and J. D. Prater: Min. Engng, 1968, vol. 1, p. 73.
17. A. Himsley: Canadian Patent 980,467, 1973.
18. R.J.M. Wyllie: World Mining, 1979, vol. 32, p. 43.
19. R.J.M. Wyllie: World Mining, 1979, vol. 32, p. 52.
20. M. Streat and R. Y. Qassim: International Symposium on Hydrometallurgy, A.I.M.E., Chicago, Chap. 18, p. 475, 1973.

MODELING OF LIQUID-MEMBRANE EXTRACTION PROCESSES

W. S. Ho
Exxon Research and Engineering Company
Clinton Township, New Jersey 08801

and

Norman N. Li
UOP Inc.
Des Plaines, Illinois 60016

A general mathematical model containing overall mass transfer coefficients for extraction and leakage has been developed for extractions by liquid surfactant membranes (LM). In this model, two extraction cases are considered--one has chemical reagent in the external feed phase to convert the leaking species into an extractable form for re-extraction into the LM emulsion, the other does not have reagent in the external phase. Most LM extraction systems can be classified into these two cases. Mass transfer equations to calculate solute recovery, contact time, extractor size, and chemical reagent consumption in relation to overall mass transfer coefficients have been derived and can be used to predict the extraction performance of a continuous LM process from a simple batch extraction experiment. It is shown from the model that the reagent for chemical conversion in the external phase can enhance extraction greatly and can minimize the effect of leakage on extraction significantly. This model has been applied successfully to experimental data in LM extractions, such as separations of chromium and phenol from their aqueous solutions.

INTRODUCTION

Liquid surfactant membranes, discovered just over 15 years ago[1], have been the subject of great interest because of their potential uses as new and effective techniques for separation and purification.[2] These uses, to name a few, include the fractionation of hydrocarbons[3-10], the recovery or enrichment of heavy metal ions[11-20], the removal of trace contaminants from waste water[10,21-26], and a number of diverse applications in the biochemical and biomedical fields.[27-36] Liquid membranes also have potential utility as membrane reactors incorporating simultaneous separation and reaction processes.[37-39]

Liquid surfactant membranes are usually formed by first making an emulsion from two immiscible phases and then dispersing this emulsion into a third phase (continuous phase) by agitation. The liquid membrane phase is the phase that separates the encapsulated internal-phase droplets in the emulsion from the external, continuous phase, as shown schematically in Fig. 1. In general, the droplets of the internal phase in the emulsion are very small, in the range

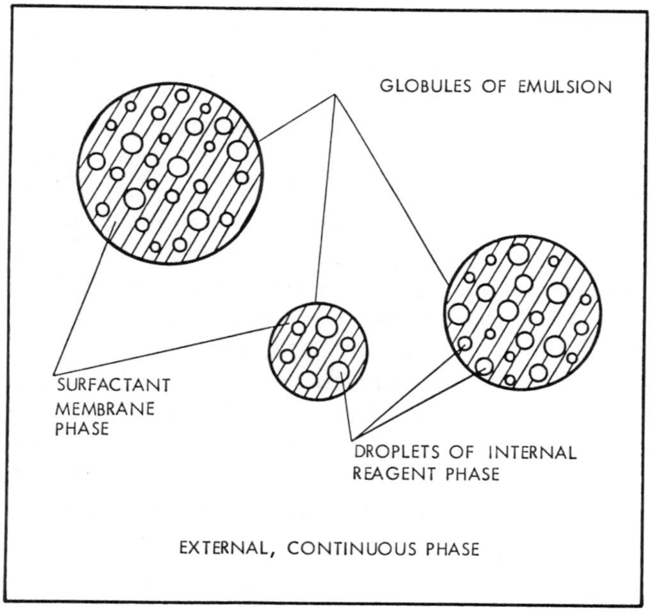

Fig. 1. Schematic diagram of a liquid surfactant membrane system.

1-10 μm diameter whereas the globules of the emulsion phase dispersed in the external phase are in 0.1-2 mm (100-2000μm) diameter. The separation of a mixture or removal of a solute in the continuous phase can occur by the selective transfer of the solute through the membrane phase to the internal phase. Surfactants and additives are normally included in the membrane phase formulation to control the stability, permeability, and selectivity of the membrane. Upon completion of the solute removal, the emulsion phase can be separated from the external phase. Then, breaking the emulsion separates the internal phase from the membrane phase.

In liquid membrane (LM) extraction, solute can transport from the external feed phase into the membrane phase because of solute solubility in the membrane phase. The solute then diffuses through the membrane phase and enters into the internal phase. This kind of LM extraction is exemplified by the fractionation of hydrocarbons with liquid membranes. The effectiveness of LM extraction can be enhanced by the use of a facilitated transport mechanism to maximize both the flux through the membrane phase and the capacity of the receiving phase for the diffusing species. Matulevicius and Li[40] and Li[41] have identified two mechanisms which are called Type 1 and Type 2 facilitations. In Type 1 facilitation, the concentration gradient of the membrane-soluble permeate is maximized by an irreversible reaction of the solute with the chemical reagent in the receiving phase, thereby maintaining the permeate concentration effectively zero in this phase. It is desirable that the reaction products be insoluble in the membrane or incapable of diffusing back through the membrane. In Type 2 facilitation, a complexing agent incorporated in membrane phase "carries" the diffusing species across the membrane to the receiving phase. This is commonly known as "carrier mediated" transport. In both cases, the diffusing species can be immobilized at the expense of some consumable reagent. For instance, phenol can react with NaOH to form the oil-insoluble sodium phenolate and the extraction of the cupric ion can be maintained by the counter transport of hydrogen ions.

Although the mathematical models for LM extractions are available in the literature, the need for an improved model still exists. In general, the available models often deal with specific extraction cases and are not general enough to take care of most LM extraction processes. Most of the models have not taken the breakage of the internal phase into account. For example, Matulevicius and Li[40] suggested that the diffusion of phenol in an LM emulsion globule be modeled mathematically by a single drop. They formulated and solved the unsteady state equations for the hollow sphere model, in which phenol diffuses from the surface of the globule to some fixed, interior position where it is removed by the reaction with the internal reagent. In assuming that an LM globule can be represented by a single drop, an effective membrane thickness needs to be assigned to account for the mass transfer resistance in the globule. Kopp et al.[42] proposed to describe the mass transfer process in the LM globule in terms of a moving boundary at which a reaction between a solute and the internal reagent occurs. This boundary moves toward the globule center as the internal reagent is consumed. They presented the solution of the mass transfer model for a planar geometry with an essentially constant solute concentration at the globule interface between the membrane and external phases. They then used that solution to represent the transport in a spherical geometry. Unfortunately, the use of that solution limits the range of applicability of their work. Recently, Ho et al.[43] solved the mass transfer process with a moving boundary for a spherical geometry. Their model also takes care of the concentration changes at the globule interface and in the external phase. However, these models mentioned all assume negligible breakage. The breakage, however, can become significant in the cases of weak membranes and stringent mixing conditions. Boyadzhiev et al.[8] took breakage into account in their modeling of hydrocarbon extractions by liquid membranes. Their model is limited to the special case of the extractions without facilitated transport mechanism. Besides, the model does not treat the leakage of extracted solute from the internal phase, via complexing with a carrier in the membrane phase, to the external phase.

In this paper, we describe a general mathematical model containing overall mass transfer coefficients for extraction and leakage in LM extraction processes. In this model two extraction cases are considered--one has chemical reagent in the external feed phase to convert the leaking species into an extractable form for re-extraction into the LM emulsion and the other does not have the reagent in the external phase. Most LM extraction systems can be classified into these two cases. This model takes care of the leakage mechanisms of solute, from the internal phase to the external phase, not only by breakage but also by complexing with a carrier in the membrane phase. This model provides a usable method for the analysis of batch extraction data, from which the overall mass transfer coefficients can be obtained. In turn, these coefficients can be used via the model for the prediction of continuous process performance and the design of LM separation processes.

THE MODEL OF LIQUID MEMBRANE EXTRACTION

Mass Transfer Mechanism

Solute can transfer from the internal phase of a liquid membrane (LM) system to the external phase by two mechanisms: (1) diffusional transport and (2) breakage. But, solute can transfer from the external phase to the internal phase only by diffusion mechanism. These mechanisms are shown schematically in Fig. 2, in which the globule of the LM emulsion is represented by a hollow sphere with the membrane phase separating the internal phase from the external phase.

1. Diffusional Transport. If the concentration driving force of solute is higher for the internal phase than for the external phase, the direction of the diffusional transport will be from the internal phase to the external phase. This diffusional transport includes five steps: (1) mass transfer from the internal phase to the interface adjacent to the internal phase, (2) interfacial mass transfer across this interface, (3) diffusion in the membrane phase from this interface to the second interface, adjacent to the external phase, (4) interfacial mass transfer across the second interface, and (5) mass transfer from the second interface to the bulk of the external phase. The mass transfer rate j_k can be expressed in the following equations with the five individual mass

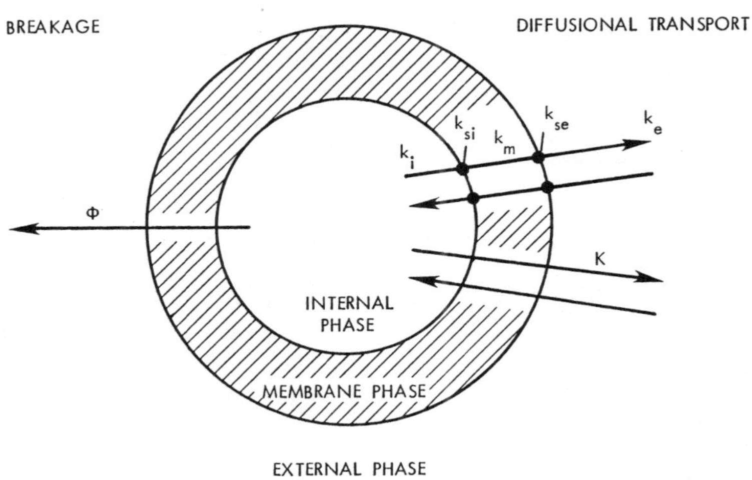

Fig. 2. Schematic of simplified mass transfer mechanism.

transfer coefficients k_i, k_{si}, k_m, k_{se}. and k_e for these five steps, respectively.

$$j_k = k_i A_i (C_i - C_{si}) \tag{1}$$

$$= k_{si} A_i (C_{si} - \frac{C_{sim}}{H_i}) \tag{2}$$

$$= k_m A_e (C_{sim} - C_{sem}) \tag{3}$$

$$= k_{se} A_e (\frac{C_{sem}}{H_e} - C_{se}) \tag{4}$$

$$= k_e A_e (C_{se} - C_e) \tag{5}$$

Rearrangement of these equations leads to the following equation:

$$j_k = \frac{A_i A_e k_i k_{si} k_m k_{se} k_e H_i (C_i - \frac{H_e}{H_i} C_e)}{A_i k_i k_{si} (k_{se} k_e + k_m k_e H_e + k_m k_{se} H_e) + A_e k_m k_{se} k_e H_i (k_{si} + k_i)} \tag{6}$$

As described above, the encapsulated droplets of the internal phase are considerably smaller in size (1-10μm in diameter) than the emulsion globules themselves, which usually have diameters in the range of 0.1 to 2 mm. And each emulsion globule contains many encapsulated droplets. Consequently, the internal phase mass transfer area, A_i, is much larger than the external phase mass transfer area, A_e. Thus, Eq. (6) reduces to the following equation:

$$j_k = [(\frac{k_m k_{se} k_e H_i}{k_{se} k_e + k_m k_e H_e + k_m k_{se} H_e}) A_e] (C_i - \frac{H_e}{H_i} C_e) \tag{7}$$

Hence, the mass transfer rate for the diffusional transport can be represented as follows:

$$j_k = K' a (V_m + V_i)(C_i - \frac{H_e}{H_i} C_e) = K (V_m + V_i)(C_i - \frac{H_e}{H_i} C_e) \tag{8}$$

where K' is the overall mass transfer coefficient based on mass transfer area per unit volume of the emulsion, V_m the membrane phase volume, V_i the total volume of the internal phase, and K the overall mass transfer coefficient. If the concentration driving force is higher for the external phase than for the internal phase, the five individual mass transfer steps described above will be reversed.

2. Breakage. The breakage of the internal phase can be assumed to be proportional to the internal phase volume under a given mixing condition.[8]

$$-\frac{dV_i}{dt} = \phi \, V_i \tag{9}$$

where ϕ is the breakage coefficient and t the extraction time. The mass transfer rate due to the breakage, j_ϕ, is

$$j_\phi = \phi \, V_i C_i \tag{10}$$

Integration of Eq. (9) gives the internal phase volume as a function of time:

$$V_i = V_{i,o} \, e^{-\phi t} \tag{11}$$

where $V_{i,o}$ is the initial volume of the internal phase. The breakage decreases the internal phase volume but increases the external phase volume, V_e, which is given in the following equation:

$$V_e = V_o - V_i = V_o - V_{i,o} \, e^{-\phi t} \tag{12}$$

where $V_o = V_{e,o} + V_{i,o}$, V_o is the sum of the external and internal phase volumes, and $V_{e,o}$ the initial external phase volume.

Case I of the Model

In Case I of the model, a chemical reagent in the external feed phase converts the leaking species into an extractable form. This model is shown schematically in Fig. 3. Solute A in the external feed phase is extracted, via the diffusion mechanism associated with the mass transfer coefficient K_A, into the internal phase where the solute reacts with the reagent in the internal phase to become Solute B. B can transfer from the internal phase, via the diffusion mechanism associated with the mass transfer coefficient K_B and the breakage mechanism associated with the coefficient ϕ, to the external phase where the external reagent converts B into the extractable solute A. In this case, A exists only in the external phase whereas B exists only in the internal phase. For example, phenol, the solute A, can be extracted into the internal phase where it reacts with the internal reagent NaOH to become sodium phenolate, the solute B. The phenolate can leak out from the internal phase, via breakage (the inonic species B can not diffuse through an oil-type membrane, i.e., K_B=0), into the external

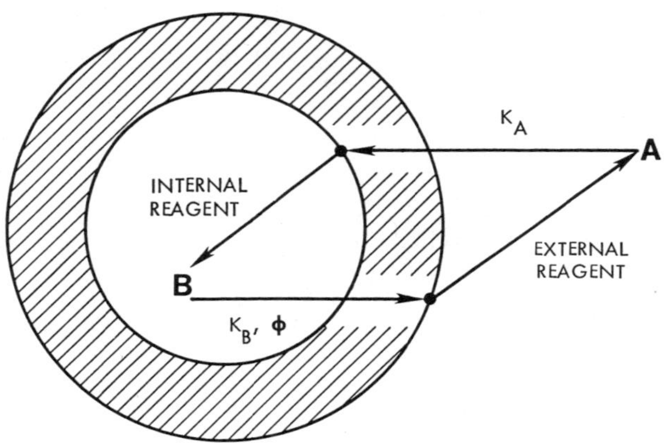

Fig. 3. Schematic of Case I of the model.

phase where H_2SO_4, the external reagent, converts phenolate into phenol, which can then be re-extracted.

The governing differential equation for Case I in a batch extraction operation is as follows:

$$\frac{d(V_e C_{eA})}{dt} = K_{A1}(V_m+V_i)(C_{iA} - \frac{H_{eA}}{H_{iA}} C_{eA}) + K_B(V_m+V_i)(C_{iB} - \frac{H_{eB}}{H_{iA}} C_{eB}) + \phi V_i C_{iB} \quad (13)$$

Since the concentration of A in the internal phase, C_{iA}, and the concentration of B in the external phase, C_{eB}, are zero, this equation reduces to Eq. (14).

$$\frac{d(V_e C_{eA})}{dt} = -K_A^\circ C_{eA} + (K_B^\circ + \phi V_i)C_{iB} \quad (14)$$

where $$K_A^\circ = K_A(V_m+V_i) \quad (15)$$

$$K_B^\circ = K_B (V_m+V_j) \quad (16)$$

$$K_A = K_{A1} \frac{H_{eA}}{H_{iA}} \quad (17)$$

K_A and K_B are the overall mass transfer coefficients for A and B, respectively, and C_{eA} and C_{iB} are the concentrations for A in the external phase and for B in the internal phase, respectively. The initial conditions for Eq. (14) are given in Eqs. (18) and (19).

$$C_{eA} = C_{eA,o} \qquad t = 0 \tag{18}$$

$$C_{iB} = C_{iB,o} \qquad t = 0 \tag{19}$$

With the assumption that solute accumulation in the membrane phase is negligible in comparison with the solute amounts in the external and internal phases, C_{iB} is given in the following equation from the conservation of mass.

$$C_{iB} = \frac{1}{V_i}(V_{e,o}C_{eA,o} + V_{i,o}C_{iB,o}) - \frac{V_e}{V_i}C_{eA} \tag{20}$$

Equation (14) can be solved for two cases: (1) small breakage and (2) large breakage.

1. Small Breakage. For small breakage, i.e., $\phi=0$ or $\phi \leq 1.4\times10^{-5}$ 1/s, the change of the internal phase volume is less than 5% as determined from Eq. (11) for one hour extraction time. Thus, V_i and V_e can be assumed constant and approximated by their initial volumes, respectively. Substituting Eq. (20) into Eq. (14) and integrating the resultant equation with the initial condition shown in Eq. (18), we obtain the solution for C_{eA}.

$$C_{eA} = \left[\frac{(V_eC_{eA,o} + V_iC_{iB,o})}{V_e}\,\frac{(K_{B\phi}/V_i)}{(K_A/V_e + K_{B\phi}/V_i)}\right] \left[1 + \frac{1}{(K_{B\phi}/V_i)}\left(\frac{K_AC_{eA,o} - K_{B\phi}C_{iB,o}}{V_eC_{eA,o} + V_iC_{iB,o}}\right)e^{-\left(\frac{K_A}{V_e} + \frac{K_{B\phi}}{V_i}\right)(V_m + V_i)t}\right] \tag{21}$$

where

$$K_{B\phi} = K_B + \phi\left(\frac{V_i}{V_m + V_i}\right) \tag{22}$$

We can then obtain the solution for C_{iB} from Eqs. (20) and (21). Rearrangement of these two equations results in the following equation:

$$\left(\frac{K_A}{V_e} + \frac{K_{B\phi}}{V_i}\right) = \frac{1}{(V_m+V_i)t}\ln\left(\frac{K_AC_{eA,o} - K_{B\phi}C_{iB,o}}{K_AC_{eA} - K_{B\phi}C_{iB}}\right) \tag{23}$$

2. Large Breakage. For large breakage, both V_i and V_e change with time in a manner that V_i and V_e can be expressed by Eqs. (11) and (12), respectively. Substituting these two equations and Eq. (20) into Eq. (14), we arrive at the solution for C_{eA}.

$$C_{eA} = \frac{(V_{e,o}C_{eA,o} + V_{i,o}C_{iB,o})}{V_o} \tag{24}$$

$$+ \frac{V_{i,o}(C_{eA,o}-C_{iB,o})}{V_o} \left(\frac{V_o-V_{i,o}}{V_o e^{\phi t}-V_{i,o}}\right)^{(1+\frac{K_A°}{\phi V_o})} e^{-\frac{K_B°}{\phi V_{i,o}}(e^{\phi t}-1)}$$

$$- \frac{E(K_A°-K_B°)(V_{e,o}C_{eA,o}+V_{i,o}C_{iB,o})}{\phi V_o(V_o-V_{i,o})} \left(\frac{V_o-V_{i,o}}{V_o e^{\phi t}-V_{i,o}}\right)^{(1+\frac{K_A°}{\phi V_o})} e^{-\frac{K_B°}{\phi V_{i,o}}e^{\phi t}}$$

where

$$E = \int_1^{e^{\phi t}} \left(\frac{V_o X-V_{i,o}}{V_o-V_{i,o}}\right)^{\frac{K_A°}{\phi V_o}} e^{\frac{K_B°}{\phi V_{i,o}}X} dX \tag{25}$$

3. Special Case: No Diffusional Transport of Solute B. A special case in the Model of Case I is that the solute B can not diffuse through the membrane phase, i.e., $K_B=0$. For example, phenolate, an ionic species, can not diffuse through an oil-type membrane phase. In general, ionic species can not diffuse through this type of membranes unless the membranes contain carriers that can complex with the ionic species. For $K_B=0$, Eq. (21) in the case of small ϕ values reduces to the following equation:

$$C_{eA} = \frac{(V_e C_{eA,o}+V_i C_{iB,o})}{(K_A°+\phi V_e)} \left[\phi+\left(\frac{K_A° C_{eA,o}-\phi V_i C_{iB,o}}{V_e C_{eA,o}+V_i C_{iB,o}}\right) e^{-\left(\frac{K_A°}{V_e}+\phi\right)t}\right] \tag{26}$$

In the case of large ϕ values, Eq. (24) reduces to the following equation:

$$C_{eA} = \frac{\phi(V_e C_{eA,o}+V_{i,o}C_{iB,o})}{(K_A°+\phi V_o)}$$

$$+ \left[\frac{(K_A°+\phi V_{i,o})C_{eA,o}-\phi V_{i,o}C_{iB,o}}{K_A°+\phi V_o}\right] \left(\frac{V_o-V_{i,o}}{V_o e^{\phi t}-V_{i,o}}\right)^{(1+\frac{K_A°}{\phi V_o})} \tag{27}$$

4. Chemical Reagent Consumption. The total consumption amount of the chemical reagent (such as H_2SO_4 for the conversion of phenolate into phenol) in the external phase, W_e, for the conversion of the leaking species B into the extractable solute A and for the reaction with the leaking internal reagent can be calculated from

the following equation:

$$W_e = \int_o^t [(K_B^\circ + \phi V_i) C_{iB} + \phi V_i C_{ir}] dt \tag{28}$$

where C_{ir} is the concentration of the internal reagent. The differential equation for the material balance of the internal reagent is as follows:

$$\frac{d(V_i C_{ir})}{dt} = - K_A^\circ \, C_{eA} - \phi V_i \, C_{ir} \tag{29}$$

The initial condition for C_{ir} is :

$$C_{ir} = C_{ir,o} \qquad t = 0 \tag{30}$$

Substitution of Eq. (21) into Eq. (29) and integration of the resultant equation give the solution of C_{ir} for small ϕ values.

$$C_{ir} = C_{ir,o} e^{-\phi t} - \left[\frac{(V_e C_{eA,o} + V_e C_{iB,o})}{V_e} \frac{(K_{B\phi}/V_i)}{(K_A/V_e + K_{B\phi}/V_i)}\right]$$

$$\left[\frac{K_A(V_m+V_i)}{\phi V_i} (1-e^{-\phi t})\right] + \left[\frac{(K_A C_{eA,o} - K_{B\phi} C_{iB,o})}{(K_A V_i + K_{B\phi} V_e)}\right]$$

$$\left[\frac{K_A(V_m+V_i)}{[(K_A/V_e + K_{B\phi}/V_i)(V_m+V_i) - \phi}\right] \left[e^{-(\frac{K_A}{V_e} + \frac{K_{B\phi}}{V_i})(V_m+V_i)t} - e^{-\phi t}\right] \tag{31}$$

Substituting Eqs. (20), (21), and (31) into Eq. (28) and integrating the resultant equation, we obtain the total consumption amount of the external reagent for small ϕ values.

$$W_e = V_i C_{ir,o}(1-e^{-\phi t}) + \left[\frac{K_A(V_m+V_i)(1-e^{-\phi t})}{V_e(K_A/V_e + K_{B\phi}/V_i)}\right]$$

$$\left[\frac{K_{B\phi}(V_e C_{eA,o} + V_i C_{iB,o})}{\phi V_i} - \frac{(K_A C_{eA,o} - K_{B\phi} C_{iB,o})}{(K_A/V_e + K_{B\phi}/V_i)(V_m+V_i) - \phi}\right]$$

$$+ \left[\frac{(K_A C_{eA,o} - K_{B\phi} C_{iB,o})}{(K_A/V_e + K_{B\phi}/V_i)^2}\right] \left[\frac{K_{A\phi}/V_e}{(K_A/V_e + K_{B\phi}/V_i)(V_m+V_i) - \phi} - \frac{K_{B\phi}}{V_i}\right]$$

$$\left[1-e^{-(\frac{K_A}{V_e} + \frac{K_{B\phi}}{V_i})(V_m+V_i)t}\right] \tag{32}$$

The total consumption amount of the chemical reagent (such as N_aOH for the conversion of phenol into phenolate) in the internal phase, W_i, for the conversion of Solute A into Solute B and for the loss due to breakage can be computed from the following equation:

$$W_i = \int_o^t (K_A{}^\circ C_{eA} + \phi V_i C_{ir})dt \tag{33}$$

Substitution of Eqs. (21) and (31) into Eq. (33) and integration of the resultant equation lead to the result of W_i for small ϕ values:

$$W_i = V_i C_{ir,o}(1-e^{-\phi t}) + [\frac{K_A(V_m+V_i)(1-e^{-\phi t})}{V_e(K_A/V_e+K_{B\phi}/V_i)}]$$
$$[\frac{K_{B\phi}(V_e C_{eA,o}+V_i C_{iB,o})}{\phi V_i} - \frac{(K_A C_{eA,o}-K_{B\phi}C_{iB,o})}{(K_A/V_e+K_{B\phi}/V_i)(V_m+V_i)-\phi}]$$
$$+ [\frac{K_A(V_m+V_i)(K_A C_{eA,o}-K_{B\phi}C_{iB,o})}{V_e(K_A/V_e+K_{B\phi}/V_i)[(K_A/V_e+K_{B\phi}/V_i)(V_m+V_i)-\phi]}]$$
$$[1-e^{-(\frac{K_A}{V_e}+\frac{K_{B\phi}}{V_i})(V_m+V_i)t}] \tag{34}$$

Case II of the Model

In Case II of the model, there is no chemical reagent in the external phase for the conversion of the leaking species into an extractable form. This model is shown schematically in Fig. 4. Solute A in the external phase is extracted, via the diffusion mechanism associated with the mass transfer coefficient K_A, into the internal phase where the solute reacts with the reagent in the internal phase to become Solute B. B can transfer from the internal phase, via both the diffusion mechanism associated with the mass transfer coefficient K_B and the breakage mechanism associated with the coefficient ϕ, to the external phase where B remains as a solute. In this case, both A and B can exist in the external phase whereas B can also exist in the internal phase, but not A. The internal reagent leaking out to the external phase can convert A into B in the external phase. A can exist in the external phase until it is exhausted owing to both extraction and internal reagent leakage. For example, phenol, the solute A, can be extracted into the internal phase where it reacts with the internal reagent NaOH to become sodium phenolate, the solute B. The phenolate can leak out from the internal phase, via breakage (the ionic species B can

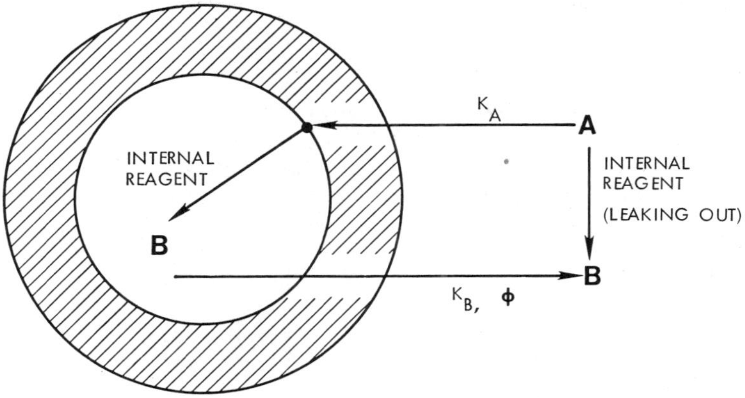

Fig. 4. Schematic of Case II of the model.

not diffuse through an oil-type membrane, i.e., K_B=0), into the external phase where it remains as phenolate. NaOH leaking out from the internal phase to the external phase can convert phenol into phenolate in the external phase. Phenol can exist in the external phase until it is exhausted owing to both extraction and NaOH leakage.

The governing equations for Case II in a batch extraction operation are as follows:

$$\frac{d(V_eC_{eA})}{dt} = -K_A^\circ C_{eA} - \phi V_i C_{ir} \qquad 0 \leq t \leq t_1 \tag{35}$$

$$\frac{d(V_eC_{eB})}{dt} = K_B^\circ\left(C_{iB} - \frac{H_{eB}}{H_{iB}} C_{eB}\right) + \phi V_i C_{iB} + \phi V_i C_{ir} \qquad 0 \leq t \leq t_1 \tag{36}$$

$$\frac{d(V_iC_{ir})}{dt} = -K_A^\circ C_{eA} - \phi V_i C_{ir} \qquad 0 \leq t \leq t_1 \tag{37}$$

$$C_{iB} = \frac{1}{V_i}(V_{e,o}C_{eA,o} + V_{e,o}C_{eB,o} + V_{i,o}C_{iB,o}) - \frac{V_e}{V_i}(C_{eA} + C_{eB}) \qquad 0 \leq t \leq t_1 \tag{38}$$

$$Ce_A = 0 \qquad t \geq t_1 \tag{39}$$

$$\frac{d(V_eC_{eB})}{dt} = K_B^\circ\left(C_{iB} - \frac{H_{eB}}{H_{iB}} C_{eB}\right) + \phi V_i C_{iB} \qquad t \geq t_1 \tag{40}$$

$$\frac{d(V_iC_{ir})}{dt} = -\phi V_i C_{ir} \qquad t \geq t_1 \tag{41}$$

$$C_{iB}=\frac{1}{V_i}(V_{e,o}C_{eA,o}+V_{e,o}C_{eB,o}+V_{i,o}C_{iB,o})-\frac{V_e}{V_i}C_{eB} \qquad t\geq t_1 \tag{42}$$

where t_1 is the time at which Solute A in the external phase is just exhausted as expressed in Eq. (39), H_{eB} is the distribution coefficient for the solute B between the membrane and external phases at equilibrium, and H_{iB} is the distribution coefficient for B between the membrane and internal phases at equilibrium. The initial conditions are given in Eqs. (43-46).

$$C_{eA}=C_{eA,o} \qquad t=o \tag{43}$$

$$C_{eB}=C_{eB,o} \qquad t=o \tag{44}$$

$$C_{iB}=C_{iB,o} \qquad t=o \tag{45}$$

$$C_{ir}=C_{ir,o} \qquad t=o \tag{46}$$

Substituting Eq. (35) into Eq. (37) and integrating the resultant equation, we obtain the following relationship:

$$V_iC_{ir}=V_eC_{eA}+(V_{i,o}C_{ir,o}-V_{e,o}C_{eA,o}) \qquad o\leq t\leq t_1 \tag{47}$$

Equations (35-42) can be solved for two cases: (1) small breakage and (2) large breakage.

1. Small Breakage. For small breakage, i.e., $\phi=0$ or $\phi\leq 1.4\times 10^{-5}$ 1/s, V_i and V_e can be assumed constant and approximated by their initial volumes. Substitution of Eq. (47) into Eq. (35) and integration of the resultant equation lead to the solution for C_{eA}.

$$C_{eA}=\frac{(K_A^\circ C_{eA,o}+\phi V_iC_{ir,o})}{(K_A^\circ+\phi V_e)}e^{-\left(\frac{K_A^\circ}{V_e}+\phi\right)t}-\frac{\phi(V_iC_{ir,o}-V_eC_{eA,o})}{(K_A^\circ+\phi V_e)} \qquad 0\leq t\leq t_1 \tag{48}$$

The time t_1 is obtained from the above equation by letting $C_{eA}=0$.

$$t_1=\frac{V_e}{K_A^\circ+\phi V_e}\ln\left(\frac{K_A^\circ C_{eA,o}+\phi V_iC_{ir,o}}{\phi(V_iC_{ir,o}-V_eC_{eA,o})}\right) \tag{49}$$

Substitution of Eq. (48) into Eq. (47) results in the solution for C_{ir}.

$$C_{ir} = \frac{V_e(K_A^\circ C_{eA,o}+V_i C_{ir,o})}{V_i(K_A^\circ+\phi V_e)} e^{-\left(\frac{K_A^\circ}{V_e}+\phi\right)t} + \frac{K_A^\circ(V_i C_{ir,o}-V_e C_{eA,o})}{V_i(K_A^\circ+\phi V_e)} \qquad 0 \leq t \leq t_1 \tag{50}$$

Substituting Eqs. (38), (48), and (50) into Eq. (36) and integrating the resultant equation, we obtain the solution for C_{eB}:

$$C_{eB} = (C_{eB,o}-f-g)e^{-\left(\frac{K_B^\circ H_{eB}}{V_e H_{iB}}+\frac{K_B^\circ}{V_i}+\phi\right)t} + fe^{-\left(\frac{K_A^\circ}{V_e}+\phi\right)t} + g \qquad 0 \leq t \leq t_1 \tag{51}$$

where

$$f = \frac{K_B^\circ V_e(K_A^\circ C_{eA,o}+V_i C_{ir,o})}{(K_A^\circ+\phi V_e)\left[K_A^\circ V_i - K_B^\circ V_e - K_B^\circ V_i(H_{eB}/H_{iB})\right]} \tag{52}$$

$$g = \left\{\frac{1}{\left[(K_B^\circ V_i H_{eB}/H_{iB})+K_B^\circ V_e+\phi V_i V_e\right]}\right\}$$

$$\left[\frac{\phi(V_i C_{ir,o}-V_e C_{eA,o})(K_A^\circ V_i+K_B^\circ V_e+\phi V_i V_e)}{(K_A^\circ+\phi V_e)} + \right.$$

$$\left.(K_B^\circ+\phi V_i)(V_e C_{eA,o}+V_e C_{eB,o}+V_i C_{iB,o})\right] \tag{53}$$

For $t \geq t_1$, i.e., $C_{eA}=0$, the solutions for C_{ir} and C_{eB} can be derived from Eqs. (40-42). Integration of Eq. (41) leads to the solution for C_{ir}:

$$C_{ir} = C_{ir}|_{t_1} \qquad t \geq t_1 \tag{54}$$

where $C_{ir}|_{t_1}$ is the concentration of the internal reagent at the time t_1. This concentration can be obtained from Eq. (50) by letting $t=t_1$. Equation (54) shows the fact that C_{ir} will not change for $t \geq t_1$ since there is no solute A to be extracted for the reaction with the internal reagent.

Substituting Eq. (42) into Eq. (40) and integrating the resultant equation, we obtain the solution for C_{eB}:

$$C_{eB} = \left[C_{eB}|_{t_1} - \frac{(K_B^\circ+\phi V_i)(V_e C_{eA,o}+V_e C_{eB,o}+V_i C_{iB,o})}{(K_B^\circ V_i H_{eB}/H_{iB})+K_B^\circ V_e+\phi V_e V_i}\right]$$

$$\left[e^{-\left(\frac{K_B^\circ H_{eB}}{V_e H_{iB}} + \frac{K_B^\circ}{V_i} +\phi\right)(t-t_1)}\right] + \frac{(K_B^\circ+\phi V_i)(V_e C_{eA,o}+V_e C_{eB,o}+V_i C_{iB,o})}{(K_B^\circ V_i H_{eB}/H_{iB})+K_B^\circ V_e+\phi V_e V_i}$$

$$t \geq t_1 \qquad (55)$$

where $C_{eB}|_{t_1}$ is the value of C_{eB} at t_1, which can be obtained from Eq. (51) by letting $t = t_1$.

2. Large Breakage. As mentioned above, for large breakage, V_i and V_e can be expressed by Eqs. (11) and (12), respectively. Substituting these two equations and Eq. (47) into Eq. (35) and integrating the resultant equaiton, we obtain the solution for C_{eA}.

$$C_{eA} = \frac{(K_A^\circ C_{eA,o}+\phi V_{i,o} C_{ir,o})}{(K_A^\circ+\phi V_o)} \left(\frac{V_o-V_{i,o}}{V_o e^{\phi t}-V_{i,o}}\right)^{\left(1+\frac{K_A^\circ}{\phi V_o}\right)}$$

$$- \frac{\phi(V_{i,o}C_{ir,o}-V_{e,o}C_{eA,o})}{(K_A^\circ+\phi V_o)} \qquad o \leq t \leq t_1 \qquad (56)$$

The time t_1 is obtained from the above equation by letting $C_{eA}=0$.

$$t_1 = \frac{1}{\phi} \ln\left[\frac{V_{i,o}}{V_o} + \left(1-\frac{V_{i,o}}{V_o}\right)\left(\frac{K_A^\circ C_{eA,o}+\phi V_{i,o} C_{ir,o}}{\phi V_{i,o} C_{ir,o}-\phi V_{e,o} C_{eA,o}}\right)^{\left(\frac{\phi V_o}{K_A^\circ+\phi V_o}\right)}\right]$$

$$(57)$$

Substitution of Eq. (56) into Eq. (47) gives the solution for C_{ir}.

$$C_{ir} = \frac{V_{e,o}(K_A^\circ C_{eA,o}+\phi V_{i,o} C_{ir,o})}{V_{i,o}(K_A^\circ+\phi V_o)} \left(\frac{V_o-V_{i,o}}{V_o e^{\phi t}-V_{i,o}}\right)^{\left(\frac{K_A^\circ}{\phi V_o}\right)}$$

$$+ \frac{(V_{i,o}C_{ir,o}-V_{e,o}C_{eA,o})}{(K_A^\circ+\phi V_o)} \left(\frac{K_A^\circ}{V_{i,o}} e^{\phi t} + \phi\right) \qquad o \leq t \leq t_1 \qquad (58)$$

Substituting Eqs. (38), (56), and (58) into Eq. (36) and integrating the resultant equation, we obtain the solution for C_{eB}.

$$C_{eB} = h+\Big[(C_{eB,o}-h)V_{e,o}^{\left(1+\frac{K_B^\circ H_{eB}}{\phi V_o H_{iB}}\right)} e^{\frac{K_B^\circ}{\phi V_{i,o}}} + Gh\left(1-\frac{H_{eB}}{H_{iB}}\right)\frac{K_B^\circ}{\phi}$$

$$-F\left(\frac{K_B^\circ V_{e,o}}{\phi V_{i,o}}\right)\frac{(K_A^\circ C_{eA,o}+\phi V_{i,o}C_{ir,o})}{(K_A^\circ+\phi V_o)} + G(K_A^\circ-K_B^\circ)\frac{(V_{i,o}C_{ir,o}-V_{e,o}C_{eA,o})}{(K_A^\circ+\phi V_o)}\Big]$$

$$\left[(V_o e^{\phi t}-V_{i,o})^{-\left(1+\frac{K_B^\circ H_{eB}}{\phi V_o H_{iB}}\right)} e^{-\frac{K_B^\circ}{\phi V_{i,o}}e^{\phi t}}\right] \qquad o\leqslant t\leqslant t_1 \tag{59}$$

where

$$h = \frac{(V_{e,o}C_{eA,o}+V_{e,o}C_{eB,o}+V_{i,o}C_{iB,o})}{V_o} + \frac{\phi(V_{i,o}C_{ir,o}-V_{e,o}C_{eA,o})}{(K_A^\circ+\phi V_o)} \tag{60}$$

$$F = \int_1^{e^{\phi t}} \left(\frac{V_o-V_{i,o}}{V_oX-V_{i,o}}\right)^{\frac{K_A^\circ}{V_o}} (V_oX-V_{i,o})^{\left(\frac{K_B^\circ H_{eB}}{\phi V_o H_{iB}}\right)} e^{\frac{K_B^\circ}{\phi V_{i,o}}X}\, dX \tag{61}$$

$$G = \int_1^{e^{\phi t}} (V_oX-V_{i,o})^{\left(\frac{K_B^\circ H_{eB}}{\phi V_o H_{iB}}\right)} e^{\frac{K_B^\circ}{\phi V_{i,o}}X}\, dX \tag{62}$$

For $t\geqslant t_1$, i.e., $C_{eA}=0$, the solution for C_{ir} is $C_{ir}|_{t_1}$ (Eq. 54), which can be obtained from Eq. (58) by using t_1 given by Eq. (57). Substitution of Eqs. (11), (12), and (42) into Eq. (40) and integration of the resultant equation lead to the solution for C_{eB}.

$$C_{eB} = \frac{(V_{e,o}C_{eA,o}+V_{e,o}C_{eB,o}+V_{i,o}C_{iBo})}{V_o}$$

$$+[C_{eB}|_{t_1} - \frac{(V_{e,o}C_{eA,o}+V_{e,o}C_{eB,o}+V_{i,o}C_{iB,o})}{V_o}][(\frac{V_o e^{\phi t_1}-V_{i,o}}{V_o e^{\phi t}-V_{i,o}})^{(1+\frac{K_B^\circ H_{eB}}{\phi V_o H_{iB}})}]$$

$$[e^{-\frac{K_B^\circ}{\phi V_{i,o}}(e^{\phi t}-e^{\phi t_1})}] + [I(1-\frac{H_{eB}}{H_{iB}})\frac{K_B^\circ(V_{e,o}C_{eA,o}+V_{e,o}C_{eB,o}+V_{i,o}C_{iBo})}{\phi V_o}]$$

$$[(V_o e^{\phi t}-V_{i,o})^{-(1+\frac{K_B^\circ H_{eB}}{\phi V_o H_{iB}})} e^{-(\frac{K_B^\circ}{\phi/v_{i,o}})e^{\phi t}}] \qquad t>t_1 \qquad (63)$$

where

$$I = \int_{e^{\phi t_1}}^{e^{\phi t}} (V_o X-V_{i,o})^{\frac{K_B^\circ H_{eB}}{\phi V_o H_{iB}}} e^{\frac{K_B^\circ}{\phi V_{i,o}}X} dX \qquad (64)$$

3. Special Case: No Diffusional Transport of Solute B. Similar to Case I model, a special case in the model of Case II is that the solute B can not diffuse through the membrane phase, i.e., $K_B=0$. For ionic species, such as phenolate and NH_4^+, generally can not diffuse through oil-type membranes unless the membranes contain carriers that can complex with the ionic species. For $K_B=0$, Eqs. (51) and (55) in the case of small ϕ values reduce to Eqs. (65) and (66), respectively.

$$C_{eB} = \frac{1}{V_e}(V_e C_{eB,o}+V_i C_{iB,o}+V_i C_{ir,o}) - \frac{V_i}{V_e}(C_{iB,o}+C_{ir,o})e^{-\phi t} \qquad o\leq t\leq t_1 \qquad (65)$$

$$C_{eB} = \frac{1}{V_e}(V_e C_{eA,o}+V_e C_{eB,o}+V_i C_{iB,o}) - \frac{V_i}{V_e}(C_{iB,o}+C_{ir,o})e^{-\phi t}$$

$$+ \frac{1}{V_e}(V_i C_{ir,o}-V_e C_{eA,o})e^{-\phi(t-t_1)} \qquad t\geq t_1 \qquad (66)$$

where t_1 is given by Eq. (49). In the case of large ϕ values, Eqs. (59) and (63) reduce to Eqs. (67) and (68), respectively.

$$C_{eB} = \frac{1}{(V_o - V_{i,o}e^{\phi t})}\left[(V_{e,o}C_{eB,o} + V_{i,o}C_{iB,o} + V_{i,o}C_{ir,o}) - V_{i,o}(C_{iB,o} + C_{ir,o})e^{-\phi t}\right] \quad o \leqslant t \leqslant t_1 \tag{67}$$

$$C_{eB} = \frac{(V_{e,o}C_{eA,o} + V_{e,o}C_{eB,o} + V_{i,o}C_{iB,o})}{(V_o - V_{i,o}e^{-\phi t})} + \frac{(V_{i,o}C_{ir,o} - V_{e,o}C_{eA,o})e^{\phi t_1} - V_{i,o}(C_{iB,o} + C_{ir,o})}{(V_o e^{\phi t} - V_{i,o})} \quad t \geqslant t_1 \tag{68}$$

where t_1 is given by Eq. (57).

4. Special Case: No Chemical Conversions in the Internal and External Phases. The other special case in the model of Case II is that there are no chemical conversions both in the internal and external phases. That is, the solute B is not related to the other solute A; then the solute A concentration does not enter into the picture for the material balance of the solute B. Because of no chemical conversion in the internal phase, the internal reagent is not needed. Mathematically, this special case is given by $K_A=0$, $C_{eA}=0$, and $C_{ir}=0$. Extractions of copper (Cu^{2+}) and hydrocarbons are examples for this case. In this case, the solution of C_{eB} for small ϕ values is obtained from Eq. (55) by letting $t_1=0$.

$$C_{eB} = \left(C_{eB,o} - \frac{(K_B^\circ + \phi V_i)(V_e C_{eB,o} + V_i C_{iB,o})}{(K_B^\circ V_i H_{eB}/H_{iB}) + K_B^\circ V_e + \phi V_e V_i}\right) e^{-\left(\frac{K_B^\circ H_{eB}}{V_e H_{iB}} + \frac{K_B^\circ}{V_i} + \phi\right)t} + \frac{(K_B^\circ + \phi V_i)(V_e C_{eB,o} + V_i C_{iB,o})}{(K_B^\circ V_i H_{eB}/H_{iB}) + K_B^\circ V_e + \phi V_e V_i} \tag{69}$$

Similarly, the solution of C_{eB} for large ϕ values is obtained from Eq. (63) by letting $t_1 = 0$.

$$C_{eB} = \frac{(V_{e,o}C_{eB,o} + V_{i,o}C_{iB,o})}{V_o} + \left[C_{eB,o} - \left(\frac{V_{e,o}C_{eB,o} + V_{i,o}C_{iB,o}}{V_o}\right)\right]$$

$$\left[\left(\frac{V_o-V_{i,o}}{V_o e^{\phi t}-V_{i,o}}\right)^{\left(1+\frac{K_B^\circ H_{eB}}{\phi V_o H_{iB}}\right)} e^{-\frac{K_B^\circ}{\phi V_{i,o}}(e^{\phi t}-1)}\right]$$

$$+\left[G\left(1-\frac{H_{eB}}{H_{iB}}\right)\frac{K_B^\circ(V_{e,o}C_{eB,o}+V_{i,o}C_{iB,o})}{\phi V_o}\right]$$

$$\left[(V_o e^{\phi t}-V_{i,o})^{-\left(1+\frac{K^\circ_B H_{eB}}{\phi V_o H_{iB}}\right)} e^{-\frac{K_B^\circ}{\phi V_{i,o}}e^{\phi t}}\right] \tag{70}$$

where G is expressed by Eq. (62). For hydrocarbon extractions, the ratio of H_{eB} to H_{iB} normally equals unity, which reduces Eq. (70) to the following equation.

$$C_{eB} = \frac{(V_{e,o}C_{eB,o}+V_{i,o}C_{iB,o})}{V_o}$$

$$+\left\{\left[C_{eB,o}-\frac{(V_{e,o}C_{eB,o}+V_{i,o}C_{iB,o})}{V_o}\right]\left(\frac{V_o-V_{i,o}}{V_o e^{\phi t}-V_{i,o}}\right)^{\left(1+\frac{K_B^\circ}{\phi V_o}\right)}\right.$$

$$\left. e^{-\frac{K_B^\circ}{\phi V_{i,o}}(e^{\phi t}-1)}\right\} \tag{71}$$

Equation (71) further reduces to the solution given by Boyadzhiev et al.[8] for $C_{eB,o}=0$, Eq. (17) in their paper.

5. Chemical Reagent Consumption. The total consumption amount of the chemical reagent in the internal phase, W_i, for the conversion of Solute A into Solute B and for the loss due to breakage can be calculated from Eq. (33). For small ϕ values, substitution of Eqs. (48) and (50) into Eq. (33) and integration of the resultant equation give the result of W_i for $0 \leq t \leq t_1$, where t_1 is given by Eq. (49).

$$W_i = \frac{V_e(K_A^\circ C_{eA,o}+\phi V_i C_{ir,o})}{(K_A^\circ+\phi V_e)}\left[1-e^{-\left(\frac{K_A^\circ}{V_e}+\phi\right)t}\right] \qquad 0 \leq t \leq t_1 \tag{72}$$

For $t \geq t_1$, the consumption of the internal reagent is due to breakage only since $C_{eA}=0$. The total consumption amount from $t=0$ to $t \geq t_1$ can be expressed by the following equation:

$$W_i = W_i\Big|_{t_1} + \int_{t_1}^{t} \phi V_1 C_{ir}\Big|_{t_1} dt \qquad t \geq t_1 \tag{73}$$

where $W_i\Big|_{t_1}$ and $C_{ir}\Big|_{t_1}$ can be obtained from Eqs. (72) and (50), respectively. Hence, Eq. (73) leads to the following result:

$$W_i = \frac{V_e(K_A^\circ C_{eA,o}+\phi V_i C_{ir,o})}{(K_A^\circ+\phi V_e)} + \frac{\phi K_A^\circ(V_i C_{ir,o}-V_e C_{eA,o})}{(K_A^\circ+\phi V_e)}(t-t_1)$$
$$+ \frac{V_e(K_A^\circ C_{eA,o}+\phi V_i C_{ir,o})}{(K_A^\circ+\phi V_e)} e^{-(\frac{K_A^\circ}{V_e}+\phi)t_1}[\phi(t-t_1)-1] \qquad t \geq t_1 \tag{74}$$

Prediction of the Performance of a Continuous Extraction Operation from a Batch Experiment

From a batch extraction experiment, the mass transfer coefficients, K_A, K_B, and ϕ, can be obtained from Eq. (21) with a small ϕ value and Eq. (24) with a large ϕ value for Case I extraction or from Eqs. (48), (51), and (55) with a small ϕ value and Eqs. (56), (59), and (63) with a large ϕ value for Case II extraction. For the special case of Case I extraction, the mass transfer coefficients can be obtained from Eqs. (26) and (27) whereas, for special cases of Case II extraction, the mass transfer coefficients can be obtained from Eqs. (65) to (71). From the mass transfer coefficients, the performance of a continuous extraction operation (under the same mixing condition as that for the batch experiment) can be predicted. This is illustrated with Case I extraction with small ϕ values. Two types of continuous operations, namely, (1) n-stage mixer-settlers and (2) mechanically agitated columns, are considered in the following. In each type, both cocurrent and countercurrent modes of operation are to be included.

1. Extraction with n-Stage Mixer-Settlers. (a) Cocurrent Mode. Figure 5(a) shows the schematic for the cocurrent mode of the continuous extraction operation with n-stage mixer-settlers of the mixer volume V each stage. The volumetric flow rates for the external, internal, and membrane phases are v_e, v_i, and v_m, respectively. The flow rates are assumed constant in the case under consideration. The material balance for the first stage can be expressed in the following two equations.

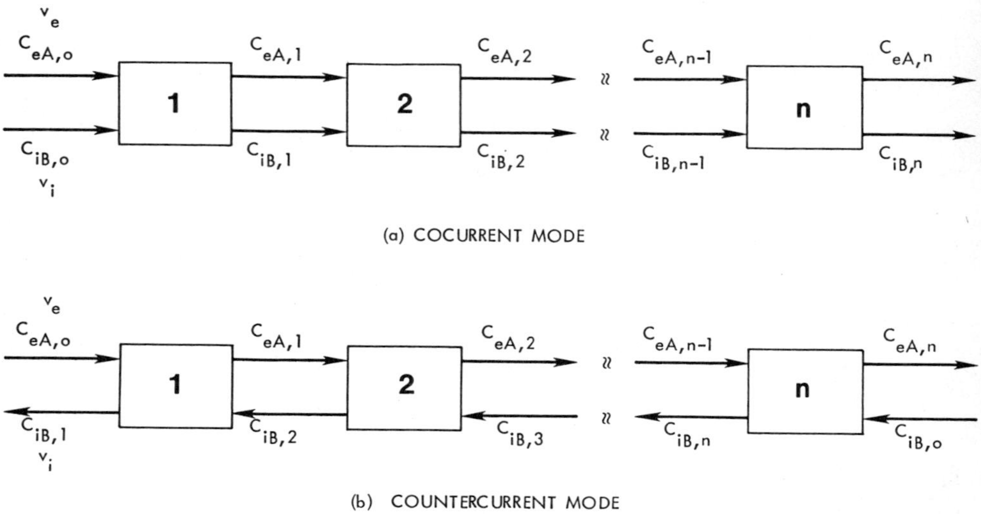

Fig. 5. Schematic of continuous extraction operations with n-stage mixer-settlers.

$$v_e(C_{eA,0}-C_{eA,1}) = v_i(C_{iB,1}-C_{iB,0}) \tag{75}$$

$$v_e(C_{eA,0}-C_{eA,1}) = K_A V(\frac{v_m+v_i}{v_e+v_m+v_i})C_{eA,1} - K_{B\phi} V(\frac{v_m+V_i}{v_e+v_m+v_i})C_{iB,1} \tag{76}$$

Rearrangement of Eq. (76) leads to the following equation:

$$C_{eA,0} = \frac{V}{v_e}(\frac{v_m+v_i}{v_e+v_m+v_i})(K_A C_{eA,1}-K_{B\phi}C_{iB,1})+C_{eA,1} \tag{77}$$

Substitution of Eq. (76) into Eq. (75) gives the following equation:

$$C_{iB,o} = -\frac{V}{v_i}(\frac{v_m+v_i}{v_e+v_m+v_i})(K_A C_{eA,1}-K_{B\phi}C_{iB,1})+C_{iB,1} \tag{78}$$

Multiplications of Eq. (77) by K_A and Eq. (78) by $K_{B\phi}$ and subtraction of one resultant equation from the other give the following equation.

$$\frac{(K_A C_{eA,0}-K_{B\phi}C_{iB,1})}{(K_A C_{eA,1}-K_{B\phi}C_{iB,1})} = V(\frac{v_m+v_i}{v_e+v_m+v_i})(\frac{K_A}{v_e}+\frac{K_{B\phi}}{v_i}) + 1 \tag{79}$$

The same type of equation can be derived for the other stages. Multiplication of these equations together results in the following equation:

$$\frac{(K_A C_{eA,0} - K_{B\phi} C_{iB,0})}{(K_A C_{eA,n} - K_{B\phi} C_{iB,n})} = \left[V\left(\frac{v_m + v_i}{v_e + v_m + v_i}\right)\left(\frac{K_A}{v_e} + \frac{K_{B\phi}}{v_i}\right) + 1\right]^n \tag{80}$$

This equation can be used to calculate the outlet concentrations, $C_{eA,n}$ and $C_{iB,n}$ or the recovery of solute for a given extractor volume V each stage with the aid of the overall material balance given by the following equation.

$$v_e(C_{eA,0} - C_{eA,n}) = v_i(C_{iB,n} - C_{iB,0}) \tag{81}$$

Rearrangement of Eq. (80) allows the calculation of the total extractor volume of the n-stage mixers, V_t, from given inlet and outlet concentrations or a given solute recovery.

$$V_t = nV = \frac{n\left[\left(\frac{K_A C_{eA,o} - K_{B\phi} C_{iB,o}}{K_A C_{eA,n} - K_{B\phi} C_{iB,n}}\right)^{\frac{1}{n}} - 1\right]}{\left(\frac{v_m + v_i}{v_e + v_m + v_i}\right)\left(\frac{K_A}{v_e} + \frac{K_{B\phi}}{v_i}\right)} \tag{82}$$

This equation allows the calculation of the total residence time or contact time for extraction, t, given by the following equation.

$$t = \frac{V_t}{(v_e + v_m + v_i)} \tag{83}$$

(b) Countercurrent Mode. Figure 5(b) shows the schematic for the countercurrent mode of the continuous extraction operation with n-stage mixer-settlers of the mixer volume V each stage. The material balance for the first stage can be expressed in the following two equations.

$$v_e(C_{eA,0} - C_{eA,1}) = v_i(C_{iB,1} - C_{iB,2}) \tag{84}$$

$$v_e(C_{eA,0} - C_{eA,1}) = K_A V\left(\frac{v_m + v_i}{v_e + v_m + v_i}\right) C_{eA,1} - K_{B\phi} V\left(\frac{v_m + v_i}{v_e + v_m + v_i}\right) C_{iB,1} \tag{85}$$

Following the similar algebric procedure described in the cocurrent mode leads to the following equation.

$$\frac{(K_A C_{eA,0} - K_{B\phi} C_{iB,1})}{(K_A C_{eA,n} - K_{B\phi} C_{iB,0})} = \left(\frac{V\left(\frac{v_m + v_i}{v_e + v_m + v_i}\right)\frac{K_A}{v_e} + 1}{V\left(\frac{v_m + v_i}{v_e + v_m + v_i}\right)\frac{K_{B\phi}}{v_i} + 1}\right)^n \tag{86}$$

This equation allows the calculation of the outlet concentrations, $C_{eA,n}$ and $C_{iB,1}$ or the solute recovery for a given extractor volume V each stage with the aid of the overall material balance given by the following equation.

$$v_e(C_{eA,0}-C_{eA,n}) = v_i(C_{iB,1}-C_{iB,0}) \tag{87}$$

Rearrangement of Eq. (86) allows the calculation of the total extractor volume of the n-stage mixers from given inlet and outlet concentrations or a given solute recovery.

$$V_t = \frac{n\left[\left(\dfrac{K_A C_{eA,0}-K_{B\phi}C_{iB,1}}{K_A C_{eA,n}-K_{B\phi}C_{iB,0}}\right)^{\frac{1}{n}} - 1\right]}{\left(\dfrac{v_m+v_i}{v_e+v_m+v_i}\right)\left|\dfrac{K_A}{v_e} - \dfrac{K_{B\phi}}{v_i}\left(\dfrac{K_A C_{eA,0}-K_{B\phi}C_{iB,1}}{K_A C_{eA,n}-K_{B\phi}C_{iB,0}}\right)^{\frac{1}{n}}\right|} \tag{88}$$

Equations (83) and (88) can be used to calculate the total residence time or contact time for extraction.

2. Extraction with a Mechanically Agitated Column. (a) Cocurrent Mode. Figure 6(a) shows the schematic for the cocurrent mode of the continuous extraction operation with a mechanically agitated column of the total extractor volume V_t. The material balance for the element with a height of Δz in the column gives the following equation for constant flow rates.

$$v_e \frac{d\,C_{eA,z}}{d\,z} = -K_A S\left(\frac{v_m+v_i}{v_e+v_m+v_i}\right)C_{eA,z}+K_{B\phi}S\left(\frac{v_m+v_i}{v_e+v_m+v_i}\right)C_{iB,z} \tag{89}$$

where S is the cross-sectional area of the column, $C_{eA,z}$ and $C_{iB,z}$ are the concentrations at the position z for the external and internal phases, respectively, and z is the vertical distance from the bottom of the column. The external phase and the emulsion are introduced at the bottom of the column. The material balance from the bottom of the column to the position z results in the following equation.

$$v_e(C_{eA,0}-C_{eA,z}) = v_i(C_{iB,z}-C_{iB,0}) \tag{90}$$

Substitution of Eq. (90) into Eq. (89) for $C_{iB,z}$ and integration of the resultant equation for the entire column of the height L lead to the following equation:

$$\frac{(K_A C_{eA,0} - K_{B\phi} C_{iB,0})}{(K_A C_{eA} - K_{B\phi} C_{iB})} = e^{V_t \left(\frac{v_m + v_i}{v_e + v_m + v_i}\right)\left(\frac{K_A}{v_e} + \frac{K_{B\phi}}{v_i}\right)} \tag{91}$$

where C_{eA} and C_{iB} are the outlet concentrations for the external and internal phases, respectively. This equation allows the calculation of these outlet concentrations or the solute recovery for a given total extractor volume V_t with the aid of Eq. (90), in which z equals L, i.e., $C_{eA,z}=C_{eA}$ and $C_{iB,z}=C_{iB}$. Rearrangement of Eq. (91) gives the expression for the calculation of the total extractor volume from given inlet and outlet concentrations or a given solute recovery.

$$V_t = \frac{1}{(K_A/v_e + K_{B\phi}/v_i)} \left(\frac{v_e + v_m + v_i}{v_m + v_i}\right) \ln\left(\frac{K_A C_{eA,o} - K_{B\phi} C_{iB,0}}{K_A C_{eA} - K_{B\phi} C_{iB}}\right) \tag{92}$$

Equations (83) and (92) can be used to calculate the total residence time or contact time for extraction.

(b) Countercurrent Mode. Figure 6(b) shows the schematic for the countercurrent mode of the continuous extraction operation with a mechanically agitated column. Similar to the cocurrent mode

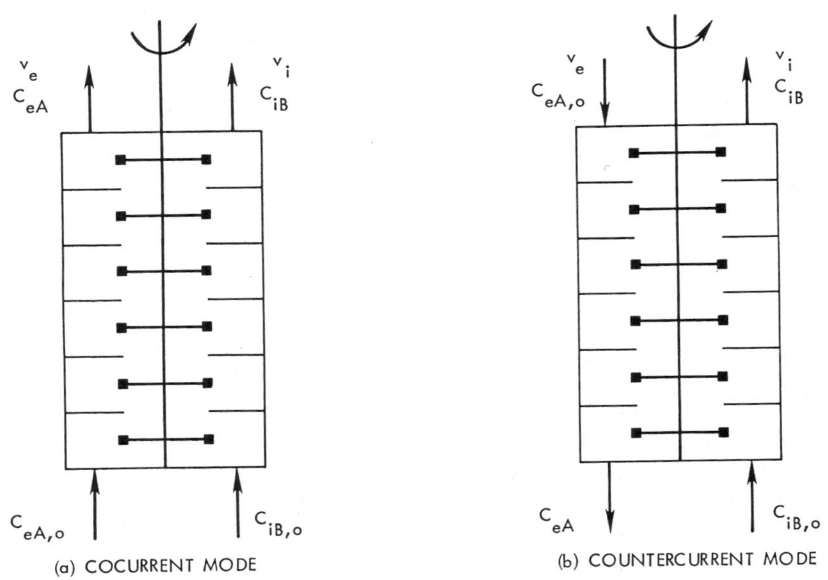

Fig. 6. Schematic of continuous extraction operations with mechanically agitated columns.

described above, the material balance gives the following two equations:

$$v_e \frac{d\, C_{eA,z}}{d\, z} = K_A S \left(\frac{v_m+v_i}{v_e+v_m+v_i}\right) C_{eA,z} - K_{B\phi} S \left(\frac{v_m+v_i}{v_e+v_m+v_i}\right) C_{iB,z} \qquad (93)$$

$$v_e(C_{eA,0}-C_{eA,z}) = v_i(C_{iB}-C_{iB,z}) \qquad (94)$$

Substituting Eq. (94) into Eq. (93) for $C_{iB,z}$ and integrating the resultant equation for the entire column, we arrive at the following result:

$$\frac{(K_A C_{eA,0}-C_{B\phi} C_{iB})}{(K_A C_{eA}-K_{B\phi} C_{iB,0})} = e^{V_t \left(\frac{v_m+v_i}{v_e+v_m+v_i}\right)\left(\frac{K_A}{v_e} - \frac{K_{B\phi}}{v_i}\right)} \qquad (95)$$

where C_{eA} and C_{iB} are the outlet concentrations for the external and internal phases, respectively. This equation allows the calculation of these outlet concentrations or the solute recovery for a given total extractor volume V_t with the aid of Eq. (94) corresponding to the overall material balance for the entire column. Rearrangement of Eq. (95) leads to the expression for the calculation of the total extractor volume from given inlet and outlet concentrations or a given solute recovery.

$$V_t = \frac{1}{(K_A/v_e - K_{B\phi}/v_i)} \left(\frac{v_e+v_m+v_i}{v_m+v_i}\right) \ln\left(\frac{K_A C_{eA,0}-K_{B\phi} C_{iB}}{K_A C_{eA}-K_{B\phi} C_{iB,0}}\right) \qquad (96)$$

Equations (83) and (96) can be employed to calculate the total residence time or contact time for extraction.

RESULTS AND DISCUSSION

With the model of liquid membrane extraction, we can obtain the mass transfer coefficients from a batch extraction experiment and predict the performance of a continuous extraction operation based on the mass transfer coefficients. In this section we discuss the applications of the model to the extractions of chromium and phenol, classification of liquid membrane extraction cases, effects of leakage and external reagent on liquid membrane extraction, chemical reagent consumption, comparison of countercurrent and cocurrent continuous extraction operations, and sizing extractor volumes.

Applications of the Model to the Extractions of Chromium and Phenol

To illustrate the applications of the model, the extraction of chromium from its acidic solution is used as the example for Case I of the model whereas the extraction of phenol from waste water is employed as the example for Case II of the model.

1. Case I: Chromium Extraction. The application of Case I of the model for the batch extraction of chromium is shown in Fig. 7. The experimental data are those reported by Hochhauser

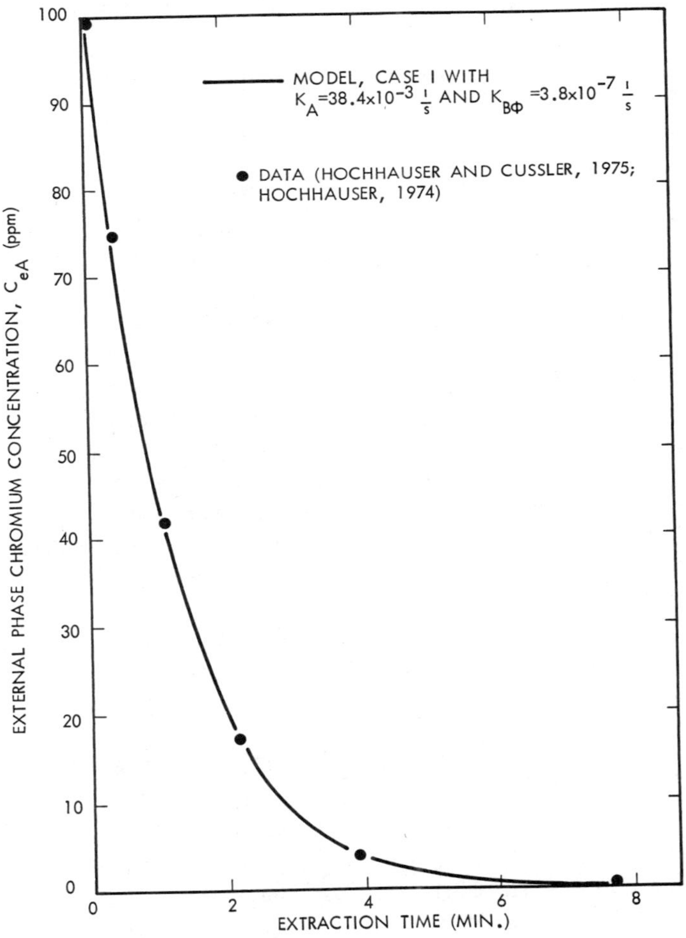

Fig. 7. Application of the model for the extraction of chromium.

Table 1. Experimental Conditions for the Extraction of Chromium[12,44]

Initial Chromium Concentration (as $H_2Cr_2O_7$)	$C_{eA,0}$ = 0.00385 N (100 ppm Cr^{6+})
Initial NaOH Concentration	$C_{ir,0}$ = 0.1 N (0.4% NaOH)
pH in the External Phase Adjusted with H_2SO_4	pH = 1.6
Emulsion	$V_{i,o}/(V_m+V_{i,o})$ = 0.286
Treat Ratio	$V_{e,o}/(V_m+V_{i,o})$ = 2.86
Surfactant System	1% Span 80 in light mineral oil
Complexing Carrier	0.05 M tridodecylamine

and Cussler[12] and Hochhauser.[44] Their experimental conditions for the chromium extraction are summarized in Table 1. In the external aqueous phase whose pH was adjusted with H_2SO_4 to 1.6, the chromium presumably existed as dichromate such as $H_2Cr_2O_7$, which is the solute A in Case I of the model. The initial concentration of $H_2Cr_2O_7$ in the external phase was 3.85×10^{-3} N, i.e., 100 ppm Cr^{6+}. The chromium was extracted with the liquid membrane emulsion containing tridodecylamine in the oil phase, which complexed with $H_2Cr_2O_7$. The complex was then stripped and the dichromate converted to CrO_4^{2-} by the internal reagent NaOH in the internal phase. In this case CrO_4^{2-} is the solute B in Case I of the model. The solid curve in Fig. 7 is the result calculated from Eq. (21) of Case I model with $K_A = 38.4 \times 10^{-3}$ 1/s and $K_{B\phi} = 3.8 \times 10^{-7}$ 1/s, which were obtained by least square fitting of this equation to the experimental data. This figure shows the fit of the model to the data is very good.

2. Case II. Phenol Extraction. The application of Case II of the model for phenol extraction is shown in Fig. 8, in which the experimental data are those reported by Li and Shrier.[21] Their experimental conditions for the extraction of phenol from waste water are summarized in Table 2. In the external phase, the initial phenol concentration was 0.01064 N, i.e., 1000 ppm. Phenol was extracted from the external phase into the globules of the liquid membrane emulsion where the solute diffused through the oil-type

Table 2. Experimental Conditions for the Extraction of Phenol[21]

Initial Phenol Concentration	$C_{eA,0}$ = 0.01064 N
Initial NaOH Concentration	$C_{ir,o}$ = 0.125 N (0.5% NaOH)
Emulsion	$V_{i,0}/(V_m+V_{i,o})$ = 0.333
Treat Ratio	$V_{e,o}/(V_m+V_{i,o})$ = 0.833
Mixing Rate	100 RPM
Temperature	25°C
Surfactant System	Span 80 in S100N Oil

membrane phase and reacted with the internal reagent NaOH to become sodium phenolate. The phenolate leaking out from the internal phase, via breakage, to the external phase remained as the same species since there was no external reagent for the conversion of the phenolate into phenol. In this case phenol is the solute A and phenolate the solute B for Case II model, in which K_B=0 since phenolate can not diffuse through the oil-type membrane phase. Li and Shrier's data shown in Fig. 8 indicate strong effect of surfactant (Span 80, sorbitan monooleate manufactured by ICI Americas) concentration on extraction results. The extraction for 0.1% surfactant concentration was very poor. The fit of Eqs. (56), (67) and (68) to the extraction data for this given surfactant concentration gives the phenol mass transfer coefficient, K_A=2.4x10^{-3} 1/s and the breakage coefficient, ϕ=0.3x10^{-3} 1/s. This breakage coefficient is very large; the breakage of the internal phase given by Eq. (11) is more than 92% for an extraction time of 143 min. With complete breakage, the total concentration in the external phase (the phenolate concentration since presumably all the phenol is converted to phenolate) should be diluted to 714 ppm because of the volume increase on the external phase due to the breakage. As shown in this figure, at the large time the model approaches this concentration value. The fit of the model to the extraction data for 0.5% surfactant concentration results in K_A=50x10^{-3} 1/s and ϕ=0.085x10^{-3} 1/s. The ϕ value is much lower than the one mentioned above but is still quite high. The breakage of the internal phase

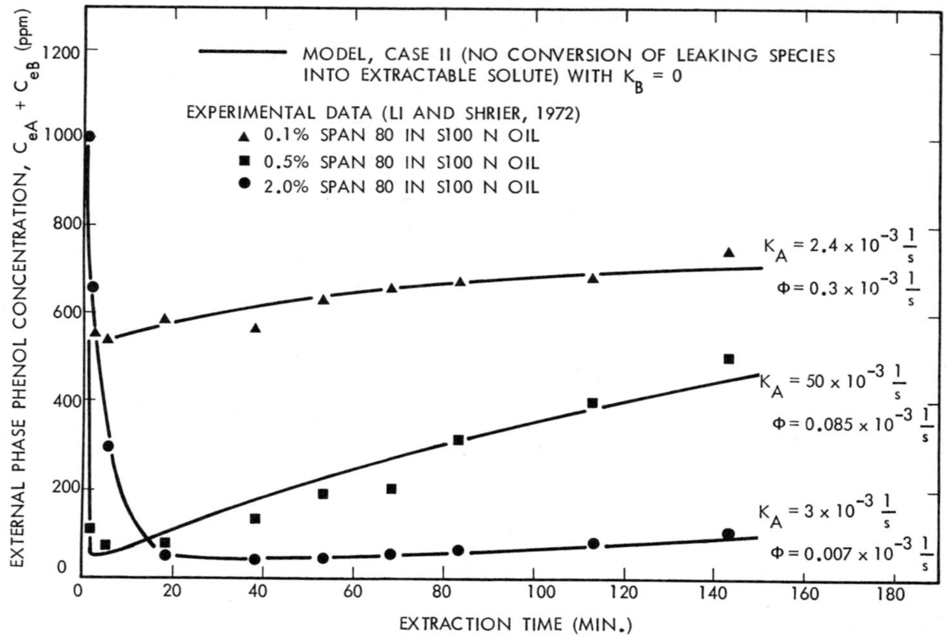

Fig. 8. Application of the model for the extraction of phenol.

given by Eq. (11) for this ϕ value is 52% at the extraction time of 143 min. The extraction result for 2% surfactant concentration is quite good. The fit of the model to the extraction data gives $K_A=3\text{x}10^{-3}$ 1/s and $\phi=0.007\text{x}10^{-3}$ 1/s. The ϕ value is small; the breakage calculated from Eq. (11) is less than 6% at the extraction time of 143 min. This figure shows the fit of the model to the experimental data is good for a wide range of breakage.

Classification of Liquid Membrane Extraction Cases

In addition to the examples just discussed above, some examples in classification of liquid membrane extraction cases are given in Table 3. In Case I, phenol, acetic acid, benzoic acid, or hydrogen sulfide as the solute A in the external phase can be extracted into the internal phase where the solute A reacts with the internal reagent, a base, and become ionic species, ϕO^-, CH_3COO^-, ϕCOO^-, or S^{2-}, respectively, as the solute B. The ionic species leaking out from the internal phase, via breakage, to the external phase can be converted to the original solute A by the external reagent, an

Table 3. Examples in Classification of Liquid Membrane Extraction Cases

Case	Extraction	External Phase			Internal Phase		Mass Transfer Coefficients
		External Reagent	Solute A	Solute B	Internal Reagent	Solute B	
I	Chromium	Acid	$H_2Cr_2O_7$	None	Base	$C_rO_4^{2-}$	$K_B \cong 0$
	Phenol	Acid	ϕOH	None	Base	ϕO^-	$K_B=0$
	Acetic Acid	Acid	CH_3COOH	None	Base	CH_3COO^-	$K_B=0$
	Benzoic Acid	Acid	$\phi COOH$	None	Base	ϕCOO^-	$K_B=0$
	Hydrogen Sulfide	Acid	H_2S	None	Base	S^{2-}	$K_B=0$
	Ammonia	Base	NH_3	None	Acid	NH_4^+	$K_B=0$
II	Phenol	None	ϕOH	ϕO^-	Base	ϕO^-	$K_B=0$
	Acetic Acid	None	CH_3COOH	CH_3COO^-	Base	CH_3COO^-	$K_B=0$
	Benzoic Acid	None	$\phi COOH$	ϕCOO^-	Base	ϕCOO^-	$K_B=0$
	Hydrogen Sulfide	None	H_2S	S^{2-}	Base	S^{2-}	$K_B=0$
	Ammonia	None	NH_3	NH_4^+	Acid	NH_4^+	$K_B=0$
	Copper	None	None	Cu^{2+}	Acid	Cu^{2+}	$K_A=0$
	Zinc	None	None	Zn^{2+}	Acid	Zn^{2+}	$K_A=0$
	Nickel	None	None	Ni^{2+}	Acid	Ni^{2+}	$K_A=0$
	Cobalt	None	None	Co^{2+}	Acid	Co^{2+}	$K_A=0$
	Hydrocarbons	None	None	Hydrocarbons	None	Hydrocarbons	$K_A=0$

acid. The solute A can then be re-extracted. On the other hand, a base, such as ammonia, as the solute A in the external phase can be extracted into the internal phase where it reacts with the internal reagent, an acid, and becomes an ionic species, such as NH_4^+, as the solute B. The ionic species leaking out from the internal phase, via breakage, to the external phase can be converted back to the original solute A by the external reagent, a base. The solute B, an ionic species, can not diffuse through an oil-type membrane phase and thus K_B can be considered zero unless B can complex with a carrier in the membrane phase.

In Case II, phenol[10,21,26,40], acetic acid[26], benzoic acid[45], hydrogen sulfide, or ammonia[31,46] as the solute A can be extracted, in a way similar to Case I, into the internal phase where the solute A is converted into ionic species as the solute B, which can not diffuse through the membrane phase, i.e., K_B=0. But the solute B leaking out into the external phase remains as an ionic species since there is not external reagent for the conversion of B into A. Other examples belonging to Case II are some hydrometallurgical extractions, such as copper[13-18,45], zinc[45], nickel[17,39] and cobalt[17], and hydrocarbon extractions.[3-5,8,9] These examples belong to the special case of Case II model in which there are no solute A related to B and no internal reagent, i.e., K_A=0, C_{eA}=0 and C_{ir}=0. The difference between hydrocarbon and hydrometallurgical extractions is that the ratio of distribution coefficients, H_{eB}/H_{iB}, is normally equal to unity for hydrocarbon extractions but not for hydrometallurgical extractions.

Effects of Leakage and External Reagent on Liquid Membrane Extraction

As discussed above, breakage decreases the recovery of solute such as phenol in Case II of liquid membrane extraction, in which there is no external reagent for the conversion of the species, phenolate, leaking out into the extractable form, phenol, in the external phase. However, an external reagent in Case I of liquid membrane extraction can play a significant role on the solute recovery. The solute recovery, ε, for the extraction of solute from the external phase into the internal phase is defined, in Case II, in the following equation.

$$\varepsilon = 1 - \frac{V_e(C_{eA}+C_{eB})}{V_{e,o}(C_{eA,o}+C_{eB,o})} \qquad (97)$$

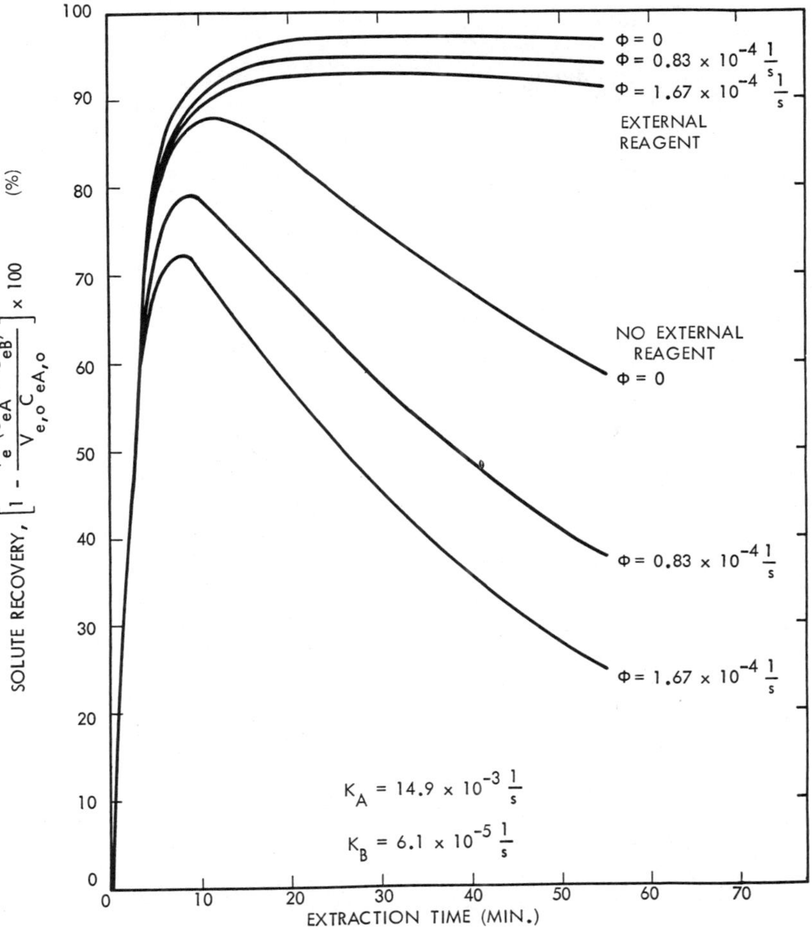

Fig. 9. Effects of leakage and external reagent on liquid membrane extraction.

In Case I, i.e., C_{eB}=0, this equation reduces to the following equation:

$$\varepsilon = 1 - \frac{V_e C_{eA}}{V_{e,o} C_{eA,o}} \tag{98}$$

The effects of leakage and an external reagent on the recovery of solute in liquid membrane extraction are shown in Fig. 9. The curves for the extraction with an external reagent are obtained from Eqs. (21) and (24) of Case I of the model whereas the curves for the

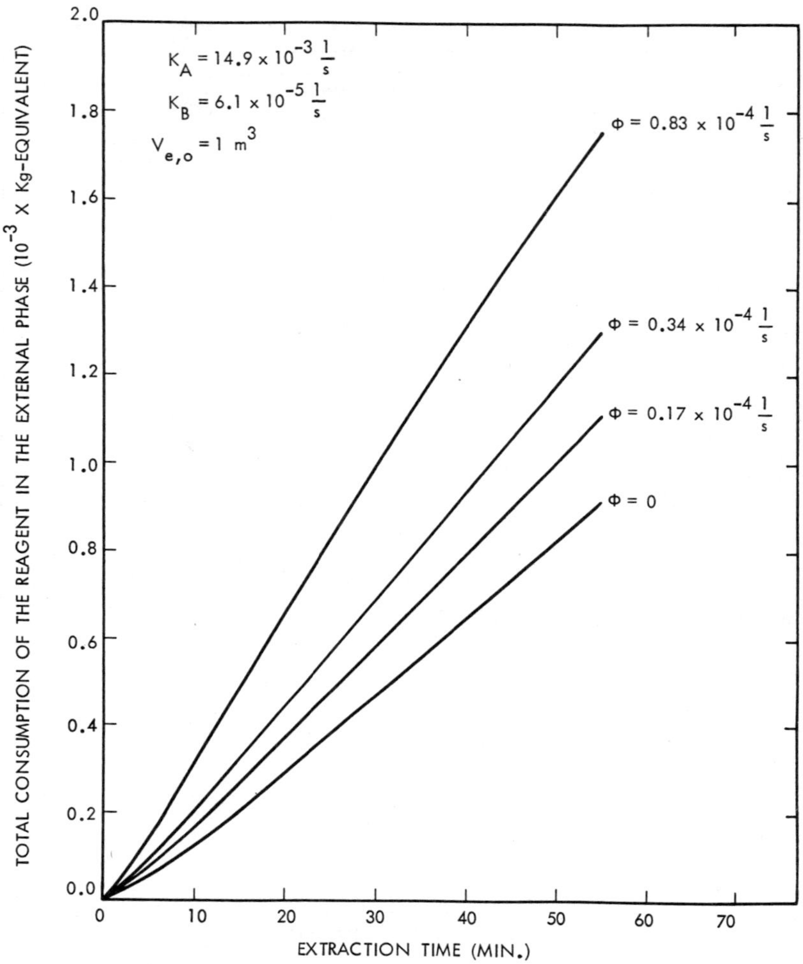

Fig. 10. Consumption of the external reagent for the conversion of leaking species into extractable solute.

extraction with no external reagent are calculated results from Eqs. (48), (51), (55), (56), (59) and (63) of Case II of the model. In the calculation the extraction conditions used are K_A=14.9x10^{-3} 1/s, K_B=6.1x10^{-5} 1/s, $V_{e,o}/(V_m+V_{i,o})$=3, $V_{i,o}/(V_m+V_{i,o})$=0.333, $C_{eA,o}$= 1.68x10^{-3} N, $C_{eB,o}$=0, $C_{iB,o}$=0, $C_{ir,o}$=3.6x10^{-2} N, and H_{eB}/G_{iB}=1. This figure shows that the effect of breakage on the solute recovery is not very significant for the extraction with an external reagent. However, for the extraction without an external reagent, the solute recovery is poor and the breakage influences the extraction significantly. Even for the cases without breakage, the leakage due to diffusional transport (K_B) can decrease the solute recovery greatly for the extraction without an external reagent but not for the extraction with an external reagent. Thus, an external reagent can effect the liquid membrane extraction greatly in enhancing the solute recovery and minimizing the effect of leakage including breakage.

Chemical Reagent Consumption

Although an external reagent can minimize leakage effect and enhance solute recovery, the cost of the external reagent should be considered. The total consumption amount of the external reagent can be calculated from Eqs. (28) and (32). Figure 10 shows the calculated results of the total consumption amount, in which the extraction conditions used in the calculation are the same as those mentioned above. The external reagent consumption increases with increasing extraction time and breakage.

Comparison of Countercurrent and Cocurrent Continuous Extraction Operations

As described above, the performance of a continuous extraction operation can be predicted by the use of mass transfer coefficients, which can be obtained from a batch extraction experiment. Figures 11 and 12 show the predicted results for the continuous extraction operations, in the case with an external reagent (Case I), with 2-stage mixer-settlers and with mechanically agitated columns, respectively. In Fig. 11 the curve for the 2-stage mixer-settler extraction operation with the countercurrent mode is obtained from Eqs. (86) and (87) whereas the curve for the cocurrent mode is calculated from Eqs. (80) and (81). In the calculations the breakage coefficient (ϕ) used is 1.52×10^{-4} 1/s and the other extraction conditions used are the same as those just mentioned above (in Figs. 9 and 10). That is, $K_A=14.9\times10^{-3}$ 1/s, $K_{B\phi}= 1.12\times10^{-4}$ 1/s, and the ratios of the volumetric flow rates in the continuous processes are the same as the ratios of the volumes in the batch extraction. Figure 11 shows that the maximum solute recovery for the 2-stage cocurrent process is limited to 94%, which is also indicated from Fig. 9 for the batch extraction with an external reagent. This limited recovery is due to the leakage problem. However, by the use of the 2-stage countercurrent process, this leakage problem can be overcome and the solute recovery higher than 94% can be achieved as shown in Fig. 11. The countercurrent process is much more effective than the cocurrent process. At 90% solute recovery, the effectiveness expressed as the

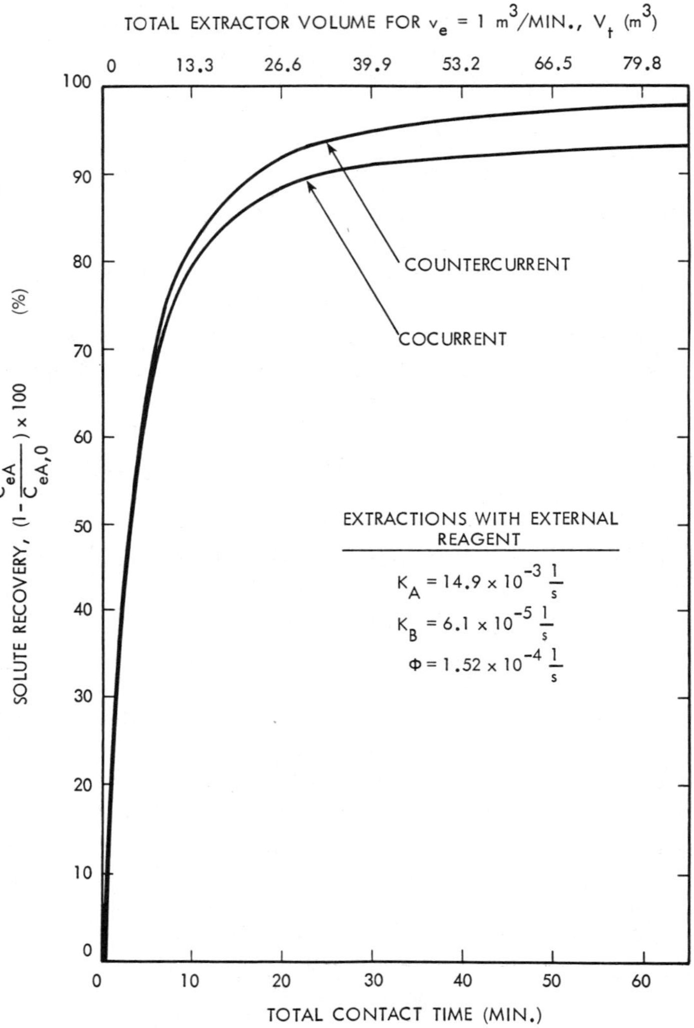

Fig. 11. Comparison of the countercurrent and cocurrent continuous extraction operations with 2-stage mixer-settlers.

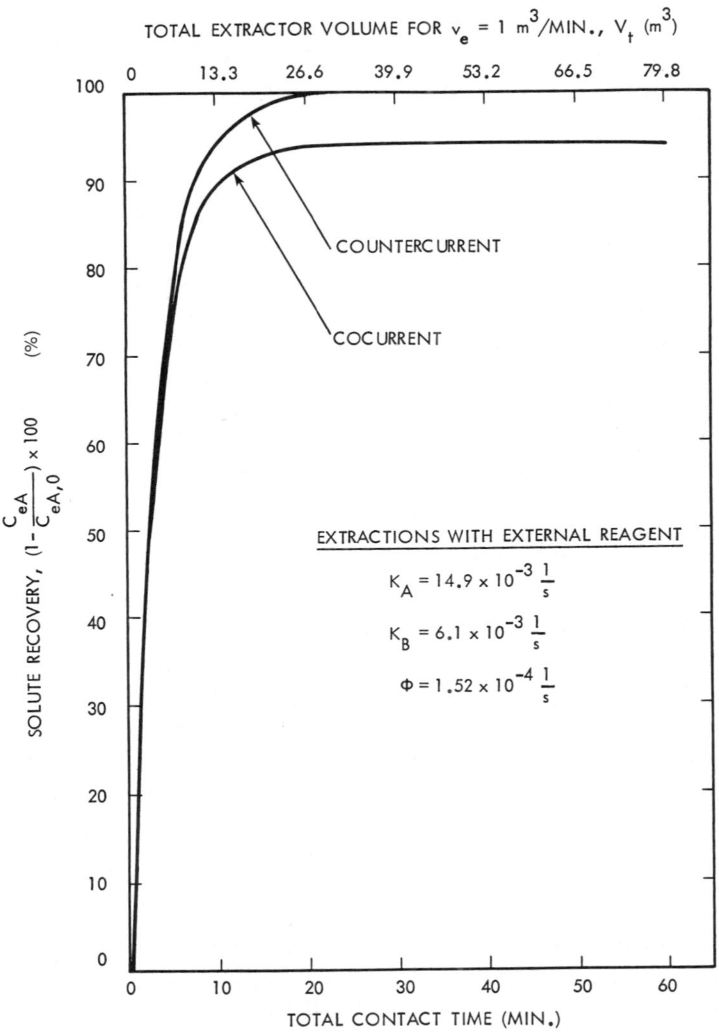

Fig. 12. Comparison of the countercurrent and cocurrent continuous extraction operations with mechanically agitated columns.

ratio of the contact time for the cocurrent process to the contact time for the countercurrent process is 1.5. The effectiveness increases to 2.8 at 93% recovery and becomes even greater as the recovery approaches 94%.

In Fig. 12 the curve for the countercurrent extraction process with a mechanically agitated column and the curve for the cocurrent extraction process are calculated from Eqs. (95) and (91), respectively. Similarly, in the calculations the breakage coefficient and the other extraction conditions used are the same as those described above for the mixer-settler operations. This figure shows that the maximum solute recovery for the cocurrent process is again limited to 94% because of the leakage problem. This leakage problem can be overcome effectively and the solute recovery approaching 100% achieved by the use of the countercurrent process. This countercurrent process can be more effective than the countercurrent process with 2-stage mixer-settlers as indicated from Figs. 11 and 12.

Sizing Extractor Volumes

The model can be used for sizing the extractor volumes for continuous liquid-membrane extraction processes. As described above, the extractor volumes for the countercurrent process and the cocurrent process with n-stage mixer settlers can be calculated from Eqs. (88) and (82), respectively. And the extractor volumes for the countercurrent process and the cocurrent process with mechanically agitated columns can be computed from Eqs. (96) and (92), respectively. The extractor volumes for an external-phase flow rate of 1 m^3/min. are also shown in Fig. 11 for 2-stage mixer-settler processes and in Fig. 12 for mechanically agitated column processes.

CONCLUSIONS

A general model containing overall mass-transfer coefficients for extraction and leakage has been developed for extractions by liquid surfactant membranes (LM). In this model two extraction cases have been considered--one has chemical reagent in the external feed phase to convert the leaking species into an extractable form for re-extraction into the LM emulsion and the other does not have the reagent in the external phase. Most LM extraction systems can be classified into these two cases. The general model is readily reduced to be applicable to special extraction cases, such as hydrocarbon and hydrometallurgical extractions, and removal of trace contaminants from waste streams. The utility of the model has been demonstrated successfully to experimental data in the extractions of chromium and phenol from their aqueous solutions. The model can fit satisfactorily extraction data with a wide range of breakage.

Mass transfer equations to calculate solute recovery, contact time, extractor size, and chemical reagent consumption in relation to overall mass transfer coefficients have been derived and can be used to predict the extraction performance of a continuous LM process from a simple batch extraction experiment. It is shown from the model that the reagent for chemical conversion in the external phase can enhance extraction greatly and can minimize the effect of leakage, including breakage, on extraction significantly. The leakage effect can also be minimized significantly via countercurrent continuous extraction, which is indicated from the model to be more effective than cocurrent continuous extraction.

NOTATION

a = mass transfer area per unit volume of the emulsion, 1/m
A = solute existing only in the external phase
A_e = total mass transfer area of the emulsion globules in the external phase, m^2
A_i = total mass transfer area of the internal phase, m^2
B = only permeate in the internal phase
C_e = solute concentration in the external phase, N
C_{eA} = Solute A concentration in the external phase, N
$C_{eA,o}$ = initial or inlet Solute A concentration in the external phase, N
$C_{eA,1}$ = outlet Solute A concentration in the external phase from Stage 1 mixer-settler, N
$C_{eA,n}$ = outlet Solute A concentration in the external phase from Stage n mixer-settler, N
C_{eB} = Solute B concentration in the external phase, N
$C_{eB,o}$ = initial or inlet Solue B concentration in the external phase, N
C_i = solute concentration in the internal phase, N
C_{iB} = Solute B concentration in the internal phase, N
$C_{iB,o}$ = initial or inlet Solute B concentration in the internal phase, N
$C_{iB,1}$ = outlet Solute B concentration in the internal phase from Stage 1 mixer-settler, N
$C_{iB,n}$ = outlet Solute B concentration in the internal phase from Stage n mixer-settler, N
C_{ir} = concentration of the reagent in the internal phase, N

$C_{ir,o}$ = initial concentration of the reagent in the internal phase, N
C_{se} = solute concentration, in the external phase, at the interface between the external and membrane phases, N
C_{sem} = solute concentration, in the membrane phases, at the interface between the external and mnembrane phases, N
C_{si} = solute concentration, in the internal phase, at the interface beween the internal and membrane phases, N
C_{sim} = solute concentration, in the membrane phase, at the interface between the internal and membrane phases, N
H_e = distribution coefficient of the solute between the membrane phase and the external phase at equilibrium
H_{eB} = distribution coefficient of Solute B between the membrane phase and the external phase at equilibrium
H_i = distribution coefficient of the solute between the membrane phase and the internal phase at equilibrium
H_{iB} = distribution coefficient of Solute B between the membrane phase and the internal phase at equilibrium
j_k = mass transfer rate due to diffusional transport, Kg-equivalent/s
j_ϕ = mass transfer rate due to breakage, Kg-equivalent/s
k_e = individual mass transfer coefficient in the external phase, m/s
k_i = individual mass transfer coefficient in the internal phase, m/s
k_m = individual mass transfer coefficient in the membrane phase, m/s
k_{se} = interfacial mass transfer coefficient at the interface adjacent to the external phase, m/s
k_{si} = interfacial mass transfer coefficient at the interface adjacent to the internal phase, m/s
K = overall mass transfer coefficient, 1/s
K_A = overall mass transfer coefficient for Solute A, 1/s
K_A° = $K_A(V_m+V_i)$, Equation (15), m^3/s
K_B = overall mass transfer coefficient for Solute B, 1/s
K_B° = $K_B(V_m+V_i)$, Equation (16), m^3/s
$K_{B\phi}$ = $K_B + \phi(\frac{V_i}{V_m+V_i})$, Equation (22), overall leakage coefficient for solute B, 1/s
L = height of a mechanically agitated column, m
n = number of mixer-settler stages
S = cross-sectional area of a mechanically agitated column, m^2
t = extraction time or total contact time, s
v_e = volumetric flow rate for the external phase, m^3/s
v_i = volumetric flow rate for the internal phase of the emulsion, m^3/s

v_m = volumetric flow rate for the membrane phase of the emulsion, m^3/s
V = extractor volume each stage in n-stage mixer-settlers, m^3
V_e = volume of the external phase, m^3
V_i = total volume of the internal phase, m^3
$V_{i,o}$ = initial, total volume of the internal phase, m^3
V_o = $V_{e,o}+V_{i,o}=V_e+V_i$, Equation (12), m^3
V_m = volume of the membrane phase, m^3
V_t = total extractor volume, m^3
W_e = total consumption amount of the reagent in the external phase, Kg-equivalent
W_i = total consumption amount of the reagent in the internal phase, Kg-equivalent
z = vertical distance from the bottom of a mechanically agitated column, m

Greek Letters

ε = solute recovery
ϕ = breakage coefficient, 1/s

REFERENCES

1. N. N. Li: "Separating Hydrocarbons with Liquid Membranes," U.S. Patent 3,410,794 (1968).
2. T. H. Maugh: Science, 1976, vol. 193, p. 134.
3. N. N. Li: AIChE J., 1971, vol. 17, p. 459.
4. N. N. Li: Ind. Eng. Process Des. Develop., 1971, vol. 10, p. 215.
5. N. D. Shah and T. C. Owens: Ind. Eng. Chem. Prod. Res. Develop., 1972, vol. 11, p. 58.
6. R. P. Cahn and N. N. Li: J. Membrane Sci., 1976, vol. 1, p. 129.
7. R. P. Cahn and N. N. Li: in Membrane Separation Processes, P. Meares, Ed., Chap. 9, pp. 327-349, Elsevier Scientific Publishing Co., Amsterdam-New York (1976).
8. L. Boyadzhiev, T. Sapundzhiev, and E. Bezenshek: Sep. Sci., 1977, vol. 12, p. 541.
9. P. Alessi, I. Kikic, and M. Orlandini-Visalberghi: Chem. Eng. J., 1980, vol. 19, p. 221.
10. W. Halwachs, E. Flaschel, and K. Schugerl: J. Membrane Sci., 1980, vol. 6, p. 33.
11. N. N. Li, R. P. Cahn, D. Naden, and R.W.M. Lai, Hydrometallurgy, 1983, vol. 9, p. 277.

12. A. M. Hochhauser and E. L. Cussler: AIChE Symp. Ser., 71, p. 136 (1975).
13. T. P. Martin and G. A. Davies: Hydrometallurgy, 1976/1977, vol. 2, p. 315.
14. K. H. Lee, D. F. Evans, and E. L. Cussler: AIChE J., 1978, vol. 24, p. 860.
15. K. Kondo, K. Kita, I. Koida, J. Irie, and E. Nakashio: J. Chem. Eng. Japan, 1979, vol. 12, p. 203.
16. W. Volkel, W. Halwachs, and K. Schugerl: J. Membrane Sci., 1980, vol. 6, p. 19.
17. J. Strzelbicki and W. Charewicz: Hydrometallurgy, 1980, vol. 5, 243.
18. J. W. Frankenfeld, R. P. Cahn, and N. N. Li: Sep. Sci. Tech., 1981, vol. 16, p. 385.
19. R. P. Cahn, J. W. Frankenfeld, N. N. Li, D. Naden, and K. N. Subramanian: in Recent Developments in Separation Science, N. N. Li, Ed., Vol. 6, CRC Press, Boca Raton, Florida (1981).
20. H. C. Hayworth, W. S. Ho, W. A. Burns, and N. N. Li: Sep. Sci. Tech., 1983, vol. 18, p. 493.
21. N. N. Li and A. L. Shrier: in Recent Developments in Separation Science, N. N. Li, Ed., Vol. 1, p. 163, CRC Press, Boca Raton, Florida (1972).
22. R. P. Cahn and N. N. Li: Sep. Sci., 1974, vol. 9, p. 505.
23. N. N. Li, R. P. Cahn, and A. L. Shrier: "Liquid Membrane Process for the Separation of Aqueous Mixtures," U.S. Patent 3,779,907 (1973).
24. J. W. Frankenfeld and N. N. Li: in Recent Developments in Separation Science, N. N. Li, Ed., Vol. 3, pp. 285-292, CRC Press, Boca Raton, Florida (1977).
25. T. Kitagawa, Y. Nishikawa, J. W. Frankenfeld, and N. N. Li: Environ. Sci. Technol., 1977, vol. 11, p. 602.
26. R. E. Terry, N. N. Li, and W. S. Ho: J. Membrane Sci., 1982, vol. 10, p. 305.
27. S. W. May and N. N. Li: Biochem. Biophys. Res. Commun., 1972, vol. 47, p. 1179.
28. S. W. May and N. N. Li: in Enzyme Engineering, E. K. Pye and L. B. Wingard, Eds., Vol. 1, p. 77, Plenum Press, New York (1974).
29. R. R. Mohan and N. N. Li: Biotechnol. Bioeng., 1974, vol. 16, p. 513.
30. R. R. Mohan and N. N. Li: Biotechnol. Bioeng., 1975, vol. 17, p. 1137.
31. W. J. Asher, K. C. Bovee, J. W. Frankenfeld, R. W. Hamilton, L. W. Henderson, P. G. Holtzapple, and N. N. Li: Kidney Int., 1975, vol. 7, S-409.

32. W. J. Asher, T. C. Vogler, K. C. Bovee, P. G. Holtzapple, and R. W. Hamilton: Kidney Int., 1976, vol. 10, S-254.
33. W. J. Asher, T. C. Vogler, K. C. Bovee, P. G. Holtzapple, and R. W. Hamilton: Trans. Am. Soc. Artif. Intern. Organs, 1977, vol. 23, p. 673.
34. N. N. Li and W. J. Asher: in Chemical Engineering in Medicine, ACS (Advances in Chemistry Series), pp. 1-14 (1973).
35. S. W. May and N. N. Li: in Biomedical Applications of Immobilized Enzymes and Proteins, T. M. S. Chang, Ed., Vol. 1, p. 171, Plenum Press, New York (1977).
36. J. W. Frankenfeld, W. J. Asher, and N. N. Li: in Recent Developments in Separation Science, N. N. Li, Ed., Vol. 4, pp. 39-50, CRC Press, Boca Raton, Florida (1978).
37. D. F. Ollis, J. B. Thompson, and E. T. Wolynic: AIChE J., 1972, vol. 18, p. 457.
38. E. T. Wolynic and D. F. Ollis: CHEMTECH, 1974, p. 111.
39. E. L. Cussler and D. F. Evans: J. Membrane Sci., 1980, vol. 6, p. 113.
40. E. S. Matulevicius and N. N. Li: Sep. Purif. Methods, 1975, vol. 4, p. 73.
41. N. N. Li: J. Membrane Sci., 1978, vol. 3, p. 265.
42. A. G. Kopp, R. J. Marr, and F. E. Moser, Instn. Chem. Eng. Symp. Ser., vol. 54, p. 279 (1978).
43. W. S. Ho, T. A. Hatton, E. N. Lightfoot, and N. N. Li: AIChE J. 1982, vol. 28, p. 662.
44. A. M. Hochhauser: Ph.D. Thesis, Carnegie-Mellon University, Pittsburgh, PA. (1974).
45. R. Marr and A. Kopp: Chem. Ing. Tech., 1980, vol. 52, p. 399.
46. H. H. Downs and N. N. Li: J. Sep. Proc. Technol. (U.K.), 1981, vol. 2, p. 19.

CHARACTERIZING EFFECTS OF NOVEL HYDROMETALLURGICAL PROCESS CHEMISTRY ON ELECTROWINNING OPERATIONS

Thomas W. Chapman

Chemical Engineering Department
University of Wisconsin
Madison, Wisconsin, U.S.A.

Cell polarization, current distribution, and current efficiency in an electrowinning cell all depend in a complex manner on thermodynamic properties, electrode kinetics, and ionic transport effects. Each of these in turn is profoundly affected by changes in system chemistry. Careful analysis and experimentation are required to determine the characteristics of the electrowinning step of a novel hydrometallurgical process.

This paper reviews the basic theory that forms the foundation for analyzing electrowinning cells and then indicates within this framework how modifications of system chemistry manifest themselves in process performance. Emphasis is placed on design and analysis of experimental work to characterize new systems.

Three examples are given that illustrate complications that arise when a new or modified electrowinning system is considered. The first is an analysis of copper chloride reduction in the presence of excess chloride, with or without iron. Chloride complexing stabilizes cuprous in solution and affects significantly cathode potential and current efficiency as well as deposit structure. The second example describes nickel electrowinning from acid solution in the absence of a buffer. Current efficiency decreases as stirring rates are increased, and cathodic passivation can be encountered at high polarization. The third example is a study of anode depolarization with sulfur dioxide and a demonstration of the use of an additional catalyst such as iodide or iron.

These three examples illustrate how solution chemistry interact with ionic transport to affect electrode behavior.

There are several electrowinning processes, such as copper winning from acidic sulfate solutions, that have been practiced industrially for so many years that their appropriate design parameters and optimal operating conditions have been identified by experience. On the other hand, new hydrometallurgical processes may require the design of an electrowinning operation for a solution that exhibits unusual properties and has not been worked with before Under these circumstances there are many variables to be considered, and efficient developemnt requires that a systematic experimental program be conducted to identify critical features of the system behavior and to determine the chemical parameters that are needed to formulate a realistic and useful process model.

In this paper the basic equations needed to simulate an electrolysis cell are reviewed, and the information that must be determined experimentally for a specific system is identified. Examples of three systems in which interesting chemical interactions are encountered are described. These are the copper chloride system in which excess chloride ions stabilize the cuprous ion by complexing, the unbuffered nickel chloride system in which cathodic hydrogen evolution may occur, and the sulfite-sulfate system in which the anodic oxidation of sulfite can be promoted by a catalyst. Each of these systems arises in a potential hydrometallurgical process, but their behavior must be elucidated before the processes can be thoroughly analyzed.

ELECTROCHEMICAL REACTOR SIMULATION

An electrowinning cell should be viewed as an electrochemical reactor in which electrical energy is used to convert reactants into products. A rational approach to reactor design involves simulation of cell performance and especially variations in cell performance with respect to various design and operating variables as well as system parameters. The particular characteristics of a cell that one wishes to simulate with a mathematical model are the following:

1. the cell polarization curve, or the relation between total current I and applied voltage V;
2. the current efficiency CE, the fraction of the total cell current that is effective in producing a desired electrode reaction;
3. the power consumption P and the specific energy consumption per unit mass of product;

4. the conversion and selectivity of the cell expressed as Δc_i, the changes in solute species concentrations from inlet to outlet; and
5. the current distribution or the nonuniformity of the electrode reactions.

The first basic equation that provides a framework for a mathematical model of a cell is the macroscopic mass balance written for each solute species

$$V_s \frac{dc_{i,ave}}{dt} = Q(c_{i,in} - c_{i,out}) + \sum_{\substack{j \\ \text{electrodes}}} W_{i,j}^{(m)} + R_{i,tot} \qquad (1)$$

which is written in the notation of Bird et al.[1] The accumulation rate of species i in the solution volume V_s results from a difference between the rates of flow in and out in streams of flowrate Q, the total rate of production by homogeneous reaction $R_{i,tot}$ and the net rate of production at the electrodes.

The total mass transfer rate of species i into the solution from electrode j may be written as

$$W_{i,j}^{(m)} = \iint_{A_j} N_{io}\, dA \qquad (2)$$

where N_{io} is the molar flux of species i out of the electrode surface and A is electrode area. For several simultaneous electrode reactions, represented by

$$\sum_{\substack{i \\ \text{species}}} s_{i,k} M_i^{z_i} \rightarrow n_k e^- \qquad (3)$$

where $s_{i,k}$ is the stoichiometric coefficient for substance M in reaction k, Faraday's law may be written as

$$N_{io} = - \sum_{\substack{k \\ \text{reactions}}} \frac{s_{i,k}}{n_k F} i_k \qquad (4)$$

where i_k is the local current density on the electrode surface resulting from reaction k and F is Faraday's constant. Thus, the mass balance Equation 1 can be used to relate concentration changes, flowrates, and electrode areas if the local current densities i_k can be estimated. (If bulk-phase homogeneous reactions occur in the cell, the kinetics and bulk mixing effects must also be introduced via the $R_{i,tot}$ term as they are in the analysis of conventional chemical reactors.)

The total cell voltage V is composed of a number of contributions:

$$V = E + \Delta\phi_{ohm} + \eta_{s,a} + \eta_{c,a} - \eta_{s,c} - \eta_{c,c} \quad (5)$$

where E is the equilibrium cell potential, $\Delta\phi_{ohm}$ is the ohmic loss from the solution resistance, and η_s and η_c are the surface and concentration overpotentials, respectively, arising at each electrode, anode and cathode. Except for the equilibrium potential, which is based on bulk phase compositions, each of the contributions to the operating cell voltage depends on total current and the current distribution. The overpotentials in particular depend on local current densities.

The surface overpotential is the additional potential in excess of the local equilibrium electrode potential that is required to drive a particular electrode reaction at a finite rate i_k. The magnitude of this quantity is determined by the kinetics of the electrode reaction, which must be determined experimentally.

The concentration overpotential arises when surface concentrations differ from the bulk concentrations because of a finite mass transfer resistance. In the notation of Reaction 3 the concentration overpotential may usually be estimated by the expression

$$\eta_c = -\frac{RT}{nF}\sum_i s_i \ln (c_{io}/c_{i\infty}) \quad (6)$$

where c_{io} and $c_{i\infty}$ are solute concentrations at the electrode surface and in the bulk solution, respectively. The difference between these depends on the flux of species i and the local mass transfer coefficient. Because the flux is given by Eq. 4, this quantity also depends on the local current density.

The problem of simulating a particular cell is to determine the current distribution, and particularly the distribution of individual electrode reaction rates, for a specified value of V or I. If V is specified, the local reaction rates i_k must adjust themselves such that the total voltage drop between any two points on the anode and on the cathode, calculated from Eq. 5, equals V. If the total current I is specified, the local current densities again must adjust themselves to maintain a uniform value of V between anode and cathode and also satisfy the constraint:

$$I = \iint_{A_a} \sum_k i_k dA = - \iint_{A_c} \sum_k i_k dA \qquad (7)$$

If this current distribution problem can be solved, one can investigate the various design factors listed above. A cell polarization curve can then be generated and the power consumption computed from

$$P = I\ V \qquad (8)$$

Local current efficiency is given directly by the relative magnitudes of the local current densities, and the overall or average current efficiency for an electrode can be computed from

$$CE_k = \iint_{A_j} i_k dA/I \qquad (9)$$

where reaction k is the desired reaction on electrode j.

Because the surface overpotential is a local surface property and the concentration overpotential usually arises in a thin mass-transfer boundary layer, the characteristics of the anode and of the cathode may usually be studied separately. This separation is implied by the form of Eq. 5. The anode and cathode interact only through the charge balance, Eq. 7, and the distributions of current, potential, and concentrations in the bulk solution. It is therefore possible to develop a cell model by determining the effects of the system chemistry on the anode and cathode individually and then to incorporate the individual electrode characteristics in the equations outlined above to arrive at a working cell model. The formulation of the general current distribution problem has been described by Newman[2] and an efficient solution algorithm has been presented by Caban and Chapman.[3]

In the equations presented above the critical information that is needed for each electrode is the relation between local conditions of solution composition, electrode surface properties, mass-transfer properties, and potential and the local fluxes or current densities. This is the relationship that varies from one system to another and depends so critically on specific chemical interactions. It should be studied in an experimental system that offers a simple geometry and known mass-transfer conditions so that the chemical factors can be identified most easily. The rotating disk electrode is an excellent tool for this purpose.[4] The three studies that are described in the subsequent sections used the RDE to identify the local dependence of electrode behavior on system chemistry, composition, and operating conditions.

COPPER CHLORIDE ELECTROLYSIS

A number of workers have proposed copper processes based on chloride solutions because strong chloride solutions provide good conductivity and high copper solubility. They also offer the possibility of maintaining copper in solution as cuprous which requires only one electron to reduce. Moreover, if oxidation of cuprous to cupric is the anodic reaction, the cell voltage as well as the total current levels can be much less than those needed in copper sulfate electrowinning.

Trainham[5] carried out a theoretical analysis of copper chloride electrolysis in which he used the RDE as a hypothetical model of an electrode with a mass-transfer boundary layer. He computed the effects of bulk concentrations and electrode potential on current density and current efficiency.

The effect of excess chloride on a copper solution is to conver most of the cupric ions to the complex $CuCl^+$ ion and most of the cuprous ions to the complex $CuCl_3^=$. The large equilibrium constant of the latter form in particular stabilizes cuprous in solution so that cuprous chloride solubility becomes significant; formal equilibrium electrode potentials for the complex ions are quite differen from the standard potentials with the free ions as shown in Table 1.

Because of the shift in the order of the reduction potentials at high chloride levels, two separate copper-reduction reactions must be considered: the reduction of cupric species to cuprous

$$CuCl^+ + 2Cl^- + e^- \rightarrow CuCl_3^= \tag{10}$$

and the reduction of cuprous to copper

$$CuCl_3^= + e^- \rightarrow Cu + 3Cl^- \tag{11}$$

Table 1. Standard electrode potentials of copper reactions with and without complexing by chloride, at 25°C.

	E°, volts	
Reaction	Noncomplexed	With chloride
copper/cuprous	0.521	0.233
copper/cupric	0.337	0.3355
cuprous/cupric	0.153	0.438

Trainham considered a one-dimensional transport model in which all copper existed as the two predominant complex ions, and he took the two electrode reactions to be reversible. He found that the effect of ionic migration in the boundary layer was small, enhancing limiting currents by only a few percent at concentrations of practical interest. It is interesting, however, that the transport of free chloride near the cathode causes the local electrostatic potential to go through a maximum so that migration does in fact enhance the transport of the anionic complex to the cathode, an effect opposite to what might be expected.

Figure 1 shows the calculated current efficiency for cuprous reduction as a function of cathode potential at various degrees of oxidation of the bulk solution copper, which might be caused by the presence of air or by mixing with anolyte. As might be expected, the predicted current efficiency drops to 50% when the copper is all present in the cupric form. The steepness of the curves results from the assumption that the electrode kinetics are very fast.

An extension of this analysis has been published by White et al.[6] who considered the case where significant levels of ferric iron are also present in the electrolyte. They assumed that the homogeneous reaction of ferric and cuprous was so fast and irreversble that those two species could not coexist at any point in the solution. Realistic electrode reaction kinetics were assumed for

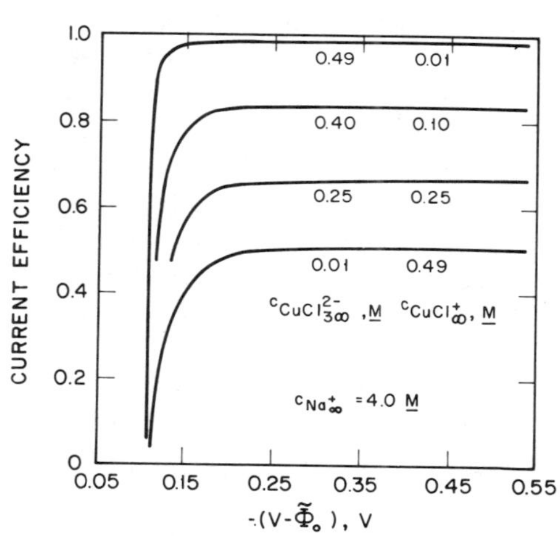

Fig. 1. Current efficiency for the reduction of cuprous to copper as a function of cathode potential at varying bulk concentrations of cuprous and cupric species.

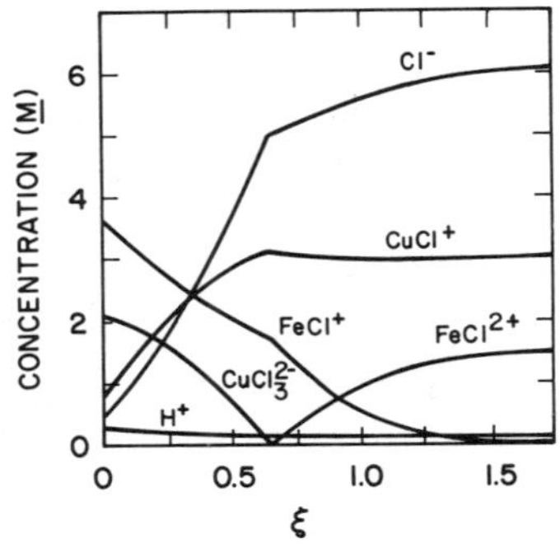

Fig. 2. Calculated concentration profiles near the cathode during reduction of a ferric chloride-cupric chloride solution. ξ is a dimensionless distance from the electrode surface.[6]

all possible electrode reactions, including reduction of ferric, ferrous, and hydrogen ions. The boundary layer mass transfer problem was solved at various specified potentials to compute concentration profiles such as shown in Fig. 2.

The most striking feature of this system is that a homogeneous reaction plane forms parallel to the electrode at some very small distance into the solution. Outside the reaction plane the metal species present are the complexed forms of ferric, ferrous, and cuprous. Close to the electrode the species are the cupric, cuprous, and ferrous complexes. At moderate cathodic potentials the only significant electrode reaciton is reduction of cupric to cuprous. As more cuprous is formed by increasing cathodic polarization, the reaction plane moves out into the boundary layer and reduces more ferric homogeneously. At more negative electrode potentials some of the cuprous formed is reduced to copper, and the reaction plane moves back in toward the electrode. The regeneration of cupric at the reaction plane increases the limiting current at the cathode above that obtained for cupric in the absence of ferric ion.

This model allows investigation of various features of the chemical system that would be difficult to observe directly. For example, calculated surface concentrations predict where cuprous chloride solubility limits may be exceeded. The model also predicts at what potential significant codeposition of iron might begin to contaminate the copper deposit.

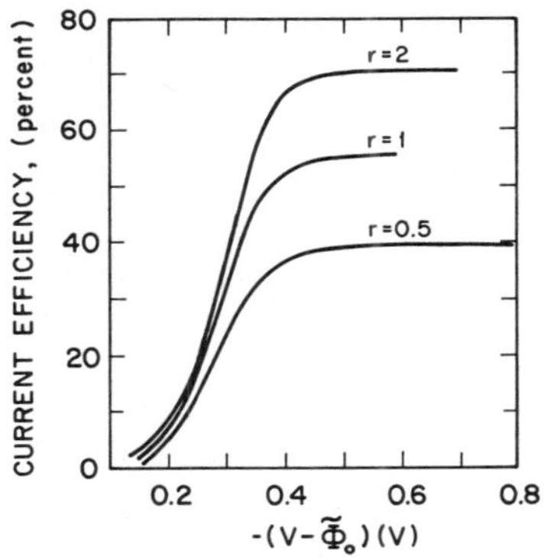

Fig. 3. Current efficiency for the reduction of cupric as a function of cathode potential at various levels of iron in the bulk solution.[6] $r = C_{CuC\ell^+,\infty}/C_{FeC\ell_2^+,\infty}$.

Figure 3 gives current efficiencies calculated for cupric reduction to copper at various iron levels as a function of cathode potential. In this case the expected current efficiency is not entirely obvious because of the interaction between the convective mass transfer and the location of the homogeneous reaction zone.

An experimental study of this system on a RDE would be interesting in that it would check the several assumptions that have been made to generate the predictions and determine the quality of copper that can be deposited from chloride solutions containing iron.

NICKEL CHLORIDE ELECTROLYSIS

A second example of an unconventional hydrometallurgical system requiring study and analysis is the nickel chloride system. Normally hydrometallurgical processing of nickel involves electrowinning from sulfate solutions. If one were considering a chloride route for processing a new ore, such as deep-sea manganese nodules, then direct electrowinning of nickel from a chloride solution might be of interest. Wruck[7] studied the deposition of nickel from chloride solutions on a RDE in order to determine the effects of solution composition, temperature, and stirring conditions on cathode polarization and current efficiency.

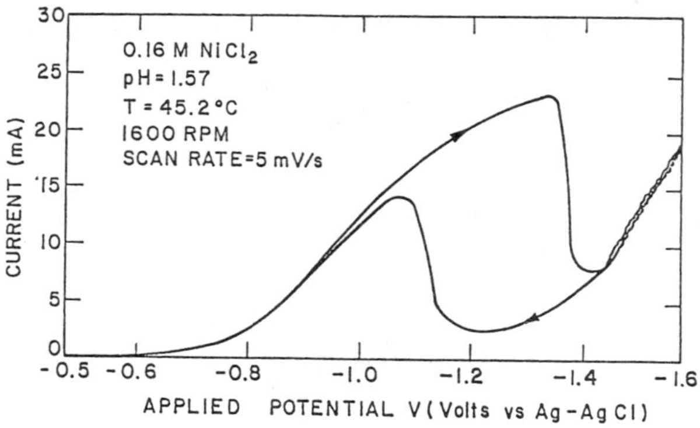

Fig. 4. Experimental cathodic polarization curves for nickel reduction from chloride solution on a rotating disk electrode. $C_{NaC\ell} = 2.0M$; $A = 0.1\ cm^2$.

An experimental polarization curve for the nickel cathode is shown in Fig. 4, which presents a potentiostatic scan on the RDE. As the potential is scanned to more cathodic values, a passivation phenomenon is observed where the current peaks at a maximum value and drops to a low level before increasing again. At slow scan rates and lower pH levels this behavior is reversible with reversed scan direction. At higher scan rates and higher pH there is increasing hysteresis in the current-voltage curve. At higher temperatures and higher rotation speeds the critical passivation point moves to higher currents and more cathodic potentials. Clearly, this passivation behavior must be avoided in any practical electrowinning operation.

The nickel-deposition experiments can be interpreted by means of a mass-transfer model which accounts for hydrogen transport and the existence of $Ni(OH)^+$, which appears to be the electroactive nickel species. The cathodic passivation occurs when H^+ reaches a limiting current such that the surface pH rises and $Ni(OH)_2$ precipitates.

The boundary-layer mass-transfer model for this system involves a continuity equation for the total concentration of nickel species, Ni^{+2} and $Ni(OH)^+$, as well as a similar equation for the net transport of all hydrogen/hydroxide species, $Ni(OH)^+$, H^+, and OH. The boundary conditions at the electrode surface require that the total nickel flux and the total hydrogen flux match the current densities of the two reduction reactions as given by their respective electrode kinetics expressions. Individual species concentration profiles are determined by solving the ionic equlibrium expressions at every point in the boundary layer simultaneously with the combined-species transport equations.

The electrode kinetics expression for nickel deposition was assumed to be of the form

$$i_{Ni} = i^{0}_{NI}\ C^{(1-\alpha/2)}_{NiOH^+}\ C^{\alpha/2}_{H^+}\ \left[\exp\left(\frac{(2-\alpha)F}{RT}\ \eta_{Ni}\right) - \exp\left(\frac{-\alpha F}{RT}\eta_{Ni}\right)\right] \tag{12}$$

which is derived from a mechanism that assumes that a one-electron discharge of $Ni\ (OH)^+$ to an adsorbed species is the rate-determining step. For hydrogen reduction the form of the kinetics was assumed to be

$$i_H = i^{0}_{H}\ C_{H^+}\ \left[\exp\left(\frac{(1-\beta)F}{RT}\eta_H\right) - \exp\left(\frac{\beta F}{RT}\ \eta_H\right)\right] \tag{13}$$

In both of these rate expressions the species concentrations that appear are those at the electrode surface.

Wruck's model of the nickel electrode introduces five parameters that must be determined experimentally, i^{0}_{Ni}, i^{0}_{H}, α, β, and D_{Ni}. All other parameters, such as stability constants and hydrogen diffusivity, were available from the literature or other experiments. Wruck measured cathode voltage and current efficiency under constant current electrolysis for a range of solution concentrations, current densities, temperatures, and disk rotation speeds. Observed current efficiencies ranged from seven to 98%. Wruck fit his data by using a nonlinear regresssion analysis to estimate the five unknown parameters of his model. The values obtained in this way were in the ranges that would be expected from earlier reports in the literature.

Figure 5 shows the species concentration profiles computed for a representative experiment. One observes in particular the increased pH near the surface.

Figure 6 compares measured and computed polarization curves for two sets of conditions. The last computed point in each series, that is, those at the highest potentials, correspond to calculated surface concentrations at which $Ni(OH)_2$ would be expected to precipitate. The location of these points is in good qualitative agreement with the observed onset of passivation.

The effect of bulk solution pH on current efficiency is shown in Fig. 7, which compares computed with experimental values. In

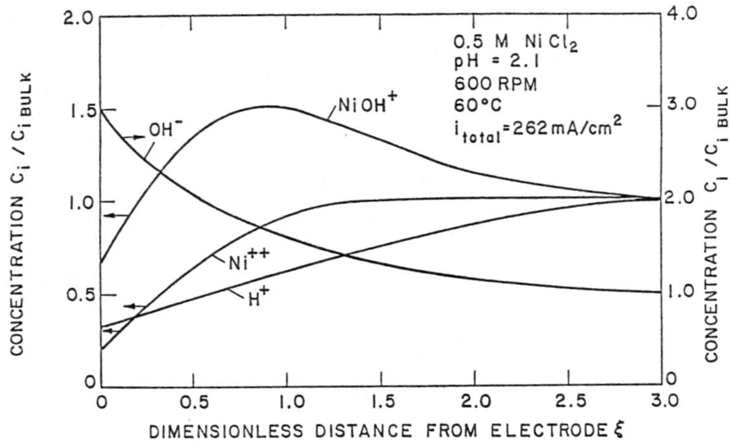

Fig. 5. Calculated concentration profiles near the rotating disk electrode during reduction of nickel from chloride solution. $C_{NaC\ell}$ = 2.0M; CE = 95%.

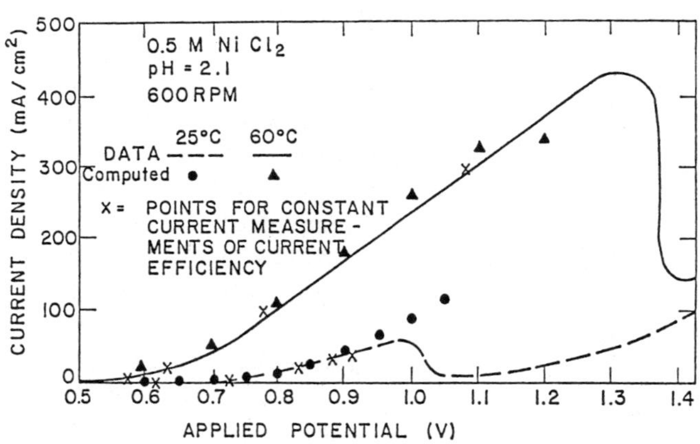

Fig. 6. Cathodic polarization behavior of the nickel chloride electrode at two temperatures.

general the model was quite successful in predicting the observed trends in current efficiency. Perhaps the most important findings of this study were that current efficiency increased with temperature and current density but that it decreased with improved stirring conditions. This latter result makes sense in the context of the transport model, but it may mean that efforts to improve operating

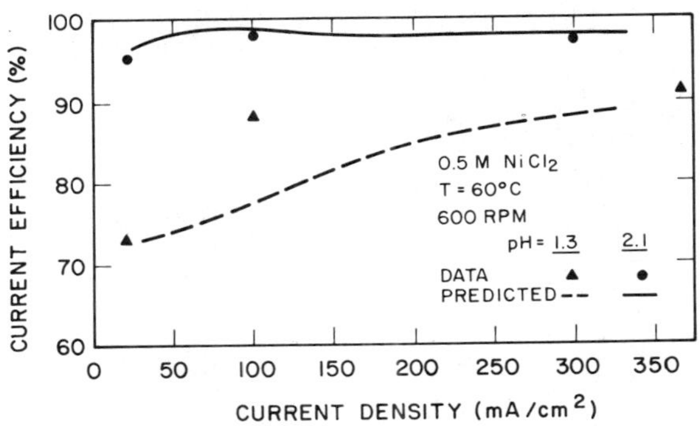

Fig. 7. Effect of current density and pH on the current efficiency for nickel deposition from chloride solution.

rates by improving stirring conditions may not be advantageous in a system such as this where hydrogen discharge is close to its limiting current.

This example demonstrates the usefulness of both controlled experiments on the RDE as well as theoretical analysis of the electrode processes. The complex phenomena occurring in this system would be very difficult to sort out with data available only from a pilot-scale electrowinning cell.

THE SULFITE ANODE

The third example of a potentially useful electrode that exhibits complex chemical behavior involves not a cathode but an anode. Figure 8 shows several polarization curves for a graphite anode in 0.2 M sulfuric acid at 50°C. Curve 1 represents the evolution of oxygen from the pure acid solution. Curve 2 is the anodic polarization curve for the same solution with 0.1 M Na_2SO_3 added. The potential with sulfite oxidation is considerably lower than that for oxygen evolution, which indicates that introduction of sulfur dioxide as an anodic reactant would reduce cell voltage and save energy in the electrowinning of metals from acidic sulfate solutions.

The potential of the sulfite oxidation is, however, still several hundred millivolts more anodic than its equilibrium electrode potential, which indicates that the electrode kinetics are slow and

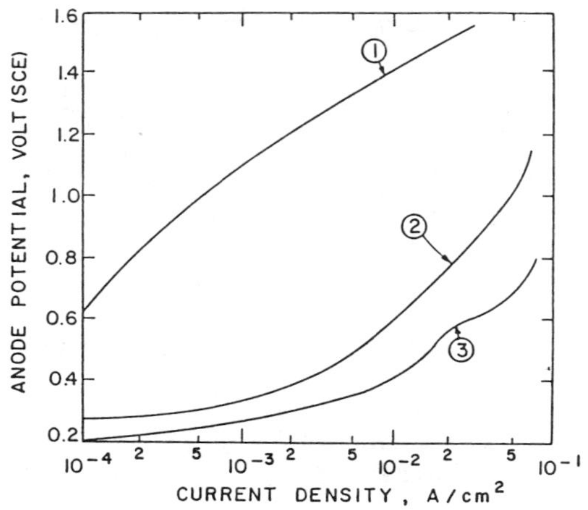

Fig. 8. Anodic polarization curves for graphite in sulfuric acid (0.2M). A = 17.5 cm^2, T = 50°C, stirred solution. Curve 1: Sulfuric acid only. Curve 2: 0.1M Na_2SO_3 added to 1. Curve 3: 1.0 g/ℓ NaI added to 2.

there is a large surface overpotential contributing to the cell voltage.

Curve 3 in Fig. 8 shows the sulfite polarization curve in the presence of 1.0 g/l NaI added as a catalyst. The addition of the iodide lowers the anode potential by as much as 300 mV at high current densities. Yen[8] has investigated the mechanism of this anode-depolarization process by studying the oxidation of sulfite on a rotating disk electrode.

Four experimental polarization curves taken on a rotating pyrolytic graphite disk electrode are shown in Fig. 9. These curves show not only the depolarization by sulfite but also the effect of iodide concentration on the degree of depolarization of the anodic reaction.

The mechanism of the sulfite-iodide anode is analogous to that of the ferric-cupric chloride system described above. While the electrochemical oxidation of sulfite on platinum or graphite is quite irreversible, the anodic oxidation of iodide is fairly fast. The thermodynamics and kinetics of the system are such that the product iodine reacts in solution with sulfite rapidly and almost completely. As a first approximation one may take both the anodic oxidation and

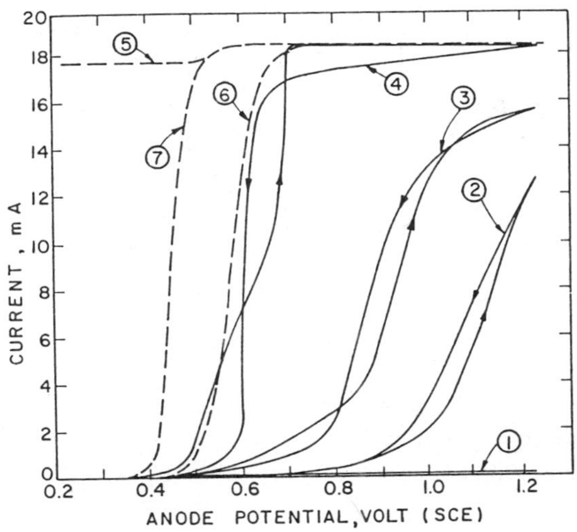

Fig. 9. Anodic polarization curves on a rotating pyrolytic graphite disk electrode. 1600 RPM, 1mV/sec scan rate, A = 0.484 cm^2. Curve 1: 1M H_2SO_4. Curve 2: 19mM Na_2SO_3, 1M H_2SO_4. Curve 3: 0.2mM NaI, 19mM Na_2SO_4, 1M H_2SO_4. Curve 4: 1mM NaI, 19mM Na_2SO_3, 1M H_2SO_4. Curves 5, 6 and 7 computed.

the homogeneous reduction of iodine by sulfite to be infinitely fast. These assumptions lead to the computed curve 5 on Fig. 9 that corresponds to conditions of curve 4. A limiting current corresponding to the bulk sulfite concentration is predicted between the reversible sulfite potential, around zero volts relative to calomel, and the reversible iodide potential, around 0.4 volts on this scale. At more positive potentials the limiting current is predicted to be the sum of the iodide and sulfite limiting currents. The model for this case is very similar to that used in the iron-copper problem.

Comparison of the computed curve with experimental curve 4 in Fig. 9 shows that the fast-reaction model is capable of reproducing the limiting current in the two-reactant system, but it is not accurate in predicting the potential at which an anodic current flows. In fact, the fast-reaction model shows very little sensitivity to iodide concentration at low levels in contrast to the observed dependence. To explain the strong dependence of the half-wave potential on iodide level, one must consider the finite rates of the homogeneous reaction and of the anodic iodide oxidation.

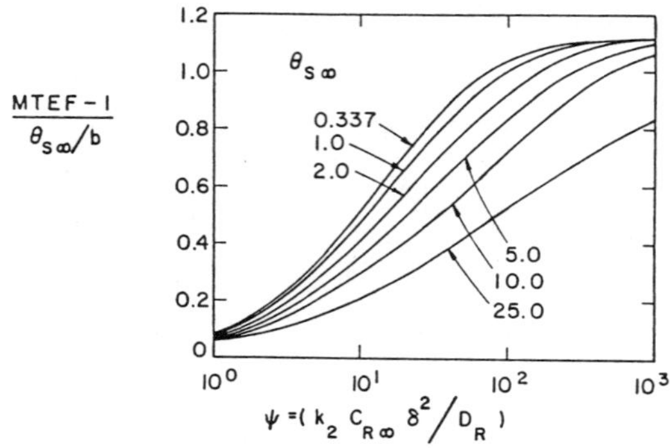

Fig. 10. Mass transfer enhancement factor (MTEF) for limiting current on a rotating disk electrode with second-order homogeneous reaction. For R = I^- and S = SO_2, b = 1/2.

Yen experimentally determined the electrode kinetics of the iodide reaction in sulfuric acid in the absence of sulfite, and he also measured the direct anodic oxidation of sulfite to confirm that the latter electrode reaction is not significant in the potential range of interest. He then developed a theoretical transport model for the RDE boundary layer in which the product of the anodic oxidation reacts in the boundary layer with a bulk-phase solute by a second-order homogeneous reaction. The transport model was solved by the collocation method of Caban and Chapman[9,10] and resulted in the plot shown in Fig. 10.

Figure 10 presents the predicted variation of the mass-transfer enhancement factor (MTEF), the factor by which the limiting current of the electroactive species is increased by the occurrence of the homogeneous reaction. The MTEF is a function of the ratio of reactant concentrations in the bulk, defined in this case as

$$\Theta_{S\infty} = C_{SO_2,\infty}/C_{I^-,\infty} \tag{14}$$

and of the dimensionless reaction-rate constant

$$\psi = \frac{k_2\, c_{I^-,\infty}}{D_{I^-}}\, \delta^2 \tag{15}$$

where k_2 is the second-order rate constant for the sulfite-iodine reaction and δ is the mass-transfer boundary-layer thickness which depends on the rotation speed and solution properties.[4,6]

Yen measured iodide limiting currents in this system at two different sulfite levels and at rotation speeds from 400 to 4900 RPM. Comparison of the observed MTEF with Fig. 10 yielded a second-order reaction rate constant of 7.5×10^9 cm^3/mole-sec, which corresponds to a fast reaction but is small enough to cause finite-rate effects at low iodide levels.

With the experimentally determined rate parameters Yen solved his transport model for currents below limiting current and generated curve 6 in Fig. 9 for conditions corresponding to experimental curve 4. Curve 7 was obtained by using the determined second-order rate constant and assuming that the iodide electrode reaction is reversible. It is seen that the theoretical transport-reaction model, with the necessary parameters determined, does a good job of predicting the anode potential under depolarizing conditions as well as fitting the limiting current.

There are other chemical systems that can be used to achieve electrode depolarization by this mechanism. The ferrous-ferric couple has a redox potential in the proper range to depolarize the sulfite anode, but the homogeneous reaction rate of ferric with sulfite is much slower than is that of iodine. Therefore most of the reaction would take place in the bulk solution, and a high iron concentration would be required to attain current densities of acceptable magnitude.

SUMMARY

The three examples that have been outlined here each involve unusual aspects of their system chemistry that affect the relationships among current density, electrode potential, solution composition, and mass-transfer conditions. The electrochemical behavior of none of these systems is straightforward or sufficiently simple that one could predict it even qualitatively with any confidence. Quantitative determination certainly requires careful experimental investigation.

The suggestion of this paper is that useful and meaningful models of such complicated systems can be developed within the framework of electrochemical mass-transfer theory and that the models can be combined with carefully designed experiments to determine the relevant chemical parameters. After the electrode characteristics are well understood, it should be possible to simplify the electrode model so that it can be used in a macroscopic electrowinning-cell design calculation.

REFERENCES

1. R. B. Bird, W. E. Stewart, and E. N. Lightfoot: Transport Phenomena, John Wiley, New York, 1960, Chapter 22.
2. J. Newman: Electrochemical Systems, Prentice-Hall, Englewood Cliffs, N.J. (1973).
3. R. Caban and T. W. Chapman: J. Electrochem. Soc., 1976, vol. 123, p. 1036.
4. A. C. Riddiford: Adv. Electrochem. Engr., 1966, vol. 4, p. 47.
5. J. A. Trainham: Master's Degree Report, Chemical Engineering Department, University of Wisconsin, Madison, WI, 1977.
6. R. White, J. A. Trainham, J. Newman, and T. W. Chapman: J. Electrochem. Soc., 1977, vol. 124, p. 669.
7. W. J. Wruck: M. S. Thesis, University of Wisconsin, Madison, WI, 1978.
8. S. C. Yen: Preliminary Ph.D. Examination Report, Chemical Engineering Department, University of Wisconsin, Madison, WI, 1982.
9. R. Caban and T. W. Chapman: Chem. Eng. Sci., 1981, vol. 36, p. 849.
10. S. C. Yen and T. W. Chapman: J. Electroanal. Chem., 1982, vol. 135, p. 305.

METAL RECOVERY FROM DILUTE AQUEOUS SOLUTIONS

BY VARIOUS ELECTROCHEMICAL REACTORS

Roland Kammel

Institut für Metallurgie - Metallhüttenkunde-
Technische Universität Berlin
Berlin, Federal Republic of Germany

Interest in the recovery of metals from dilute aqueous solutions is steadily increasing because of environmental concerns and potential economic rewards. Since conventional electrowinning cell systems with parallel plate electrodes are unsuitable for an efficient and economic treatment of effluents, many novel electrolytic methods with two-, semithree- and three-dimensional electrodes have been developed and few cell designs are already successfully applied on an industrial scale. The most important characteristic of these cell systems to ensure high current efficiencies and high production rates is improved mass transfer by use of extended cathode surfaces and vigorous agitation of the electrolyte. In competition with other effluent treatment methods such as adsorption, evaporation, precipitation, cementation, ion exchange, liquid-liquid extraction and reverse osmosis, the electrolytic processes offer the possibility of recovering the metal values from dilute solutions as a marketable metal product in a one step process. This review is focused on industrial applied cell designs and their special field of application. Their economics and cost savings in comparison with conventional treatment techniques are outlined and discussed.

INTRODUCTION

To meet the challenges of increasingly stringent environmental directives for emission of metal concentrations in industrial discharges, effluents, or waste waters, improved technologies for metal recovery have to be developed and introduced in a short period of time.

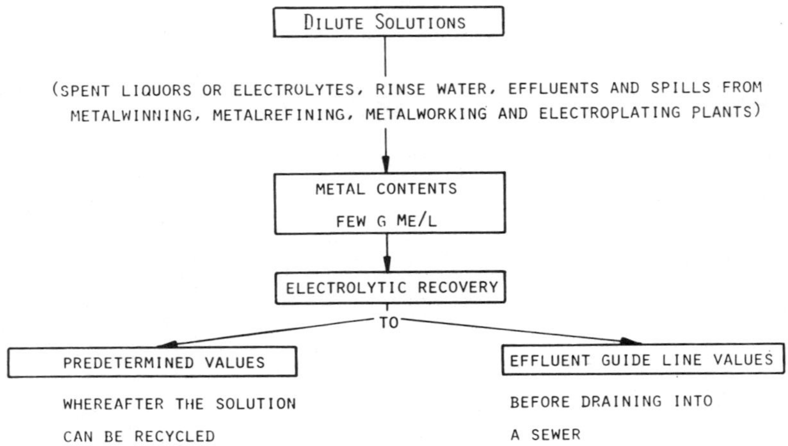

Fig. 1. Electrolytic treatment of dilute solutions.

The main sources of metal containing dilute solutions include: mine waters or leach liquors from natural ore leaching; solutions from primary ore as well as mine dump or heap leaching procedures; spent electrolytes from hydrometallurgical operations; echants and pickling solutions from metal working processes; and rinse waters, spills and effluents from metal finishing and electroplating industries. In most cases, these solutions contain less than a few grams of metal per liter.

As shown in Fig. 1, these low metal contents have to be reduced either to predetermined values after which the solution can be recycled into the process or to residual values of few ppm or less before draining into a sewer to meet effluent guidelines or other environmental regulations. In Table 1 are summarized some of the residual concentrations of waste water to be discharged as proposed by the US Environmental Protection Agency for 1982.[1]

For the recovery of the metal values from these dilute solutions many physical, chemical, and electrolytic methods have been developed which involve various separation and concentration procedures as well as the production of marketable compounds or metals.[2,3,4,5] In Table 2 is a comparison of the advantages and disadvantages of some of the most frequently applied methods for effluent treatment in industry.[6] The obvious attractions for considering electrolytic metal extraction method for the treatment of dilute solutions are that it is a one step process, it does not create further waste, and it recovers metal directly as a marketable product.

Table 1. Pretreatment Regulations Proposed by Environmental Protection Agency (USA for Waste Water).

PROPOSED PRETREATMENT REGULATIONS
COMPLIANCE DATE - SOMETIME IN 1982

FOR COMPANIES DISCHARGING LESS THAN 10.000 GAL/DAY

POLLUTANT	CONCENTRATION (MG/L.) 1-DAY MAX	30-DAY AVERAGE
CYANIDE AMENABLE TO CHLORINE DESTRUCTION	2.0	0.8
HEXAVALENT CHROMIUM	0.25	0.09
LEAD	0.8	0.4
CADMIUM	1.0	0.5
PH	ADJUSTED TO 7.5 TO 10.0	

FOR COMPANIES DISCHARGING MORE THAN 10.000 GAL/DAY

POLLUTANT	1-DAY MAX	30-DAY AVERAGE
CYANIDE AMENABLE TO CHLORINE DESTRUCTION	0.2	0.08
TOTAL CYANIDE	0.64	0.24
HEXAVALENT CHROMIUM	0.25	0.09
TOTAL CHROMIUM	4.2	1.6
COPPER	4.6	2.0
NICKEL	3.6	1.8
ZINC	3.4	1.5
SILVER	1.0	0.34
LEAD	0.8	0.4
CADMIUM	1.0	0.5
TOTAL METALS*	7.5	3.9

*TOTAL OF CU, NI, ZN AND TOTAL CR. DOES NOT INCLUDE CD AND PB.

The idea to use electrolytic process for recovering metals from dilute liquors is not new[7] and the first industrial trials were carried out with the Siemens-Halske gold electrowinning process in 1898 and later with silver recovery cells of the type shown in Fig. 2.

The data in Fig. 3 show that the Siemens-Halske gold extraction from dilute leach liquors in conventional cells with vertical cathode plates can only be carried out with very low current densities.[9] These results further reveal that economic operation in conventinal cells cannot be achieved due to low current efficiency caused by poor mass transfer using vertical cathode plates.[10] Therefore, improved mass transfer conditions are necessary to ensure sufficient current efficiencies and high deposition rates per unit of cell volume down to very low final metal concentrations in the ppm range.

This paper sets itself the following tasks: to introduce briefly some of the theoretical aspects for an efficient recovery of metal

Table 2. Comparison of Effluent Treatment Methods.

Methods	Quality of Treated Effluent	Cost of Effluent Treatment (a)	Selectivity for Individual Metals	Feasibility of Metals Recovery for Salable Product	Cost of Metals Recovery (b)	Other Factors
Precipitation o.g. as hydroxide	Very good	Low/moderate (depends on solid liquid separation)	None	Poor	High	Additional treatment to destroy complexing anions
Ion exchange	Very good	High	High or moderate (depends on pH etc)	Good	Moderate	Resin may be affected by oxidising agents
Electrolysis	Usually suitable for closed loop operation only	High	High (with control of potential)	Very good	Low (included in a)	
Reverse osmosis	Low-high	Depends on desired effluent quality	None	Good for in-plant solution recycle	Low (included in a)	Restriction on pH and Eh of effluents
Evaporation	No effluent	High	None	Good for in-plant solution recycle	Low (included in a)	
Liquid-liquid Extraction	Low-high (organic as well as residual metals in raffinate)	Moderate to high (depends on solvent loss)	High or moderate (depends on pH and extractant)	Good	Low for in-plant recycle otherwise moderate	Truly continuous countercurrent operation

Dr. D. S. Flett

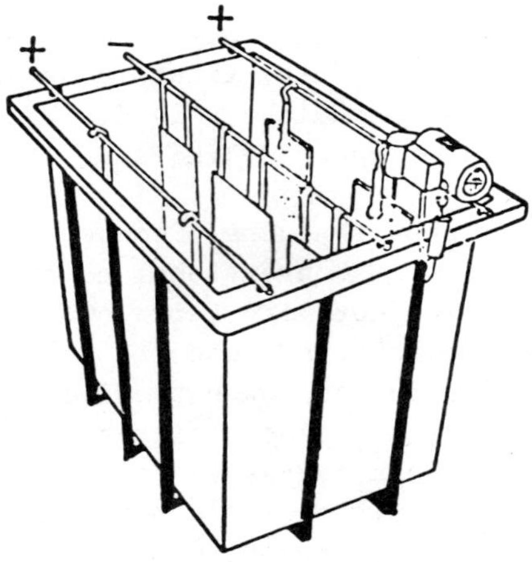

Fig. 2. Silver recovery cell.

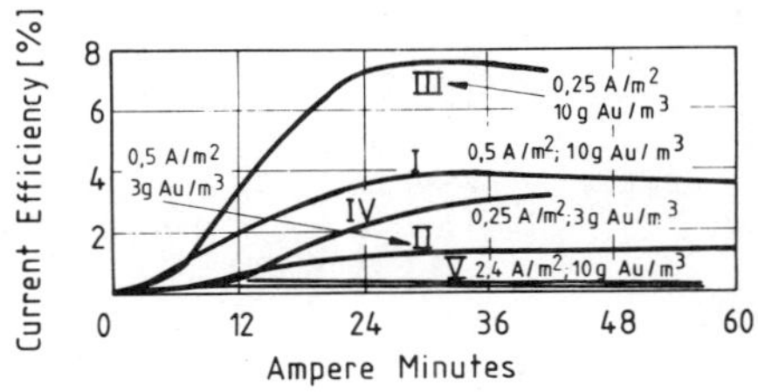

Fig. 3. Current efficiency for electrowinning of gold from dilute cyanide solution (Siemens-Halske-process).

values from dilute solutions; to review the variety of cell systems now available for electrolytic metal recovery; to classify the basic concepts used in the various electrolytic devices; to outline data and results of these novel electrolytic recovery systems; and to compare the electrolytic method with other effluent treatment processes.

THEORETICAL CONSIDERATIONS

The electrowinning of metals from dilute solutions is not favored thermodynamically. This is shown in Fig. 4 where the reversible potential data were calculated from the Nernst equation for some uncomplexed ions.[11] The presence of complexing agents in the solutions will often shift the reversible potentials several hundred millivolts more cathodic. In a sufficiently dilute conductive solution, mass transfer of the metal ions to be discharged is controlled in the vicinity of the cathode surface mainly by diffusion. Experimental results for dilute electrolytes with ions of low valency follows Eq. (1):

$$-\frac{dm}{dt} = D\,q\,\frac{(c_o - c_e)}{\delta N} \qquad (1)$$

where the mass transfer dm/dt is dependent on the surface area q of the cathode and the decrease in concentration $(c_o - c_e)$ divided by the boundary layer thickness δN. D is the diffusion coefficient of the metal ion to be recovered. The concentration profile is presented in Fig. 5. The velocity profile of the adjacent electrolyte film at a vertical cathode under natural convection conditions, which are associated with differences of specific gravity of the solution within the diffusion layer, are also shown in Fig. 5.[12] In a dilute solution, the difference in specific gravity between the bulk solution and the solution within the boundary layer is extremely small and, therefore, mass transfer by natural convection can be neglected.

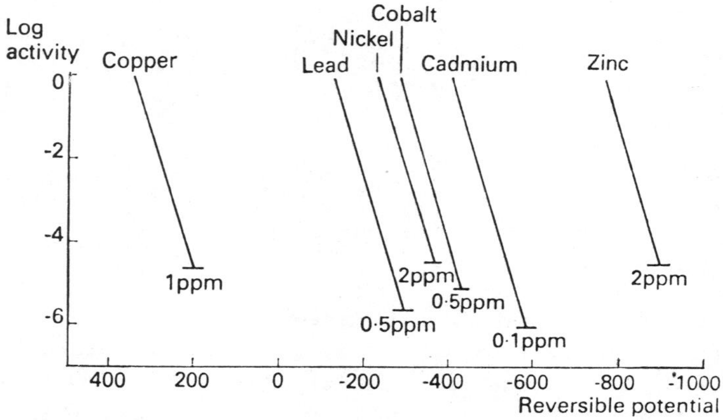

Fig. 4. Reversible potentials for some uncomplexed metal ions as a function of activity

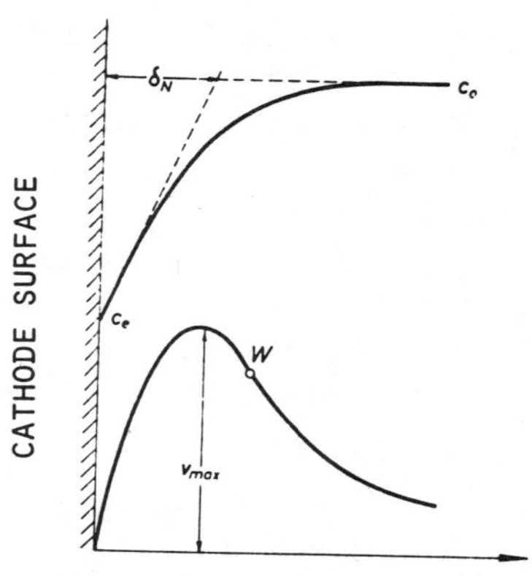

Fig. 5. Concentration profile of dischargable metal ions in the cathodic diffusion layer; velocity profile of the solution due to natural convection. S_N = thickness of diffusion layer (model by Nernst); C_O = bulk concentration; C_E = concentration at the electrode; V_{max} = maximum of velocity of hydrodynamic flow; W = point of inflection.

The thickness of the diffusion layer δN increases under potentiostatic conditions with the square root of the time, t, of the electrolytic metal deposition according to Eq. 2:

$$\delta N = k_1 \sqrt{t} \tag{2}$$

thus bringing about a continuous decrease of mass transfer. The factor k_1 is mainly defined by the diffusion coefficient. Under galvanostatic conditions, however, the concentration c of dischargeable ions at the cathode surface is dropping to zero within a so called transition time τ according to Eq. (3)

$$\tau = k_2 \frac{c^2}{i^2} \tag{3}$$

The transition time τ can be decreased by diluting the solution and by increasing the current density at the cathode. The factor k_2 is mainly defined by the diffusion coefficient and the valency of the ions to be discharged. A transition time τ of about 100 seconds has been determined in a cell with vertical cathode plates[13] using a conventional copper refining solution containing 40 g/l copper at a current density of 400 A/m^2. Therefore, the transition time τ will be very short for dilute solutions at high current densities.

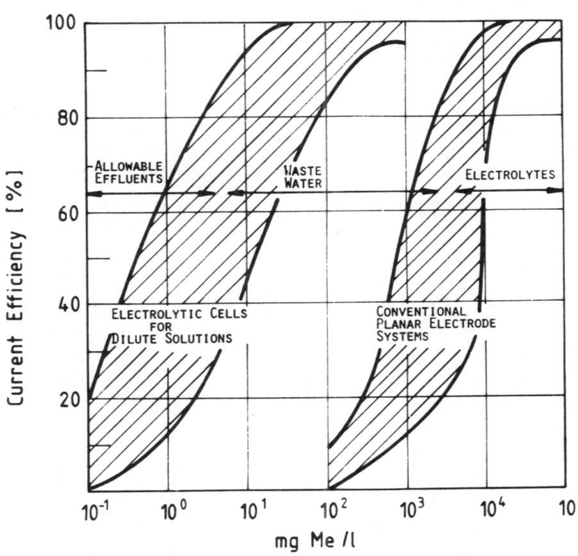

Fig. 6. Current efficiency as a function of metal concentrations for conventional and recovery electrolysis systems.

Mass transfer in connection with electrolytic metal deposition can also be expressed by Eq. (4):

$$- \frac{dm}{dt} = k_3 \, I \, \eta \tag{4}$$

with the electrolytic current I and the current efficiency η. The factor k_3 is defined by the electrochemical equivalent weight of the metal ions to be discharged. If mass transfer of the electrolytic process is mainly controlled by diffusion, expressions (1) and (4) can be combined to yield Eq. 5:

$$D \, q \, \frac{c_o - c_e}{\delta_N} = k_3 \, I \, \eta \tag{5}$$

which leads to Eq. (6) for the current

$$I = \frac{D}{k_3} \, q \, \frac{c_o - c_e}{\delta N} \tag{6}$$

Equation (6) reveals that a high space-time yield or a large electrolytic current, respectively, can only be achieved by increasing the surface area q of the cathode and/or by decreasing the thickness δ_N of the cathodic boundary layer, because the concentration c_o of dilute metal containing solutions will always be comparatively small. Figure 6 clearly shows that new types of electrowinning cell design must be adapted for the purpose of economic metal recovery from very dilute solutions and industrial waste waters.[14]

Electrolytic Metal Recovery Cell Designs

Over the last decades a large number of electrolytic cell designs for the treatment of dilute solutions have been suggested. Only a few have been introduced in the industry while others have remained unused or have never even been tried in a laboratory scale.

Table 3 classifies some of the technically important electrorecovery cell systems for dilute solutions into 2-dimensional, semi-3-dimensional and 3-dimensional systems[15]. The changes in design with regard to mass transport and surface area compared with conventional electrowinning cell systems are marked + for medium, ++ for major and 0 for no improvements.

The various cell designs for metal recovery from dilute solutions shown in Table 4 have been classified as concentrator cells and direct recovery cells.[16] The concentrator cells produce a concentrated solution (electrodialysis) or a metal deposit on cathodes

Table 3. Characteristics of Technically Important Electrolytic Recovery Cells.

Cell Type	Improved Mass Transfer	Extended Electrode Surface	Electrode Type
Eco	++	0	2-dimensional
Chemlec	++	0	
Impact Rod	+	+	
Swiss Roll	+	++	semi-2-dimensional
ESE	+	++	
Fixed Bed	+	++	3-dimensional
Fluidized Bed	++	++	
Rotating Tubular	++	++	

Effect: 0 low + medium ++ high (Kreysa)

Table 4. Cell Designs for Electrorecovery of Metals from Dilute Solutions.

Type	Max. anode current density A/m^2 for a dilute (250 ppm) cond. solution	Effective concentration limit (level where c.e is 25 %)	Energy consumption kWh/kg Cu from steady state run at 150 ppm Cu, 100 g/l H_2SO_4 (1)	Comments
Concentrator Cells				
Electrodialysis		30	20	Feasible mid 1960's w/development of improved IX membranes
"Swiss-roll"cell	500	0.5	8 - 9 (3)	Robertson, ETH Zürich. Relatively high pumping costs " double up when metal "re-won
Dupont ESE cell	500	0.5	8 - 9 (3)	Keating, Du Pont, USA. As per swiss roll
H.S.A.	2000	0.1	7 - 8 (3)	Fleet, Das Gupta, HSA, Toronto. Pat. UK, Aug. 1974
Direct Recovery Cells				
2D Cathode (Mesh)	50 - 100	10	3.5 - 4.5	Cooke, Holland: ICI, UK 1968 Pat. 1268182
Rotating cathode: coherent deposit	300	10	4.5 - 6.0	Surfleet, ESRC, UK. Trans. Inst.Met.Fin. 5 (1972) 227-32
Rotating cathode: powder production	2000	< 1	6.5 - 7.5	Holland, Ecological Engineering, UK.
Fluidized bed cell: non-cond. particles	100	10	3.5 - 5.0 (4.0 - 5.5) (4)	Lopez-Cacicedo, ECRS, UK. Chemelec cell. Pat.1423369, Sept. 1973
Fluidized bed cell: conducting particles	1500 (2)	< 1	6.0 - 7.0	Goodridge, Backhurst, Fleischmann, Plimley-NRDC, UK; CJBD; AKZO. Master Pat.: 1 194 181, May 1966
Particulate bed cell: conducting particles (recirculating bed cell)	1500	<1	5.0 - 6.5	Moergli, Parel SA (CJED, Anglo Am. Diamond Shamrock Corp., Pat. 1974 US 3974049
Rotating tubular bed reactor	<500	10	ca. 5.0	Kammel, Lieber, GOEMA, Stuttgart, USP 4159235
Impact rod reactor	<800	10	ca. 3.0	Kammel, Lieber, GOEMA, Stuttgart, USP 4172780

(1) AC + 1.1 DC
(2) Difficult to quantify since rod anodes, cylindrical membranes are used
(3) Includes 1.8 kWh/kg for winning from the concentrated solution
(4) When inert cathode is used and metal is refined

Dr. Geoffry M. Roberts
completed by new data

(Swiss Roll, ESE-, Fixed Bed-Cells) from which the metal is recovered either by chemical dissolution to yield a concentrated solution or anodically stripped and cathodically deposited as in electrorefining procedures. The direct recovery cells produce marketable metal deposits either as cathodic metal plates (Chemelec Cell), as powder (Eco Cell), as particles (fluidized bed cell), and as flakes (Rotating Tubular Bed- and Impact Rod Reaktor of the GOECOMET® type).

Recovery Cells with Extended Cathodic Surface Area

A large variety of cell constructions has been suggested for extended cathode surface areas of very different shape and dimensions.[2] Metal strip in straight lines as well as wound up as a spiral or helix or in a cylindrical configuration, cathodes of the disc type compiled with narrow spacings, expanded metal sheets as single electrodes or multi-mesh arrangements, and cathodes consisting of woven or knitted metal wire or screen meshing of different aperture have been used for electrolytic treatment of dilute metal containing solutions.

The "ESE-Cell" (Extended Surface Electrolysis) by Du Pont and the "Swiss Roll" by Robertson, ETH Zürich[11] were developed at the same time. From the photograph and the schematic drawing in Fig. 7 it can be seen that the Swiss-roll which represents an electrode configuration of the helix type, is constructed by rolling a flexible sandwich of electrodes and separators around an axis and

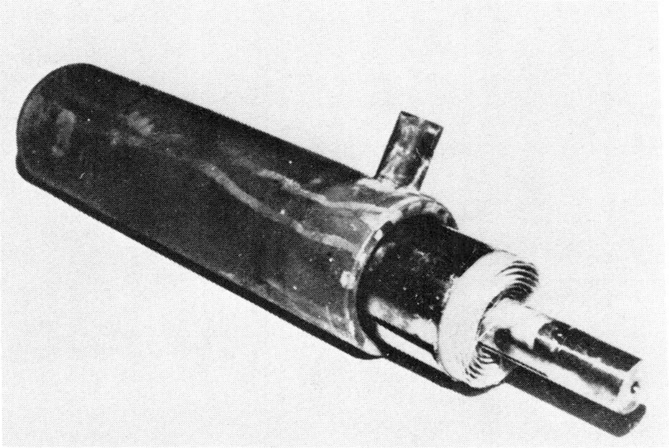

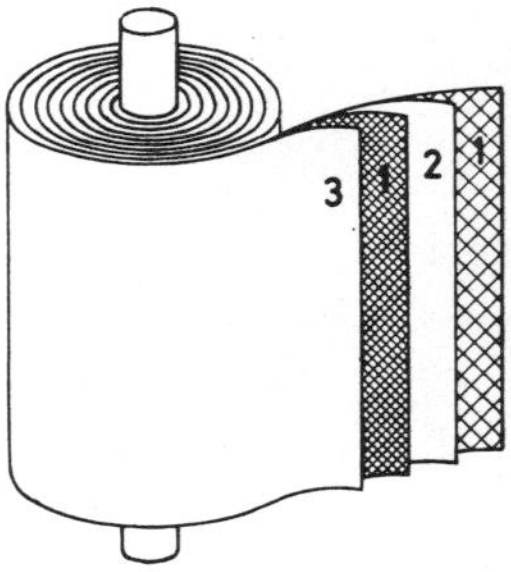

Fig. 7. Swiss roll cell.

inserted in a pipe. Due to the limited volume available for the recovered metal the cathode deposits have to be stripped periodically either by leaching or anodically. Despite the fact that both cell types have shown very promising results in laboratory and field trials, they have yet to be used in industry for the treatment of effluents.

Cells with Vigorous Agitation of the Liquid

A relative motion between the cathode surface and the solution can be achieved by vibration, rotation or mechanical agitation of the electrodes, by circulating the electrolyte using pumps or by air injection and by hydraulic agitation of nonconductive particles within the space between the electrodes.

Only two types of these cell designs have up until now found commercial use in waste water treatment, namely the "Eco-Cell" developed by Ecological Engineering Ltd.[17,18] and the "Chemelec Cell" marketed by BEWT (British Engineers Water Treatment) under license of the British Electricity Council.[19,20] The characteristic cell designs of the "ECO-Cell" shown in Fig. 8 is a fast rotating cylinder cathode in a membrane divided cell. The metal deposits are scraped

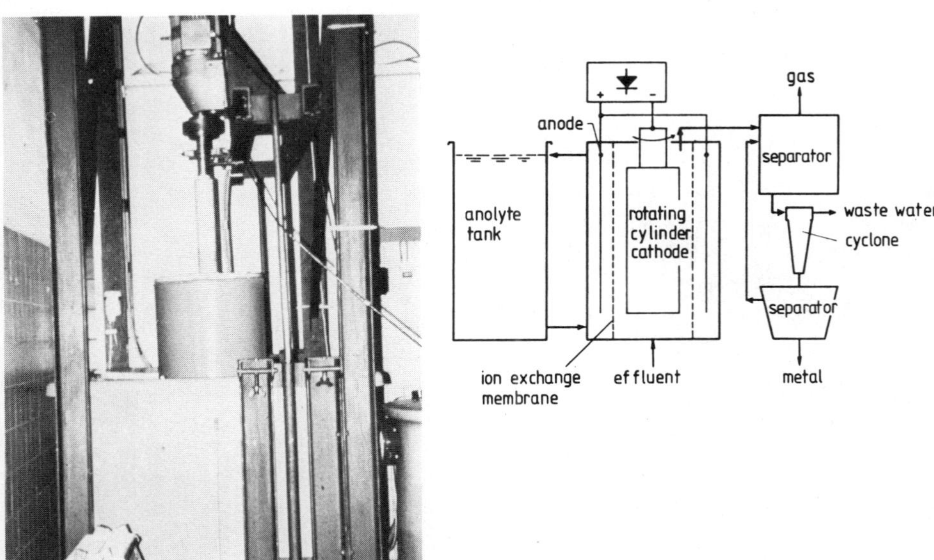

Fig. 8. Eco cell

off the cathode in powder form and continuously removed from the cell and then separated from the liquid. Under these operating conditions overall mass transfer rates have been observed which are up to 1000 times greater than those in conventional planar cell systems. The first cell installation with a cascade effect has been operated since 1977 for decopperizing effluents (400 ppm down to 2 ppm) from the manufacture of copper phtalo-cyanine and other organic compounds. Cells of this type have been also applied for silver recovery.

In the "Chemelec Cell" shown in Fig. 9 mass transfer is improved by a combination of mesh electrodes and a non-conducting fluidized bed as turbulence promotor. This cell is used for the treatment of a variety of waste waters, especially in the UK. From Table 4 it can be seen that the "Chemelec Cell" will not readily produce final effluents for discharge. A cost comparison of the "Chemelec Cell" with other waste water treatment methods is given in Table 5 for copper recovery from effluents containing 20 tons copper per year.[19]

Cells with Large Cathode Areas and Intense Agitation

Typical examples of industrially used direct recovery cells with extended cathode areas and intense agitaton listed in Tables 3 and 4 include the "Fixed Bed"-, the "Fluidized Bed"-, the "Rotating Tubular"- and the "Impact Rod"-Reactor (GOECOMET®).

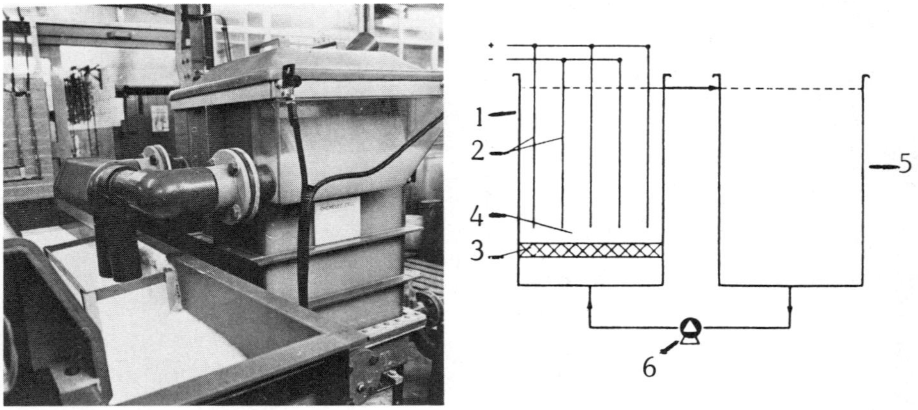

Fig. 9. Chemelec cell. 1) Cell; 2) Expanded Mesh Electrodes; 3) Distributor; 4) Fluidized Glass Bead; 5) Rinse Tank; 6) Pump.

Table 5. Cost Comparison for Copper Recovery from Effluents with Chemelec Cell (20 t Cu/per year)[19].

System	Capital £	Cost	
		Running £/tonne	Total £/tonne
Reverse osmosis	60,000	250	550
Ion exchange	40,000	100	300
Electrolytic conventional (Plate type cathodes)	15,000	160	235
Solvent extraction	25,000	80	205
"Chemelec" cells (Extended surface cathodes)			
Cu mesh	10,000	75	125
Ti mesh	10,000	55	105

C. L. Lopez-Cacicedo

Fixed Bed Cell Systems

A characteristic example for concentrator cells with large 3-dimensional cathode surface areas of the fixed bed or packed particulate bed cell type with a special tetragonal shape to ensure low residual concentrations is shown in Fig. 10. Cell types for industrial application have been developed as particulate bed ("Enviro-Cell", Carbone A.G., Germany), as graphite particulate bed (National Institute for Metals, SA), as knitted mesh electrodes (Heraeus, Germany), as particulate coke bed (Kennecott Corp., USA), and with packed carbon fibers of 5-10 μm (HSA-Reactors Ltd., Canada). Some of these cells have achieved commerical operations. From all these concentrator cells, however, the deposited metal has to be periodically removed in a second step either by anodic or by chemical dissolution.

Fluidized Cell Systems

Another important route for electrolytic treatment of effluents is to use the various types of fluidized bed electrodes. The cell consists of cathode and anode compartments which are separated from each other by a membrane or diaphragm. This is shown in Fig. 11. Their principle difference in comparison with fixed bed is the hydraulic agitation of each single particle in the cathode compactment by the waste water to be treated. Fluidized bed cells with

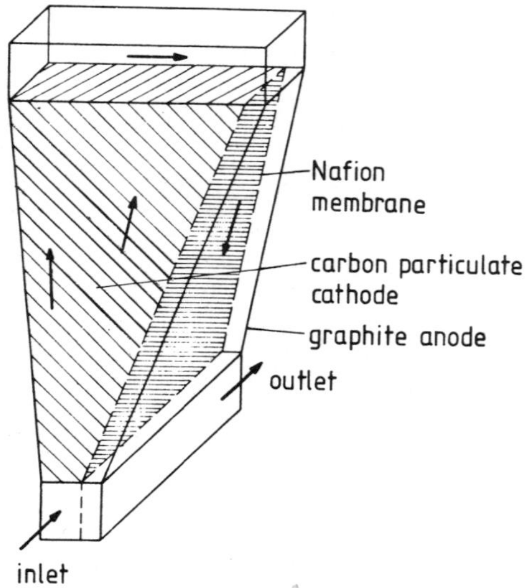

Fig. 10. Fixed bed recovery electrolysis cell.

their excellent mass transfer conditions and highly extended surface areas have shown in short term performances with various laboratory scale cell designs that high recovery rates at favorable current efficiencies can be achieved even at very low residual metal values in the effluents.[21, 22, 23, 24, 25]

However, it has been difficult to scale up these systems from pilot size to commercial units due to problems associated with the diaphragm, with potential distribution in the bed, the steady hydraulic fluidization of the bed, with metal deposits on the feeders and bridging effects, with very complicated cell construction and necessary regulating systems. These problems have stopped the further development of the "Parel SA" cell developed by the Diamond Schamrock Corp.[16] In this version the fluidized bed is circulating rapidly along the ion-permeable diaphragm and slowly descending on the cell wall.

The only fluidized bed system which up until now operates in industrial scale has been developed by Akzo Zout Chemie, Holland. From the schematic view of the Akzo cell shown in Fig. 12, it can be seen that the design consists of a cylindrical fluidized cathode bed in which are enclosed cylinderical diaphragm containing the anodes.[26] Two of these units have been installed at the Enka Co. in

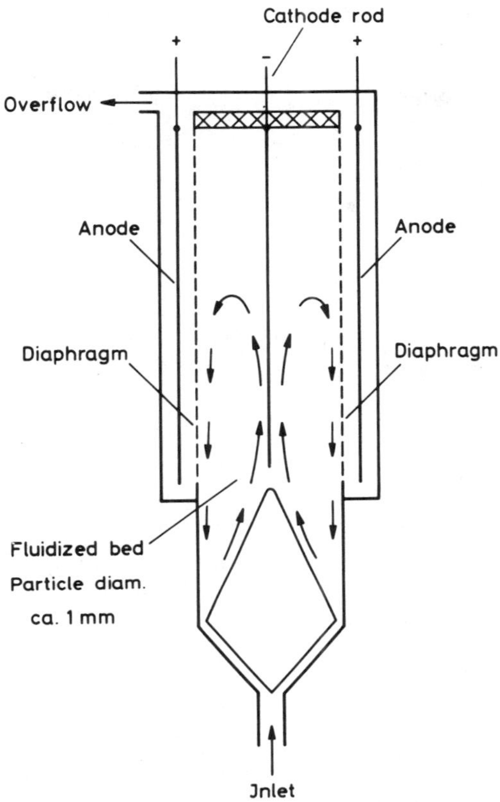

Fig. 11. Fluidized bed electrolysis cell.

Wuppertal, Germany, for copper removal from sulphuric acid solutions (100 g H_2SO_4/1). A schematic flow diagram and some data of the commercial operation are given in Table 6.[27] Further development of this cell system has been recently abandoned by Akzo Zout Chemie.

Rotating Cylinder and Impact Electrodes

Comprehensive investigations at the Techische Universitat and Technische Fachhochschule Berlin, employing quite different electrodes, resulted in the development of two new cell types with specific features of construction. The rotating cathode cylinder shown in Fig. 13 (GOECOMET® Au 80 x)[28] consists of two perforated concentric tubes of non-conductive material, the interspace of which is filled to a great extent with conductive particles. Outside of the perfo-

Table 6. Industrial Copper Removal by Fluidized Bed Cell of Akzo Zout Chemie.

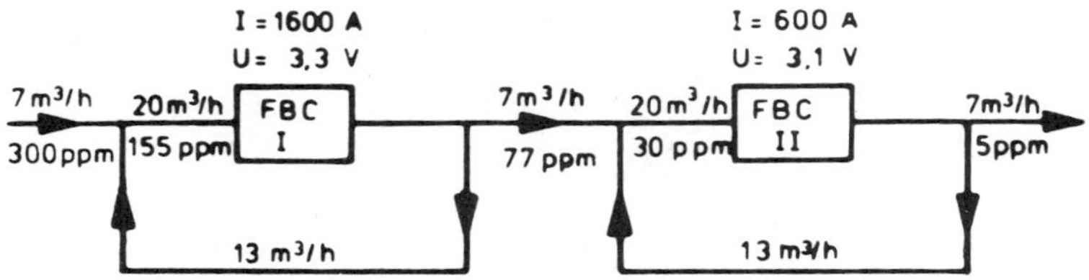

operation data

	Cell I	Cell II
cell volume	192 l	192 l
initial concentration	300 ppm	77 ppm
final concentration	77 ppm	5 ppm
flow rate	7 m^3/h	7 m^3/h
current	1600 A	600 A
voltage	3.3 V	3.1 V
current efficiency	0.82	0.71
power consumption	0.75 kWh/m^3	0.27 kWh/m^3

rated rotating casing, mesh metal anodes are arranged, and the electrolyte is circulated through the spacings between the agitated particles by forced flow. In this way a large cathodic surface is combined with a fast relative motion between the cathode and the solution. Furthermore, all particles are continuously in contact with the cathode bus bar in spite of their agitation, and the thickness of the solution adherent to the surface of the particles is frequently reduced leading to improved mass transfer. This mechanical effect can be easily intensified to such a degree that the metal deposit can be peeled off after a thickness of some ten micrometers has been attained. Under this condition, the rotating cylinder electrode can be operated continuously.

The electrodes of the "Impact Rod Reactor" (GOECOMET® Ag 80)[29,30] consisting of single metal rods, the ends of which are agitated by cathodic guide rails in such a way that the rods are frequently impacting each other as shown in Fig. 14. This steady interrupted movement of the electrodes helps to improve the mass transfer. The metal deposits on the rods are easily peeled off and can be collected from the bottom of the cell system. The cathode

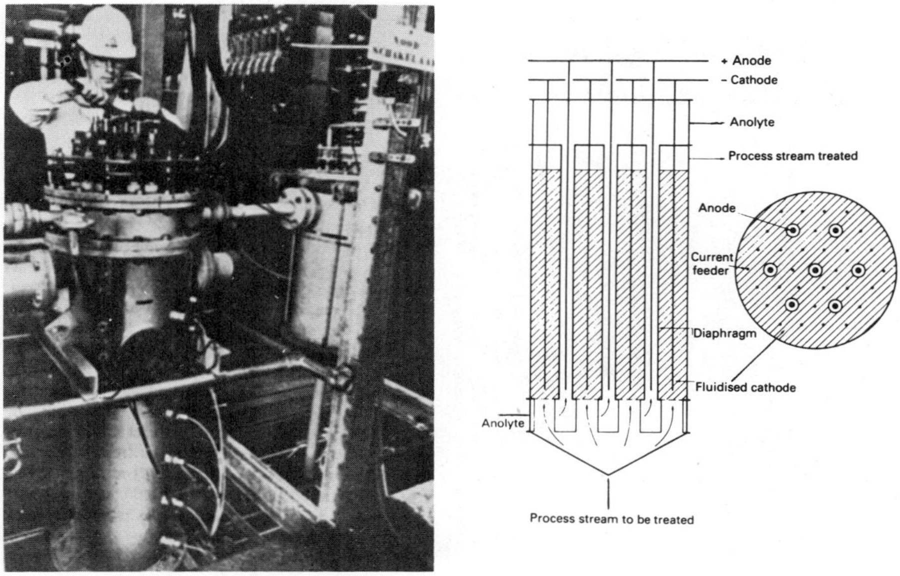

Fig. 12. Akzo Zout Chemie fluidized bed electrolysis cell.

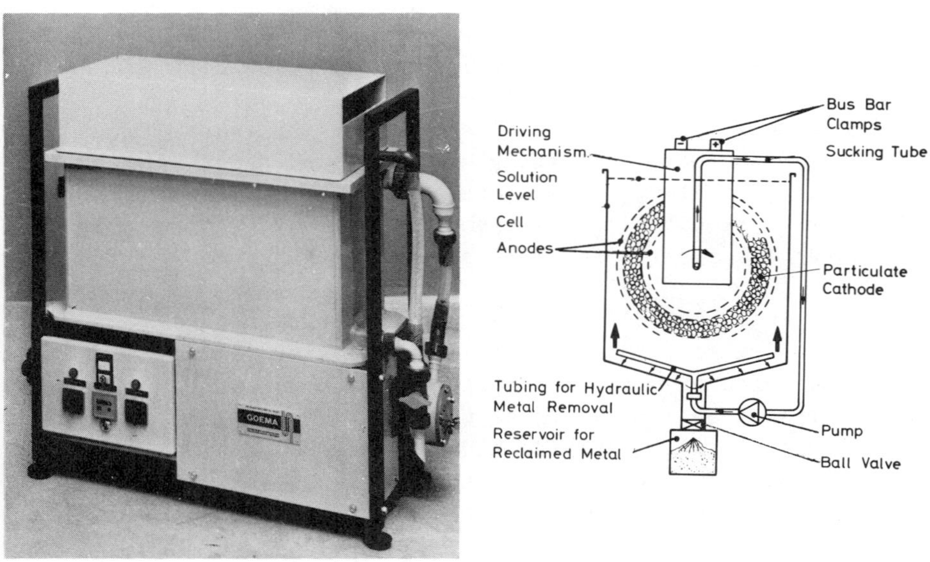

Fig. 13. Rotating tubular bed electrolysis cell (GOECOMET® Au 80 x)

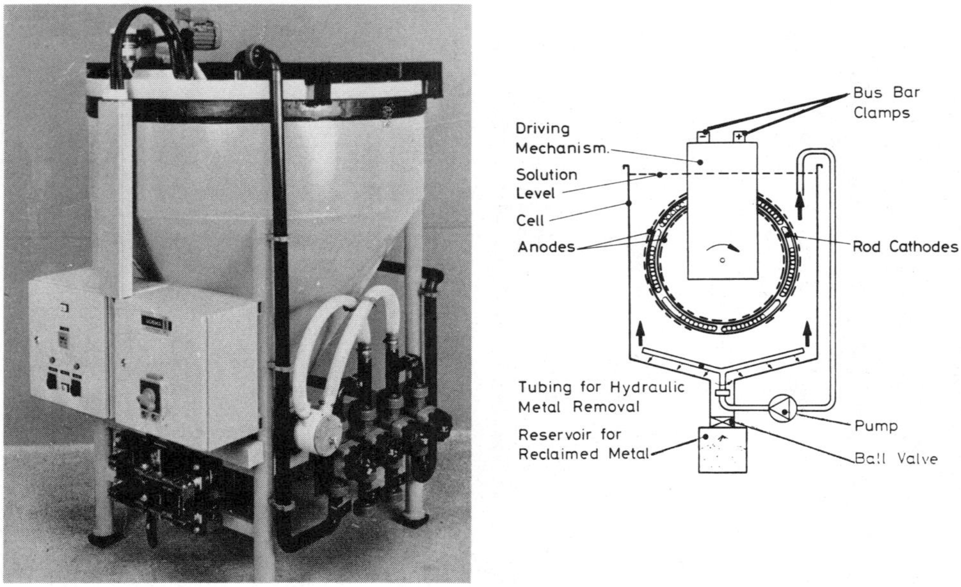

Fig. 14. Impact rod electrolysis cell (GOECOMET® Ag 80)

rods are encased by knitted mesh metal anodes, and the electrode spacing can be kept quite narrow leading to a low cell voltage.

These cell types have been tested over the years using a large variety of rinse waters, plating effluents and spent solutions, containing zinc, cadmium, nickel, copper, silver and gold as well as gold alloys. Precious metals have been recovered in industrial scale using these cells for several years.

The results shown in Fig. 15 indicate that the recovery of gold from various rinse waters in the electroplating industry can be carried out with high current efficiencies down to final concentrations of less than 1 ppm. Rinse waters from pure gold solutions contain only potassium cyanide and gold potassium cyanide. Hard gold effluents consist of weak acid solutions of gold potassium cyanide with certain amounts of cobalt. Gold alloy solution are prepared from gold potassium cyanide with copper and cadmium cyanide salt additions. The technical and economic data compiled in Table 7 from an industrial operation shows that the recovery cost of gold is nearly negligible.[31]

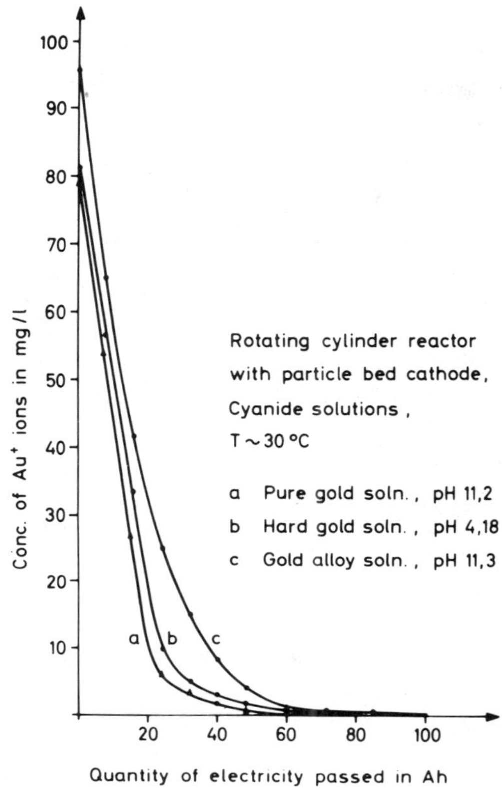

Fig. 15. Electrolytic recovery of gold from rinse water.

The impact rod reactor has been operated successfully in several industrial silver plating installations for silver recovery as well as for the treatment of solutions from photographic processing. Fig. 16 shows the fast decrease of the silver content during electrolytic treatment of electroplating rinse water with the impact rod reactor. The deposited silver is peeled off from the rod cathodes in the form of small flakes which are easily removed from the bottom of the cell. The technical and economic data for electrolytic silver recovery given in Table 8 indicate also that the investment costs for reclaiming precious metals from dilute solutions using a direct electrolytic recovery process are rapidly repaid and that this cell system can be economically operated down to very low final metal values. It should be noted that the simultaneous anodic oxidation of cyanide reduces the cyanide waste treatment operation and the savings for sodium hypochlorite readily covers a substantial amount of the recovery cost.

Table 7. Technical and Economic Data for Electrolytic Recovery of Gold from Rinse Water.

100 l Gold Containing Solution With
0,2 g/l Au = 20 g Au

Current:	10	A
Cell Voltage:	4,5	V
Treatment Time:	16	h
Energy Demand:		
Electrolysis:	1	kWh
Driving Motor:	0,5	kWh
Concentrations:		
Starting:	0,2	g/l
Final:	0,2	ppm
Weight of Recovered Gold:	20	g
Value of Dissolved Gold:	600,00	DM
Value of Recovered Gold:	540,00	DM
Recovery Costs:		
Energy Costs:	0,20	DM
Handling (0,5 h):	1,00	DM
Maintenance:	8,00	DM
Surplus per Charge:	530,00	DM

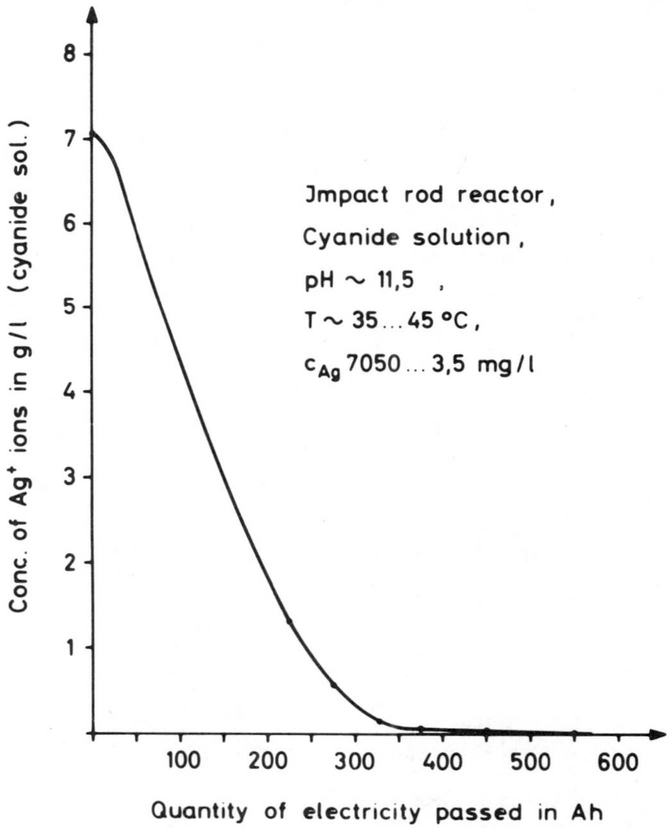

Fig. 16. Electrolytic recovery of silver from rinse water.

Table 8. Technical and Economic Data for Electrolytic Recovery of Silver from Rinse Water.

200 L SILVER CONTAINING SOLUTION WITH
5 G/L AG = 1000 G AG

CURRENT:	125	A
CELL VOLTAGE:	4,5	V
TREATMENT TIME:	16	H
ENERGY DEMAND:		
ELECTROLYSIS:	14	KWH
DRIVING MOTOR:	1	KWH

CONCENTRATIONS:

	CYANIDE:	SILVER:
STARTING:	20 G/L	5 G/L
FINAL:	8 G/L	2 PPM

WEIGHT OF RECOVERED SILVER:	1	KG
CYANIDE DESTRUCTION:	2,4	KG
VALUE OF DISSOLVED SILVER:	700,-	DM
VALUE OF RECOVERED SILVER:	630,-	DM
RECOVERY COSTS:		
ENERGY COSTS:	2,-	DM
HANDLING (0,5 H):	10,-	DM
MAINTENANCE:	17,-	DM
SAVINGS BY ANODIC DETOXICATION OF 2,4 KG CYANIDE:	11,-	DM
SURPLUS PER CHARGE:	610,-	DM

Table 9. Weak Acid Cation Exchange Resins for Cadmium Recovery (According to G. Spanier).

Properties	Carboxylic Acid Resin	Chelating Resin
Matrix of Resin	Acrylate, macroporous	Polystyrene, macroporous
Groups effecting ion exchange	Carboxylic Acid -COOH	Iminodiacetate $-N(CH_2COOH)_2$
Loading Form	Mono-sodium -COONa + -COOH	Mono-sodium $-N(CH_2COONa)(CH_2COOH)$
pH Range of operation	4 - 7 (...9)	2 - 6 (...9) depending from type of metal
Usable capacity	1 - 3 Equivalents per liter of resin (depending from pH in lower range) (the equivalent weight of Cd = 56.2 g)	depending from type of metal e.g. 40-50 g Cu, 30-40 g Ni or Zn, or 30-40 g of different metals like Cd per liter of Resin
Selectivity	$H^+ >> Cu^{2+} > Fe^{2+} > Zn^{2+} > Ni^{2+} > Cd^{2+} > Ca^{2+}$	$H^+ >> Cu^{2+} > Ni^{2+} > Zn^{2+} > Dc^{2+} > Fe^{2+} >> Ca^{2+}$

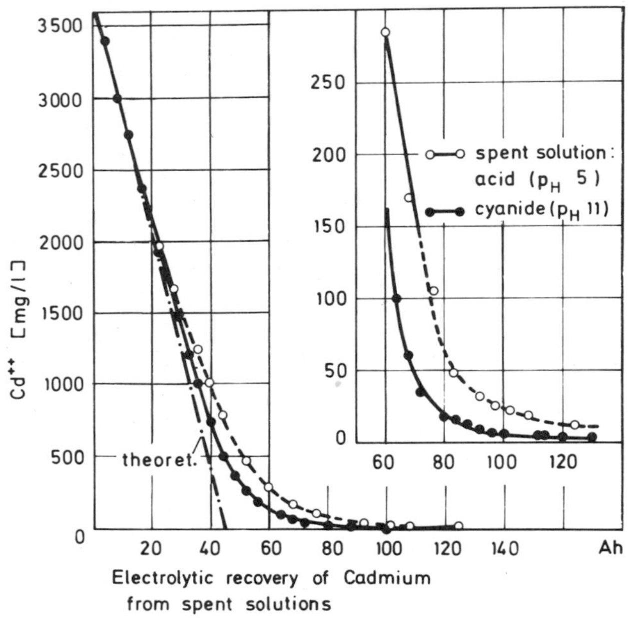

Fig. 17. Electrolytic recovery of cadmium from spent solutions.

From an environmental aspect, avoiding cadmium containing sludges by electrolytic recovery of this toxic metal has become a very important technology during the last few years. In electroplating shops for the deposition of cadmium layers, acid solutions of cadmium sulfate and alkaline solutions with a cadmium cyanide complex, respectively, are commonly used. The effluents of counter current rinse tanks contain several grams of cadmium per liter or less. From these solutions cadmium can be recovered by use of the rotating cylinder or impact rod reactor as indicated in Fig. 17. An evident decrease of current efficiency is only observed in the ppm range.[32,33]

Results from combined electrolytic and various ion exchange resin studies indicate that a combination of electrolysis and ion exchange will result in a most economic and reliable operation. Special weak acid cation exchange resins have proven to be successful for the recovery of metals from very dilute solutions.[34] By using two equal ion exchange columns in series, a most economic and reliable operation can be achieved. Table 9 shows that the

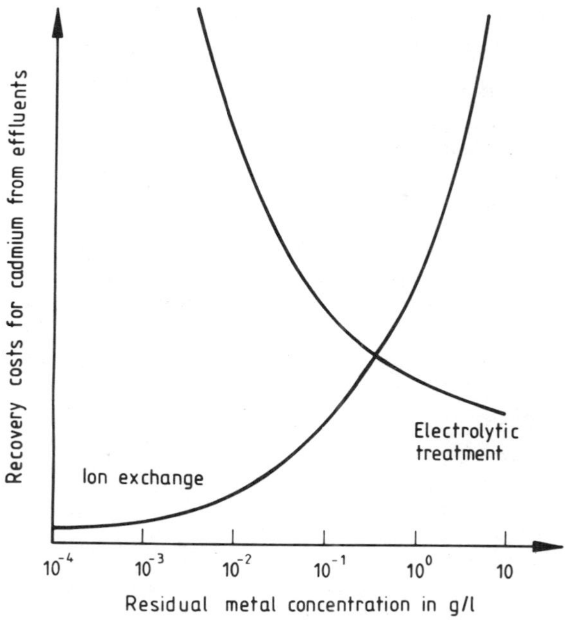

Fig. 18. Costs for recovery of cadmium by electrolytic treatment and ion exchange as a function of the cadmium content of the effluent

capacity of an appropriate resin amounts to at least 80 grams of cadmium per liter of resin. However, the recovery costs are increasing with increasing metal content of the waste water.

On the other hand, direct electrolytic recovery of metallic cadmium is most advantageous with increasing metal content of the solution, as can be seen from Fig. 17. The costs of both methods of operation are shown in Fig. 18, which leads to a minimum, dependent on the plant conditions, costs of energy, chemicals and labor, throughput, etc.[35] The cadmium content can be reduced to about 0.05 ppm by this effluent treatment method.

If cadmium from a cyanide solution is to be recovered, most of the cyanide will be oxidized anodically. The remaining cyanide can be easily oxidized by hypochlorite in this alkaline solution, and afterwards adjust the pH value to about 5.5. The ion exchange resin can be favorably regenerated by sulfuric acid and conditioned with sodium hydroxide. The cadmium containing regenerated solution is returned to the electrolytic recovery cell, thus improving the conductivity of the solution and saving electric energy. By this procedure cadmium containing sludges can be avoided. Savings in cost

Table 10. Costs for Electrolytic and Conventional Treatment of Cadmium Effluents.

Acid effluents (1 m^3)		
Cadmium content	2 g/l Cd	
Free sulfuric acid	4 g/l H_2SO_4	
Electrolytic treatment:		
Energy cost (20 kWh à 0,15 DM/kWh)		3,00 DM
Neutralization of acid		4,00 DM
Regeneration of ion exchange resin (6 liters)		6,00 DM
Draining of treated waste water		1,00 DM
Total cost	about	14,00 DM
Conventional treatment:		
Value of metal (2 kg Cd à 10,- DM/kg)		20,00 DM
Neutralization and precipitation		4,00 DM
Filtration of sludge		2,00 DM
Transportation of sludge		1,00 DM
Dumping cost		2,00 DM
Draining of treated waste water		1,00 DM
Total cost	about	30,00 DM

in using electrolytic cadmium recovery compared with conventional treatment by detoxification and precipitation have been calculated and summarized in Tables 10 and 11. The cost of installation and labor have not been taken into consideration because these figures will depend on the individual operating condition of each plant. The data presented are valid for 1 m^3 of effluent. Due to the detoxification of cyanide at the anodes, electrolytic treatment of cyanide solutions can bring about higher savings compared with acid effluents.[36]

CONCLUSIONS

Special electrolytic cell systems have been developed from laboratory and pilot scale demonstration units to full scale industrial operation for the recovery of metals from dilute aqueous solutions. These cells are equipped with extended electrode surface area and are operated with vigorous agitation of the electrolyte to improve mass transfer conditions. These novel electrowinning cell designs make possible the extraction of metal values efficiently to very low residual concentrations in the ppm range or even less than 1 ppm.

Table 11. Costs for Electrolytic and Conventional Treatment of Cadmium Effluents.

CYANIDE EFFLUENTS (1 M^3)		
CADMIUM CONTENT 3,75 G/L CD		
CYANIDE CONTENT 12 G/L NACN		
ALKALI CONTENT 5 G/L NAOH		
ELECTROLYTIC TREATMENT:		
ENERGY COST (18 KWH À 0,15 DM/ KWH)	ABOUT	3,00 DM
DETOXICATION OF RESIDUAL CYANIDE		16,00 DM
NEUTRALIZATION OF ALKALI		7,00 DM
REGENERATION OF ION EXCHANGE RESIN (6 LITERS)		6,00 DM
DRAINING OF TREATED WASTE WATER		1,00 DM
TOTAL COST	ABOUT	33,00 DM
CONVENTIONAL TREATMENT:		
VALUE OF METAL (3,75 KG CD À 10,- DM/KG)		37,50 DM
DETOXICATION WITH HYPOCHLORITE		60,00 DM
NEUTRALIZATION AND PRECIPITATION		3,50 DM
FILTRATION OF SLUDGE		2,00 DM
TRANSPORTATION OF SLUDGE		2,00 DM
DUMPING COST		4,00 DM
DRAINING OF TREATED WASTE WATER		1,00 DM
TOTAL COST	ABOUT	110,00 DM

A comparison of several electrolytic cells for waste water treatment have been reported by Kreysa[15] using new data and are summarized in Tables 12 and 13. The results show that some types of cells allow high volumetric thoughputs (V_D) with favorable current efficiencies (η). However, some of these advantageous appearing cells have been either tested only on a laboratory scale and short term pilot scale or are facing problems in commercial applications due to their complicated design and control systems. Further improvements with regard to an optimum compromise between the best technological design and simplicity are therefore necessary.

Electrolysis results indicate that favorable metal recoveries can be attained not only for electropositive metals but also for metals as electronegative as cadmium. A combination of electrowinning and ion exchange can make possible a more economic treatment procedure for the recovery of the electronegative metals. It seems likely that as more practical experience is gained and further extension of their applications are developed, these electrolytic processes will be increasingly accepted in industrial operations.

Table 12. Operational Data of Different Cells for Electrolytic Waste Water Treatment.

Type	Scale	Removed metal	C_o ppm	C_1 ppm	V_R l	V_D l/h	I A	i A/cm^2	U_Z V	η	V_v
Eco cell	Ind.	Cu	100	2	275	8000	1000	0,059	12	0,65	0,2
Impact rod cell	Ind.	Ag	<5000	1	450	25	150	20.10^{-3}	4,5	0,25	1500
Swiss-roll cell	Lab.	Cu	380	25	0,3	0,7	0,55	$0,46.10^{-3}$	1,56	0,38	0,8
Fixed bed cell	Ind.	Cu	50	0,1	4,8	50	3,2	$0,76.10^{-3}$	1,9	0,66	0,1
Fluidized bed cell	Ind.	Cu	77	5	192	7000	600	-	3,1	0,71	0,15
Rotating tubular bed cell	Ind. Lab.	Au Cu	80 2340	0,2 5	50 50	3,1 2,5	10 8	1.10^{-3} 27.10^{-3}	4,5 3,0	0,01 0,35	1250 4,50

C_o, C_1: starting resp. final concentration; V_R: cell volume; V_D: throughput; I: cell current; i: current density; U_Z: cell voltage; η : current efficiency; V_v: value of metal per volume of solution

Figures from Kreysa completed by new data

Table 13. Comparison of Specific Data of Different Cells for Electrolytic Waste Water Treatment.

Type	Scale	Removed metal	E_s^e kWh/m^3	E_s^t kWh/m^3	η_e	g_n^s l/l.h	V_v $/m^3
Eco cell	Ind.	Cu	1,5	3,0	0,05	20	0,2
Impact rod cell	Ind.	Ag	30	32	0,02	2 - 5	1500
Swiss-roll cell	Lab.	Cu	1,23	-	0,22	20	0,80
Fixed bed cell	Ind.	Cu	0,12	-	0,31	28	0,1
Fluidized bed cell	Ind.	Cu	0,27	-	0,20	30	0,15
Rotating tubular bed reactor	Ind. Lab.	Au Cu	15 10	17 12	0,001 0,30	0,4 0,2	1250 4,50

E_s^e, E_s^t: energy demand for electrolysis resp. electrolysis and agitation; η_e: energy efficiency; g_n^s: volume of waste water treated per hour and cell volume for recovery rate of 90 percent; V_v: value of metal per volume of solution

Figures from Kreysa completed by new data

ACKNOWLEDGEMENTS

The author acknowledges most thankfully the long time cooperation with Profesor Dr. Ing. H. W. Lieber, Labor für Galvanotechnik, Technische Fachhochschule Berlin, and the experimental contribution of Dr. Ing. E. Avci, Dr. Ing. A. Atlam, and Dipl. Ing. H. G. Eran. The pilot plant and industrial scale tests were kindly supported by Dr. K. G. Götzelmann of Fa. GOEMA, D-7000 Stuttgart 1.

REFERENCES

1. I. B. Ishan: Plating & Surface Finishing, 1979, no. 3, pp. 18-19
2. R. Kammel and H.-W. Lieber: Z. Galvanotechnik, 1977, vol. 68, pp 57-63, pp. 123-130, pp. 413-418, pp. 710-715, pp. 789-794, pp. 883-886; 1978, vol. 69, pp. 317-324, pp. 624-630, pp. 687-696.
3. R. Kammel and H.-W. Lieber: (Japanese Translator Ch. Yamauchi and T. Oki), Jour. of the Metal Finishing Soc. of Japan, 1979, vol. 30, no. 6, pp. 304-316; no. 7, pp. 327-378; no. 8, pp. 421-427; no. 9, pp. 477-489.
4. R. Kammel and H.-W. Liever: Z. Metall. 1980, vol. 34, pp. 258-261.
5. R. Kammel and H.-W. Lieber: Z. Galvanotechnik, 1977, vol. 68, pp. 241-251.
6. D. S. Flett: Electrolytic Recovery of Metals from Dilute Aqueous Solutions, lecture, Warren Spring Laboratory, Stevenage, Herts.
7. V. Tafel: Lehrbuch der Metallhüttenkunde, vol. 1, p. 16, S. Hirzel-Verlag, Leipzig, 1951.
8. V. Tafel: Lehrbuch der Metallhüttenkunde, vol. 1, pp. 45-46, S. Hirzel-Verlag, Leipzig, 1951.
9. B. Neumann: Z. Elektrochemie, 1906, vol. 12, pp. 569-578.
10. N. Ibl: Z. Erzmetall., 1969, vol. 22, pp. B87-B98.
11. P. M. Robertson, B. Scholder, G. Theis and N. Ibl: Chemistry and Industry, 1978, no. 1, July, pp. 459-465.
12. H.-W. Lieber: Beitrag zur Kenntnis der kathodischen Abscheidungs-bedingungen bei der Raffinationselektrolyse des Kupfers, Ph.D. Thesis, Technical University, Berlin, Federal Republic of Germany, 1966.
13. H. Wöbking and H. Wörz: Grundlagen des PCR-Verfahrens und Ergebnisse bei dessen Anwendung in der Raffinationselektrolyse der Montanwerke Brixlegg, lecture, GDMB-Kupferausschuß, Köln, Federal Republic of Germany, March 20, 1980.
14a. K. B. Keating and J. M. Williams: Resource, Recovery and Conservation, 1976, vol. 2, pp. 39-55.
14b. G. Kreysa: Z. Metalloberfläche, 1981, vol. 35, pp. 211-217.
15. G. Kreysa: Personal Communication.

16. G. M. Roberts: The Role of Electrolytic Processes for Metal Winning from Dilute Solutions, paper presented at Symposium, Energy Considerations in Electrolytic Processes, University of New Castle-upon-Tyne, July 1980.
17. F. S. Holland: U.S. Patent no. 4,028,199, 1977.
18. F. S. Holland: Chemistry and Industry, 1978, no. 1, July, pp. 453-458.
19. C. L. Lopez-Casiedo: Trans. Inst. Metal Finish, 1975, vol. 53, no. 2, pp. 74-77.
20. F. C. Walsh and D. R. Gabe: Surface Technology, 1981, vol. 12, pp. 25-37.
21. F. Goodridge: Chem. Process Eng., 1968, February, pp. 93-55, 100.
22. J. A. E. Wilkinson: Trans. Inst. Metal Finishing, 1971, vol. 49, p. 16.
23. D. S. Flett: Chemistry and Industry, 1972, no. 24, pp. 983-988.
24. H.-D. Steppke and R. Kammel: Z. Erzmetall. 1973, vol. 26, pp. 533-538.
25. G. Kreysa and E. Heitz: Chemie-Ing.-Technik, 1976, vol. 48, pp. 852-860.
26. C. M. S. Raats, U. F. Boon and G. Heiden: Chemistry and Industry, 1978, no. 13, pp. 465-468.
27. G. Kreysa: personal communication.
28. R. Kammel and H.-W. Lieber: Method and Apparatus for Treating Metal Containing Waste Water, U.S. Patent No. 4,159,235, June 26, 1979
29. R. Kammel and H.-W. Lieber: Apparatus for Treating Metal Containing Waste Waters, U.S. Patent no. 4,172,780 October 30, 1979.
30. Leaflet, Firma Dr. Götzelmann KG, P. O. Box 997, D-7000 Stuttgart 1.
31, R. Kammel and H.-W. Lieber: Jour. Metals, 1981, vol. 33, no. 10, pp. 45-48.
32. R. Kammel: Process Developments to Meet Environmental Challenges in Metal Finishing, invitation lecture, 10th World Congress on Metal Finishing, Interfinish '80, Metal Finishing Society of Japan, 1980, pp. 17-21.
33. R. Kammel, H.-W. Lieber and E. Avci: Z. Wissenschaft und Umwelt, 1981, no. 2, pp. 47-53.
34. G. Spanier: Z. Galvanotechnik, 1977, vol. 68, pp. 705-709.
35. R. Kammel and H.-W. Lieber: Electrolytic Treatment of Dilute Metal Containing Solutions, TMS paper selection no. A82-24, AIME Meeting, February 1982, Dallas, TX.
36. S. Bustos, R. Kammel and H.-D. Steppke: Z. Metall. 1980, vol. 9, pp. 843-847.

THE PRECIPITATION OF HEMATITE FROM IRON-LOADED VERSATIC ACID SOLUTIONS BY HYDROLYTIC STRIPPING

A. J. Monhemius and L. A. C. Teixeira
Department of Metallurgy and Materials Science
Imperial College of Science and Technology
London SW7 2BP UNITED KINGDOM

and

G. Thorsen
Department of Chemical Engineering
The Norwegian Institute of Technology
University of Trondheim
7034 Trondheim - NTH, Norway

Solutions of Versatic acid containing ferric iron precipitate hematite when reacted with water at temperatures above about 130°C. Factors affecting the rates of precipitation have been investigated. It is shown that precipitation can initiate either by a process of homogeneous self-nucleation or by heterogeneous reaction at the surface of hematite particles in solution. The former process is favored by high loadings of iron in the organic phase and by increases in reaction temperature. It is suggested that differences in the structures of the iron-versatic complexes in the organic phase determine the course of the reaction. The effects of the age of the organic solutions on the rates of precipitation are discussed.

INTRODUCTION

Hydrolytic stripping is the name given to the process whereby metal oxides or hydroxides are directly precipitated from metal-loaded carboxylic acid solutions by reaction with water at elevated temperatures. It has been shown that solutions of Versatic acid loaded with iron, nickel or copper, when reacted with water at temperatures in the range 150-200°C, precipitate Fe_2O_3, $Ni(OH)_2$ and Cu_2O respectively.[1,2] The type of reaction involved is analogous to the hydrolysis of aqueous metal salts and can be represented, for the case of iron, by the following reaction:

$$2R_3Fe_{(org)} + 3H_2O = Fe_2O_{3(s)} + 6RH_{(org)}$$

where R represents the carboxylate anion. This equation is highly simplified since it is known that iron in carboxylic acid solutions can exist in a range of complexes and the free carboxylic acid itself undergoes dimerisation in non-polar aliphatic diluents. Nevertheless the equation serves to illustrate the essential features of the reaction. Iron is precipitated as hematite and the carboxylic acid is regenerated in the acid form. A similar reaction can be written in simplified form to describe the precipitation of nickel as its hydroxide:

$$NiR_{2(org)} + 2H_2O = Ni(OH)_{2(s)} + 2RH_{(org)}$$

From the hydrometallurgical point of view, hydrolytic stripping of iron is potentially the most interesting application of the new process.[3] The great affinity of carboxylic acid extractants for ferric iron allows very pure iron-loaded organic phases to be produced by conventional solvent extraction techniques. The fact that pure water is the reactant used in hydrolytic stripping means that the hematite produced is uncontaminated by other inorganic impurities. In principle this will enable further use to be made of the hematite in iron-making processes or even in more specialized applications, such as pigments. This is in contrast with the iron residues produced currently by conventional aqueous phase hydrolysis of iron-bearing solutions, where the products are not suitable for further use and furthermore can give rise to severe disposal problems. Possible applications of the hydrolytic stripping of iron in the electrolytic production of zinc[4] and in the treatment of metal-bearing mine waters[5] have been discussed.

One of the major factors that will affect the viability of the hydrolytic stripping process is the thermal stability of the organic extractant - this must be capabable of withstanding the elevated temperatures required for the process so that it can be recycled between extraction and hydrolytic stripping. In this respect it has been reported previously[1] that Versatic acid, which is a tertiary carboxylic acid, is thermally stable under the conditions used for the hydrolytic stripping of iron up to 200°C. This is not the case however for some other liquid cation exchange extractants, which in principle could undergo similar hydrolytic reactions. For example, it has been found that di-2-ethylhexyl phosphoric acid does not show the necessary thermal stability, decomposing under hydrolytic strip-ping conditions. Di-nonylnapthalene sulphonic acid has also been found to be unsuitable.

This paper reports the preliminary results obtained from a detailed investigation into the kinetics of the hydrolytic stripping of iron-loaded Versatic acid solutions. The data presented illustrate the major features of the kinetic behavior of this system.

EXPERIMENTAL

Preparation of Iron-loaded Versatic Acid Solutions

Iron-loaded Versatic acid solutions were prepared immediately prior to each hydrolytic stripping experiment by exchange extraction from aqueous ferric chloride solutions with organic solutions containing the calcium salt of Versatic acid.

Calcium loading was achieved by dissolving CaO powder into solutions of Versatic 10 in Escaid 110 (33 V/V %), in a ratio of 50 g CaO per litre organic phase. The oxide was added gradually through a sieve to the stirred organic solution, maintained at about 50°C. After one hour, the organic solution was vacuum filtered through glass-fibre paper to remove residual solid.

If necessary the composition of the organic phase was adjusted to that required for the hydrolytic stripping experiment by mixing the required volume of calcium-loaded Versatic with further unloaded Versatic acid solution and/or Escaid diluent. Iron loading of the organic phase was then brought about by exchange extraction with ferric chloride solution containing 10 g/l Fe^{3+}. The phase ratio was selected so that extraction of all the iron contained in the aqueous phase would give the required concentration of iron in the organic phase. Exchange extraction was carried out by vigorous shaking of the phases in a separating funnel for about one minute. The extraction proceeded via an intermediate stage of partial precipitation of finely dispersed ferric hydroxide. This normally rapidly redissolved into the organic phase, which changed from colorless to dark brown. Following phase separation, the iron-loaded Versatic solution was scrubbed with dilute HCl to a final pH of about 4 to remove residual calcium from the organic phase. Finally the organic phase was washed twice with distilled water to remove traces of chloride and calcium and then filtered through silicone-coated phase separation paper to remove entrained water. The organic phase was then used without delay for the hydrolytic stripping experiment.

Hydrolytic Stripping Experiments

Kinetic studies on the rates of the hydrolytic stripping reaction were carried out in a stainless-steel autoclave of 300 ml internal volume. The autoclave was equipped with a variable speed Magnedrive stirrer (Autoclave Engineers), a sampling system and a heated reservoir of 60 ml volume, from which water could be injected into the autoclave under pressure.

A kinetic run consisted of charging the autoclave with the required volume of freshly-prepared iron-Versatic solution (usually 210 ml) plus hematite seed material, if used. The hematite powder used as a seed was 99% Fe_2O_3, obtained from BDH, and had a surface area of 8.3 m^2/g. The reservoir was also filled with the required volume of water (usually 10 ml). The autoclave was flushed with nitrogen and heated to the reaction temperature. A 10 ml sample of organic was then withdrawn to give a zero time iron analysis and the experiment was started by injecting the water from the reservoir, which had also been heated to the reaction temperature. Following water injection, there was an instantaneous temperature drop in the autoclave of about 6°C, due to the expansion of water vapor. Recovery of the temperature to the set point occurred within one minute and for the majority of the kinetic runs, the effect of this short temperature excursion on the measured rate could be safely ignored. The rate of precipitation was followed by taking a series of samples (2-3 ml each) of the organic phase at timed intervals. These were immediately filtered under vacuum and stored for subsequent analysis.

Following each run, the autoclave was emptied and the whole system cleaned out with acetone. Prior to unseeded (self-nucleating) experiments, the autoclave was also chemically cleaned with hot dilute oxalic acid to remove all traces of residual iron oxides.

Analytical Methods

Iron in the organic phase was determined by UV-visible spectrophotometry using the absorption shoulder at 404 nm. The molar extinction coefficient at this wavelength was 323±4 $(mol\ cm)^{-1}$ and the absorbance was linearly related to iron concentration to at least 0.14 g/l Fe^{3+} (2.5×10^{-3} M). Versatic-Escaid solutions without iron do not absorb at this wavelength. Organic samples taken from the autoclave were diluted as appropriate with absolute alcohol prior to absorbance measurements. The time zero sample was always also analyzed by volumetric analysis with potassium dichromate, after stripping the iron from the organic sample with HCl. This served as an internal standard for the absorbance measurements of the subsequent samples from the experiment.

Data Treatment

The kinetic data were evaluated by the method of initial rates. Experiments were usually continued until at least 95% precipitation of iron had occurred and the initial rates were obtained by polynomial regression analysis on the iron concentration-time profile obtained for each run. Adequate fitting of the data was obtained with a second degree polynomial of the form

$$[Fe]_t = [Fe]_o - r_o t + r_1 t^2$$

where $[Fe]_t$ is the iron concentration at time t, $[Fe]_o$ is the initial iron concentration, r_o is the initial rate of precipitation and r_1 is a positive constant. Curve fitting was done using the MINITAB statistical computer package. The goodness of fit of the polynomial was judged by the squared correlation coefficient, which was usually better than 99% and the t-ratios of the coefficients, particularly of r_o, the initial rate. The latter was generally higher than 10. The agreement between the $[Fe]_o$ value obtained from the fitted polynomial and the analyzed iron concentration of the time zero sample was generally better than 1.5%. A typical precipitation profile together with the fitted polynomial curve is shown in Fig. 1.

RESULTS

The kinetic variables investigated were temperature, total concentration of Versatic 10 in the organic phase (V), initial organic iron concentration (Fe) and the weight of hematite powder added as a seed material (S).

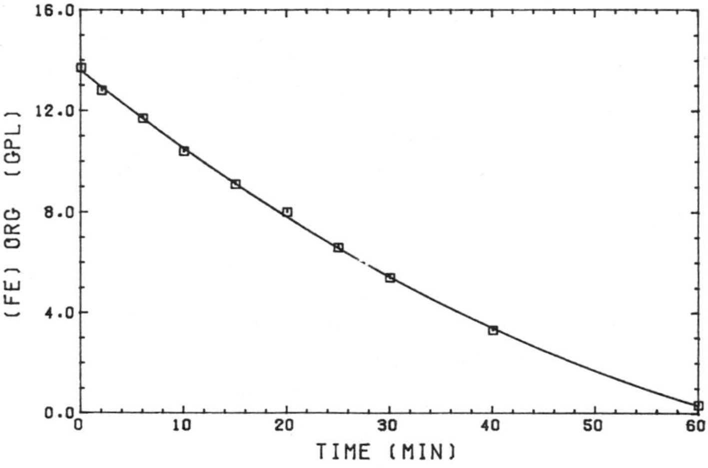

Fig. 1. Typical precipitation data with fitted polynomial curve.
Experimental conditions: T = 135°C, V = 33% V/V,
Fe = 14 g/l (nominal, 13.7 g/l actual)
S = 69 g/l Fe_2O_3 powder
Fitted polynomial $[Fe]_t = 13.6 - 0.322t + 0.0017t^2$

Initial experiments established that the water to organic phase ratio had no effect on the rate of precipitation over the range 1:10 to 1:40. All subsequent experiments were carried with a phase ratio of 1:20 (10 ml water: 200 ml organic phase). The rate of precipitation was also found to be independent of the stirring speed over the range 800 to 1200 rpm. A stirring speed of 1000 rpm was used as standard.

A two-level, three variable factorial design strategy was followed to investigate the effects of the kinetic variables. The levels of the variables were selected in order to explore a range of experimental conditions which were both realistic from an industrial point of view and sufficiently extensive to provide information of their effects on the kinetic behavior of the system.

The two levels of Versatic concentration selected were 25 and 33% V/V. The upper iron concentration used was 14 g/l, which is close to the maximum loading readily achievable with 25% Versatic solutions. The lower level of iron concentration was 7 g/l. The levels chosen for the weight of added hematite powder seed were 23 and 46 g/l. These weights corresponded approximately to one and two times the weight of hematite that would be precipitated from an

Table 1. Results of Factorial Experiment.

Minimum and maximum levels of variables (T = 135°C)

Variable	Min	Max	Units
Versatic concentration (V)	25	33	% V/V
Organic Fe concentration (Fe)	7	14	g/1
Wt. of Fe_2O_3 seed (S)	23	46	g/1

Experimental conditions, results and factorial analysis

V	Fe	S	Initial rate		Effect	Mean diff.	t ratio
(% V/V)	(g/1)	(g/1)	g/1 min.	ln(molal/min)			
25	7	23	0•12	- 5•95	Mean	-5•34	-
33	7	23	0•07	- 6•50	V	-1•05	9•27*
25	14	23	0•78	- 4•07	Fe	+0•97	8•61*
33	14	23	0•14	- 5•80	VFe	-0•45	4•00*
25	7	46	0•28	- 5•10	S	+0•49	4•31*
33	7	46	0•15	- 5•74	VS	+0•09	0•82
25	14	46	0•74	- 4•13	FeS	+0•32	2•81
33	14	46	0•21	- 5•40	VFeS	+0•14	1•22

Standard error of ln rate (molal/min) = 0•113

t-ratio for 99% significance = 3•71 (6 d.f.)

* statistically significant effects.

organic phase initially containing 14 g/1 Fe. The conditions that were fixed during the factorial experiment were as follows: temperature 135°C; stirring speed 1000 rpm; water to organic phase ratio 1:20; nigrogen atmosphere. The response used to analyze the results of the factorial experiment was the natural logarithm of the initial rate, obtained from the fitted polynomial and expressed in molal min^{-1} units. Certain experiments within the design were replicated to obtain an estimate of experimental error. The results of the factorial experiment, together with estimates of the main effects and interactions of the variables, are given in Table 1.

Subsequently, other experiments were carried out to further investigate the effect of seed material on the reaction. These experiments included some in which no seed was added and others in which additions were made of up to double the amount of seed material used in the factorial experiments. The results are reported in Table 2. The results of a third series of experiments, in which the reaction temperature was increased up to 170°C, are given in Table 3.

Table 2. Effect of Seed on Initial Rate of Reaction (T = 135°C).

V	Fe	S	Initial rate	
(% V/V)	(g/l)	(g/l)	g/l. min	molal/min x 10^3
33	14	92	0·45	9·70
33	14	69	0·32	6·89
33	14	0	0	0
33	7	0	0	0
25	14	0	0·79	17·20
25	7	0	0	0

Table 3. Effect of Temperature on Initial Rate of Reaction. (V = 33% V/V, Fe = 14 g/l)

Temperature	S	Initial rate	
(°C)	(g/l)	g/l. min	molal/min x 10^3
135	(0	0	0
	(23	0·14	3·02
140	(0	0	0
	(23	0·22	4·74
145	(0	0·10	2·15
	(23	0·37	7·97
150	23	0·69	14·86
155	(0	0·36	7·75
	(23	0·76	16·37
160	23	1·23	26·49
170	0	3·00	64·61

DISCUSSION

The results of the experiments carried out at 135°C and reported in Tables 1 and 2 are displayed in graphical form in Figure 2 as plots of initial reaction rate against weight of added hematite seed material. A number of the kinetic features of this system are made evident by this diagram.

The effect of seeding on the rate of reaction is clearly illustrated. For three of the solution compositions investigated, namely (33/14), (33/7) and (25/7), where the numbers in brackets signify

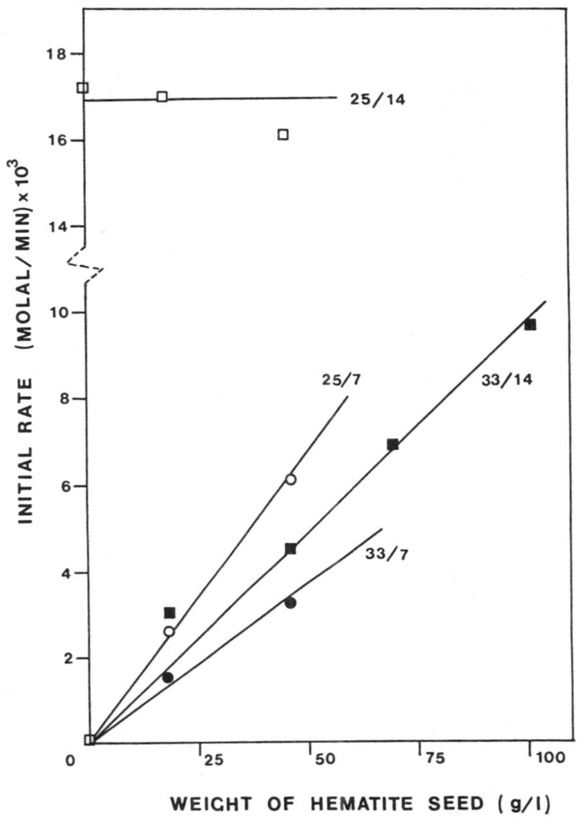

Fig. 2. Initial rate of precipitation as a function of weight of hematite seed added.
T = 135°C. Solution compositions indicated: V/Fe.

the Versatic concentration in % V/V and initial organic iron concentration in g/l, respectively, there is no reaction in the absence of seed and the initial rate of reaction is directly proportional to the weight of hematite added as a seed. In the case of the fourth solution composition (25/14), the behaviour is quite different. There is a very fast rate of reaction at zero seed and the initial rate is more or less unaffected by additions of hematite powder. Thus in this type of solution, precipitation of hematite occurs spontaneously and the rate of precipitation is rapid and is independent of the surface area of solid present in the system. This type of behaviour is referred to subsequently as self-nucleating precipitation.

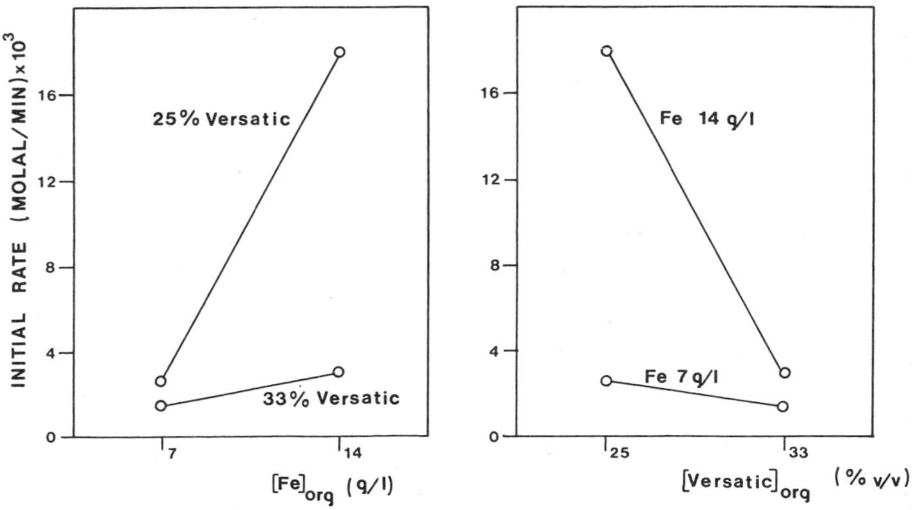

Fig. 3. Initial rate of precipitation as a function of initial iron concentration at constant Versatic concentration and vice versa. (T = 135°C)

The mode of precipitation exhibited by the other solutions, referred to as heterogeneous precipitation, evidently requires the presence of active solid surface to catalyze the reaction and the initial rate of reaction is directly proportional to the weight, and hence the area of solid surface available.

The effects of the other main variables, Versatic 10 concentration and iron concentration, are also illustrated in Fig. 2. At a fixed level of Versatic concentration, the initial rate of reaction increases with increasing initial iron concentration in the organic phase. The Versatic concentration has a negative effect on the initial rate of reaction since, for a fixed initial iron concentration, the initial rate decreases with increasing Versatic acid concentration. These increases and decreases in initial rate are however not additive, being instead dependent on the level of the fixed variable, as illustrated in Fig. 3. It may be seen that the increase in rate with iron concentration is much greater at the low level of Versatic (25%) than at the high level. Similarly the decrease in rate with Versatic concentration is more marked at high iron (14 g/l) than at low iron. This interaction between the main effects of iron and Versatic concentrations is manifested in the analysis of the factorial experiment in Table 1 by the fact that the VFe interaction is statistically significant.

Although the total concentration of Versatic acid in the organic phase is a convenient experimental variable, it is not a meaningful kinetic variable, since its value remains constant during the hydrolytic stripping reaction and therefore it should not influence the rate of reaction. The organic phase prior to reaction contains a certain amount of free (i.e. unbound) Versatic acid, plus a range of iron-Versatic complexes, which may be solvated by Versatic acid molecules and also may contain OH^- ions.[6,7] The distribution of these species will be determined by the iron concentration and the total Versatic acid concentration in the organic phase. However, the existence of the VFe interaction discussed above indicates, as might be anticipated, that the distribution of the complexes in the organic phase is not a linear function of the iron and Versatic concentrations. Furthermore the existence of two distinct modes of precipitation, namely heterogeneous and self-nucleating, which are dependent on the compositon of the organic phase, suggests that there are some organic iron complexes which are more readily hydrolysable than others. It is significant that the self-nucleating mode of precipitation is shown by solutions which are heavily loaded with iron, i.e. where the metal concentration in the organic phase is approaching saturation. In order to explore these aspects of the hydrolytic stripping reaction in more detail, information on the compositions and distributions of the iron complexes in the organic phase is necessary. There is little published information available in this area, even at ambient temperatures, and there is none available for the elevated temperatures used in this study. Work is now underway to rectify this situation, involving the use of infrared spectroscopy to determine the compositions of the organic phases at the temperatures used for hydrolytic stripping.

The effect of temperature on the initial rate of reaction over the range 135 to 170°C is illustrated in Fig. 4 in the form of an Arrhenius plot. The line plotted represents data obtained for solutions of composition (33/14), which, at 135°C, show heterogeneous precipitation behaviour - see Fig. 2. However, as the reaction temperature was increased, it was found that these solutions began to exhibit self-nucleating behaviour and could precipitate without added seed. As shown by the data in Table 3, self-nucleation became evident at about 145°C, although at this temperature the rate of precipitation in the absence of seed was slow and could be increased by adding seed. By 170°C however the reaction had become fully self-nucleating and the initial rate was unaffected by the addition of seed. In spite of this apparent change in the mechanism of precipitation with increasing temperature, it may be seen from Fig. 4 that there is no obvious change in slope of the Arrhenius plot. The

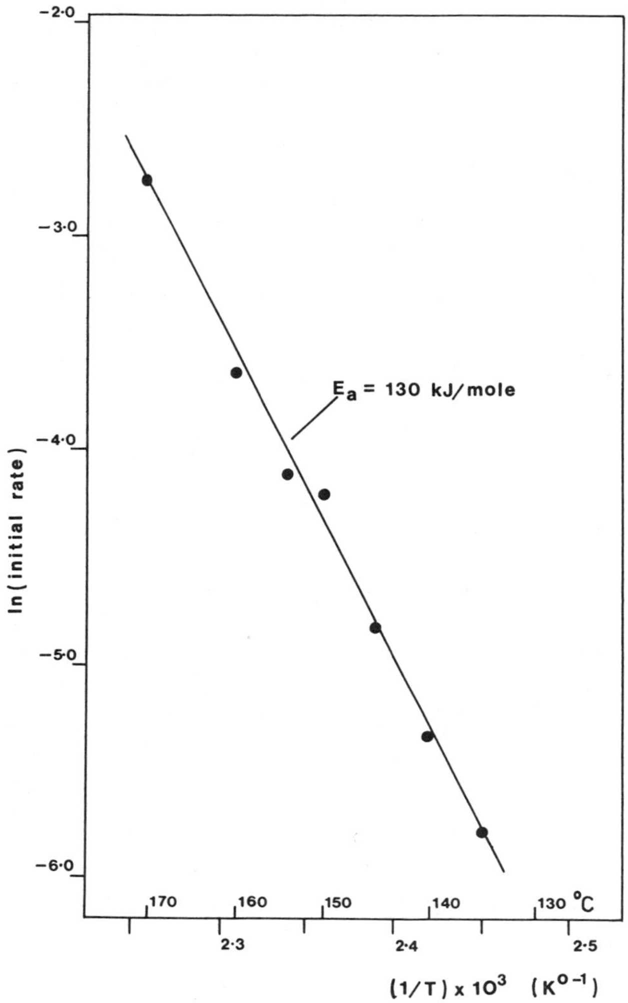

Fig. 4. Initial rate of precipitation as a function of temperature - Arrhenius plot. Solution compositions: V = 33% V/V
Fe = 14g/l

apparent activation energy obtained from this plot over the entire temperature range investigated is 130±35 kJ/mole, indicating that the reactions are chemically rate controlled.

As reported above, it was found that changing the volume phase ratio of water to organic phase had no discernable effect on the rate of reaction. This was so for both self-nucleating and heterogeneous precipitation and this strongly suggests that the reactions do not occur at the aqueous/organic interface, since the interfacial area should vary with the phase ratio. It is therefore more likely that the reactions involve water which is dissolved in the organic phase.

In the case of heterogeneous precipitation, the rate controlling step appears to be a chemical reaction at the surface of the hematite particles added as seed, because the initial rate of reaction is directly proportional to the surface area of solid available.

The so-called self-nucleating mode of precipitation, where the rate of reaction is independent of the interfacial area of both organic/water and organic/solid, is assumed to occur by a chemically rate-controlled process of homogeneous nucleation in the organic phase, involving reaction between readily hydrolysable iron-Versatic complexes and dissolved water. This reaction route is favored by high metal loading in the organic phase, i.e. where the concentration of unbound (free) Versatic acid is low.

The observed change from heterogeneous to self-nucleating behaviour with increasing temperature is thought to be due to changes in the nature and distribution of the iron-Versatic complexes in the organic phase with temperature. It is hoped that the infrared investigation mentioned above will help to elucidate these changes in the organic phase.

Ageing Effects in the Organic Phase

In the preliminary stages of this work, which preceded the experiments reported in this paper, great difficulty was experienced in obtaining reproducible kinetic data. At that time, for experimental convenience, stock solutions of iron-loaded Versatic acid, sufficient for a number of experiments, were being prepared. It became apparent that the age of the iron-Versatic solution had a marked effect on the rate of precipitation. As the solution aged the rate of precipitation slowed down, until eventually precipitation would not occur. The rate of decrease was most marked in the first few days following preparation of the solution. Thereafter the rate of precipitation decreased more slowly, but could stop entirely in solutions about two months old. Dilution of aged stock iron-Versatic solution with unloaded Versatic acid solution was found to reverse this process to a certain extent, but it was not possible by this method to obtain rates of precipitation as fast as those exhibited by freshly-prepared iron-Versatic solutions.

Further investigation revealed that the ageing effect was not confined only to the iron-Versatic solutions. It was found that the age of the calcium-Versatic solutions, prepared as precursors to the iron-Versatic solutions, also affected the rate of the hydrolytic stripping reaction, although this effect was less marked than in the case of the iron-Versatic solutions. For conditions where heterogeneous precipitation occurred, it was found that there was no measurable difference in the rates of precipitation for iron-Versatic solutions freshly prepared from calcium-Versatic solutions which were up to three weeks old, although the rates of precipitation slowed if the calcium-Versatic solution was allowed to age further. In the case of self-nucleating precipitation, however, the age of the calcium-Versatic solution had a strong effect. The rates of reaction in iron-Versatic solutions exhibiting self-nucleating precipitation, freshly prepared from calcium-Versatic solutions which had aged for only a few days were considerably slower than those obtained with similar solutions prepared from freshly made calcium-Versatic solutions.

The causes of these ageing effects are not understood, although it is suspected that slow polymerization of the metal-Versatic complexes occurs in the organic phase. Other evidence has been reported[8] which suggests that polymerization of metal carboxylate solutions can occur with ageing.

It is evident also that the changes brought about by ageing in the calcium-Versatic solutions can carry over to a certain extent to the stage where calcium is replaced by iron in the organic phase. Thus if the ageing process involves some type of structuring in the organic phase, these structures at least partially survive during the exchange extraction process.

In order to avoid these ageing problems and to obtain meaningful and reproducible kinetic data, the experimental procedure outlined in this paper was adopted. In this procedure, both the intermediate calcium-Versatic solution and the final iron-Versatic solution were prepared on the day that a hydrolytic stripping experiment was carried out.

It should be noted that results published previously[1,3] illustrating the rates of hydrolytic stripping were obtained before the effects of solution ageing were appreciated. The solutions used were prepared from aged stock solutions and therefore the rates of reaction are not representative and are slower than would be obtained with fresh solutions. It will be noted that most of the data presented in this paper were obtained at a temperature of 135°C, whereas previously it was reported that temperatures greater than 150°C were necessary to obtain significant rates of reaction. With freshly prepared solutions, it has been found that the reactions will occur readily at lower temperatures.

ACKNOWLEDGEMENTS

One of us (LACT) wishes to acknowledge the financial support received by him from the Brazilian National Research Council (CNPq).

REFERENCES

1. G. Thorsen and A. J. Monhemius: TMS Paper No. A-79-12, AIME Annual Meeting, New Orleans, 1979.
2. G. Thorsen and A. J. Monhemius: US Patent 4,282,189. Aug. 4, 1981.
3. A. J. Monhemius and G. Thorsen: Proceedings ISEC 80, vol 3, paper no. 80-91, Liege, Belgium, 1980.
4. G. Thorsen and A. Grislingas: SME Paper No 80-52, AIME Annual Meeting, Las Vegas, 1980.
5. H. F. Svendsen and G. Thorsen: Proceedings Oslo Symposium 1982, paper no. IV-26. Soc. Chem. Ind., London 1982.
6. A. J. Van der Zeeuw: Hydrometallurgy, 1979, vol. 4, pp 21-37 and 39-50.
7. M. I. Stefanakis: Ph.D. Thesis, Univ. of London, 1982.
8. A. W. Fletcher and D. S. Flett: In Solvent Extraction Chemistry of Metals. Ed. H.A.C. McKay et al. pp 359-375. Macmillan & Co. Ltd. London, 1965.

INDEX